Selected Papers of Antoni Zygmund
Volume 3

Mathematics and Its Applications (*East European Series*)

Volume 41/3

Selected Papers of Antoni Zygmund

Volume 3

edited by

A. Hulanicki
University of Wrocław, Poland

and

P. Wojtaszczyk and W. Żelazko
Institute of Mathematics,
Polish Academy of Sciences,
Warsaw, Poland

KLUWER ACADEMIC PUBLISHERS
DORDRECHT / BOSTON / LONDON

Library of Congress Cataloging in Publication Data

Zygmund, Antoni, 1900-
[Selections. 1989]
Selected papers of Antoni Zygmund / edited by A. Hulanicki and P. Wojtaszczyk and W. Żelazko.
p. cm. -- (Mathematics and its applications. East European series)
ISBN 0-7923-0474-8 (U.S. : set)
1. Harmonic analysis. 2. Zygmund, Antoni, 1900-
I. Hulanicki, Andrzej, 1933- . II. Wojtaszczyk, Przemysław, 1940- . III. Żelazko, Wiesław. IV. Title. V. Series: Mathematics and its applications (Kluwer Academic Publishers). East European series.
QA403.Z9425 1989
515'.2433--dc20 89-20033

ISBN 0-7923-0473-X
ISBN 0-7923-0474-8 (Set)

Published by Kluwer Academic Publishers,
P.O. Box 17, 3300 AA Dordrecht, The Netherlands.

Kluwer Academic Publishers incorporates
the publishing programmes of
D. Reidel, Martinus Nijhoff, Dr W. Junk and MTP Press.

Sold and distributed in the U.S.A. and Canada
by Kluwer Academic Publishers,
101 Philip Drive, Norwell, MA 02061, U.S.A.

In all other countries, sold and distributed
by Kluwer Academic Publishers Group,
P.O. Box 322, 3300 AH Dordrecht, The Netherlands.

Printed on acid-free paper

SERIES EDITOR'S PREFACE

'Et moi, ..., si j'avait su comment en revenir, je n'y serais point allé.'

Jules Verne

The series is divergent; therefore we may be able to do something with it.

O. Heaviside

One service mathematics has rendered the human race. It has put common sense back where it belongs, on the topmost shelf next to the dusty canister labelled 'discarded non-sense'.

Eric T. Bell

Mathematics is a tool for thought. A highly necessary tool in a world where both feedback and non-linearities abound. Similarly, all kinds of parts of mathematics serve as tools for other parts and for other sciences.

Applying a simple rewriting rule to the quote on the right above one finds such statements as: 'One service topology has rendered mathematical physics ...'; 'One service logic has rendered computer science ...'; 'One service category theory has rendered mathematics ...'. All arguably true. And all statements obtainable this way form part of the raison d'être of this series.

This series, *Mathematics and Its Applications*, started in 1977. Now that over one hundred volumes have appeared it seems opportune to reexamine its scope. At the time I wrote

> "Growing specialization and diversification have brought a host of monographs and textbooks on increasingly specialized topics. However, the 'tree' of knowledge of mathematics and related fields does not grow only by putting forth new branches. It also happens, quite often in fact, that branches which were thought to be completely disparate are suddenly seen to be related. Further, the kind and level of sophistication of mathematics applied in various sciences has changed drastically in recent years: measure theory is used (non-trivially) in regional and theoretical economics; algebraic geometry interacts with physics; the Minkowsky lemma, coding theory and the structure of water meet one another in packing and covering theory; quantum fields, crystal defects and mathematical programming profit from homotopy theory; Lie algebras are relevant to filtering; and prediction and electrical engineering can use Stein spaces. And in addition to this there are such new emerging subdisciplines as 'experimental mathematics', 'CFD', 'completely integrable systems', 'chaos, synergetics and large-scale order', which are almost impossible to fit into the existing classification schemes. They draw upon widely different sections of mathematics."

By and large, all this still applies today. It is still true that at first sight mathematics seems rather fragmented and that to find, see, and exploit the deeper underlying interrelations more effort is needed and so are books that can help mathematicians and scientists do so. Accordingly MIA will continue to try to make such books available.

If anything, the description I gave in 1977 is now an understatement. To the examples of interaction areas one should add string theory where Riemann surfaces, algebraic geometry, modular functions, knots, quantum field theory, Kac-Moody algebras, monstrous moonshine (and more) all come together. And to the examples of things which can be usefully applied let me add the topic 'finite geometry'; a combination of words which sounds like it might not even exist, let alone be applicable. And yet it is being applied: to statistics via designs, to radar/sonar detection arrays (via finite projective planes), and to bus connections of VLSI chips (via difference sets). There seems to be no part of (so-called pure) mathematics that is not in immediate danger of being applied. And, accordingly, the applied mathematician needs to be aware of much more. Besides analysis and numerics, the traditional workhorses, he may need all kinds of combinatorics, algebra, probability, and so on.

In addition, the applied scientist needs to cope increasingly with the nonlinear world and the extra mathematical sophistication that this requires. For that is where the rewards are. Linear models are honest and a bit sad and depressing: proportional efforts and results. It is in the non-linear world that infinitesimal inputs may result in macroscopic outputs (or vice versa). To appreciate what I am hinting at: if electronics were linear we would have no fun with transistors and computers; we would have no TV; in fact you would not be reading these lines.

There is also no safety in ignoring such outlandish things as nonstandard analysis, superspace and anticommuting integration, p-adic and ultrametric space. All three have applications in both electrical engineering and physics. Once, complex numbers were equally outlandish, but they frequently proved the shortest path between 'real' results. Similarly, the first two topics named have already provided a number of 'wormhole' paths. There is no telling where all this is leading - fortunately.

Thus the original scope of the series, which for various (sound) reasons now comprises five subseries: white (Japan), yellow (China), red (USSR), blue (Eastern Europe), and green (everything else), still applies. It has been enlarged a bit to include books treating of the tools from one subdiscipline which are used in others. Thus the series still aims at books dealing with:

- a central concept which plays an important role in several different mathematical and/or scientific specialization areas;
- new applications of the results and ideas from one area of scientific endeavour into another;
- influences which the results, problems and concepts of one field of enquiry have, and have had, on the development of another.

These three volumes of selected papers by Antoni Zygmund - of which this is the third one - are the first of their kind in this series. More are scheduled to appear fairly soon. Often, collected or selected works are treated as separate projects on their own and are not placed in any series whatever (or are placed in a series of collected works); so possibly a few words of explanation, or at least motivation, are in their place.

The interconnectedness of all things is something in which I (like Douglas Adams' creation Svlad Cjelli) firmly believe, especially in mathematics and science (though for vastly different reasons). In fact this trust in the importance of interrelations is a main driving force behind the whole Mathematics and Its Applications series.

Interconnectedness between things mathematical can be found in monographs, especially when written with that point in view, and in selected proceedings of conferences of an interspecialistic character, but it can also be found in the work of a single scientist, especially a great one, a point that needs no discussion in the present case; and, in fact, in such a case the refusal to restrict considerations to the domain of one labelled, classified, and recognized subdiscipline may take on dramatic forms.

Abel recommended to read (only) the works of the original masters. I have my own relatively modest collection of collected and selected works, and, with increasing awareness of just how valuable Abel's remark is, I find myself dipping into them more and more frequently.

In addition, the great ones often have the invigorating habit of scattering their papers far and wide; add ups and downs in the popularity and status of journals, an increasing tendency to publish important material only in proceeding volumes, and finding all the relevant stuff of a particular author on a particular topic may well become something of a serious chore in itself.

Moreover, if present trends continue, very few libraries will be able to afford (more or less) complete collections of even only the more important journals, proceedings, and other edited volumes.

All this makes judiciously chosen selected/collected works (and other ways of grouping interesting material) particularly inviting and attractive.

The shortest path between two truths in the real domain passes through the complex domain.

J. Hadamard

La physique ne nous donne pas seulement l'occasion de résoudre des problèmes ... elle nous fait pressentir la solution.

H. Poincaré

Never lend books, for no one ever returns them; the only books I have in my library are books that other folk have lent me.

Anatole France

The function of an expert is not to be more right than other people, but to be wrong for more sophisticated reasons.

David Butler

Bussum, September 1989

Michiel Hazewinkel

TABLE OF CONTENTS

BIBLIOGRAPHY OF ANTONI ZYGMUND ix

PAPER

[141] (with **A. P. Calderón**) 'A note on the interpolation of linear operations', *SM* 12 (1951), 194–204. 1

[142] 'Polish mathematics between the two wars (1919–39)', in: *Proceedings of the Second Canadian Mathematical Congress, Vancouver, 1949*, pp. 3–9, Univ. of Toronto Press, 1951. 12

[143] (with **A. P. Calderón**) 'On the existence of certain singular integrals', *Acta Math.* 88 (1952), 85–139. 19

[148] (with **R. Salem**) 'Some properties of trigonometric series whose terms have random signs', *Acta Math.* 91 (1954), 245–301. 74

[150] (with **A. P. Calderón**) 'Singular integrals and periodic functions', *SM* 14 (1954), 249–271. 131

[151] (with **R. Salem**) 'Sur un théorème de Piatetçki-Shapiro', *CRAS* 240 (1955), 2040–2042. 154

[153] (with **A. P. Calderón**) 'On a problem of Mihlin', *TAMS* 78 (1955), 209–224. 157

[164] (with **A. P. Calderón**) 'Addenda to the paper "On a problem of Mihlin"', *ibid.*, 84 (1957), 559–560. 173

[155] (with **A. P. Calderón**) 'A note on the interpolation of sublinear operations', *Amer. J. Math.* 78 (1956), 282–288. 175

[156] (with **A. P. Calderón**) 'On singular integrals', *ibid.*, 289–309. 182

[157] (with **A. P. Calderón**) 'Algebras of certain singular operators', *ibid.*, 310–320. 203

[158] 'On a theorem of Marcinkiewicz concerning interpolation of operations', *J. Math. Pures Appl.* (9) 35 (1956), 223–248. 214

[159] 'Hilbert transforms in E^n', in: *Proceedings of the International Congress of Mathematicians, 1954, Amsterdam*, Vol. 3, pp. 140–151, Erven P. Noordhoff N.V., Groningen, and North-Holland Publishing Co., Amsterdam, 1956. 240

[160] 'On the Littlewood–Paley function $g^*(\theta)$', *Proc. Nat. Acad. Sci. U.S.A.* 42 (1956), 208–212. 252

[162] (with **A. P. Calderón**) 'Singular integral operators and differential equations', *Amer. J. Math.* 79 (1957), 901–921. 257

The papers appearing in these volumes are numbered according to the comprehensive list of publications of Antoni Zygmund which are presented in chronological order beginning on p. ix.

[166] (with **M. Weiss**) 'A note on smooth functions', *Nederl. Akad. Wetensch. Proc. Ser. A* 62 = *Indag. Math.* 21 (1959), 52–58. 278

[174] (with **A. P. Calderón**) 'Local properties of solutions of elliptic partial differential equations', *SM* 20 (1961), 171–225. 285

[179] (with **E. M. Stein**) 'On the differentiability of functions', *ibid.*, 23 (1963/1964), 247–283. 340

[180] (with **A. P. Calderón**) 'On the higher gradients of harmonic functions', *ibid.*, 24 (1964), 211–226. 377

[182] (with **E. M. Stein**) 'On the fractional differentiability of functions', *Proc. London Math. Soc.* (3) 14A (1965), 249–264. 393

[187] (with **E. M. Stein**) 'Boundedness of translation invariant operators on Hölder spaces and L^p-spaces', *Ann. of Math.* (2) 85 (1967), 337–349. 409

[192] (with **M. Weiss**) 'On multipliers preserving convergence of trigonometric series almost everywhere', *SM* 30 (1968), 111–120. 422

[196] 'On certain lemmas of Marcinkiewicz and Carleson', *J. Approx. Theory* 2 (1969), 249–257. 432

[202] 'A Cantor–Lebesgue theorem for double trigonometric series', *SM* 43 (1972), 173–178. 441

[212] (with **A. P. Calderón**) 'A note on singular integrals', *ibid.*, 65 (1979), 77–87. 447

BIBLIOGRAPHY OF ANTONI ZYGMUND

Abbreviations

BIAP *Bulletin International de l'Académie Polonaise des Sciences et des Lettres, Classe des Sciences Mathématiques et Naturelles, Série A: Sciences Mathématiques*
CRAS *Comptes Rendus hebdomadaires des séances de l'Académie des Sciences (Paris)*
FM *Fundamenta Mathematicae*
MR *Mathematical Reviews*
SM *Studia Mathematica*
TAMS *Transactions of the American Mathematical Society*
Zbl. *Zentralblatt für Mathematik*
[*M*] Józef Marcinkiewicz, *Collected Papers*, PWN–Polish Scientific Publishers, Warszawa, 1964.
[*S*] Raphaël Salem, *Œuvres mathématiques*, Hermann, Paris, 1967.

Other abbreviations follow those of Mathematical Reviews.

1923

[1] 'Sur la théorie riemannienne des séries trigonométriques', *CRAS* 177, 521–523; Errata, *ibid.*, 804.
[2] 'Sur les séries trigonométriques', *ibid.*, 576–579; Errata, *ibid.*, 804.
[3] (with **W. Sierpiński**) 'Sur une fonction qui est discontinue sur tout ensemble de puissance du continu', *FM* 4, 316–318; reprinted in: Wacław Sierpiński, *Œuvres choisies*, Tome II, pp. 497–499, PWN–Polish Scientific Publishers, Warszawa, 1975.

1924

[4] (with **A. Rajchman**) 'Sur les principes et les problèmes de la théorie riemannienne des séries trigonométriques', *Ann. Soc. Polon. Math.* 3, 147–148.
[5] 'Sur un théorème de M. Marcel Riesz', *ibid.*, 153.
[6] 'Sur les séries de Fourier restreintes', *CRAS* 178, 181–182.

The papers reprinted in this three-volume work are marked with *.
Those appearing in this particular volume are marked with **.

[7] 'Sur une généralisation de la méthode de Cesàro', *ibid.*, 179, 870–872.

[8] (with **S. Saks**) 'Sur les faisceaux des tangentes à une courbe', *FM* 6, 117–121.

[9] 'O module ciągłości sumy szeregu sprzężonego z szeregiem Fouriera' (On the modulus of continuity of the sum of a series conjugate to a Fourier series), *Prace Mat.-Fiz.* 33, 125–132 [in Polish, French summary].

1925

[10] (with **S. Saks**) 'Un teorema sulle curve continue', *Boll. Un. Mat. Ital.* 4, 7–10.

*[11] 'Sur la dérivation des séries de Fourier', *BIAP*, Année 1924, 243–249.

*[12] 'Sur la sommation des séries trigonométriques conjuguées aux séries de Fourier', *ibid.*, 251–258.

[13] 'Sur la sommation des séries trigonométriques et celles de puissances par les moyennes typiques', *CRAS* 181, 1122–1123.

1926

[14] 'Sur la sommabilité des séries de Fourier des fonctions vérifiant la condition de Lipschitz', *BIAP*, Année 1925, 1–9.

*[15] (with **A. Rajchman**) 'Sur la relation du procédé de sommation de Cesàro et celui de Riemann', *ibid.*, 69–80.

*[16] 'Sur un théorème de M. Gronwall', *ibid.*, 207–217.

[17] 'Sur la sommation des séries par le procédé des moyennes typiques', *ibid.*, 265–287.

*[18] 'Remarque sur la sommabilité des séries de fonctions orthogonales', *ibid.*, Année 1926, 185–191.

*[19] 'Contribution à l'unicité du développement trigonométrique', *Math. Z.* 24, 40–46.

*[20] 'Sur la théorie riemannienne des séries trigonométriques', *ibid.*, 47–104.

[21] 'Sur un théorème de la théorie de la sommabilité', *ibid.*, 25, 291–296.

*[22] (with **A. Rajchman**) 'Sur la possibilité d'appliquer la méthode de Riemann aux séries trigonométriques sommables par le procédé de Poisson', *ibid.*, 261–273.

*[23] 'Sur les séries trigonométriques sommables par le procédé de Poisson', *ibid.*, 274–290.

*[24] 'Une remarque sur un théorème de M. Kaczmarz', *ibid.*, 297–298.

[25] 'O teorji średnich arytmetycznych" (On the theory of arithmetic means), *Mathesis Polska* 1, 75–85 and 119–129 [in Polish].

1927

*[26] 'Sur l'application de la première moyenne arithmétique dans la théorie des séries de fonctions orthogonales', *FM* 10, 356–362.

[27] 'O pewnym szeregu potęgowym' (On some power series), *Mathesis Polska* 2, 62–63 [in Polish].

1928

[28] 'Sur la convergence absolue des séries de Fourier', *Ann. Soc. Polon. Math.* 7, 266–267.

[29] 'Sur le problème d'unicité pour certains systèmes orthogonaux', *ibid.*, 267.

*[30] 'Sur la sommation des séries de fonctions orthogonales', *BIAP*, Année 1927, 295–308.

[31] 'Über einige Sätze aus der Theorie der divergenten Reihen', *ibid.*, 309–331.

[32] 'Remarque sur les sommes positives des séries trigonométriques', *ibid.*, 333–341.

*[33] 'Remarque sur un théorème de M. Fekete', *ibid.*, 343–347.

[34] 'Sur les fonctions conjuguées', *CRAS* 187, 1025–1026.

*[35] 'Remarque sur la convergence absolue des séries de Fourier', *J. London Math. Soc.* 3, 194–196.

[36] 'Über die Beziehungen der Eindeutigkeitsfragen in den Theorien der trigonometrischen Reihen und Integrale', *Math. Ann.* 99, 562–589.

1929

*[37] 'Sur les fonctions conjuguées', *FM* 13, 284–303.

[38] 'Kilka uwag o zbiorach jednoznaczności w teorji całek trygonometrycznych' (Some remarks on sets of uniqueness in the theory of trigonometric integrals), in: *Księga Pamiątkowa Pierwszego Polskiego Zjazdu Matematycznego, Lwów, 7-10.IX.1927*, p. 117, Dodatek do "Ann. Soc. Polon. Math.", Kraków [in Polish].

[39] 'Uwaga o funkcjach ciągłych nieróżniczkowalnych' (Remark on nondifferentiable continuous functions), *Mathesis Polska* 4, 1–7 [in Polish, French summary].

[40] *Równania różniczkowe. Skrypt autoryzowany. Cz. I. Równania różniczkowe zwyczajne w dziedzinie rzeczywistej* (Differential Equations. Authorized lecture notes. Part I. Ordinary Differential Equations in the Real Domain), ed. by A. Koźniewski and E. Szpilrajn, Komisja Wyd. Koła Mat.-Fiz. Słuchaczów Uniw. Warsz., Warszawa, viii + 403 pp. [in Polish].

1930

[41] 'Remarques sur les ensembles d'unicité dans quelques systèmes orthogonaux', in: *Atti del Congresso Internazionale dei Matematici, Bologna, 3–10 Settembre 1928 (VI)*, Tomo II, pp. 309–311, Nicola Zanichelli, Bologna.

*[42] 'On the convergence of lacunary trigonometric series', *FM* 16, 90–107; Errata, *ibid.*, 18 (1932), 312.

[43] 'Über einseitige Lokalisation', *Jahresber. Deutsch. Math.-Verein.* 39, 47–52.

[44] 'Sur les séries trigonométriques lacunaires', *J. London Math. Soc.* 5, 138–145.

[45] 'O pewnych nierównościach' (On some inequalities), *Mathesis Polska* 5, 83–100 [in Polish].

*[46] (with **R. E. A. C. Paley**) 'On some series of functions, (1)', *Proc. Cambridge Philos. Soc.* 26, 337–357.

*[47] (with **R. E. A. C. Paley**) 'On some series of functions, (2)', *ibid.*, 458–474.

[48] 'A note on series of sines', *Quart. J. Math. Oxford Ser.* 1, 102–107.

*[49] 'Sur la théorie riemannienne de certains systèmes orthogonaux. I', *SM* 2, 97–170.

[50] 'Un théorème sur les séries orthogonales', *ibid.*, 181–182.

*[51] (with **R. E. A. C. Paley**) 'On the partial sums of Fourier series', *ibid.*, 221–227.

1931

[52] 'Über einige Probleme und Sätze aus der Theorie der trigonometrischen Reihen', *Ann. Soc. Polon. Math.* 10, 129.

[53] 'On a theorem of Ostrowski' *J. London Math. Soc.* 6, 162–163; *Zbl.* 2, 189.

*[54] 'Quelques théorèmes sur les séries trigonométriques et celles de puissances', *SM* 3, 77–91; *Zbl.* 3, 253.

[55] 'On a theorem of Privaloff', *ibid.*, 239–247; *Zbl.* 3, 204.

1932

[56] 'On lacunary trigonometric series', *Bull. Amer. Math. Soc.* 38, 344.

*[57] 'Errata et remarques à mon travail »Sur les fonctions conjuguées«, Fund. Math. XIII, p. 284–303', *FM* 18, 312.

[58] 'O pewnych zagadnieniach teorji szeregów trygonometrycznych' (On some questions in the theory of trigonometric series), *Mathesis Polska* 7, 49–58 [in Polish].

[59] 'Sur la théorie riemannienne de certains systèmes orthogonaux. II', *Prace Mat.-Fiz.* 39, 73–117; *Zbl.* 5, 290.

*[60] (with **R. E. A. C. Paley**) 'On some series of functions, (3)', *Proc. Cambridge Philos. Soc.* 28, 190–205; *Zbl.* 6, 198.

*[61] (with **R. E. A. C. Paley**) 'A note on analytic functions in the unit circle', *ibid.*, 266–272; *Zbl.* 5, 66.

*[62] 'A remark on conjugate series', *Proc. London Math. Soc.* (2) 34, 392–400; *Zbl.* 5, 353.

[63] 'Sur un théorème de M. Pólya', in: *Verhandlungen des Internationalen Mathematiker-Kongresses, Zürich 1932*, II. Band, p. 38, Orell Füssli Verlag, Zürich.

[64] 'On lacunary trigonometric series', *TAMS* 34, 435–446; *Zbl.* 5, 63.

1933

[65] 'On continuability of power series', *Acta Litt. Sci. Szeged* 6, 80–84; *Zbl.* 7, 312.

[66] 'On an integral inequality', *J. London Math. Soc.* 8, 175–178; *Zbl.* 7, 302.

*[67] (with **R. E. A. C. Paley** and **N. Wiener**) 'Notes on random functions', *Math. Z.* 37, 647–668; *Zbl.* 7, 354.

[68] 'O zachowaniu się pewnych szeregów funkcyjnych' (On the behaviour of some function series), *Mathesis Polska* 8, 76–87 [in Polish].

1934

[69] (with **S. Saks**) 'On functions of rectangles and their application to analytic functions', *Ann. Scuola Norm. Sup. Pisa* 3, 27–32.

*[70] 'On the differentiability of multiple integrals', *FM* 23, 143–149; *Zbl.* 10, 14.

*[71] 'Some points in the theory of trigonometric and power series', *TAMS* 36, 586–617; *Zbl.* 10, 62.

1935

[72] (with **B. Jessen** and **J. Marcinkiewicz**) 'Note on the differentiability of multiple integrals', *FM* 25, 217–234; reprinted in [*M*], pp. 81–95; *Zbl.* 12, 59.

[73] 'Sur le caractère de divergence des séries orthogonales', *Mathematica (Cluj)* 9, 86–88; *Zbl.* 14, 14.

[74] *Trigonometrical Series*, Monografje Matematyczne, Vol. 5, Z Subwencji Funduszu Kultury Narodowej, Warszawa–Lwów, iv + 332 pp.

1936

[75] (with **J. Marcinkiewicz**) 'On the differentiability of functions and summability of trigonometrical series', *FM* 26, 1–43; reprinted in [*M*], pp. 125–163; *Zbl.* 14, 111.

[76] (with **J. Marcinkiewicz**) 'Mean values of trigonometrical polynomials', *ibid.*, 28, 131–166; reprinted in [*M*], pp. 260–292; *Zbl.* 17, 251.

[77] (with **J. Marcinkiewicz**) 'Some theorems on orthogonal systems', *ibid.*, 309–335; reprinted in [*M*], pp. 209–232; *Zbl.* 16, 205.

*[78] 'A remark on Fourier transforms', *Proc. Cambridge Philos. Soc.* 32, 321–327; *Zbl.* 14, 258.

1937

[79] (with **J. Marcinkiewicz**) 'Sur les fonctions indépendantes', *FM* 29, 60–90; reprinted in [*M*], pp. 233–259; *Zbl.* 16, 409.

[80] (with **J. Marcinkiewicz**) 'Remarque sur la loi du logarithme itéré', *ibid.*, 215–222; reprinted in [*M*], pp. 299–306; *Zbl.* 18, 32.

[81] 'Note on trigonometrical and Rademacher's series', *Prace Mat.-Fiz.* 44, 91–107; *Zbl.* 16, 109.

[82] (with **J. Marcinkiewicz**) 'Two theorems on trigonometrical series', *Mat. Sb.* (*N.S.*) 2 (44), 733–737 [with Russian summary]; reprinted in [*M*], pp. 293–298; *Zbl.* 18, 18.

1938

[83] 'A remark on conjugate series', *Bull. Sémin. Math. Univ. Wilno* 1, 16–18; *Zbl.* 19, 301.

[84] (with **J. Marcinkiewicz**) 'Proof of a gap theorem', *Duke Math. J.* 4, 469–472; reprinted in [*M*], pp. 424–427; *Zbl.* 19, 270.

[85] (with **J. Marcinkiewicz**) 'A theorem of Lusin', *ibid.*, 473–485; reprinted in [*M*], pp. 428–443; *Zbl.* 19, 420.

*[86] 'On the convergence and summability of power series on the circle of convergence (I)', *FM* 30, 170–196; *Zbl.* 19, 16.

[87] (with **J. Marcinkiewicz**) 'Sur les séries de puissances', *Mathematica* (*Cluj*) 14, 21–30; reprinted in [*M*], pp. 454–462; *Zbl.* 20, 231.

[88] 'Proof of a theorem of Paley', *Proc. Cambridge Philos. Soc.* 34, 125–133; *Zbl.* 19, 17.

[89] (with **J. Marcinkiewicz**) 'Quelques théorèmes sur les fonctions indépendantes', *SM* 7, 104–120; reprinted in [*M*], pp. 374–388, *Zbl.* 18, 75.

[90] (with **S. Saks**) *Funkcje analityczne. Wykłady uniwersyteckie* (Analytic Functions. University Lectures), Monografie Matematyczne, Vol. 10, Z Subwencji Funduszu Kultury Narodowej Józefa Piłsudskiego, Warszawa–Lwów–Wilno, viic + 431 pp. [in Polish].

[91] 'Fundamenta Mathematicae', *Mathesis Polska* 11, 22–26 [in Polish].

[92] 'Międzynarodowy Kongres Matematyczny w Oslo' (International Congress of Mathematicians in Oslo), *ibid.*, 28–30 [in Polish].

1939

*[93] 'Sur un théorème de M. Fejér', *Bull. Sémin. Math. Univ. Wilno* 2, 3–12; *MR* 1-9.

[94] (with **J. Marcinkiewicz**) 'Sur la derivée seconde généralisée', *ibid.*, 35–40; reprinted in [*M*], pp. 582–587; *MR* 1-8.

*[95] 'Note on the formal multiplication of trigonometrical series', *ibid.*, 52–56; *MR* 1-11.

[96] (with **J. Marcinkiewicz**) 'Quelques inégalités pour les opérations linéaires', *FM* 32, 115–121; reprinted in [*M*], pp. 541–546; *Zbl.* 21, 407.

[97] (with **J. Marcinkiewicz**) 'On the summability of double Fourier series', *ibid.*, 122–132; reprinted in [*M*], pp. 547–556; *Zbl.* 22, 18.

[98] *Trigonometricheskie ryady*, GONTI, Moskva, 329 pp. [Russian transl. of [74]].

1941

[99] (with **J. Marcinkiewicz**) 'On the behavior of trigonometric series and power series', *TAMS* 50, 407–453; reprinted in [*M*], pp. 609–654; *MR* 3-105.

1942

[100] 'Two notes on inequalities', *J. Math. Phys. Mass. Inst. Tech.* 21, 117–123; *MR* 4-135.

*[101] 'On the convergence and summability of power series on the circle of convergence (II)', *Proc. London Math. Soc.* (2) 47, 326–350; *MR* 4-76.

1943

*[102] 'Complex methods in the theory of Fourier series', *Bull. Amer. Math. Soc.* 49, 805–822; *MR* 5-119.

[103] 'A theorem on generalized derivatives', *ibid.*, 917–923; *MR* 5-175.

[104] 'A property of the zeros of Legendre polynomials', *TAMS* 54, 39–56; *MR* 5-180.

1944

*[105] (with **J. D. Tamarkin**) 'Proof of a theorem of Thorin', *Bull. Amer. Math. Soc.* 50, 279–282; *MR* 5-229.

*[106] 'On certain integrals', *TAMS* 55, 170–204; *MR* 5-230.

1945

*[107] 'On the degree of approximation of functions by Fejér means', *Bull. Amer. Math. Soc.* 51, 274–278; *MR* 6-265.

*[108] 'Proof of a theorem of Littlewood and Paley', *ibid.*, 439–446; *MR* 7-8.

*[109] 'Smooth functions', *Duke Math. J.* 12, 47–76; *MR* 7-60.

[110] 'A theorem on fractional derivatives', *ibid.*, 455–464; *MR* 7-148.

*[111] (with **R. Salem**) 'Lacunary power series and Peano curves', *ibid.*, 569–578; reprinted in [*S*], pp. 336–345; *MR* 7-378.

[112] 'The approximation of functions by typical means of their Fourier series', *ibid.*, 695–704; *MR* 7-435.

1946

[113] 'On the theorem of Fejér–Riesz', *Bull. Amer. Math. Soc.* 52, 310–318; *MR* 7-434.

[114] (with **R. Salem**) 'The approximation by partial sums of Fourier series', *TAMS* 59, 14–22; reprinted in [*S*], pp. 366–373; *MR* 7-435.

*[115] (with **R. Salem**) 'Capacity of sets and Fourier series', *ibid.*, 23–41; reprinted in [*S*], pp. 374–392; *MR* 7-434.

1947

*[116] 'On the summability of multiple Fourier series', *Amer. J. Math.* 69, 836–850; *MR* 9-235.

[117] 'A remark on characteristic functions', *Ann. Math. Statist.* 18, 272–276; *MR* 9-88.

*[118] 'On trigonometric integrals', *Ann. of Math.* (2) 48, 393–440; *MR* 9-88.

*[119] (with **R. Salem**) 'On a theorem of Banach', *Proc. Nat. Acad. Sci. U.S.A.* 33, 293–295; reprinted in [*S*], pp. 393–395; *MR* 9-88.

*[120] (with **R. Salem**) 'On lacunary trigonometric series', *ibid.*, 333–338; reprinted in [*S*], pp. 396–401; *MR* 9-181.

1948

*[121] 'On a theorem of Hadamard', *Ann. Soc. Polon. Math.* 21, 52–69; *MR* 10-186; Errata, *ibid.*, 357–358; *MR* 11-20.

[122] 'Two notes on the summability of infinite series', *Colloq. Math.* 1, 225–229; *MR* 10-446.

*[123] (with **R. Salem**) 'On lacunary trigonometric series, II', *Proc. Nat. Acad. Sci. U.S.A.* 34, 54–62; reprinted in [*S*], pp. 405–413; *MR* 9-425.

*[124] (with **R. Salem**) 'A convexity theorem', *ibid.*, 443–447; reprinted in [*S*], pp. 427–431; *MR* 10-247.

[125] 'On certain methods of summability associated with conjugate trigonometric series', *SM* 10, 97–103; *MR* 10-31.

[126] 'An example in Fourier series', *ibid.*, 113–119; *MR* 10-603.

[127] (with **M. Kac** and **R. Salem**) 'A gap theorem', *TAMS* 63, 235–243; reprinted in [*S*], pp. 414–422; *MR* 9-426.

[128] (with **S. Saks**) *Funkcje analityczne. Wykłady uniwersyteckie* (Analytic Functions. University Lectures), 2nd ed. of [90], Monografie Matematyczne, Vol. 10, Spółdzielnia Wydawnicza »Czytelnik«, xi + 431 pp. [in Polish].

1949

*[129] 'On the boundary values of functions of several complex variables, I', *FM* 36, 207–235; *MR* 12-18.

*[130] 'On a theorem of Littlewood', *Summa Brasil. Math.* 2, no. 5, 51–57; *MR* 12-88.

[131] *Fourier Series. 8 Lectures*, Canadian Mathematical Congress, The Univ. of British Columbia, 79 pp.

1950

[132] 'A remark on functions of several complex variables', *Acta Sci. Math. (Szeged)* 12, 66–68; *MR* 11-652.

*[133] (with **R. Salem**) 'La loi du logarithme itéré pour les séries trigonométriques lacunaires', *Bull. Sci. Math.* (2) 74, 209–224: reprinted in [*S*], pp. 465–480; *MR* 12-605.

*[134] (with **A. P. Calderón**) 'Note on the boundary values of functions of several complex variables', in: *Contributions to Fourier Analysis*, pp. 145–165, Ann. of Math. Stud., no. 25, Princeton Univ. Press, Princeton, N.J.; *MR* 12-19.

*[135] (with **A. P. Calderón**) 'On the theorem of Hausdorff–Young and its extensions', *ibid.*, pp. 166–188; *MR* 12-255.

[136] (with **A. P. Calderón** and **A. González-Domínguez**) 'Nota sobre los valores limites de funciones analiticas', *Rev. Un. Mat. Argentina* 14 (1949/1950), 16–19; *MR* 11-168.

[137] 'A remark on the integral modulus of continuity', *Univ. Nac. Tucumán Rev. Ser. A* 7, 259–269; *MR* 13-118.

[138] *Trigonometric Interpolation*, Univ. of Chicago, iv + 99 pp.; *MR* 11-654.

1951

[139] 'An individual ergodic theorem for non-commutative transformations', *Acta Sci. Math. (Szeged)* 14, 103–110; *MR* 13-661.

[140] 'A remark on characteristic functions', in: *Proceedings of the Second Berkeley Symposium on Mathematical Statistics and Probability, 1950*, pp. 369–372, Univ. of California Press, Berkeley–Los Angeles; *MR* 13-362.

[141] (with **A. P. Calderón) 'A note on the interpolation of linear operations', *SM* 12, 194–204; *MR* 13-754.

**[142] 'Polish mathematics between the two wars (1919–39)', in: *Proceedings of the Second Canadian Mathematical Congress, Vancouver, 1949*, pp. 3–9, Univ. of Toronto Press; *MR* 13-197.

1952

[143] (with **A. P. Calderón) 'On the existence of certain singular integrals', *Acta Math.* 88, 85–139; *MR* 14-637.

[144] (with **A. P. Calderón**) 'On singular integrals in the theory of the potential', in: *Proceedings of the International Congress of Mathematicians, Cambridge, Mass., U.S.A., August 30–September 6, 1950*, Vol. I, pp. 375–376, Amer. Math. Soc., Providence, R.I.

[145] (with **S. Saks**) *Analytic Functions*, Monografie Matematyczne, Vol. 28, Polskie Towarzystwo Matematyczne, Warszawa–Wrocław, viii + 451 pp. [English transl. of [90]]; *MR* 14-1073.

[146] *Trigonometrical Series*, reprint of [74], Chelsea Publications Co., New York, vi + 330 pp.; *MR* 17-844.

1953

[147] (with **R. Salem**) 'Sur les séries trigonométriques dont les coefficients ont des signes aléatoires', *CRAS* 236, 571–573; reprinted in [*S*], pp. 497–498; *MR* 14-1081.

1954

[148] (with **R. Salem) 'Some properties of trigonometric series whose terms have random signs', *Acta Math.* 91, 245–301; reprinted in [*S*], pp. 501–557; *MR* 16-467.

[149] (with **G. Szegö**) 'On certain mean values of polynomials', *J. Analyse Math.* 3, 225–244; *MR* 16-355.

[150] (with **A. P. Calderón) 'Singular integrals and periodic functions', *SM* 14, 249–271; *MR* 16-1017.

1955

[151] (with **R. Salem) 'Sur un théorème de Piatetçki-Shapiro', *CRAS* 240, 2040–2042; reprinted in [*S*], pp. 590–592; *MR* 17-150.

[152] (with **R. Salem**) 'Sur les ensembles parfaits dissymétriques à rapport constant', *ibid.*, 2281–2283; reprinted in [*S*], pp. 593–595; *MR* 17-150.

[153] (with **A. P. Calderón) 'On a problem of Mihlin', *TAMS* 78, 209–224; *MR* 16-816.

[154] *Trigonometrical Series*, reprint of [74], Dover Publications, New York, vii + 329 pp.; *MR* 17-361.

1956

[155] (with **A. P. Calderón) 'A note on the interpolation of sublinear operations', *Amer. J. Math.* 78, 282–288; *MR* 18-586.

[156] (with **A. P. Calderón) 'On singular integrals', *ibid.*, 289–309; *MR* 18-894.

[157] (with **A. P. Calderón) 'Algebras of certain singular operators', *ibid.*, 310–320; *MR* 19-414.

**[158] 'On a theorem of Marcinkiewicz concerning interpolation of operations', *J. Math. Pures Appl.* (9) 35, 223–248; *MR* 18-321.

**[159] 'Hilbert transforms in E^n', in: *Proceedings of the International Congress of Mathematicians, 1954, Amsterdam*, Vol. 3, pp. 140–151, Erven P. Noordhoff N.V., Groningen, and North-Holland Publishing Co., Amsterdam; *MR* 19-139.

**[160] 'On the Littlewood–Paley function $g^*(\theta)$', *Proc. Nat. Acad. Sci. U.S.A.* 42, 208–212; *MR* 17-1080.

[161] (with **R. Salem**) 'A note on random trigonometric polynomials', in: *Proceedings of the Third Berkeley Symposium on Mathematical Statistics and Probability, 1954–1955*, Vol. 2, pp. 243–246, Univ. of California Press, Berkeley–Los Angeles; reprinted in [*S*], pp. 586–589; *MR* 18-891.

1957

[162] (with **A. P. Calderón) 'Singular integral operators and differential equations', *Amer. J. Math.* 79, 901–921; *MR* 20#7196.

[163] 'On singular integrals', *Rend. Mat. e Appl.* (5) 16, 468–505; *MR* 20#2585.

[164] (with **A. P. Calderón) 'Addenda to the paper "On a problem of Mihlin"', *TAMS* 84, 559–560; *MR* 18-894.

[165] 'Problem 131', in: *The Scottish Book. A Collection of Problems*, p. 58, ed. by S. Ulam, Los Alamos Scientific Laboratory, Los Alamos, N. Mex.; reprinted in: *The Scottish Book. Mathematics from the Scottish Café*, p. 215, ed. by R. D. Mauldin, Birkhäuser, Boston–Basel–Stuttgart, 1981.

1959

[166] (with **M. Weiss) 'A note on smooth functions', *Nederl. Akad. Wetensch. Proc. Ser. A* 62 = *Indag. Math.* 21, 52–58; *MR* 21#5849.

[167] 'On the preservation of classes of functions', *J. Math. Mech.* 8, 889–895; Erratum, *ibid.*, 9(1960), 663; *MR* 22#A8277.

[168] (with **S. Saks**) *Funkcje analityczne* (Analytic Functions), 3rd ed. of [90], Monografie Matematyczne, Vol. 10, PWN–Polish Scientific Publishers, Warszawa, viii + 431 pp. [in Polish].

[169] *Trigonometric Series*, 2nd ed., Cambridge Univ. Press, London–New York, Vol. 1, xii + 383 pp., Vol. 2, vii + 354 pp.; *MR* 21#6498.

1960

[170] (with **M. Weiss**) 'On the existence of conjugate functions of higher order', *FM* 48, 175-187; *MR* 22#A3934.

[171] (with **A. P. Calderón**) 'A note on local properties of solutions of elliptic partial differential equations', *Proc. Nat. Acad. Sci. U.S.A.* 46, 1385–1389; *MR* 25#4243.

[172] 'Józef Marcinkiewicz', *Wiadom. Mat.* (2) 4, 11–41 [in Polish]; *MR* 22#A6683.

1961

[173] (with **E. M. Stein**) 'Smoothness and differentiability of functions', *Ann. Univ. Sci. Budapest. Eötvös Sect. Math.* 3–4 (1960/1961), 295–307; *MR* 24#A1982.

[174] (with **A. P. Calderón) 'Local properties of solutions of elliptic partial differential equations', *SM* 20, 171–225; *MR* 25#310.

[175] 'Sur la différentiabilité des fonctions', in: *Séminaire Pierre Lelong*, Vol. 3, pp. 1201–1208, Faculté des Sciences de Paris.

1962

[176] (with **A. P. Calderón**) 'On the differentiability of functions which are of bounded variation in Tonelli's sense', *Rev. Un. Mat. Argentina* 20, 102–121; *MR* 27#1542.

[177] 'On one-sided localization of trigonometric series', in: *Studies in Mathematical Analysis and Related Topics*, pp. 435–447, Stanford Univ. Press, Stanford, Calif.; *MR* 26#2794.

1963

[178] (with **A. P. Calderón**) 'On higher gradients of harmonic functions', in: *Outlines of the Joint Symposium on Partial Differential Equations* (Novosibirsk 1963), pp. 57–60, Acad. Sci. USSR, Siberian Branch, Moscow; *MR* 35#5647.

1964

[179] (with **E. M. Stein) 'On the differentiability of functions', *SM* 23 (1963/1964), 247–283; *MR* 28#2176.
[180] (with **A. P. Calderón) 'On the higher gradients of harmonic functions', *ibid.*, 24, 211–226; *MR* 29#4903.
[181] 'Józef Marcinkiewicz', in: [*M*], pp. 1–30 [English translation of [172]].

1965

[182] (with **E. M. Stein) 'On the fractional differentiability of functions', *Proc. London Math. Soc.* (3) 14A, 249–264; *MR* 31#1340.
[183] (with **S. Saks**) *Analytic Functions*, 2nd ed. of [145], enlarged, Monografie Matematyczne, Vol. 28, PWN–Polish Scientific Publishers, Warszawa, ix + 508 pp.; *MR* 31#4889.
[184] *Intégrales singulières*, Publ. Sémin. Math. d'Orsay, Univ. de Paris, 56 pp.
[185] *Trigonometricheskie ryady*, "Mir", Moskva, 1965, Vol. 1, 615 pp., Vol. 2, 537 pp. [Russian transl. of [169]]; *MR* 31#2554.

1966

[186] (with **M. Weiss**) 'An example in the theory of singular integrals', *SM* 26 (1965/1966), 101–111; *MR* 32#8067.

1967

[187] (with **E. M. Stein) 'Boundedness of translation invariant operators on Hölder spaces and L^p-spaces', *Ann. of Math.* (2) 85, 337–349; *MR* 35#5964.
[188] 'A note on the differentiability of integrals', *Colloq. Math.* 16, 199–204; *MR* 35#1732.
[189] (with **A. P. Calderón** and **M. Weiss**) 'On the existence of singular integrals', in: *Singular Integrals*, pp. 56–73, Proc. Sympos. Pure Math., Vol. 10, Amer. Math. Soc., Providence, R.I.; *MR* 49#3473.
[190] 'Preface', in: [*S*], pp. 15–18.
[191] (with **J.-P. Kahane**) 'Introduction', *ibid.*, pp. 19–39.

1968

[192] (with **M. Weiss) 'On multipliers preserving convergence of trigonometric series almost everywhere', *SM* 30, 111–120; *MR* 38#2526.
[193] (with **E. M. Stein**) 'On the boundary behavior of harmonic functions', in: *Orthogonal Expansions and their Continuous Analogues* (Proc. Conf., Edwardsville, Ill., 1967), pp. 127–141, Southern Illinois Univ. Press, Carbondale, Ill.; *MR* 38#2322.
[194] 'Mary Weiss: December 11, 1930 — October 8, 1966', *ibid.*, pp. xi–xviii; *MR* 37#2 6153.

[195] *Trigonometric Series*, reprint of [169] with corrections and some additions, Cambridge Univ. Press, London–New York, Vol. 1, xiv + 383 pp., Vol. 2, vii + 364 pp. (two volumes bound as one); *MR* 38#4882.

1969

**[196] 'On certain lemmas of Marcinkiewicz and Carleson', *J. Approx. Theory* 2, 249–257; *MR* 41#6022.

1970

[197] 'A theorem on the formal multiplication of trigonometric series', in: *Functional Analysis and Related Fields* (Proc. Conf. for M. Stone, Univ. Chicago, Ill., 1968), pp. 224–227, Springer-Verlag, Berlin–Heidelberg–New York; *MR* 43#800.

[198] (with **S. Saks**) *Fonctions analytiques*, Masson et Cie, Paris, ii + 389 pp. [French transl. of [183]]; *MR* 42#4708.

1971

[199] *Algunos resultados y problemas sobre diferenciabilidad de funciones* (lecture notes), Universidad de Madrid, Madrid, 24 pp.

[200] (with **S. Saks**) *Analytic Functions*, 3rd ed. of [145], enlarged, Elsevier Publishing Co., Amsterdam–London–New York, and PWN–Polish Scientific Publishers, Warsaw, xiv + 504 pp.; *MR* 50#2456.

[201] *Intégrales singulières*, Lecture Notes in Math., Vol. 204, Springer-Verlag, Berlin–Heidelberg–New York, iv+53 pp.; *MR* 57#7056.

1972

**[202] 'A Cantor–Lebesgue theorem for double trigonometric series', *SM* 43, 173–178; *MR* 47#711.

1973

[203] (with **A. P. Calderón**) 'Addendum to the paper "On singular integrals"', *SM* 46, 297–299; *MR* 49#3474; Errata, *ibid.*, 47, 305; *MR* 49#3475 [Addendum to [189]].

1974

[204] 'On Fourier coefficients and transforms of functions of two variables', *SM* 50, 189–201; *MR* 52#8788.

[205] (with **C. F. Fefferman**) 'Infinite series', in: *The New Encyclopaedia Britannica*, Vol. 13, pp. 470–475, 15th ed., Chicago.

[206] (with **C. F. Fefferman**) 'Fourier analysis', *ibid.*, pp. 509–525.

1975

[207] 'The role of Fourier series in the development of analysis', Proceedings of the American Academy Workshop on the Evolution of Modern Mathematics (Boston, Mass., 1974), *Historia Math.* 2, 591–594; *MR* 58# 78.

1976

[208] 'Notes on the history of Fourier series', in: *Studies in Harmonic Analysis* (Proc. Conf., DePaul Univ., Chicago, Ill., 1974), pp. 1–19, MAA Stud. Math., Vol. 13, Math. Assoc. Amer., Washington, D.C.; *MR* 56#12740.

1977

[209] (with **R. L. Wheeden**) *Measure and Integral. An Introduction to Real Analysis*, Pure and Appl. Math., Vol. 43, Marcel Dekker Inc., New York–Basel, x + 274 pp.; *MR* 58#11295.

[210] *Trigonometric Series*, reprint of [194], Cambridge Univ. Press, Cambridge–New York–Melbourne, Vol. 1, xiv + 383 pp., Vol. 2, vii + 364 pp. (two volumes bound as one); *MR* 58#29731.

1978

[211] (with **A. P. Calderón**) 'On singular integrals with variable kernels', *Applicable Anal.* 7 (1977/1978), 221–238; *MR* 58#23379.

1979

[212] (with **A. P. Calderón) 'A note on singular integrals', *SM* 65, 77–87; *MR* 80k:42022.

1981

[213] 'Steinhaus and the development of Polish mathematics', in: *The Scottish Book. Mathematics from the Scottish Café*, pp. 29–34, ed. by R. D. Mauldin, Birkhäuser, Boston–Basel–Stuttgart.

1982

[214] 'Stanisław Saks (1897–1942)', *Wiadom. Mat.* (2) 24, 145–156 [in Polish].

1987

[215] 'Aleksander Rajchman (1890–1940)', *Wiadom. Mat.* (2) 27, 219–231 [in Polish].

A note on the interpolation of linear operations

by

A. P. CALDERÓN and A. ZYGMUND (Chicago).

1. This note gives an extension of results previously obtained by the authors in [1]. A knowledge of the latter paper is not assumed here, though it could shorten the exposition below.

Let E be a measure space, i. e. a space in which a non-negative and countably additive measure μ is defined for a class of (measurable) sets. It is not assumed that the measure of the whole space is finite. Given any measurable function f defined on E and any number $r>0$, we shall write

$$\|f\|_{r,\mu}=\Big(\int_E |f|^r d\mu\Big)^{1/r}.$$

Correspondingly, $\|f\|_{\infty,\mu}$ will denote the essential upper bound of $|f|$, that is the least number M such that the set of points where $|f|>M$ is of μ-measure zero. The class of functions for which $\|f\|_{r,\mu}$ is finite will be denoted by $L^{r,\mu}$. Sometimes we shall simply write $\|f\|_r$ and L^r.

If $r\geqslant 1$, L^r is a vector space in which the distance

$$d(f_1,f_2)=\|f_1-f_2\|_r$$

of two points satisfies the usual requirements of distance in metric spaces. If $0<r<1$, this distance does not satisfy the triangle inequality. We may then either not require the triangle inequality or define the distance by the formula

$$d(f_1,f_2)=\|f_1-f_2\|_r^r=\int_E |f_1-f_2|^r d\mu.$$

In the latter case, the triangle inequality is restored, and L^r is again a metric space.

Reprinted from *SM* 12, 194–204 (1951).

A function f, defined on E, will be called *simple*, if it only takes a finite number of values and if it vanishes outside a set of finite measure (the latter condition is automatically satisfied if E itself is of finite measure). The set of all simple functions will be denoted by S. It is dense in every L^r for $0<r<\infty$. It is immediate that S is also dense in L^∞, if the measure of E is finite, though not otherwise.

In what follows we shall constantly use two facts, namely, Hölder's inequality

$$(1.1)\qquad \Big|\int_E fg\,d\mu\Big| \leqslant \|f\|_r\|g\|_{r'} \qquad \text{for} \qquad 1\leqslant r\leqslant\infty,\ r'=r/(r-1)$$

and the formula

$$(1.2)\qquad \|f\|_r=\sup_g \int fg\,d\mu \qquad \text{for} \qquad g\in S,\ \|g\|_{r'}=1,\ 1\leqslant r\leqslant\infty.$$

Let E_1 and E_2 be two measure spaces with measures μ and ν respectively. An operation $h=Tf$ will be called *of type* (r,s) if it is defined and additive for all $f\in L^{r,\mu}$, with h defined on E_2, and if there exists a finite constant M such that

$$(1.3)\qquad \|h\|_{s,\nu}\leqslant M\|f\|_{r,\mu}$$

for all f in L^r. The least value of M is the *norm* of the operation. If $0<r<\infty$, and if Tf is defined for all $f\in S$ and satisfies (1.3), then Tf can be extended to all $f\in L^r$, with the preservation of the M in (1.3), since S is dense in L^r.

M. Riesz, [2], has given a basic result about the operations which are simultaneously of two types (r,s). His result, in the form given in [1], can be stated as follows:

Theorem A. *Let E_1 and E_2 be two measure spaces with measures μ and ν respectively. Let $h=Tf$ be a linear operation defined for all simple functions f in E_1, with h defined on E_2. Suppose that T is simultaneously of the types $(1/\alpha_1, 1/\beta_1)$ and $(1/\alpha_2, 1/\beta_2)$, that is that*

$$\|Tf\|_{1/\beta_1}\leqslant M_1\|f\|_{1/\alpha_1}, \qquad \|Tf\|_{1/\beta_2}\leqslant M_2\|f\|_{1/\alpha_2},$$

the points (α_1,β_1) and (α_2,β_2) belonging to the square

$$0\leqslant\alpha\leqslant1, \qquad 0\leqslant\beta\leqslant1.$$

Then T is also of the type $(1/\alpha, 1/\beta)$ *for all*

$$\alpha=(1-t)\alpha_1+t\alpha_2 \qquad (0<t<1) \tag{1.5}$$
$$\beta=(1-t)\beta_1+t\beta_2$$

with

$$\|Tf\|_{1/\beta}\leqslant M_1^{1-t}M_2^t\|f\|_{1/\alpha}. \tag{1.6}$$

In particular, if $\alpha\neq 0$, *the operation T can be uniquely extended to the whole space* $L^{1/\alpha,\mu}$, *preserving* (1.6).

One of the aims of this note is to prove the following extension of this result:

Theorem A_1. *Theorem* A *holds if the points* (α_1,β_1) *and* (α_2,β_2) *belong to the strip*

$$0\leqslant\alpha\leqslant 1, \qquad 0\leqslant\beta<\infty. \tag{1.7}$$

One may ask what is the interest of this generalization if in applications we encounter, almost exclusively, operations of type (r,s), where both r and s are not less than 1.

This is the reason. If the measure of E_2 is finite, then, as Hölder's inequality shows, every operation of type (r,s) is automatically of type (r,s_1) for $0<s_1<s$, and it is natural to inquire about the behavior of the norm of the operation as a function of the point (r,s). A more serious justification of Theorem A_1 is its application to linear operations defined on the classes H^r (see below), where r is any positive number. The restriction of s in Theorem C below to values $\geqslant 1$ while r itself is assumed to be merely positive, is unnatural. Sometimes we really need an interpolation of operations of type (r,s) when both r and s are positive. Theorem C_1 below, which is the main result of this note, gives such an interpolation. Theorem A_1 (as well as Theorem B_1) will serve as a step in the proof of Theorem C_1.

2. We now pass to the proof of Theorem A_1. This proof uses the same basic idea as the proof of Theorem A (see [1]), supplemented by a simple device necessitated by the fact that the relation (1.1) and (1.2) fail for $0<r<1$.

Let (α_1,β_1) and (α_2,β_2) belong to the strip (1.7). Let $k>0$ be so small that

$$k\beta_1<1, \qquad k\beta_2<1,$$

and let (α,β) be given by (1.5). We observe that $k\beta<1$. Hence

$$\|Tf\|_{1/\beta}^{k}=\||Tf|^{k}\|_{1/\beta k}=\sup_{g}\int_{E_2}|Tf|^{k}g\,d\nu. \tag{2.1}$$

Here g is simple and $\|g\|_{1/(1-\beta k)}=1$. We may assume that $\|f\|_{1/\alpha}=1$, $g\geqslant 0$. Let us fix f and g, write

$$f=|f|e^{iu},$$

and consider the integral

$$I=\int_{E_2}|Tf|^{k}g\,d\nu. \tag{2.2}$$

Denoting by $\alpha(z)$ and $\beta(z)$ the functions (1.5), where t is replaced by z, we consider the functions

$$F_z=|f|^{\frac{\alpha(z)}{\alpha}}e^{iu},\qquad G_z=g^{\frac{1-k\beta(z)}{1-k\beta}}$$

and the integral

$$\Phi(z)=\int_{E_2}|TF_z|^{k}|G_z|\,d\nu. \tag{2.3}$$

This integral reduces to I for $z=t$ (since $g\geqslant 0$).

It is easily seen that G_z and TF_z are linear combinations of functions λ^z with $\lambda>0$ and with coefficients functions defined on E_2. Thus $|F_z|^k|G_z|$ is for every point in E_2 a continuous and subharmonic function in z, for $0\leqslant x\leqslant 1$ $(z=x+iy)$.

It is also bounded there. For let $c_1,c_2,\ldots$ and $c_1',c_2',\ldots$ be the various values taken by the functions f and g respectively, and let $\chi_1,\chi_2,\ldots$ and $\chi_1',\chi_2',\ldots$ be the characteristic functions of the sets where they are taken. Writing $c_j=|c_j|e^{iu_j}$, we have, for $0\leqslant x\leqslant 1$,

$$F_z=\sum e^{iu_j}|c_j|^{\frac{\alpha(z)}{\alpha}}\chi_j,$$

$$|TF_z|^k=|\sum e^{iu_j}|c_j|^{\frac{\alpha(z)}{\alpha}}T\chi_j|^k\leqslant \text{const}\cdot\sum|T\chi_j|^k,$$

$$|G_z|=\left|\sum c_l'^{\frac{1-k\beta(z)}{1-k\beta}}\chi_l'\right|\leqslant\text{const}\cdot\sum\chi_l',$$

$$|\Phi(z)|\leqslant\text{const}\cdot\int_{E_2}\sum|T\chi_j|^k\chi_l'\,d\nu=\text{const}\cdot\sum\int_{E_{2,l}}|T\chi_j|^k\,d\nu, \tag{2.4}$$

where $E_{2,l}$ is the subset of E_2 where $\chi_l'\neq 0$. Thus $E_{2,l}$ is of finite measure. Taking k so small that $k<1/\beta_1$ and applying Hölder's inequality so as to introduce the integrals $\int |T\chi_j|^{1/\beta_1}$, which are finite by assumption, we see that the right side of (2.4) is finite, which proves the boundedness of $\Phi(z)$.

Let us consider any z with $x=0$. The real parts of $a(z)$ and $\beta(z)$ are a_1 and β_1. An application of Hölder's inequality to (2.3) gives

$$|\Phi(z)| \leqslant \|TF_z\|_{1/\beta_1}^k \|G_z\|_{1/(1-k\beta_1)} \leqslant M_1^k \|F_z\|_{1/a_1}^k \|G_z\|_{1/(1-k\beta_1)}.$$

On account of our assumptions concerning f and g,

$$\|F_z\|_{1/a_1} = \| |f|^{a_1/a} \|_{1/a_1} = \|f\|_{1/a}^{a_1/a} = 1^{a_1/a} = 1,$$

$$\|G_z\|_{1/(1-k\beta_1)} = \| |g|^{\frac{1-k\beta_1}{1-k\beta}} \|_{1/(1-k\beta_1)} = \|g\|_{1/(1-k\beta)}^{\frac{1-k\beta_1}{1-k\beta}} = 1.$$

Hence $|\Phi(z)| \leqslant M_1^k$ on the line $x=0$. Similarly $|\Phi(z)| \leqslant M_2^k$ for $x=1$. Hence $I=\Phi(t) \leqslant M_1^{k(1-t)} M_2^{kt}$. Applying (2.1), we get (1.6).

In the foregoing argument we tacitly used the assumption that $a>0$. If $a=0$, then also $a_1=a_2=0$.

The assumption of Theorem A_1 can then be written

$$\|Tf\|_{1/\beta_j} \leqslant M_j \operatorname{ess\,sup} |f| \qquad (j=1,2),$$

and a simple application of Hölder's inequality (valid for all β_1, β_2 non-negative and finite) gives

$$\|Tf\|_{1/\beta} \leqslant M_1^{1-t} M_2^t \operatorname{ess\,sup} |f|.$$

3. Theorem B. *Let E and $E_1, E_2, \ldots, E_n$ be measure spaces with measures ν and $\mu_1, \mu_2, \ldots, \mu_n$ respectively. Let $h=T[f_1, f_2, \ldots, f_n]$ be a multilinear (i. e. linear in each f_j) operation defined for simple functions f_j on E_j $(j=1,2,\ldots,n)$. The functions h are defined on E. Suppose that T is simultaneously of the types*

$$(1/a_1^{(1)}, 1/a_2^{(1)}, \ldots, 1/a_n^{(1)}, 1/\beta^{(1)}) \quad \textit{and} \quad (1/a_1^{(2)}, 1/a_2^{(2)}, \ldots, 1/a_n^{(2)}, 1/\beta^{(2)}),$$

that is that

$$(3.1) \qquad \|T[f_1, f_2, \ldots, f_n]\|_{1/\beta^{(k)}} \leqslant M_k \|f_1\|_{1/a_1^{(k)}} \ldots \|f_n\|_{1/a_n^{(k)}} \qquad (k=1,2),$$

where

$$(3.2) \qquad 0 \leqslant \beta^{(k)} \leqslant 1, \qquad 0 \leqslant a_j^{(k)} \leqslant 1 \quad (k=1,2;\ j=1,2,\ldots,n).$$

Then T is also of the type $(1/\alpha_1, 1/\alpha_2, \ldots, 1/\alpha_n, 1/\beta)$ *for*

$$\alpha_j = (1-t)\alpha_j^{(1)} + t\alpha_j^{(2)}, \qquad \beta = (1-t)\beta^{(1)} + t\beta^{(2)} \qquad (0<t<1),$$

and satisfies the inequality

$$\|T[f_1, f_2, \ldots, f_n]\|_{1/\beta} \leqslant M_1^{1-t} M_2^t \|f_1\|_{1/\alpha_1} \cdots \|f_n\|_{1/\alpha_n}. \tag{3.3}$$

If, in addition, all the α_j *are positive, T can be extended by continuity to* $L_{1/\alpha_1} \times L_{1/\alpha_2} \times \ldots \times L_{1/\alpha_n}$, *preserving* (3.3).

This theorem was proved in [1]. Here we shall prove the following generalization:

Theorem B_1. *Theorem* B *holds if the points* $(\alpha_1^{(k)}, \alpha_2^{(k)}, \ldots, \alpha_n^{(k)}, \beta^{(k)})$ *satisfy, instead of* (3.2), *the condition*

$$0 \leqslant \alpha_j^{(k)} \leqslant 1, \qquad 0 \leqslant \beta^{(k)} < \infty \qquad (k=1,2). \tag{3.5}$$

The proof is obtained by a modification of the proof of Theorem B (see [1]), the same modification which extended Theorem A to Theorem A_1. We may be brief here. Let us assume that the numbers $\alpha_1, \alpha_2, \ldots, \alpha_n$ are all positive, and let k be a positive number, so small that both $k\beta^{(1)}$ and $k\beta^{(2)}$ are <1. Let us fix simple functions $f_1, f_2, \ldots, f_n$ with $\|f_j\|_{1/\alpha_j} = 1$ for $j=1,2,\ldots,n$, and a non-negative simple function g satisfying $\|g\|_{1/(1-k\beta)} = 1$. We fix t in (3.3), write $f_j = |f_j| e^{iu_j}$ and consider the integral

$$\Phi(z) = \int_E |T[|f_1|^{\alpha_1(z)/\alpha_1} e^{iu_1}, \ldots, |f_n|^{\alpha_n(z)/\alpha_n} e^{iu_n}]|^k g^{\frac{1-k\beta(z)}{1-k\beta}} dv, \tag{3.6}$$

which for $z=t$ reduces to

$$I = \int_E |T[f_1, f_2, \ldots, f_n]|^k g \, dv.$$

Since g and f_j are simple functions, the integrand in (3.6) is, for each point in E, a continuous subharmonic function of z. Hence $\Phi(z)$ is a subharmonic function of z, continuous and bounded in every vertical strip of finite width of the z-plane (the proof is the same as in the case of Theorem A_1). For $x=0$ Hölder's inequality gives

$$|\Phi(z)| \leqslant \| g^{\frac{1-k\beta^{(1)}}{1-k\beta}} \|_{1/(1-k\beta^{(1)})} \| |T[\ldots, |f_j|^{\alpha_j(z)/\alpha_j} e^{iu_j}, \ldots]|^k \|_{1/k\beta^{(1)}}$$

$$\leqslant 1 \cdot M_1^k \prod_j \| |f_j|^{\alpha_j^{(1)}/\alpha_j} \|^k_{1/\alpha_j^{(1)}} = M_1^k.$$

Similarly, $|\Phi(z)|\leqslant M_2^k$ for $x=1$. Hence $I=\Phi(t)\leqslant M_1^{k(1-t)}M_2^{kt}$. Since the upper bound of I for all simple g's with $\|g\|_{1/(1-k\beta)}=1$ gives $\|T[f_1,\dots,f_n]\|_{1/\beta}^k$, the inequality (3.4) follows when $\|f_j\|_{1/a_j}=1$ for all j, and so also for all simple f_j.

Let us now suppose that some of the a_j, but not all of them, are zero. The case $a_1=0$, $a_2\neq 0,\dots,a_n\neq 0$ is entirely typical. Then also $a_1^{(1)}=a_1^{(2)}=0$. For fixed $f_1, T[f_1,f_2,\dots,f_n]$ is a multilinear operation in $f_2,\dots,f_n$, and the assumption (3.1) can be written

$$T[f_1,f_2,\dots,f_n]_{1/\beta^{(k)}}\leqslant M_k'\|f_2\|_{1/a_2^{(k)}}\cdots\|f_n\|_{1/a_n^{(k)}} \qquad (k=1,2), \tag{3.7}$$

where $M_k'=M_k \operatorname{ess\,sup}|f_1|$. By the case already dealt with, the left side of (3.7) does not exceed

$$M_1'^{1-t}M_2'^{t}\|f_2\|_{1/a_1}\cdots\|f_n\|_{1/a_n}=M_1^{1-t}M_2^{t}\|f_1\|_{1/a_1}\|f_2\|_{1/a_2}\cdots\|f\|_{1/a_n}.$$

The case $a_1=a_2=\dots=a_n=0$ is disposed of similarly as for $n=1$.

It remains to show that if all the a_j are positive and if (3.4) is valid for simple f_j, then T can be extended by continuity to $L^{1/a_1}\times L^{1/a_2}\times\dots\times L^{1/a_n}$.

Suppose first that $0\leqslant\beta\leqslant 1$. Then

$$\begin{aligned}
&\|T[f_1^{(1)},f_2^{(1)},\dots,f_n^{(1)}]-T[f_1^{(2)},f_2^{(2)},\dots,f_n^{(2)}]\|_{1/\beta}\\
&\leqslant\|T[f_1^{(1)},f_2^{(1)},\dots,f_n^{(1)}]-T[f_1^{(2)},f_2^{(1)},\dots,f_n^{(1)}]\|_{1/\beta}\\
&+\|T[f_1^{(2)},f_2^{(1)},\dots,f_n^{(1)}]-T[f_1^{(2)},f_2^{(2)},\dots,f_n^{(1)}]\|_{1/\beta}\\
&+\dots\dots\dots\dots\dots\dots\dots\dots\\
&+\|T[f_1^{(2)},f_2^{(2)},\dots,f_n^{(1)}]-T[f_1^{(2)},f_2^{(2)},\dots,f_n^{(2)}]\|_{1/\beta},
\end{aligned}\tag{3.8}$$

which shows, on account of (3.4), that the left side of (3.8) is small if all $\|f_j^{(1)}-f_j^{(2)}\|_{1/a_j}$ $(j=1,2,\dots,n)$ are small and all $\|f_j^{(1)}\|_{1/a_j}$ and $\|f_j^{(2)}\|_{1/a_j}$ are $O(1)$. If $\beta>1$, we consider instead of (3.8) a similar inequality with norms $\|\dots\|_{1/\beta}$ replaced by $\|\dots\|_{1/\beta}^{1/\beta}$.

4. We are now going to discuss operations defined for the functions of a class H^r, $r>0$, that is for functions

$$F(z)=c_0+c_1z+c_2z^2+\dots$$

regular in the unit circle $|z|<1$ and such that the expression

$$\left\{\frac{1}{2\pi}\int_0^{2\pi}|F(\varrho e^{i\theta})|^r d\theta\right\}^{1/r}$$

remains bounded as $\varrho\to 1$. The limit of this expression for $\varrho\to 1$ then exists and will be denoted by $\|F\|_r$. It is very well known that

$$\|F\|_r=\left\{\frac{1}{2\pi}\int_0^{2\pi}|F(e^{i\theta})|^r d\theta\right\}^{1/r},$$

where $F(e^{i\theta})$ denotes the non-tangential boundary values of $F(z)$.

An operation

$$h=T[F]$$

will be called *of type* (r,s) if it is defined for all $F\epsilon H^r$, satisfies $T[\lambda_1F_1+\lambda_2F_2]=\lambda_1T[F_1]+\lambda_2T[F_2]$ for all constants λ_1,λ_2, and if there is an M independent of F and such that

$$\|h\|_s\leqslant M\|F\|_r. \tag{4.1}$$

Here h is supposed to belong to some fixed $L^{s,\nu}$ and $\|h\|_s=\|h\|_{s,\nu}$.

If $T[F]$ is initially defined only for all polynomials

$$p(z)=d_0+d_1z+\ldots+d_kz^k,$$

and satisfies (4.1), T can be uniquely extended to all F in H^r, with the preservation of the M in (4.1), since the set of all polynomials $p(z)$ is dense in every H^r.

The following theorem was established in [1] (see also [3] and [4]):

Theorem C. *Let* (α_1,β_1) *and* (α_2,β_2) *be two points of the strip*

$$0<\alpha<\infty,\qquad 0\leqslant\beta\leqslant 1. \tag{4.2}$$

Let T *be a linear operation defined for all polynomials* p, *whose values are measurable functions in a measurable space* E, *with measure* ν, *and such that*

$$\|Tp\|_{1/\beta_1}\leqslant M_1\|p\|_{1/\alpha_1},\qquad \|Tp\|_{1/\beta_2}\leqslant M_2\|p\|_{1/\alpha_2}. \tag{4.3}$$

Then for every point (α,β) *of the segment*

$$\alpha=\alpha_1(1-t)+\alpha_2t,\qquad \beta=\beta_1(1-t)+\beta_2t\qquad (0<t<1),$$

we have the inequality

$$\|Tp\|_{1/\beta}\leqslant KM_1^{1-t}M_2^t\|p\|_{1/\alpha}, \tag{4.4}$$

K denoting a constant depending on α_1,α_2 *only.*

In particular, T *can be extended to the whole space* $H^{1/\alpha}$ *with the preservation of* (4.4).

This result will now be generalized as follows:

Theorem C_1 *). *Theorem C holds if the strip (4.2) is replaced by the quadrant*

$$0<\alpha<\infty, \qquad 0\leqslant\beta\leqslant\infty.$$

Let us suppose that

$$\alpha_1\leqslant\alpha_2,$$

and let us fix a positive integer n so large that $\alpha_2/n<1$. Hence also $\alpha_1/n<1$.

For any system of n simple complex-valued functions $g_1, g_2,\dots,g_n$ defined on the interval $(0,2\pi)$ we set

$$T^*[g_1,g_2,\dots,g_n]=T[F_1F_2\dots F_n], \tag{4.5}$$

where

$$F_j(z)=\frac{1}{2\pi}\int_0^{2\pi}\frac{e^{it}+z}{e^{it}-z}g_j(t)\,dt \qquad (j=1,2,\dots,n). \tag{4.6}$$

Recalling the very well known fact that, for every $g\in L^r$ with $1<r<\infty$, the function

$$F(z)=\frac{1}{2\pi}\int_0^{2\pi}\frac{e^{it}+z}{e^{it}-z}g(t)\,dt \tag{4.7}$$

satisfies the inequality

$$\|F\|_r\leqslant A_r\|g\|_r, \tag{4.8}$$

we see that each F_j belongs to H^r, no matter how large is r. Hence also $F_1F_2\dots F_n$ belongs to every H^r, and in particular to both H^{1/α_1} and H^{1/α_2}. On account of (4.3), the operation T is extensible both in H^{1/α_1} and H^{1/α_2}, without increase of the norm. The extensions are the same for functions common to both classes, since these extensions are almost everywhere ordinary limits of the same sequence Tp_j. Thus

$$\|T[F_1F_2\dots F_n]\|_{1/\beta_k}\leqslant M_k\|F_1F_2\dots F_n\|_{1/\alpha_k} \qquad (k=1,2).$$

*) The proof given here of Theorem C_1 is (assuming the validity of Theorem B_1) essentially the same as the proof, given in [1], of Theorem C. We repeat the proof of Theorem C_1 here to make the present note self-contained.

Using Hölder's inequality and (4.8) we have

$$\|F_1F_2\dots F_n\|_{1/\alpha_k}=\|F_1\|_{n/\alpha_k}\dots\|F_n\|_{n/\alpha_k}$$
$$\leqslant(A_{n/\alpha_k})^n\|g_1\|_{n/\alpha_k}\dots\|g_n\|_{n/\alpha_k}.$$

Hence, from the definition of T^*,

$$(4.9)\qquad \|T^*[g_1,g_2,\dots,g_n]\|_{1/\beta_k}\leqslant M_k(A_{n/\alpha_k})^n\|g_1\|_{n/\alpha_k}\dots\|g_n\|_{n/\alpha_k}.$$

An application of Theorem B_1 gives

$$(4.10)\qquad \|T^*[g_1,g_2,\dots,g_n]\|_{1/\beta_k}\leqslant(A^{1-t}_{n/\alpha_1}A^{t}_{n/\alpha_2})^n(M_1^{1-t}M_2^t)^n\prod_j\|g_j\|_{n/\alpha}.$$

Formula (4.5) defines T^* when $g_1,g_2,\dots,g_n$ are simple. The formulae (4.9) show that T^* can be extended to $L^{n/\alpha_k}\times\dots\times L^{n/\alpha_k}$ ($k=1,2$) and that the extension satisfies (4.10). But if $g_j\in L^{n/\alpha_k}$, then the F_j in (4.6) belongs to H^{n/α_k}. Hence $F_1F_2\dots F_n$ belongs to H^{1/α_k}, which means that $T[F_1F_2\dots F_n]$ is defined. We shall show that (4.5) *holds for the extended* T.

For if the g_j belong to L^{n/α_k}, and if g_j^m are simple functions such that $\|g_j^m-g_j\|_{n/\alpha_k}\to0$ as $m\to\infty$, then

$$\|T^*[g_1^m,\dots,g_n^m]-T^*[g_1,\dots,g_n]\|_{1/\beta_k}\to0,$$

by the argument used at the end of Section 3. On the other hand, if F_j^m is derived from g_j^m by means of the formula (4.6), we have

$$\|F_j^m-F_j\|_{n/\alpha_k}\to0,\qquad \|F_j^m\|_{n/\alpha_k}\leqslant A_{n/\alpha_k}\|g_j^m\|_{n/\alpha_k}=O(1),$$

so that, as in (3.8) (or in its analogue for $\beta_k>1$) but using (4.3) in the proof,

$$\|T[F_1^m\dots F_n^m]-T[F_1\dots F_n]\|_{1/\beta_k}\to0,$$

which proves (4.5) in the case considered.

We are now going to prove that for a fixed polynomial p we have (4.4).

Let $B(z)$ be the *Blaschke product* of $p(z)$, that is the product of the factors

$$\frac{z-a_j}{1-\bar a_j z}$$

extended over all the zeros a_j of $p(z)$ situated in $|z|<1$. Thus

$$p(z)=e^{i\gamma}B(z)G(z),$$

where γ is a real constant and $G(z)$ a polynomial without zeros in $|z|<1$ and satisfying the condition $\operatorname{Im} G(0)=0$. Without loss of generality we may assume that $\gamma=0$. Hence

$$(4.11)\quad p=F_1F_2\ldots F_n,\ \text{where}\ F_1=BG^{1/n}, F_2=F_3=\ldots=F_n=G^{1\,n}.$$

All the functions F_j are bounded, and so also of the class H^{n/a_1}. Assuming as we may, that $\operatorname{Im} G^{1/n}(0)$, we see that each F_j is representable by the formula (4.6), where the g_j are of the class L^{n/a_1} and real-valued. Hence

$$Tp=T[F_1F_2\ldots F_n]=T^*[g_1,g_2,\ldots,g_n].$$

The functions g_j also belong to $L^{n/a}$ (because $a\geqslant a_1$ or simply because they belong to every L^r, $r>0$). But the formula (4.5), which was initially established for g_j simple, shows that the operation can be extended to $L^{n/a}\times L^{n/a}\times\ldots\times L^{n/a}$, with the preservation of the inequality (4.10). Combining (4.5) with (4.11) we get

$$\|Tp\|_{1/\beta}=\|T^*[g_1,g_2,\ldots,g_n]\|_{1/\beta}$$

$$\leqslant(A^{1-t}_{n/a_1}A^t_{n/a_2})^n M_1^{1-t}M_2^t\prod_j\left\{\int_0^{2\pi}|g_j(t)|^{n/a}dt\right\}^{a/n}.$$

The last product Π here does not exceed

$$\prod_j\left\{\int_0^{2\pi}|F_j(e^{it})|^{n/a}dt\right\}^{a/n}=\prod_j\left\{\int_0^{2\pi}|G(e^{it})|^{1/a}dt\right\}^{a/n}=(2\pi)^{1/a}\|p\|_{1/a},$$

which gives (4.4) with

$$K=(2\pi)^{1/a_1}\delta^n,\quad\text{where}\quad \delta=\max(A_{n/a_1},A_{n/a_2}).$$

Bibliography.

[1] A. P. Calderón and A. Zygmund, *On the theorem of Hausdorff-Young*, *Contributions to Fourier Analysis*, Annals of Mathematics Studies 25, p. 166-168, Princeton University Press (1950).

[2] M. Riesz, *Sur les maxima des formes bilinéaires et sur les fonctionnelles linéaires*, Acta Mathematica 49 (1926), p. 465-497.

[3] R. Salem and A. Zygmund, *A Convexity Theorem*, Proc. of National Acad. of Sciences 34 (1948), p. 443-447.

[4] G. O. Thorin, *Convexity Theorem*, Uppsala 1948, p. 1-57.

OHIO STATE UNIVERSITY AND THE UNIVERSITY OF CHICAGO

(Reçu par la Rédaction le 9. 4. 1951).

Zam. 38/509/51— 100 — Wrocławska Drukarnia Naukowa. — F-2-11471.

POLISH MATHEMATICS BETWEEN THE TWO WARS (1919–39)

A. Zygmund, *University of Chicago*

The Polish mathematical school of the period 1919–1939 was an interesting phenomenon: first, because of its achievements, and secondly, because of the place and circumstances in which it arose.

The Second World War was disastrous to Poland in every respect. It brought about very severe losses in the cultural life of the country through the destruction both of the human element and of scientific workshop. In this respect, the losses of Polish mathematics were particularly heavy. A large number of prominent mathematicians and of promising youth died or were killed. A number of others, dispersed by war, have not returned to Poland. Finally, those who survived and stayed in the country are faced with the problem of rebuilding academic life and will not be able, for some time to come, to devote all their time to mathematical research. Thus a certain period in Polish mathematics is definitely over and, looking backwards, already from a certain distance, one can objectively appraise its achievements.

One might say that before 1919 there had been Polish mathematicians but there was no Polish mathematical school. The rapid growth of Polish mathematics after 1919 was partly spontaneous, helped by the recent freeing of the country from foreign occupation, and partly a result of thoughtful planning.

The development of Polish mathematics is in the first place due to Janiszewski, Mazurkiewicz and Sierpinski in Warsaw, and to Banach and Steinhaus in Lwów. The role of Janiszewski here was particularly significant and unique. Born in 1888, he died in 1920, at the beginning of the period

Reprinted from *Proceedings of the Second Canadian Mathematical Congress, Vancouver, 1949*, pp. 3–9 (1951).

which interests us, and so did not live to see the fruition of his ideas, but he had been the chief planner of the Polish school. A talented mathematician (topologist) himself, he realized the difficulties of organizing good mathematical research in a country without strong and continuous mathematical tradition. His idea was that the surest and quickest way to success here would be first through concentration on a particular mathematical discipline which would be the main source of interest and of problems for a larger group of mathematicians, and secondly, through starting a mathematical publication specializing in this selected branch of mathematics. This specialization was contemplated as an initial stage only. Once a strong point was established, gradual extension of interest to other fields of mathematics was expected.

JANISZEWSKI expounded his ideas in a number of articles (see e.g., *Nauka Polska* [Polish Science], vol. I, 1917, Warszawa, pp. 11–18), and rereading these now one can but admire the sharpness of his vision and the correctness of his ideas. Incidentally, his remarks would be of interest even now, since we still refuse to see the folly of a haphazard dispersal of mathematical papers over a very large number of mathematical publications some of which are accessible with great difficulty, often for economic reasons.

At that time, the Theory of Sets, Topology, and Real Variable were attracting a number of Polish mathematicians. It was natural to make a starting point here, and in 1920 the first volume of the publication *Fundamenta Mathematicae* appeared in Warsaw. It was a success from the start. It gave an outlet to Polish mathematical production and attracted foreign papers. Before September of 1939, thirty-two volumes of the *Fundamenta* had been published.

The appearance of the *Fundamenta* contributed much to the development of Polish mathematics, and especially of the Warsaw school, which centered around MAZURKIEWICZ and SIERPIŃSKI. The main factors, however, in this development, were the personalities of Mazurkiewicz and Sierpinski themselves, and the ideas of Janiszewski, which influenced his pupils even after Janiszewski's death. The devotion of the teachers to mathematical research, their lectures and seminars, and above all personal contact, had a strong influence upon young students of mathematics in Warsaw. Among pupils of Mazurkiewicz and Sierpiński were KURATOWSKI, KNASTER, SAKS, ZALCWASSER, TARSKI, WUNDHEILER, ZARANKIEWICZ, LINDENBAUM, BORSUK, EILENBERG, SZPILRAJN-MARCZEWSKI, LUBELSKI, KERNER, and several others, including the writer of this note. One should also list here the names of NEYMAN, NIKODYM, and RAJCHMAN, though those people joined the Warsaw group as mature mathematicians, with independent interests.

With time, the interests of the group became more varied and differentiated. Some of the younger generation went in the direction of trigonometric series (mainly under the influence of RAJCHMAN), Analytic Functions, Mathematical Statistics and Probability (with NEYMAN), Linear Operations, Theory of Numbers, etc. Such a development had been foreseen by JANISZEWSKI and reassured the critics who had feared that the initial specialization in the Theory of Sets might have a detrimental effect upon the development of mathematics in the country. A few years before 1939 a new mathematical periodical, *Acta Arithmetica*, devoted to the Theory of Numbers, was started in Warsaw.

Actually the danger of overspecialization was never very great since almost simultaneously with the Warsaw school another centre of mathematical research was developing in Poland. The centre was in Lwów and the leading people were STEINHAUS and BANACH. The main interest of the school was Functional Analysis, though later strong interest developed in the Calculus of Probability (mainly through Steinhaus) and the theory of the Potential. In 1930 the group started a new publication *Studia Mathematica*, with the intention of attracting, in the first place, papers devoted to Functional

Analysis and its applications. Before 1939 eight volumes of the *Studia* had appeared. In addition to the founders of the group one should mention here their pupils: SCHAUDER, KACZMARZ, NIKLIBORC, MAZUR, AUERBACH, ORLICZ, ULAM, KAC, and many others.

Another mathematical enterprise of that period in Poland was the series of mathematical monographs, *Monografje Matematyczne*, of which ten volumes appeared before 1939. The series contains some very well known books like BANACH'S on Linear Operations, SAKS'S on Integration, KURATOWSKI'S on Topology and SIERPIŃSKI'S on the Hypothesis of the Continuum.

Though it was outside mathematics proper, the Warsaw school of mathematical logic should be mentioned here. It was led by LUKASIEWICZ, LEŚNIEWSKI and TARSKI. There was considerable interchange of ideas between the logical and the mathematical groups.

Warsaw and Lwów occupied exceptional positions in this picture of Polish mathematics because of the presence of "schools" there. However, good mathematical work was also being done at other Polish universities, in Cracow, Poznan and Wilno. Especially Cracow must be singled out here. It was the oldest Polish university (founded in 1364) and had the strongest mathematical tradition in the country, mainly through the presence of ZAREMBA. The Cracow group also comprised ROSENBLATT, WILKOSZ, WAZEWSKI, LEJA, GOLAB. The publication *Annales de la Société Polonaise de Mathématiques* was the organ of the Cracow group. In Wilno one should mention KEMPISTY, MARCINKIEWICZ, and S. K. ZAREMBA; in Poznań, SLEBODZIŃSKI and BIERNACKI.

It is not the purpose of this brief note to give an exhaustive picture of Polish mathematics between the two wars. Many names and many facts are omitted here. The question, however, naturally suggests itself, what caused this eruption of mathematical activity in a country like Poland, where there seemed to exist little mathematical tradition. The achieve-

of Polish mathematicians both in the country and abroad, but also published translations of foreign papers, gave information about mathematical developments abroad, and kept the interest in mathematics alive. To him part of the credit for later development of Polish mathematics must be given.

It remains to say a few words about the effect of the last war upon mathematics in Poland. From the point of view of personal losses, the picture is a grim one. About fifty actively working mathematicians died or were killed, half of the number in existence in 1939. These figures are taken from the article by Marczewski, "The Development of Mathematics in Poland" in Polish, with a French summary, Cracow, 1948, pp. 1–46. Of the older generation, ZAREMBA, DICKSTEIN, PRZEBORSKI, HOBORSKI, and KWIETNIEWSKI are dead; KEMPISTY was arrested and died in prison. KACZMARZ was killed in the campaign of September 1939. MARCINKIEWICZ, mobilized, was taken prisoner and disappeared without trace. SAKS was killed when trying to escape from prison. The following were killed as a result of a deliberate policy of extermination: RAJCHMAN, ZALCWASSER, KERNER, JACOB, LINDENBAUM, A. LOMNICKI, STOZEK, RUZIEWICZ, BARTEL, SCHAUDER, AUERBACH, SCHREIER, EIDELHEIT, KOLODZIEJCZYK, and many others. MAZURKIEWICZ and BANACH died immediately after the war, as a result of wartime difficulties.

Owing to territorial changes brought about by the war, Wilno and Lwów are no longer within the boundaries of Poland. The University of Wilno was transferred to Toruń, that of Lwów to Wroclaw (Breslau). The publication of the *Studia* has been resumed in the latter place, and another publication was started there, *Colloquium Mathematicum*. Warsaw has resumed publication of the *Fundamenta*, of *Mathematical Monographs*, and of *Prace Matematyczno-Fizyczne*. In Cracow again appear the *Annals of the Polish Mathematical Society*. *Acta Arithmetica* and *Wiadomości Matematyczne* have not so far been resumed. The libraries of the Mathematical Institutes in Wilno and Lwów are no longer available. The

ments of Polish mathematics in a relatively brief period of twenty years can objectively be considered as remarkable. Furthermore, though in independent Poland (after 1918) there has been a general and very strong upsurge in scientific activity, the advance of mathematics seems to be ahead of the advances in other branches of science.

A complete answer to our question is impossible to give. As in many biological phenomena, there are elements here which defy rational explanation. Partial explanation is, however, possible, and two reasons may be given here. The first is the more obvious one. A human group linked by deep interest in mathematics and by devotion to mathematical research is bound to attract young and talented people and so propagate itself in ever growing measure. The development is even more rapid if there are common mathematical problems and collaboration in their solution. This is the case of Warsaw and Lwów.

Another reason for the success of mathematics in Poland was that, in spite of appearances, there had been some mathematical tradition there. The nineteenth century, especially the second half of it, was a very painful period in the history of Poland. It is enough to say, in connection with the questions which interest us, that for almost fifty years, up to 1915, the lectures at the university of Warsaw were not given in Polish. While there were prominent Polish mathematicians in the first half of the nineteenth century (one may mention the name of WROŃSKI), the second half of it saw a change for the worse. Many scientists, among them mathematicians, in order to do their research, had to emigrate. Some of them kept contact with the old country and so preserved the scientific tradition there. In mathematics an inestimable role was played by S. DICKSTEIN. For several decades he published in Warsaw, at his own expense, two mathematical publications, *Prace Matematyczno-Fizyczne* and *Wiadomości Matematyczne*, the only mathematical periodicals in Poland at that time. Not only did he try to attract here the papers

library of the Warsaw Mathematical Seminar was completely destroyed by an incendiary bomb during an aerial attack against the city. Part of the losses in books were made good by the generous help of American mathematicians organized by J. R. KLINE.

Thus the task of rehabilitation of Polish mathematics to former prominence will not be an easy one. If, however, the successes of the past are any indication as to the possibilities of the future, this task will be accomplished. We all wish our Polish colleagues every success in their efforts.

ON THE EXISTENCE OF CERTAIN SINGULAR INTEGRALS.

By

A. P. CALDERON and A. ZYGMUND

Dedicated to Professor MARCEL RIESZ, on the occasion of his 65th birthday

Introduction.

Let $f(x)$ and $K(x)$ be two functions integrable over the interval $(-\infty, +\infty)$. It is very well known that their composition

$$\int_{-\infty}^{+\infty} f(t)\, K(x-t)\, dt$$

exists, as an absolutely convergent integral, for almost every x. The integral can, however, exist almost everywhere even if K is not absolutely integrable. The most interesting special case is that of $K(x) = 1/x$. Let us set

$$\tilde{f}(x) = \frac{1}{\pi}\int_{-\infty}^{+\infty} \frac{f(t)}{x-t}\, dt.$$

The function $\tilde{f}$ is called the conjugate of f (or the Hilbert transform of f). It exists for almost every value of x in the Principal Value sense:

$$\tilde{f}(x) = \lim_{\varepsilon \to 0} \frac{1}{\pi}\left(\int_{-\infty}^{x-\varepsilon} + \int_{x+\varepsilon}^{\infty}\right) \frac{f(t)}{x-t}\, dt.$$

Moreover it is known (See [9] or [7], p. 317) to satisfy the M. Riesz inequality

$$\text{(1)} \qquad \left[\int_{-\infty}^{+\infty} |\tilde{f}|^p\, dx\right]^{1/p} \le A_p \left[\int_{-\infty}^{+\infty} |f|^p\, dx\right]^{1/p}, \qquad 1<p<\infty,$$

where A_p depends on p only. There are substitute result for $p = 1$ and $p = \infty$. The limit $\tilde{f}$ exists almost everywhere also in the case when $f(t)\,dt$ is replaced there by $dF(t)$, where $F(t)$ is any function of bounded variation over the whole interval $(-\infty, +\infty)$. (For all this, see e.g. [7], Chapters VII and XI, where also bibliographical references can be found).

Reprinted from *Acta Math.* 88, 85–139 (1952).

The corresponding problems for functions of several variables have been little investigated, and it is the purpose of this paper to obtain some results in this direction. To indicate the problems we are going to discuss let us consider two classical examples.

Let $f(s,t)$ be a function integrable over the whole plane, and let us consider in the half-space $z>0$ the Newtonian potential $u(x,y,z)$ of the masses with density $f(s,t)$. Thus

$$u(x,y,z)=\iint f(s,t)\frac{ds\,dt}{R}, \qquad R^2=(x-s)^2+(y-t)^2+z^2,$$

the integration being extended over the whole plane. Let us also consider the partial derivatives

$$u_z=-z\iint f(s,t)\frac{ds\,dt}{R^3}, \qquad u_x=-\iint f(s,t)\frac{x-s}{R^3}\,ds\,dt.$$

Here $-(4\pi)^{-1}u_z$ is the Poisson integral of f, and it is a classical fact that it tends to $f(x,y)$ as $z\to 0$, at every point (x,y) at which f is the derivative of its indefinite integral. On the other hand, by formally replacing z by 0 in the formula for u_x we obtain the singular integral

$$\iint f(s,t)\frac{x-s}{[(x-s)^2+(y-t)^2]^{3/2}}\,ds\,dt.$$

It can be written in the form

$$\iint f(s,t)\,K(x-s,y-t)\,ds\,dt, \tag{2}$$

with

$$K(x,y)=\frac{x}{(x^2+y^2)^{3/2}}. \tag{3}$$

It is a simple matter to show that at every point (x_0,y_0) at which f is the derivative of its indefinite integral the existence of the integral is equivalent to the existence of lim u_x, as the point (x,y,z) approaches $(x_0,y_0,0)$ non-tangentially (and that both expressions have the same value), but neither fact seems to have been established unconditionally. Here again the integral (2) is taken in the principal value sense, which in two dimensions means that first the integral is taken over the exterior of the circle with center (x_0,y_0) and radius ε, and then ε is made to tend to 0.

Another example, of a somewhat similar nature, arises from considering in the plane the logarithmic potential u of masses with density $f(s,t)$. Hence

$$u(x, y) = \iint f(s, t) \log \frac{1}{r} \, ds \, dt; \qquad r^2 = (x-s)^2 + (y-t)^2.$$

If in order to avoid unnecessary complications we assume that f vanishes in a neighborhood of infinity, then in any finite circle u is the convolution of two integrable functions, and so the integral converges absolutely almost everywhere. The integral obtained by formal differentiation, say with respect to x, is

$$(4) \qquad -\iint f(s, t) \frac{x-s}{(x-s)^2 + (y-t)^2} \, ds \, dt,$$

and so, as a convolution of two integrable functions, again converges absolutely almost everywhere and represents a function integrable over any finite portion of the plane. Using this fact one proves without difficulty (see [1]) that the integral actually represents u_x. Thus u_x and u_y exist almost everywhere.

Let us, however, differentiate the integral (4) formally once more, with respect to x and with respect to y. We get the integrals of type (2) with

$$(5) \qquad K(x, y) = \frac{x^2 - y^2}{(x^2 + y^2)^2}, \qquad K(x, y) = \frac{2xy}{(x^2 + y^2)^2},$$

respectively. These two kernels are not essentially different, since one is obtained from the other through a rotation of the axes by 45°. It may also be of interest to observe that they appear respectively as the real and imaginary parts of

$$\frac{1}{z^2} = \frac{1}{(x + iy)^2}.$$

The existence almost everywhere of the integrals (2) in the cases (5) has been established by Lichtenstein for functions f which are continuous (or, slightly more generally, Riemann integrable). This result seems not to have been superseded so far, though the existence almost everywhere of u_{xx}, u_{yy}, u_{xy} together with the relation $u_{xx} + u_{yy} = -2\pi f$ was established by Lichtenstein [6] (see also [2], [8]) in the much more general case of f quadratically integrable.

The kernels (3) and (5) have one feature in common: they are of the form

$$g(\varphi)\varrho^{-2}, \quad x = \varrho \cos \varphi, \quad y = \varrho \sin \varphi,$$

where $g(\varphi)$ is a function of angle φ (actually a trigonometric polynomial) whose mean value over $(0, 2\pi)$ is zero. Several examples of kernels of this type could be considered, but we shall now state the problem in a more general form.

Suppose we have a function $f(x_1, x_2, \dots x_n)$ integrable over the whole n-dimensional space, and a kernel

$$K(x_1, x_2, \dots x_n) = \varrho^{-n} \Omega(\alpha_1, \alpha_2, \dots, \alpha_n),$$

where $x_j = \varrho \cos \alpha_j$ for all j, and $\alpha_1, \alpha_2, \dots, \alpha_n$ are the direction angles. What can be said about the existence and the properties of the integral

$$\tilde{f}(x_1, x_2, \dots, x_n) = \int f(s_1, s_2, \dots, s_n) K(x_1 - s_1, \dots, x_n - s_n)\, ds_1 \dots ds_n \,? \tag{6}$$

An answer to this problem is our main object here.

This is the plan of the paper.

In Chapter I it will be shown that, *if* $f \in L^p$, $1 < p < \infty$, *then the integral* (6) *converges, in the metric* L^p, *to a function* $\tilde{f} \in L^p$, *provided*

a) *the mean value of* Ω *over the unit sphere is zero,*

b) *the function* $\Omega(\alpha_1, \alpha_2, \dots, \alpha_n)$ *satisfies a smoothness condition* (See Chapter II).

(In the case $p = 2$ condition b) can be considerably relaxed).

The function $\tilde{f}$ satisfies the condition analogous to (1). The cases $p = 1$ and $p = \infty$ are also investigated.

The main result of Chapter II is that under conditions a) and b) the integral (6) exists almost everywhere not only for $p > 1$, but also for $p = 1$. The result holds, if $f\, ds_1 \dots ds_n$ is replaced by $d\mu$, where μ is an arbitrary mass distribution with finite total mass. If $f \in L^p$, $p > 1$, the partial integrals of the integral (6), that is the integrals over the exterior of the sphere of radius ε and center $(x_1, x_2, \dots x_n)$ are majorized by a function of L^p, independent of ε.

Chapter III is devoted to some applications of the results previously obtained to the problem of the differentiability of the potential.

Other problems connected with our main topic will be considered in an another paper.

CHAPTER I.

Mean Convergence of Singular Integrals.

Let E^n be the n-dimensional euclidean space. If P and Q are points in E^n, $(P - Q)$ will denote either the vector going from Q to P, or the point whose coordinates are the components of $(P - Q)$. The length of $(P - Q)$ will be denoted by $|P - Q|$, and Σ will stand for the surface of the sphere of radius 1 with center at the origin of coordinates, O.

We shall be concerned with kernels of the form

$$K(P-Q) = |P-Q|^{-n}\,\Omega[(P-Q)|P-Q|^{-1}],$$

where $\Omega(P)$ is a function defined on Σ and satisfying the conditions

(1) $$\int_{\Sigma} \Omega(P)\,d\sigma = 0$$

and

$$|\Omega(P)-\Omega(Q)| \leq \omega(|P-Q|),$$

where ω is an increasing function such that $\omega(t) \geq t$, and

$$\int_0^1 \omega(t)\frac{dt}{t} = \int_1^{\infty} \omega\left(\frac{1}{t}\right)\frac{dt}{t} < \infty.$$[1]

More precisely, we shall investigate the convergence of the integral

(2) $$\tilde{f}_{\lambda}(P) = \int_{E^n} K_{\lambda}(P-Q)\,f(Q)\,dQ,$$

where $f(Q)$ is a function of L^p, $p \geq 1$ in E^n, dQ is the element of volume in E^n and

$$K_{\lambda}(P-Q) = \begin{cases} K(P-Q) & \text{if } |P-Q| \geq 1/\lambda, \\ 0 & \text{otherwise.} \end{cases}$$

Using Hölder's inequality, or the boundedness of K_{λ}, we see that (2) is absolutely convergent for $1 < p < \infty$ and $p = 1$ respectively.

We shall begin by proving that *in the case when the function f in* (2) *belongs to* L_2, $\tilde{f}_{\lambda}$ *converges in the mean of order two as* $\lambda \to \infty$.

Let

$$K_{\lambda\mu} = \begin{cases} K(P-O) & \text{if } \mu \geq |P-O| \geq 1/\lambda, \\ 0 & \text{otherwise.} \end{cases}$$

As we shall see, the Fourier transform of $K_{\lambda\mu}$ converges boundedly as μ and λ tend to infinity successively, and then the desired result will follow easily.

In polar coordinates we have the following expression for the Fourier transform $\hat{K}_{\lambda\mu}$ of $K_{\lambda\mu}$,

$$\hat{K}_{\lambda\mu}(P) = \int_{E^n} K_{\lambda\mu}(Q)\,e^{ir\varrho\cos\varphi}\,dQ = \int_{1/\lambda}^{\mu} \varrho^{-1}\,d\varrho \int_{\Sigma} \Omega(Q')\,e^{ir\varrho\cos\varphi}\,d\sigma,$$

---

[1] This implies the convergence of $\int^{\infty} \omega\left(\frac{c}{t}\right)\frac{dt}{t}$ for every $c > 0$, a fact we shall use in what follows.

It may be added that the condition $\omega(t) \geq t$ is quite harmless, since we can always replace $\omega(t)$ by Max $\{\omega(t), t\}$. The case $\omega(t) = t^{\alpha}, \alpha > 0$, is, of course, the most important one.

where $r = |P - O|$, $\varrho = |Q - O|$, $Q' = (Q - O)\,|Q - O|^{-1}$, and φ is the angle between the vectors $(P - O)$ and $(Q - O)$. Introducing the variable $s = \varrho r$ we can write

$$\hat{K}_{\lambda\mu}(P) = \int_{r/\lambda}^{r\mu} \frac{ds}{s} \int_{\Sigma} \Omega(Q')\, e^{is\cos\varphi}\, d\sigma,$$

and owing to the fact that

$$\int_{\Sigma} \Omega(Q')\, d\sigma = 0,$$

we also have

$$\hat{K}_{\lambda\mu}(P) = \int_{r/\lambda}^{r\mu} \frac{ds}{s} \int_{\Sigma} \Omega(Q')\,[e^{is\cos\varphi} - e^{-s}]\, d\sigma$$

$$= \int_{\Sigma} \Omega(Q')\, d\sigma \int_{r/\lambda}^{r\mu} \frac{e^{is\cos\varphi} - e^{-s}}{s}\, ds.$$

Now, if $\varphi \neq \frac{\pi}{2}$, the inner integral in the last expression converges as λ and μ tend to infinity, and it is not difficult to verify that it never exceeds $2 \log \frac{c}{|\cos\varphi|}$ in absolute value, where $c > 1$ is a constant. But $\Omega(Q')$ is a bounded function and the integral

$$\int_{\Sigma} \log \frac{c}{|\cos\varphi|}\, d\sigma$$

is finite, and therefore $\hat{K}_{\lambda\mu}$ is bounded and converges, as μ and λ tend to infinity successively. Therefore, if $\hat{K}_{\lambda}$ is the Fourier transform of $K_{\lambda}(P - O) \in L^2$, we have

$$\lim_{\mu \to \infty} \hat{K}_{\lambda\mu} = \hat{K}_{\lambda},$$

and $\hat{K}_{\lambda}$ converges boundedly to a function $\hat{K}$ as $\lambda \to \infty$.

Let now

$$\tilde{f}_{\lambda\mu}(P) = \int_{E^n} K_{\lambda\mu}(P - Q)\, f(Q)\, dQ.$$

Then, if $\hat{\tilde{f}}_{\lambda\mu}$ is the Fourier transform of $\tilde{f}_{\lambda\mu}$, we have

$$\hat{\tilde{f}}_{\lambda\mu} = \hat{K}_{\lambda\mu} \hat{f},$$

and since $\hat{K}_{\lambda\mu} \to \hat{K}_{\lambda}$ boundedly as $\mu \to \infty$, $\hat{\tilde{f}}_{\lambda\mu}$ converges in the mean to $\hat{K}_{\lambda} \hat{f}$. On the other hand, $\tilde{f}_{\lambda\mu}$ converges to $\tilde{f}_{\lambda}$ as $\mu \to \infty$ and therefore we have

(2 a) $$\hat{\tilde{f}}_{\lambda} = \hat{K}_{\lambda} \hat{f}.$$

Letting now λ tend to infinity, $\hat{K}_\lambda$ will converge boundedly to $\hat{K}$, and $\hat{\tilde{f}}_\lambda$ will converge in the mean to $\hat{K}\hat{f}$. Therefore $\tilde{f}_\lambda$ will converge in the mean to the Fourier transform of $\hat{K}\hat{f}$. This completes the argument.

Remark. In the above argument we used the fact that $\Omega(P')$ was merely bounded. Actually the only property of Ω we need (except for (1)) is the uniform boundedness of $\int\limits_{\Sigma} |\Omega(Q')| \log \frac{c}{|\cos\varphi|}\, d\sigma$. This condition is certainly satisfied if $|\Omega| \log^+ |\Omega|$ is integrable.

Before we pass to the general case we shall prove some lemmas which will be needed also in a later section.

Given a non-negative function $f(P)$ not identically zero in E^n, we shall denote by $f^*(t)$, $0 < t < \infty$, any non-increasing function equimeasurable with $f(P)$. If f belongs to L^p, $1 \le p < \infty$, in E^n, then $f^*(t)$ belongs to L^p in $0 < t < \infty$ and thus is integrable over every finite interval. In this case we introduce also the function

$$y = \beta_f(x) = \frac{1}{x}\int\limits_0^x f^*(t)\, dt; \quad x > 0,$$

which is continuous and either strictly decreasing or possibly constant in an interval $(0, x_0)$ and strictly decreasing for $x \ge x_0$. In both cases we have $\beta_f(x) \to 0$ as $x \to \infty$. The function inverse to $y = \beta_f(x)$ will be denoted by $x = \beta^f(y)$. If $\beta_f(x)$ tends to infinity as x tends to zero, $\beta^f(y)$ is well defined for $y > 0$. If $\beta_f(x)$ is bounded, $\beta^f(y)$ is well defined for all y less than the least upper bound y_0 of $\beta_f(x)$. In this case we extend the domain of $\beta^f(y)$ by defining $\beta^f(y) = 0$ for $y > y_0$ and $\beta^f(y_0) = \overline{\lim\limits_{y \to y_0}}\, \beta^f(y)$.

Thus we have

$$\beta^f[\beta_f(x)] \ge x, \quad \beta_f[\beta^f(y)] \le y$$

$$\lim_{y \to 0} \beta^f(y) = \infty, \quad \lim_{y \to \infty} \beta^f(y) = 0.$$

We now have the following:

Lemma 1.[1] *Given an $f(P) \ge 0$ of L^p, $p \ge 1$ and any number $y > 0$, there is a sequence of non-overlapping cubes I_k such that*

---

[1] In the one-dimensional case this lemma is contained in a lemma by F. Riesz (See [7], page 242).

$$y \leq \frac{1}{|I_k|} \int_{I_k} f(P)\, dP \leq 2^n y; \quad (k = 1, 2, \ldots),$$

and $f(P) \leq y$ *almost everywhere outside* $D_y = \bigcup_k I_k$. *Moreover* $|D_y| \leq \beta^f(y)$ *and*

$$y \leq \frac{1}{|D_y|} \int_{D_y} f(P)\, dP \leq 2^n y.$$

Proof. On account of the properties of $\beta_f(x)$, for the given y we can find an x such that $\beta_f(x) < y$. Then over any cube I of measure x we have

$$\frac{1}{|I|} \int_I f(P)\, dP \leq \frac{1}{x} \int_0^x f^*(t)\, dt \leq \beta_f(x) < y.$$

Divide now E^n into a mesh of cubes of measure x and carry out the following process: divide each cube into 2^n equal cubes and select those where the average of the function $f(P)$ is larger than or equal to y. Then divide the remaining ones again in 2^n equal cubes and select those where the average of the function is larger than or equal to y. Continuing this process we obtain a sequence of cubes I_k which we shall show have the required properties. First of all, we obviously have

$$\frac{1}{|I_k|} \int_{I_k} f(P)\, dP \geq y.$$

Moreover, since every selected cube I_k was obtained from dividing a cube I where the average of the function $f(P)$ was less than y, we also have

$$\int_{I_k} f(P)\, dP \leq \int_I f(P)\, dP \leq |I|\, y = |I_k|\, 2^n y,$$

and therefore

$$\frac{1}{|I_k|} \int_{I_k} f(P)\, dP \leqq 2^n y.$$

Now, every point outside $D_y = \bigcup I_k$ is contained in arbitrarily small cubes over which the average of $f(P)$ is less than y. Therefore the derivative of the indefinite integral of f cannot exist and be larger than y, and since $f(P)$ is almost everywhere equal to the derivative of its indefinite integral, we conclude that $f(P) \leq y$ almost everywhere outside D_y.

Finally we have

$$y|I_k| \le \int_{I_k} f(P)\,dP \le 2^n y |I_k|,$$

and hence

$$y\sum_1^m |I_k| \le \int_{\bigcup_1^m I_k} f(P)\,dP \le 2^n y \sum_1^m |I_k|,$$

or

$$y \le \frac{1}{\sum_1^m |I_k|} \int_{\bigcup_1^m I_k} f(P)\,dP \le 2^n y.$$

Since $|\bigcup_1^m I_k| = \sum_1^m |I_k|$, it follows that

$$\frac{1}{\sum_1^m |I_k|} \int_{\bigcup_1^m I_k} f(P)\,dP \le \beta_f\left(\sum_1^m |I_k|\right).$$

Therefore

$$y \le \beta_f\left(\sum_1^m |I_k|\right),$$

and

$$\sum_1^m |I_k| \le \beta^f(y),$$

and letting m tend to infinity we get $|D_y| \le \beta^f(y)$. Therefore $|D_y|$ is finite, and repeating the argument above, replacing now $\sum_1^m |I_k|$ by $\sum_1^\infty |I_k| = |D_y|$, we shall finally get

$$y \le \frac{1}{|D_y|} \int_{D_y} f(P)\,dP \le 2^n y.$$

This completes the proof.

Lemma 2. *Let $f \ge 0$ belong to L^p, $1 \le p \le 2$, in E^n, and let E_y be the set of points where the function*

$$\tilde{f}_\lambda(P) = \int_{E^n} K_\lambda(P-Q) f(Q)\,dQ$$

exceeds y in absolute value. Then

$$|E_y| \le \frac{c_1}{y^2} \int_{E^n} [f(P)]_y^2\,dP + c_2 \beta^f(y), \tag{3}$$

where $[f(P)]_y$ denotes the function equal to $f(P)$ if $f(P) \le y$ and equal to y otherwise, and c_1 and c_2 are constants independent of λ.

Proof. In order to simplify notation, every constant depending only on the dimension n and the function Ω will be denoted by c simply.

Let D_y be the set of Lemma 1 and define

$$h(P) = \begin{cases} \dfrac{1}{|I_k|}\displaystyle\int_{I_k} f(Q)\,dQ, & \text{if } P \in I_k; \\ f(P) \text{ otherwise.} \end{cases}$$

Then $f(P) = h(P) + g(P)$, with $g(P) = 0$ outside D_y, and

$$\int_{I_k} g(P)\,dP = 0, \quad k = 1, 2, \ldots .$$

Define now

$$\tilde{h}_\lambda(P) = \int_{E^n} K_\lambda(P - Q)\,h(Q)\,dQ,$$

$$\tilde{g}_\lambda(P) = \int_{E^n} K_\lambda(P - Q)\,g(Q)\,dQ,$$

and denote by E_1 the set of points where $|\tilde{h}_\lambda(P)| \geq y/2$, and by E_2 that where $|\tilde{g}_\lambda(P)| \geq y/2$.

As we have already shown (see (2 a)),

(4)
$$\int_{E^n} |\tilde{h}_\lambda(P)|^2\,dP \leq c \int_{E^n} h(Q)^2\,dQ,$$

where c is a constant independent of λ. From this it easily follows that

$$|E_1| \leq \frac{4c}{y^2} \int_{E^n} h(Q)^2\,dQ.$$

Now, on account of the definition of h, we have $h(P) = f(P) \leq y$ outside D_y and therefore $h(P) = [f(P)]_y$ outside D_y; moreover $h(P) \leq 2^n y$ in D_y. Therefore, denoting by D'_y the complement of D_y, we have

$$\int_{E^n} h(Q)^2\,dQ = \int_{D_y} h(Q)^2\,dQ + \int_{D'_y} h(Q)^2\,dQ \leq 2^{2n} y^2 |D_y| + \int_{E^n} [f(P)]_y^2\,dP,$$

and

$$|E_1| \leq \frac{c}{y^2} \int_{E^n} [f(P)]_y^2\,dP + c\,|D_y|.$$

To estimate the measure of E_2 we proceed as follows. Denote by S_k the sphere with the same center as I_k, and radius equal to the diameter of I_k, and call $\bar{D}_y = \bigcup_1^\infty S_k$ and $\bar{D}'_y$ its complement. Then $|\bar{D}_y| \leq c\,|D_y|$ and

$$|E_2| \leq |\bar{D}_y| + |E_2 \cap \bar{D}'_y| \leq c\,|D_y| + |E_2 \cap \bar{D}'_y|.$$

Since $g(P) = 0$ outside D_y, we have

$$\tilde{g}_\lambda(P) = \sum_k \int_{I_k} g(Q) K_\lambda(P-Q)\, dQ.$$

Let us now estimate the integral of $|g_\lambda(P)|$ over $\bar{D}'_y$. Suppose that P belongs to $\bar{D}'_y$ and consider one of the cubes I_k. If I_k has no points in common with the sphere with center at P and radius $1/\lambda$ we have

$$\int_{I_k} g(Q) K_\lambda(P-Q)\, dQ = \int_{I_k} g(Q) K(P-Q)\, dQ$$

since $K_\lambda = K$ outside that sphere. Since the integral of g over I_k is zero we have, furthermore,

$$\int_{I_k} g(Q) K_\lambda(P-Q)\, dQ = \int_{I_k} g(Q)[K(P-Q) - K(P-Q_k)]\, dQ,$$

where Q_k is the center of I_k. Now, if P is outside S_k and Q is in I_k, from the continuity properties of Ω and, by an elementary geometrical argument, we deduce that

$$|K(P-Q) - K(P-Q_k)| \leq c|P-Q_k|^{-n}\, \omega[c|I_k|^{1/n}|P-Q_k|^{-1}],$$ [1]

and therefore

$$\left|\int_{I_k} g(Q) K_\lambda(P-Q)\, dQ\right| \leq c|P-Q_k|^{-n}\, \omega[c|I_k|^{1/n}|P-Q_k|^{-1}] \int_{I_k} |g(Q)|\, dQ.$$

On the other hand, if I_k intersects the sphere of radius $1/\lambda$ and center at P, and P is outside S_k, I_k is entirely contained in the sphere of radius $3/\lambda$, and center at

[1] Since this argument is going to be used repeatedly, we shall give it here. Let us denote by R and S the projections of Q and Q_k on the unit sphere with center at P. Then $K(P-Q) - K(P-Q_k)$ can be written

$$\frac{\Omega(R)}{|P-Q|^n} - \frac{\Omega(S)}{|P-Q_k|^n} = \frac{\Omega(R) - \Omega(S)}{|P-Q_k|^n} + \left\{\frac{1}{|P-Q|^n} - \frac{1}{|P-Q_k|^n}\right\}\Omega(R).$$

The second term on the right is numerically

$$\leq \frac{c}{|P-Q_k|^n}\,\frac{|I_k|^{1/n}}{|P-Q_k|} \leq \frac{c}{|P-Q_k|^n}\,\omega\left(\frac{|I_k|^{1/n}}{|P-Q_k|}\right).$$

For the first term on the right, we have

$$|R-S| \leq c|I_k|^{1/n}|P-Q_k|^{-1},$$

and thus

$$|\Omega(R) - \Omega(S)| \leq \omega(|R-S|) \leq \omega(c|I_k|^{1/n}|P-Q_k|^{-1}).$$

Collecting the results we obtain the desired inequality.

P, so that, if $\gamma(t)$ is the characteristic function of the interval (0,3) and c is a bound for Ω, we have

$$|K_\lambda(P-Q)| \le c\lambda^n,$$

or, for all Q in I_k,

$$|K_\lambda(P-Q)| \le c\lambda^n \gamma[\lambda|P-Q|].$$

From this it follows that

$$\Big|\int_{I_k} g(Q) K_\lambda(P-Q)\, dQ\Big| \le c\lambda^n \int_{I_k} \gamma(\lambda|P-Q|)|g(Q)|\, dQ,$$

and this combined with the estimate above gives

$$|\tilde{g}_\lambda(P)| \le \sum_k \Big\{c|P-Q_k|^{-n}\omega[c|I_k|^{1/n}|P-Q_k|^{-1}]\int_{I_k}|g(Q)|\,dQ + \\ + c\lambda^n\int_{I_k}\gamma(\lambda|P-Q|)|g(Q)|\,dQ\Big\}$$

or

$$|\tilde{g}_\lambda(P)| \le c\lambda^n\int_{D_y}\gamma(\lambda|P-Q|)|g(Q)|\,dQ + \\ + \sum_k\Big\{c|P-Q_k|^{-n}\omega[c|I_k|^{1/n}|P-Q_k|^{-1}]\int_{I_k}|g(Q)|\,dQ\Big\}.$$

Integrating this over the complement $\bar{D}'_y$ of $\bar{D}_y$ we get (denoting by S'_k the complement of S_k)

$$\int_{\bar{D}'_y}|\tilde{g}_\lambda(P)|\,dP \le c\int_{D_y}|g(Q)|\,dQ\int_{E^n}\lambda^n\gamma(\lambda|P-Q|)\,dP + \\ + \sum_k\Big\{c\int_{S'_k}|P-Q_k|^{-n}\omega[c|I_k|^{1/n}|P-Q_k|^{-1}]\,dP\int_{I_k}|g(Q)|\,dQ\Big\}.$$

Now, on account of the properties of $\omega(t)$, the integrals with respect to P inside the summation sign are easily seen to be less than a constant, and the inner integral in the first integral on the right is a constant, regardless of the values of λ and P. Thus the last inequality reduces to

$$\int_{\bar{D}'_y}|\tilde{g}_\lambda(P)|\,dP \le c\int_{D_y}|g(Q)|\,dQ.$$

Now, according to the definitions of g and h, we have

$$|g(P)| \le f(P) + h(P),$$

and

$$\int_{D_y}|g(P)|\,dP \le \int_{D_y}[f(P)+h(P)]\,dP = 2\int_{D_y}f(P)\,dP,$$

and by Lemma 1 the last integral does not exceed $2^n y|D_y|$. Therefore

$$\int_{D'_y} |\tilde{g}_\lambda(P)|\, dP \leq c\, y\, |D_y|$$

and

$$|E_2 \cap \bar{D}'_y| \leq c\, |D_y|.$$

Collecting all estimates we get

$$|E_1| + |E_2| \leq \frac{c}{y^2} \int_{E^n} [f(P)]_y^2\, dP + c\, |D_y|.$$

Since $E_y \subset E_1 \cup E_2$ and $|D_y| \leq \beta'(y)$, Lemma 2 follows from the preceeding inequality.

Theorem 1. *Let $f(P)$ belong to L^p, $1 < p < \infty$, in E^n, then the function*

$$\tilde{f}_\lambda(P) = \int_{E^n} K_\lambda(P - Q)\, f(Q)\, dQ.$$

also belongs to L^p, and

$$\Big[\int_{E^n} |\tilde{f}_\lambda(P)|^p\, dP\Big]^{1/p} \leq A_p \Big[\int_{E^n} |f(P)|^p\, dP\Big]^{1/p},$$

where A_p is a constant independent of λ and f.

Proof. Without loss of generality we may assume that $f(P) \geq 0$. We shall start with the case $1 < p < 2$. According to Lemma 2,

$$|E_y| \leq \frac{c_1}{y^2} \int_{E^n} [f(P)]_y^2\, dP + c_2\, \beta'(y),$$

where E_y is the set of points where $|\tilde{f}_\lambda(P)|$ exceeds y, and c_1 and c_2 are absolute constants.

We have

$$\int_{E^n} |\tilde{f}_\lambda(P)|^p\, dP = p \int_0^\infty |E_y|\, y^{p-1}\, dy,$$

and replacing on the right $|E_y|$ by its estimate we get

$$\int_{E^n} |\tilde{f}_\lambda(P)|^p\, dP \leq c_1 \int_0^\infty \frac{p\, y^{p-1}}{y^2} \int_{E^n} [f(P)]_y^2\, dP\, dy + c_2 \int_0^\infty \beta'(y)\, p\, y^{p-1}\, dy.$$

For the first integral on the right we have

$$c_1 \int_0^\infty \frac{p\, y^{p-1}}{y^2} \int_{E^n} [f(P)]_y^2\, dP\, dy = c_1 \int_{E^n} dP \int_0^\infty p\, y^{p-1} \frac{[f(P)]_y^2}{y^2}\, dy,$$

and since $[f(P)]_y = y$ for $y \leq f(P)$, and $[f(P)]_y = f(P)$ for $y \geq f(P)$, the right hand side of the last expression can be replaced by

$$c_1 \int_{E_n} d\,P \left[\int_0^{f(P)} p\,y^{p-1}\,d\,y + \int_{f(P)}^{\infty} p\,y^{p-3} f^2(P)\,d\,y \right] =$$

$$= c_1 \int_{E^n} \left[f(P)^p + \frac{p}{2-p} f(P)^p \right] d\,P = c_1 \frac{2}{2-p} \int_{E^n} f(P)^p\,d\,P.$$

To estimate the second integral, we set $y = \beta_f(x)$ and get

$$\int_0^{\infty} \beta^f(y)\,p\,y^{p-1}\,d\,y = -\int_0^{\infty} x\,d\,\beta_f^p(x).$$

Now

$$x\,\beta_f^p(x) = \frac{1}{x^{p-1}} \left[\int_0^x f^*(t)\,d\,t \right]^p \leq \int_0^x f^*(t)^p\,d\,t,$$

and since $f(p)$ belongs to L^p so does $f^*(t)$, and $x\,\beta_f^p(x) \to 0$ as $x \to 0$. Therefore we have

$$-\int_0^{\infty} x\,d\,\beta_f^p(x) \leq \int_0^{\infty} \beta_f^p(x)\,d\,x$$

and by a familiar theorem of Hardy (See [7], p. 72) the last expression does not exceed

$$\left(\frac{p}{p-1}\right)^p \int_0^{\infty} f^*(t)^p\,d\,t = \left(\frac{p}{p-1}\right)^p \int_{E^n} f(P)^p\,d\,P.$$

Collecting all inequalities, we finally get

$$\int_{E^n} |\tilde{f}_\lambda(P)|^p\,d\,P \leq \left[\frac{2\,c_1}{2-p} + c_2 \left(\frac{p}{p-1}\right)^p \right] \int_{E^n} f(P)^p\,d\,P. \tag{5}$$

In the case when f belongs to L^p with $p > 2$ let g be any function belonging to L^q $(1/p + 1/q = 1)$ and vanishing outside a bounded set. Then

$$\int_{E^n} g(P)\,\tilde{f}_\lambda(P)\,d\,P = \int_{E^n} g(P)\,d\,P \int_{E^n} K_\lambda(P-Q)\,f(Q)\,d\,Q,$$

and inverting the order of integration, which is justified since the double integral is absolutely convergent,

$$\int_{E^n} g(P)\,\tilde{f}_\lambda(P)\,d\,P = \int_{E^n} f(Q)\,d\,Q \int_{E^n} K_\lambda(P-Q)\,g(P)\,d\,P = \int_{E^n} f(-Q)\,\tilde{g}'_\lambda(Q)\,d\,Q,$$

where $g'(P) = g(-P)$.

Therefore

$$\left|\int_{E^n} g(P)\tilde{f}_\lambda(P)\,dP\right| = \left|\int_{E^n} f(-Q)\tilde{g}'_\lambda(Q)\,dQ\right| \leq \left[\int_{E^n} |f|^p\,dP\right]^{1/p}\left[\int_{E^n} |\tilde{g}'_\lambda|^q\,dQ\right]^{1/q},$$

and since $q < 2$ we may replace the last integral by the corresponding integral of $|g(Q)|$ times A_q^q, and we get

$$\left|\int_{E^n} g(P)\tilde{f}_\lambda(P)\,dP\right| \leq A_q\left[\int_{E^n} |f(Q)|^p\,dQ\right]^{1/p}\left[\int_{E^n} |g(Q)|^q\,dQ\right]^{1/q},$$

which implies that

$$\left[\int_{E^n} |\tilde{f}_\lambda(P)|^p\,dP\right]^{1/p} \leq A_q\left[\int_{E^n} |f(P)|^p\,dP\right]^{1/p}.$$

This completes the proof.

Remark. The inequality (5) leads to a very crude estimate for the least value A_p^* of A_p, namely

$$A_p^* = O\left(\frac{1}{p-1}\right) + O\left(\frac{1}{2-p}\right); \quad 1 < p < 2. \tag{6}$$

This can easily be improved to

$$A_p^* = O\left(\frac{1}{p-1}\right); \quad 1 < p \leq 2. \tag{7}$$

For, anyway, A_p^* is finite, and so, using instead of (4) the inequality

$$\int_{E^n} |\tilde{h}_\lambda|^4\,dP \leq A_4 \int_{E^n} h^4\,dP,$$

and repeating the proof of Lemma 2, we obtain instead of (3) the inequality

$$|E_y| \leq \frac{c_1}{y^4}\int_{E^n} [f(P)]_y^4\,dP + c_2\beta^f(y)$$

for all $f \in L^p$, $1 \leq p \leq 4$ which, by an argument similar to the one used in the preceding theorem, leads to (7). Since $A_p = A_q$ for $q = \dfrac{p}{p-1}$, we have

$$A_q = O(q); \quad q \geq 2.$$

Another way of obtaining (7) would be to apply the theorem of M. Riesz on the interpolation of linear operations (See [7], p. 198) to the two exponents $p < 2 < q$.

Theorem 2. *Let $f(P)$ be a function such that*

$$\int_{E^n} |f(P)|(1+\log^+|f(P)|)\,dP < \infty.$$

Then $\tilde{f}_\lambda$ is integrable over any set S of finite measure and

$$\int_S |\tilde{f}_\lambda|\,dP \le c\int_{E^n}|f|\,dP + c\int_{E^n}|f|\ \log^+\left(|S|^{\frac{n+1}{n}}|f|\right)dP + c|S|^{-\frac{1}{n}},$$

where c is a constant independent of S and λ.

Proof. We may assume, without loss of generality, that $f(P)\ge 0$. Let E_y be the set of points where $|\tilde{f}_\lambda(P)| > y$ and $E'_y = E_y \cap S$. Then

$$\int_S |\tilde{f}_\lambda|\,dP = \int_0^\infty |E'_y|\,dy.$$

Now $|E'_y| \le |E_y|$ and $|E'_y| \le |S|$, and therefore we may write

$$\int_S |\tilde{f}_\lambda|\,dP \le \int_0^{y_0}|S|\,dy + \int_{y_0}^\infty |E_y|\,dy = |S|\,y_0 + \int_{y_0}^\infty |E_y|\,dy,$$

y_0 being any positive number.

According to Lemma 3, we have

$$|E_y| \le \frac{c_1}{y^2}\int_{E^n}[f(P)]_y^2\,dP + c_2\beta^f(y),$$

and from this it follows that

$$\int_{y_0}^\infty |E_y|\,dy \le c_1\int_0^\infty \frac{1}{y^2}\,dy \int_{E^n}[f(P)]_y^2\,dP + c_2\int_{y_0}^\infty \beta^f(y)\,dy.$$

Now, in the proof of Theorem 1 we have shown that the first integral on the right does not exceed a constant multiple of the integral

$$\int_{E^n} f(P)\,dP.$$

On the other hand, if we select $y_0 = \beta_f(|S|)$, the integral on the right reduces, after introducing the variable $x = \beta^f(y)$ and integrating by parts, to

$$\int_{y_0}^{\infty} \beta^f(y)\,dy = \int_{|S|}^{0} x\,d\beta_f(x) \le \int_0^{|S|} \beta_f(x)\,dx = \int_0^{|S|} \frac{dx}{x} \int_0^x f^*(t)\,dt = \int_0^{|S|} f^*(t) \log \frac{|S|}{t}\,dt.$$

Now, the convex functions $\Phi(x) = x \log^+ (|S|^{\frac{n+1}{n}} x)$ and

$$\Psi(y) = \begin{cases} y\,|S|^{-\frac{n+1}{n}} & \text{for } 0 \le y \le 1, \\ e^{y-1}\,|S|^{-\frac{n+1}{n}} & \text{for } 1 \le y, \end{cases}$$

are conjugate in the sense of Young[1], so that Young's inequality gives

$$\int_0^{|S|} f^*(t) \log \frac{|S|}{t}\,dt = 2 \int_0^{|S|} f^*(t) \frac{1}{2} \log \frac{|S|}{t}\,dt \le 2 \int_{E^n} f \log^+ (|S|^{\frac{n+1}{n}} f)\,dP +$$

$$+ 2\,|S|^{-\frac{n+1}{n}} \int_0^{|S|} \left(\frac{|S|}{t}\right)^{1/2} dt = 2 \int_{E^n} f \log^+ (|S|^{\frac{n+1}{n}} f)\,dP + 4\,|S|^{-1/n}.$$

Finally, collecting results and observing that

$$|S|\,y_0 = |S|\,\beta_f(|S|) = \int_0^{|S|} f^*(t)\,dt \le \int_{E^n} f\,dP,$$

we establish our assertion.

Theorem 3. *Let f be integrable in E^n. Then if S is a set of finite measure we have*

$$\int_S |\tilde{f}_\lambda(P)|^{1-\varepsilon}\,dP \le \frac{c}{\varepsilon}\,|S|^{\varepsilon} \left[\int_{E^n} |f(P)|\,dP\right]^{1-\varepsilon},$$

where c is a constant independent of ε, S, λ and f.

Proof. Again we shall only consider the case when $f \ge 0$. We have

$$|E_y| \le \frac{c_1}{y^2} \int_{E^n} [f(P)]_y^2\,dP + c_2\,\beta^f(y).$$

From this it follows that

$$y\,|E_y| \le c_1 \int_{E^n} [f(P)]_y \frac{[f(P)]_y}{y}\,dP + c_2\,y\,\beta^f(y).$$

---

[1] See [7], p. 64.

Since $\frac{1}{y}[f(P)]_y \leq 1$, and since for $\beta^f(y) = x$ we have

$$y\,\beta^f(y) = x\,\beta_f(x) = \int_0^x f^*(t)\,dt \leq \int_{E^n} f(P)\,dP,$$

we get

$$y\,|E_y| \leq c \int_{E^n} f(P)\,dP.$$

If we write $E'_y = E_y \cap S$, we have

$$|E'_y| \leq |S|, \qquad |E'_y| \leq |E_y|$$

and

$$\int_S |\tilde{f}_\lambda(P)|^{1-\varepsilon}\,dP = -\int_0^\infty y^{1-\varepsilon}\,d\,|E'_y| \leq (1-\varepsilon)\int_0^\infty \frac{|E'_y|}{y^\varepsilon}\,dy \leq$$

$$\leq (1-\varepsilon)\int_0^{y_0} \frac{|S|}{y^\varepsilon}\,dy + (1-\varepsilon)\,c\left[\int_{E^n} f(P)\,dP\right]\int_{y_0}^\infty \frac{dy}{y^{1+\varepsilon}}.$$

If we set here

$$y_0 = |S|^{-1}\,c \int_{E^n} f(P)\,dP,$$

our assertion follows. This completes the proof.

Theorem 4. *Let $\mu(P)$ be a mass-distribution that is a completely additive function of Borel set in E^n, and suppose that the total variation V of μ in E^n is finite. Then if*

$$\tilde{f}_\lambda(P) = \int_{E^n} K_\lambda(P-Q)\,d\mu(Q),$$

over every set S of finite measure we have

$$\int_S |\tilde{f}_\lambda(P)|^{1-\varepsilon}\,dP \leq \frac{c}{\varepsilon}\,|S|^\varepsilon\,V^{1-\varepsilon}.$$

Proof. This theorem is a straightforward consequence of the preceding one.

Let $H(P)$ be a non negative continuous function vanishing outside a bounded set and such that

$$\int_{E^n} H(P)\,dP = 1.$$

Then it is known that (see e.g. Lemma 1 in Chapter II)

$$\tilde{f}_\lambda(P) = \lim_{k\to\infty} k^n \int_{E^n} H[k(P-Q)]\,\tilde{f}_\lambda(Q)\,dQ$$

almost everywhere. But

$$k^n \int_{E^n} H[k(P-Q)]\tilde{f}_\lambda(Q)\,dQ = k^n \int_{E^n} H[k(P-Q)]\,dQ \int_{E^n} K_\lambda(Q-R)\,d\mu(R) =$$

$$= \int_{E^n} K_\lambda(P-Q)\Big[k^n \int_{E^n} H[k(Q-R)]\,d\mu(R)\Big]\,dQ,$$

and thus from the preceding theorem it follows that

$$\int_S \Big| k^n \int_{E^n} H[k(P-Q)]\tilde{f}_\lambda(Q)\,dQ\Big|^{1-\varepsilon} dP \le \frac{c}{\varepsilon}|S|^\varepsilon \Big[\int_{E^n}\Big|\int_{E^n} k^n H[k(Q-R)]\,d\mu(R)\Big|\,dQ\Big]^{1-\varepsilon}.$$

It is now readily seen that the last integral on the right does not exceed V. Therefore, substituting V on the right and applying Fatou's Lemma to the left-hand side we get the theorem.

Theorem 5. *Let $f(P)$ be a function in E^n such that*

$$\int_{E^n} |f(P)|(1 + \log^+|P-O| + \log^+|f(P)|)\,dP < \infty;$$

then for $\lambda \ge 1$ the function

$$\tilde{F}_\lambda(P) = \tilde{f}_\lambda(P) - K_1(P-O)\int_{E^n} f(Q)\,dQ$$

is integrable and

$$\int_{E^n} |\tilde{F}_\lambda(P)|\,dP \le c\int_{E^n} |f(P)|(1 + \log^+|P-O| + \log^+|f(P)|)\,dP + c,$$

where c is a constant independent of λ and f.

Proof. For the sake of simplicity of notation we shall denote any constant by c. Let

$$f_0(P) = f(P) \text{ if } |P-O| \le 1,$$

and

$$f_0(P) = 0$$

otherwise, and $f_k(P) = f(P)$ if $2^{k-1} < |P-O| \le 2^k$, $f_k(P) = 0$ otherwise, $k = 1, 2, \ldots$.

Let

$$\tilde{f}_{k\lambda}(P) = \int_{E^n} K_\lambda(P-Q) f_k(Q)\,dQ$$

and

$$\tilde{F}_{k\lambda} = \tilde{f}_{k\lambda} - K_1(P-O)\int_{E^n} f_k(Q)\,dQ.$$

Now, if $k \geq 1$ and S_k denotes the sphere $|P - O| \leq 2^{k+1}$, then $|S_k| = c\, 2^{(k+1)n}$, and Theorem 2 gives

$$\int_{S_k} |\tilde{f}_{k\lambda}|\, dP \leq c \int_{E^n} |f_k|\, dP + c \int_{E^n} |f_k| \log^+ (|S_k|^{\frac{n+1}{n}} |f_k|)\, dP + c\, |S_k|^{-\frac{1}{n}} \leq$$

$$\leq c \int_{E^n} |f_k| (1 + \log^+ |P - O| + \log^+ |f_k|)\, dP + c\, 2^{-k-1},$$

since $|P - O| \geq 2^{k-1}$ wherever $f_k(P) \neq 0$. As easily seen, this inequality, with suitable c, also holds for $k = 0$. On the other hand,

$$\int_{S_k} |K_1(P - O)|\, dP \leq c \log 2^{k+1},$$

so that

$$\left| \int_{S_k} K_1(P - O)\, dP \int_{E^n} f_k(Q)\, dQ \right| \leq c \log 2^{k+1} \int_{E^n} |f_k(Q)|\, dQ \leq$$

$$\leq c \int_{E^n} (1 + \log^+ |P - O|)\, |f_k(Q)|\, dQ.$$

This, together with the estimate for the integral of $|\tilde{f}_{k\lambda}(P)|$, gives

$$\int_{S_k} |\tilde{F}_{k\lambda}|\, dP \leq c \int_{E^n} |f_k| (1 + \log^+ |P - O| + \log^+ |f_k|)\, dP + c\, 2^{-k-1}.$$

Since for $\lambda \geq 1$ and $|P - Q| \geq 1$ we have $K_\lambda(P - Q) = K(P - Q)$, and since $f_k(P)$ vanishes outside S_{k-1}, for P outside S_k we have

$$\tilde{F}_{k\lambda}(P) = \int_{S_{k-1}} [K_\lambda(P - Q) - K_1(P - O)]\, f_k(Q)\, dQ =$$

$$= \int_{S_{k-1}} [K(P - Q) - K(P - O)]\, f_k(Q)\, dQ.$$

Now, an argument already used (see footnote to Lemma 2) shows that, on account of the continuity condition satisfied by $\Omega(P)$, for every P outside S_k and Q inside S_{k-1} the following inequality holds:

$$|K(P - Q) - K(P - O)| \leq c\, |P - O|^{-n}\, \omega(c\, 2^{k+1} |P - O|^{-1}).$$

Thus, if S'_k denotes the complement of S_k, we obtain

$$\int_{S'_k} |\tilde{F}_{k\lambda}|\, dP \leq \int_{S'_k} dP \int_{E^n} c\, |P - O|^{-n}\, \omega(c\, 2^{k+1} |P - O|^{-1})\, |f_k(Q)|\, dQ =$$

$$= c \int_{E^n} |f_k(Q)|\, dQ \int_{2^{k+1}}^{\infty} r^{-n}\, \omega(c\, 2^{k+1} r^{-1})\, r^{n-1}\, dr = c \int_{E^n} |f_k(Q)|\, dQ,$$

and collecting the results we have

$$\int_{E^n} |\tilde{F}_{k\lambda}|\, dP \le c \int_{E^n} |f_k| (1 + \log^+ |P - O| + \log^+ |f_k|)\, dP + c\, 2^{-(k+1)}.$$

Since $\tilde{F}_\lambda(P) = \sum_0^\infty \tilde{F}_{k\lambda}(P)$, the theorem follows by adding the above inequalities.

This result can be worded in a different manner. Since the functions

$$\Phi(x) = x \log^+ \alpha x$$

and

$$\Psi(y) = \begin{cases} y\alpha^{-1} & \text{for } 0 \le y \le 1 \\ e^{y-1}\alpha^{-1} & \text{for } y \ge 1 \end{cases}$$

are conjugate in the sense of Young, setting $x = |f(P)|$, $\alpha = 1 + |P - O|^{n+1}$ and $y = \frac{1}{2} \log^+ |P - O|$, for $y \ge 1$, Young's inequality gives

$$\tfrac{1}{2} |f(P)| \log^+ |P - O| \le |f(P)| \log^+ [(1 + |P - O|^{n+1}) |f|] + \\ + |P - O|^{1/2} (1 + |P - O|^{n+1})^{-1}$$

so that if $|f| \log^+ [(1 + |P - O|^{n+1}) |f|]$ is integrable the same is true for the product $|f| \log^+ |P - O|$, and since $|f| \le |f| \log^+ |P - O|$ for $|P - O| \ge e$, and $|f| \le 1 + |f| \log^+ |f|$ for $|P - O| < e$, it follows that

$$\int_{E^n} |f| (1 + \log^+ |P - O| + \log^+ |f|)\, dP \le c \int_{E^n} |f| \log^+ [(1 + |P - O|^{n+1}) |f|]\, dP + c,$$

and we have the following:

Corollary. The function $\tilde{F}_\lambda(P)$ of the preceding theorem satisfies the inequality

$$\int_{E^n} |\tilde{F}_\lambda(P)|\, dP \le c \int_{E^n} |f(P)| \log^+ [(1 + |P - O|^{n+1}) |f(P)|]\, dP + c.$$

If the integral of f extended over the whole space is zero, then in the last inequality we can replace $\tilde{F}_\lambda$ by $\tilde{f}_\lambda$. For $n = 1$ this result reduces to a known theorem about Hilbert transforms of functions on the real line [5].

Theorem 6. *Let $f(P)$ be a function bounded in E^n and $|f(P)| \le M$. Then the integral*

$$\tilde{F}_\lambda(P) = \int_{E^n} [K_\lambda(P - Q) - K_1(O - Q)] f(Q)\, dQ$$

is absolutely convergent, and

$$\int_{E^n} \Psi[c^{-1} M^{-1} |\tilde{F}_\lambda(P)|,\ P]\, dP \le 1,$$

where c is a constant independent of f and $\lambda \ge 1$, and $\Psi(y, P)$ is the function defined by $\Psi(y, P) = y\alpha^{-1}$ for $0 \le y \le 1$, $\Psi(y, P) = e^{y-1}\alpha^{-1}$ for $y \ge 1$, $\alpha = 1 + |P - O|^{n+1}$.

Proof. First we observe that for fixed P and λ the function

$$K_\lambda(P-Q)-K_1(O-Q)$$

is bounded. Moreover as Q tends to infinity this function is of the order

$$|Q-O|^{-n}\,\omega[c|P-O|\,|Q-O|^{-1}],$$

and thus is absolutely integrable, and the integral of its absolute value is a function of P bounded on every bounded set. Consider now the functions

$$\Phi(x,P)=x\log^+[(1+|P-O|^{n+1})x]$$

and

$$\Psi(y,P)=\begin{cases} y(1+|P-O|^{n+1})^{-1} & \text{for } 0\le y\le 1\\ e^{y-1}(1+|P-O|^{n+1})^{-1} & \text{for } y\ge 1,\end{cases}$$

which, for fixed P, are conjugate in the sense of Young, and let $g(P)$ be a function vanishing outside a bounded set and such that

$$\int_{E^n}\Phi(|g(P)|,P)\,dP\le 1.$$

Then we have

$$\int_{E^n} g(P)\tilde F_\lambda(P)\,dP=\int_{E^n} g(P)\,dP\int_{E^n}[K_\lambda(P-Q)-K_1(O-Q)]f(Q)\,dQ,$$

and since the double integral is absolutely convergent we may invert the order of integration and write

$$\int_{E^n} g(P)\tilde F_\lambda(P)\,dP=\int_{E^n} f(Q)\,dQ\int_{E^n}[K_\lambda(P-Q)-K_1(O-Q)]g(P)\,dP.$$

But, according to the corollary of Theorem 5,

$$\int_{E^n}\Big|\int_{E^n}[K_\lambda(P-Q)-K_1(O-Q)]g(P)\,dP\Big|\,dQ\le c$$

and therefore, if $|f(P)|\le M$, then

$$\Big|\int_{E^n} g(P)\tilde F_\lambda(P)\,dP\Big|\le cM.$$

The same conclusion holds if we multiply g by any function of absolute value 1; therefore we also have the stronger inequality

$$\text{(7 a)}\qquad \int_{E^n}|g(P)|\,|\tilde F_\lambda(P)|\,dP\le cM.$$

Let us now define

$$F_k(P) = \begin{cases} |\tilde{F}_\lambda(P)|, \text{ if } |\tilde{F}_\lambda(P)| \leq k \text{ and } |P - O| \leq k, \\ 0 \quad \text{otherwise,} \end{cases}$$

and denoting the function $\frac{d}{dx}\Psi(x, P)$ by $\Psi'(x, P)$ let us also define

$$g_k(P) = \begin{cases} \Psi'[c^{-1}\beta M^{-1} F_k(P), P] \text{ for } F_k(P) \neq 0, \\ 0 \text{ for } F_k(P) = 0, \end{cases}$$

where c is the same as in (7 a) and where, assuming that $F_k \not\equiv 0$, we select the constant β in such a way that

$$\int_{E^n} \Phi(g_k(P), P)\, dP = 1.$$

Then (see [7], p. 64) Young's inequality degenerates into equality,

$$g_k(P) \cdot [c^{-1}\beta M^{-1} F_k(P)] = \Phi(g_k, P) + \Psi(c^{-1}\beta M^{-1} F_k, P),$$

and integrating with respect to P we get

$$c^{-1}\beta M^{-1} \int_{E^n} g_k(P) F_k(P)\, dP = \int_{E^n} \Phi(g_k, P)\, dP + \int_{E^n} \Psi(c^{-1}\beta M^{-1} F_k, P)\, dP =$$

$$= 1 + \int_{E^n} \Psi[c^{-1}\beta M^{-1} F_k(P), P]\, dP.$$

But we also have

$$\int_{E^n} g_k(P) F_k(P)\, dP = \int_{E^n} g_k(P)\, |\tilde{F}_\lambda(P)|\, dP \leq c M.$$

Thus we get

$$1 + \int_{E^n} \Psi[c^{-1}\beta M^{-1} F_k(P), P]\, dP \leq \beta.$$

This implies, first of all, that $\beta \geq 1$ and secondly that

$$\int_{E^n} \frac{1}{\beta} \Psi[c^{-1}\beta M^{-1} F_k(P), P]\, dP \leq 1.$$

Now, since $\Psi(x, P)$ is convex, increasing and vanishes for $x = 0$, and since $\beta \geq 1$, we have

$$\frac{1}{\beta} \Psi[c^{-1}\beta M^{-1} F_k(P), P] \geq \Psi[c^{-1} M^{-1} F_k(P), P],$$

and from this and the inequality above it follows that

$$\int_{E^n} \Psi[c^{-1}M^{-1}F_k(P), P] \leq 1,$$

a relation which also holds for $F_k \equiv 0$. Finally since $F_k(P) \to |\tilde{F}_\lambda(P)|$ as $k \to \infty$, an application of Fatou's lemma establishes our assertion.

Theorem 7. *Let $f(P)$ belong to L^p, $1 < p < \infty$ then*

$$\tilde{f}_\lambda(P) = \int_{E^n} K_\lambda(P-Q) f(Q)\, dQ$$

converges in the mean of order p as $\lambda \to \infty$, to a function $\tilde{f}(P)$ of L^p in E^n.

If $f(P)$ is such that

(7 b) $$\int_{E^n} |f| \log^+ [(1 + |P-O|^{n+1}) |f|]\, dP < \infty,$$

then $\tilde{F}_\lambda(P)$ converges in the mean of order 1 to a function $\tilde{F}(P)$ integrable in E^n.

Proof. If $g(P)$ is a function with continuous first derivatives and vanishing outside a bounded set, then

$$\tilde{g}_\lambda(P) = \int_{E^n} K_\lambda(P-Q)\, g(Q)\, dQ$$

converges uniformly to a function $\tilde{g}(P)$ and moreover, outside a bounded set, $\tilde{g}_\lambda(P) = \tilde{g}(P)$ for $\lambda \geq 1$. This is easy to verify on account of the properties of $K_\lambda(P-Q)$, of the differentiability of g, and of the fact that

$$\tilde{g}_\lambda(P) = \int_{E^n} K_\lambda(P-Q)[g(Q) - g(P)]\, dQ.$$

Therefore, not only $\tilde{g}_\lambda(P) \to \tilde{g}(P)$ but also

$$\int_{E^n} |\tilde{g}_\lambda(P) - \tilde{g}(P)|^p\, dP \to 0$$

as $\lambda \to \infty$, for any $p \geq 1$.

Let now f be a function of L^p, $1 < p < \infty$. Given any $\varepsilon > 0$ there exists a function g with continuous first derivatives and vanishing outside a bounded set such that

$$\Big[\int_{E^n} |f(P) - g(P)|^p\, dP\Big]^{1/p} < \varepsilon.$$

Then, if $h = f - g$,

$$\tilde{f}_\lambda = \tilde{g}_\lambda + \tilde{h}_\lambda,$$

and

$$\Big[\int_{E^n} |\tilde{f}_\lambda - \tilde{f}_\mu|^p \, dP\Big]^{1/p} \le \Big[\int_{E^n} |\tilde{g}_\lambda - \tilde{g}_\mu|^p \, dP\Big]^{1/p} + \Big[\int_{E^n} |\tilde{h}_\lambda - \tilde{h}_\mu|^p \, dP\Big]^{1/p}.$$

Now, since

$$\Big[\int_{E^n} |h|^p \, dP\Big]^{1/p} < \varepsilon,$$

Theorem 1 gives

$$\Big[\int_{E^n} |\tilde{h}_\lambda|^p \, dP\Big]^{1/p} < A_p\, \varepsilon,$$

and thus we get

$$\Big[\int_{E^n} |\tilde{f}_\lambda - \tilde{f}_\mu|^p \, dP\Big]^{1/p} < \Big[\int_{E^n} |\tilde{g}_\lambda - \tilde{g}_\mu|^p \, dP\Big]^{1/p} + 2 A_p\, \varepsilon.$$

As λ and μ tend to infinity, the integral on the right tends to zero; therefore for λ and μ large we shall have

$$\Big[\int_{E^n} |\tilde{f}_\lambda - \tilde{f}_\mu|^p \, dP\Big]^{1/p} < 3 A_p\, \varepsilon,$$

and, since ε is arbitrary, the first part of the theorem is established.

For the second part we shall begin by showing that, given any $\varepsilon > 0$, there exists a function g with continuous first derivatives and vanishing outside a bounded set, such that

$$\int_{E^n} \left|\frac{f-g}{\varepsilon}\right| \log^+ \left[(1 + |P-O|^{n+1}) \frac{|f-g|}{\varepsilon}\right] dP \le 1. \tag{8}$$

For let S be a sphere with center at O and so large that

$$\int_{S'} \left|\frac{f}{\varepsilon}\right| \log^+ \left[(1 + |P-O|^{n+1}) \left|\frac{f}{\varepsilon}\right|\right] dP \le \tfrac{1}{2}. \tag{9}$$

For the points P inside S we shall have

$$x \log^+ [(1 + |P-O|^{n+1})\, x] \le c\, x^2,$$

for all $x \ge 0$ and a suitable c.

We now select k so large that

$$\int_S \left|\frac{f-[f]_k}{\varepsilon/2}\right| \log^+ \left[(1 + |P-O|^{n+1}) \left|\frac{f-[f]_k}{\varepsilon/2}\right|\right] dP \le \tfrac{1}{4},$$

and then g in such a way that $g = 0$ outside S and

$$\int_S \left|\frac{g-[f]_k}{\varepsilon/2}\right|^2 dP \le \frac{1}{2c}.$$

Then

$$\int_S \left|\frac{g-[f]_k}{\varepsilon/2}\right| \log^+ \left[(1+|P-O|^{n+1}) \left|\frac{g-[f]_k}{\varepsilon/2}\right|\right] dP \le \tfrac{1}{2}$$

and, applying Jensen's inequality,

$$\int_S \left|\frac{f-g}{\varepsilon}\right| \log^+ \left[(1+|P-O|^{n+1}) \left|\frac{f-g}{\varepsilon}\right|\right] dP \le \tfrac{1}{2},$$

which in conjunction with (9) gives (8).

Let now $h(P) = f(P) - g(P)$; then

$$\int_{E^n} |\tilde{F}_\lambda - \tilde{F}_\mu| \, dP \le \int_{E^n} |\tilde{G}_\lambda - \tilde{G}_\mu| \, dP + \int_{E^n} |\tilde{H}_\lambda - \tilde{H}_\mu| \, dP,$$

where

$$\tilde{F}_\lambda(P) = \int_{E^n} K_\lambda(P-Q) f(Q)\, dQ - K_1\,(P-O) \int_{E^n} f(Q)\, dQ,$$

and similarly for $\tilde{G}_\lambda$ and $\tilde{H}_\lambda$.

Now

$$\int_{E^n} |\tilde{G}_\lambda - \tilde{G}_\mu| \, dP \to 0$$

as λ and μ tend to infinity. On the other hand, since

$$\int_{E^n} \left|\frac{h}{\varepsilon}\right| \log^+ \left[(1+|P-O|^{n+1}) \left|\frac{h}{\varepsilon}\right|\right] dP \le 1,$$

by the corollary of Theorem 5 we have

$$\int_{E^n} \left|\frac{\tilde{H}_\lambda}{\varepsilon}\right| dP \le 2c;$$

therefore for λ and μ large we shall have

$$\int_{E^n} |\tilde{F}_\lambda - \tilde{F}_\mu| \, dP \le 4c\varepsilon,$$

and since ε is arbitrary the theorem is established.

Remarks 1°. Under the assumptions of Theorem 2, the function $\tilde{f}_\lambda(P)$ converges to a limit $\tilde{f}(P)$, in the mean of order 1, over every set of finite measure. Under the assumption (7 b), this mean convergence holds over E^n, but, unless $\int_{E^n} f\, dP = 0$ (or $K \equiv 0$), neither $\tilde{f}_\lambda$ nor $\tilde{f}$ are of the class L.

2°. Under the assumptions of Theorem 3, the function $\tilde{f}_\lambda(P)$ converges to $\tilde{f}(P)$, in the mean of order $1-\varepsilon$, over any set of finite measure.

3°. If $f(P)$ satisfies the assumptions of Theorem 6 and, in addition, vanishes in the neighborhood of infinity, the function $\exp\{c^{-1}M^{-1}\tilde{f}(P)\}$ is integrable over any bounded set S. If f is also continuous, $\exp k|f|$ is integrable over S for any $k > 0$.

CHAPTER II.

The pointwise convergence of the singular integrals.

In this section we shall investigate the convergence of the singular integrals at individual points. In the case where $f(P)$ belongs to L^p, $p > 1$, we shall prove that the singular integrals converge almost everywhere and that moreover they are dominated by a function of L^p, uniformly in λ. On the other hand, we shall show that the pointwise limit still exists almost everywhere even if the function $f(P)$ is replaced by a completely additive function of Borel set of finite total variation.

We shall begin by proving two lemmas.

Lemma 1. *Let $N(P)$ be a function in E^n and suppose that*

$$|N(P)| \leq \varphi(|P-O|),$$

where $\varphi(x)$ is a decreasing function of x such that

$$\int_{E^n} \varphi(|P-O|)\,dP < \infty.$$

Then, if $f(P)$ is a function of L^p, $1 \leq p < \infty$ we have

$$\lim_{\lambda\to\infty} \lambda^n \int_{E^n} N[\lambda(P-Q)]\,f(Q)\,dQ = f(P)\int_{E^n} N(Q)\,dQ$$

at every point P of the Lebesgue set of $f(P)$.

(One says that P is a Lebesgue point for f if the derivative of $\int |f(P)-f(Q)|\,dQ$ is equal to zero at P. This implies in particular that the derivative of $\int f(Q)\,dQ$ at P is $f(P)$).

Proof. Let P be a point of the Lebesgue set of $f(P)$ and let $I(\varrho)$ be the integral of $|f(P)-f(Q)|$ over the sphere with center at P and radius ϱ. Then $I(\varrho)\varrho^{-n}$ is a bounded function and tends to zero as ϱ tends to zero.

We have

$$\lambda^n \int_{E^n} N[\lambda(P-Q)] f(Q)\, dQ = \lambda^n \int_{E^n} N[\lambda(P-Q)][f(Q)-f(P)]\, dQ +$$
$$+ f(P) \int_{E^n} \lambda^n N[\lambda(P-Q)]\, dQ$$

and the second integral on the right is equal to $f(P) \int_{E^n} N(Q)\, dQ$ so that if we show that the first integral tends to zero as $\lambda \to \infty$ the lemma will be established. Now, this integral is in absolute value less than or equal to $\lambda^n \int_0^\infty \varphi(\lambda\varrho)\, dI(\varrho)$, and thus it suffices to show that the latter tends to zero.

Since $\varphi(\varrho)$ is decreasing and

$$J = \int_0^\infty \varphi(\varrho)\varrho^{n-1}\, d\varrho < \infty$$

$\varphi(\varrho)\varrho^n$ tends to zero as $\varrho \to 0$ and $\varrho \to \infty$. On the other hand, we have $I(\varrho)\varrho^{-n} \le c$ where c is a constant, or $I(\varrho) \le c\varrho^n$, so that integrating by parts we have

$$\lambda^n \int_0^\infty \varphi(\lambda\varrho)\, dI(\varrho) = -\lambda^n \int_0^\infty I(\varrho)\, d\varphi(\lambda\varrho),$$

and since φ is a decreasing function, if $c(\delta)$ denotes the least upper bound of $I(\varrho)\varrho^{-n}$ in $0 \le \varrho \le \delta$ we can write

$$-\lambda^n \int_0^\infty I(\varrho)\, d\varphi(\lambda\varrho) \le -\lambda^n c(\delta) \int_0^\delta \varrho^n\, d\varphi(\lambda\varrho) - \lambda^n c \int_\delta^\infty \varrho^n\, d\varphi(\lambda\varrho) \le$$
$$\le n\, c(\delta) \int_0^\delta \lambda^n \varphi(\lambda\varrho)\varrho^{n-1}\, d\varrho + n\, c \int_\delta^\infty \lambda^n \varphi(\lambda\varrho)\varrho^{n-1}\, d\varrho + \lambda^n c\, \delta^n \varphi(\lambda\delta) =$$
$$= n\, c(\delta) \int_0^{\lambda\delta} \varrho^{n-1}\varphi(\varrho)\, d\varrho + n\, c \int_{\lambda\delta}^\infty \varrho^{n-1}\varphi(\varrho)\, d\varrho + c\,\lambda^n \delta^n \varphi(\lambda\delta).$$

Now as $\lambda \to \infty$ the two last terms tend to zero and the first remains less than $n\, c(\delta) J$. Therefore

$$\overline{\lim_{\lambda\to\infty}}\, \lambda^n \int_0^\infty \varphi(\lambda\varrho)\, dI(\varrho) \le n\, c(\delta) J,$$

and since $c(\delta) \to 0$ as $\delta \to 0$, we have

$$\lim_{\lambda\to\infty} \lambda^n \int_0^\infty \varphi(\lambda\varrho)\, dI(\varrho) = 0,$$

and the lemma is established.

Lemma 2. *Under the assumptions of Lemma 1, the function*

$$\lambda^n \int_{E^n} N[\lambda(P-Q)] f(Q)\, dQ$$

converges to $f(P) \int_{E^n} N(Q)\, dQ$ *in the mean of order* p.

Proof. Let us denote by $\hat{f}_\lambda$ the integral in question and by $\hat{f}$ its pointwise limit. If f is bounded and vanishes outside a sphere of radius r with center at 0, $\hat{f}_\lambda$ is bounded for $|P-O| < 2r$, and less than $c\lambda^n \varphi\left(|P-O|\frac{\lambda}{2}\right)$ for $|P-O| \geq 2r$, where c is a constant. Inside the sphere $|P-O| < 2r$, $\hat{f}_\lambda$ converges to $\hat{f}$ boundedly and almost everywhere. Thus over the sphere $|P-O| < 2r$ we have

$$\lim_{\lambda \to \infty} \int |\hat{f}_\lambda - \hat{f}|^p\, dP = 0.$$

Over the exterior of the sphere we have

$$\int |\hat{f}_\lambda|^p\, dP \leq c\lambda^{np} \int \varphi\left(|P-O|\frac{\lambda}{2}\right)^p dP \leq c[\lambda^n \varphi(r\lambda)]^{p-1} \lambda^n \int \varphi\left(|P-O|\frac{\lambda}{2}\right) dP,$$

and the last expression tends to zero as $\lambda \to \infty$.

To extend the result to general functions we observe first that, with $1/p + 1/q = 1$,

$$\int_{E^n} |\hat{f}_\lambda|^p\, dP = \int_{E^n} dP \left| \int_{E^n} \lambda^n N[\lambda(P-Q)] f(Q)\, dQ \right|^p \leq$$

$$\leq \int_{E^n} dP \left| \int_{E^n} \{\lambda^n \varphi(\lambda|P-Q|)\}^{1/q} \{\lambda^n \varphi(\lambda|P-Q|)\}^{1/p} |f(Q)|\, dQ \right|^p \leq$$

$$\leq \int_{E^n} dP \left\{ \int_{E^n} \lambda^n \varphi(\lambda|P-Q|)\, dQ \right\}^{p/q} \left\{ \int_{E^n} \lambda^n \varphi(\lambda|P-Q|) |f|^p\, dQ \right\},$$

and since

$$\int_{E^n} \lambda^n \varphi(\lambda|P-Q|)\, dP = \int_{E^n} \lambda^n \varphi(\lambda|P-Q|)\, dQ = \int_{E^n} \varphi(|P-O|)\, dP,$$

we obtain

$$\left[\int_{E^n} |\hat{f}_\lambda(P)|^p\, dP\right]^{1/p} \leq c\left[\int_{E^n} |f(Q)|^p\, dQ\right]^{1/p},$$

where $c = \int_{E^n} \varphi(|P-O|)\, dP$.

Thus, given $f \in L^p$, we may split f into two functions, $f = g + h$, where g is bounded and vanishes outside a bounded set and $\left[\int_{E^n} |h|^p\, dP\right]^{1/p} < \varepsilon$. Then we have $\hat{f}_\lambda = \hat{h}_\lambda + \hat{g}_\lambda$ and

$$\Big[\int_{E^n} |\hat{f}_\lambda - f|^p \, dP\Big]^{1/p} \le \Big[\int_{E^n} |\hat{h}_\lambda - h|^p \, dP\Big]^{1/p} + \Big[\int_{E^n} |\hat{g}_\lambda - g|^p \, dP\Big]^{1/p}.$$

As $\lambda \to \infty$, the last integral tends to zero, and since

$$\Big[\int_{E^n} |\hat{h}_\lambda - h|^p \, dP\Big]^{1/p} \le \Big[\int_{E^n} |\hat{h}_\lambda|^p \, dP\Big]^{1/p} + \Big[\int_{E^n} |h|^p \, dP\Big]^{1/p} \le (c+1)\,\varepsilon,$$

for λ sufficiently large we shall have

$$\Big[\int_{E^n} |\hat{f}_\lambda - f|^p \, dP\Big]^{1/p} \le (c+2)\,\varepsilon,$$

and the assertion follows from the fact that ε is arbitrary.

Lemma 3. *Let $N_1(P)$ be equal to 1 in the sphere of volume 1 and center at 0, and zero elsewhere; and let $\bar{f}(P)$ be defined by*

$$\bar{f}(P) = \sup_\lambda \lambda^n \int_{E^n} N_1[\lambda(P-Q)] \, \big|f(Q)\big| \, dQ.$$

Then, if f belongs to L^p, $1 < p < \infty$, the same is true of $\bar{f}(P)$ and

$$\int_{E^n} \bar{f}(P)^p \, dP \le c \int_{E^n} |f(P)|^p \, dP,$$

where c is a constant depending on p only. If $|f| \log^+ |f|$ is integrable then $\bar{f}(P)$ is locally integrable, and over every set S of finite measure we have

$$\int_S \bar{f}(P)\, dP \le c \int_S |f(P)|\, dP + c \int_S \log^+ \Big[|S|^{\frac{n+1}{n}} |f(P)|\Big] dP + c\,|S|^{-\frac{1}{n}}.$$

In general, under the assumptions of Lemma 1 we have

$$\sup_\lambda \Big| \lambda^n \int_{E^n} N[\lambda(P-Q)] f(Q)\, dQ \Big| \le \bar{f}(P) \int_{E^n} \varphi(|P-O|)\, dP,$$

where $\bar{f}(P)$ is the function defined above.

Proof. Without loss of generality we may assume that $f(P) \ge 0$. Denote by $\bar{D}_y$ the set of points where $\bar{f}(P) > y$. The sets $\bar{D}_y$ are closely related to the sets D_y of Lemma 1 of the preceding chapter and we shall refer part of the argument to that lemma and Theorems 1 and 2 in that section.

Let $P \in \bar{D}_y$. On account of the definition of $\bar{D}_y$ then there exists a sphere with center at P over which the average of $f(Q)$ is larger than y. Suppose that r is the radius of this sphere and consider a subdivision of the space in equal cubes of edge not less than $r/2$ and less than r, as in Lemma 1 of the preceding chapter. Consider all the cubes in the subdivision intersecting the sphere. Their total measure or volume does not exceed a fixed positive number $1/\alpha$ (depending only on the dimension of the space) times the volume of the sphere. Therefore the average of $f(P)$ over the union of those cubes is larger than αy, and hence there exists at least one cube where the average of the function exceeds αy, and which is therefore contained in $D_{\alpha y}$. If we define D_y^* to be the set obtained from D_y by enlarging five times the edges of the cubes whose union is D_y, while keeping their centers and orientation fixed, it will turn out that the center of our sphere, that is P, is contained in $D_{\alpha y}^*$. But since P is an arbitrary point of $\bar{D}_y$ it follows that $\bar{D}_y \subset D_{\alpha y}^*$ and thus

$$|\bar{D}_y| \le |D_{\alpha y}^*| \le 5^n |D_{\alpha y}| \le 5^n \beta^f(\alpha y).$$

From this the assertion on the class of $\bar{f}(P)$ would follow as in Theorems 1 and 2 of the preceding chapter. We omit the argument because it would be a mere repetition.

In the general case we have

$$\Big|\int_{E^n} \lambda^n N[\lambda(P-Q)] f(Q)\,dQ\Big| \le \int_{E^n} \lambda^n \varphi(\lambda|P-Q|)\,|f(Q)|\,dQ,$$

and, denoting by $I(\varrho)$ the integral of $|f(Q)|$ over the sphere with center at P and radius ϱ, the last integral can be written as follows

$$\int_0^\infty \lambda^n \varphi(\lambda\varrho)\,dI(\varrho).$$

Now, if v_n denotes the volume of the sphere of radius 1, we have $I(\varrho) \le v_n \varrho^n \bar{f}(P)$, and since $\varphi(\lambda\varrho)\varrho^n \to 0$ as ϱ tends to zero or infinity, we can integrate the last integral by parts and write

$$\int_0^\infty \lambda^n \varphi(\lambda\varrho)\,dI(\varrho) = -\lambda^n \int_0^\infty I(\varrho)\,d\varphi(\lambda\varrho) \le -\lambda^n \int_0^\infty v_n \varrho^n \bar{f}(P)\,d\varphi(\lambda\varrho) =$$

$$= \bar{f}(P) \int_0^\infty \lambda^n \varphi(\lambda\varrho)\,d v_n \varrho^n = \bar{f}(P) \int_{E^n} \varphi(|P-O|)\,dP.$$

This completes the argument.

Remark. The first part of Lemma 3 concerning N, and for $n = 1$ is the very well known result of Hardy and Littlewood (see [7], p. 244). In the general case spheres with center at P can be replaced by cubes with center at P (which also follows from the result concerning the function N). In this case, and for $p > 1$, the case of general n can easily be deduced from the original Hardy-Littlewood result by induction and the cubes can even be replaced by arbitrary n-dimensional intervals with fixed orientation. (See [4]). However, the case of $|f| \log^+ |f|$ integrable seems to require a special treatment.

An alternative proof for the latter case was communicated to us by Professor B. Jessen. He pointed out that it is enough to prove the result for differentiation with respect to a net of cubes, and that in this case the result for general n is deducible from the result for $n = 1$ by a measure preserving mapping of E^n onto E^1, which transforms the cubes of the net into intervals of E^1.

Theorem 1. *If $f(P)$ belongs to L^p, $1 < p < \infty$ then*

$$\tilde{f}_\lambda(P) = \int_{E^n} K_\lambda(P-Q) f(Q)\, dQ$$

converges almost everywhere to a function $\tilde{f}(P)$ as $\lambda \to \infty$. Moreover the function $\sup_\lambda |\tilde{f}_\lambda(P)|$ belongs to L^p and

$$\int_{E^n} \sup_\lambda |\tilde{f}_\lambda(P)|^p\, dP \leq c \int_{E^n} |f(P)|^p\, dP,$$

c being a constant which depends on p and on the kernel K_λ only.

Proof. In the preceding chapter we have shown that $\tilde{f}_\lambda$ converges in the mean of order p to a function $\tilde{f}$ of L^p.

Let $H(P)$ be a non negative continuous function with everywhere continuous first derivatives, vanishing outside the sphere with center at 0 and radius 1, and such that

$$\int_{E^n} H(P)\, dP = 1.$$

Then the function

$$\hat{f}_\mu(P) = \mu^n \int_{E^n} H[\mu(P-Q)]\, \tilde{f}(Q)\, dQ$$

converges almost everywhere to $\tilde{f}$ as $\mu \to \infty$ and moreover

$$\int_{E^n} \sup_\mu |\hat{f}_\mu|^p\, dP \leq c \int_{E^n} |\tilde{f}|^p\, dP \leq c' \int_{E^n} |f|^p\, dP,$$

c and c' being two constants independent of f. Since $\tilde{f}_\lambda$ converges in the mean of order p to $\tilde{f}$ and $H[\mu(P-Q)]$ belongs to all classes L^p we have

$$\hat{f}_\mu(P) = \lim_{\lambda\to\infty} \int_{E^n} \mu^n H[\mu(P-Q)]\tilde{f}_\lambda(Q)\,dQ =$$

$$= \lim_{\lambda\to\infty} \int_{E^n} \mu^n H[\mu(P-Q)] \int_{E^n} K_\lambda(Q-S) f(S)\,dS,$$

or, interchanging the order of integration,

$$\hat{f}_\mu(P) = \lim_{\lambda\to\infty} \int_{E^n} f(S)\,dS\Big[\int_{E^n} \mu^n H[\mu(P-Q)] K_\lambda(Q-S)\,dQ\Big].$$

Now, since $H[\mu(P-Q)]$ belongs to all classes L^p and has continuous first derivatives, as $\lambda\to\infty$ the inner integral converges pointwise and in the mean of order $q = \dfrac{p}{p-1}$, so that we can pass to the limit under the integral sign and write

$$\hat{f}_\mu(P) = \int_{E^n} f(S)\Big[\lim_{\lambda\to\infty} \int_{E^n} \mu^n H[\mu(P-Q)] K_\lambda(Q-S)\,dQ\Big]\,dS.$$

Since $K_\lambda(Q-S) = \lambda^n K_1[\lambda(Q-S)]$, introducing the variable $\mu(P-Q) = R$ the inner integral can be written as

$$\int_{E^n} \mu^n H(R)\frac{\lambda^n}{\mu^n} K_1\Big\{\frac{\lambda}{\mu}[(P-S)\mu - R]\Big\}\,dR$$

and, setting

$$\tilde{H}(P) = \lim_{\lambda\to\infty} \int_{E^n} \lambda^n K_1[\lambda(P-Q)] H(Q)\,dQ,$$

we have

$$\lim_{\lambda\to\infty} \int_{E^n} \mu^n H[\mu(P-Q)] K_\lambda(Q-S)\,dQ = \mu^n \tilde{H}[\mu(P-S)]$$

and

$$\hat{f}_\mu(P) = \int_{E^n} \mu^n \tilde{H}[\mu(P-S)] f(S)\,dS.$$

Now $H(P)$ has continuous first derivatives and vanishes outside the sphere with center at 0 and radius 1 and therefore $\tilde{H}(P)$ is bounded and, for $|P-O| \geq 2$,

$$\tilde{H}(P) = \int_{E^n} K_1(P-Q) H(Q)\,dQ.$$

On the other hand, since

$$\int_{E^n} H(P)\,dP = 1,$$

for $|P-O| \geq 2$ we have also

$$\tilde{H}(P) - K_1(P-O) = \int_{E^n} [K_1(P-Q) - K_1(P-O)]\, H(Q)\, dQ,$$

and, on account of the conditions satisfied by the function Ω in the definition of $K_\lambda(P-Q)$ it follows that, for large $|P-O|$ and $|Q-O| \le 1$,

$$|K_1(P-Q) - K_1(P-O)| \le c|P-O|^{-n}\,\omega(|P-O|^{-1})$$

and thus

$$|\tilde{H}(P) - K_1(P-O)| \le c|P-O|^{-n}\,\omega(|P-O|^{-1}),$$

where c is a constant. Now $K_1(P-O)$ is bounded, as well as $\tilde{H}(P)$, so that for all P the inequality

$$|\tilde{H}(P) - K_1(P-O)| \le c \min\{1;\, |P-O|^{-n}\,\omega(|P-O|^{-1})\}$$

will hold, c being again a constant.

Now

$$\hat{f}_\mu(P) - \tilde{f}_\mu(P) = \int_{E^n} \mu^n \{\tilde{H}[\mu(P-Q) - K_1[\mu(P-Q)]\}\, f(Q)\, dQ.$$

Thus from Lemmas 1 and 3 it follows that $\hat{f}_\mu(P) - \tilde{f}_\mu(P)$ converges almost everywhere and that

$$\int_{E^n} \sup_\mu |\hat{f}_\mu(P) - \tilde{f}_\mu(P)|^p\, dP \le c \int_{E^n} |f(P)|^p\, dP.$$

Since

$$\lim_{\mu\to\infty} \hat{f}_\mu(P) = \tilde{f}(P)$$

almost everywhere, and

$$\int_{E^n} \sup_\mu |\hat{f}_\mu(P)|^p\, dP \le c' \int_{E^n} |f(P)|^p\, dP$$

and $\tilde{f}_\mu \to \tilde{f}$ in mean of order p, the theorem follows.

Theorem 2. *Let $\mu(P)$ be a mass distribution, that is a completely additive function of Borel set in E^n and suppose that the total variation V of $\mu(P)$ in E^n is finite. Then the integral*

$$\tilde{f}_\lambda(P) = \int_{E^n} K_\lambda(P-Q)\, d\mu(Q)$$

has a limit $\tilde{f}$ almost everywhere as λ tends to infinity, and over every set S of finite measure we have

$$\int_S |\tilde{f}|^{1-\varepsilon}\, dP \le \frac{c}{\varepsilon} |S|^\varepsilon V^{1-\varepsilon}.$$

Proof. We may assume that $\mu(P) \geq 0$, and we shall show that, given any sphere of finite radius and an $\varepsilon > 0$, the integral converges in that sphere outside a subset of measure less than ε. We shall begin with the following observation. Let P be a point, I_ϱ a cube with center at P and edge equal to ϱ, and D any set contained in I_ϱ such that $|D| \geq \alpha |I_\varrho|$, α being a fixed positive number. Then

$$\lim_{\varrho \to 0} \frac{\mu(D)}{|D|}$$

exists and

$$\sup_\varrho \frac{\mu(D)}{|D|}$$

is finite for almost every P in E^n.

Let now S be an arbitrary sphere and $\bar{S}$ the sphere with the same center as S and radius twice as large. Fix $\alpha = 2^{-2n}$. Given an $\varepsilon > 0$ choose y so large that the set of points P in $\bar{S}$ such that

$$\sup_\varrho \frac{\mu(D)}{|D|} > y$$

be of measure less than $\varepsilon\, 2^{-n}\, n^{-n/2}$, and let A be an open set covering this set and the set of measure zero carrying the singular part of μ in $\bar{S}$, and such that also $|A| \leq \varepsilon\, 2^{-n}\, n^{-n/2}$. Now call $x_1, x_2, \ldots, x_n$ the coordinates of a point in E^n and cover A by means of half open cubes

$$\frac{m_i}{2^k} \leq x_i < \frac{m_i + 1}{2^k}; \qquad (i = 1, 2, \ldots, n)$$

where m_i and k are integers, in the following manner: first let k_0 be the smallest value of k for which there is a cube of the above form entirely contained in A, and take all such cubes contained in A; then let $k = k_0 + 1$ and take all the new cubes contained in the remaining part of A and so on. Thus we shall obtain A as the union of non-overlapping half-open cubes which we shall denote by I_k with the property that every I_k is contained in a cube with edge twice as long and containing a point outside A.

Denote by A_1 the union of all those I_k intersecting S. Then it is clear that, if ε is sufficiently small, every I_k in A_1 will be contained in a cube with edge twice as long and containing a point P outside A and in $\bar{S}$. Thus from the definitions of A and of α it follows that $\mu(I_k) \leq y |I_k|$ for every $I_k \in A_1$. Moreover, outside A and in $\bar{S}$, and therefore also outside A_1 and in S, the function $\mu(P)$ is absolutely continuous and its derivative does not exceed y. Let finally A_2 be the union of all spheres S_k with center at the centers of I_k and radius equal to the diameter of $I_k \subset A_1$. Clearly, $|A_2| \leq n^{n/2} 2^n |A_1| < \varepsilon$. We shall show that outside A_2 and in S the integral converges almost everywhere. For this purpose let $g(P)$ be the function equal to the derivative of $\mu(P)$ outside A_1 and in S, equal to $\mu(I_k)\,|I_k|^{-1}$ in every cube I_k of A_1, and equal to zero elsewhere. Let also $\nu(P)$ be equal to $\mu(P)$ minus the indefinite integral of $g(P)$. Then $\nu(I_k) = 0$, $g(P)$ is less than or equal to y, and

$$\int_{I_k} |d\nu(P)| \leq 2y|I_k|.$$

Let now P be a point interior to S, outside A_2, and where the density of A_1 is zero. Then, denoting by a prime the complement of a set,

$$\int_{E^n} K_\lambda(P-Q)\,d\mu(Q) = \int_{(A_1 \cup S)'} K_\lambda(P-Q)\,d\mu(Q) + \\ + \int_{E^n} K_\lambda(P-Q)\,g(Q)\,dQ + \int_{A_1} K_\lambda(P-Q)\,d\nu(Q).$$

Since P is interior to S, its distance to the set $(A_1 \cup S)'$ is positive and therefore the first integral converges. Moreover since $g(P)$ is a bounded function which vanishes outside a bounded set the second integral converges almost everywhere, so that the whole problem reduces to showing the convergence of the last integral.

We have

$$\int_{A_1} K_\lambda(P-Q)\,d\nu(Q) = \sum\nolimits_1 \int_{I_k} K_\lambda(P-Q)\,d\nu(Q) + \sum\nolimits_2 \int_{I_k} K_\lambda(P-Q)\,d\nu(Q),$$

where the first sum is extended over the cubes of A_1 intersecting the sphere with center at P and radius $1/\lambda$ and the second over those entirely outside this sphere. Now, since P is outside A_2, if I_k intersects the sphere with center at P and radius $1/\lambda$ it follows that I_k is contained in a sphere with center at P and radius equal to $3/\lambda$. Therefore we have

$$\left|\sum\nolimits_1 \int_{I_k} K_\lambda(P-Q)\,d\nu(Q)\right| \leq \lambda^n c \sum\nolimits_1 \int_{I_k} |d\nu(Q)| \leq 2y\lambda^n c \sum\nolimits_1 |I_k|.$$

But since all the cubes in $\sum_1$ are contained in the sphere of radius $3/\lambda$ and P is a point of density zero of A_1, we have

$$\lim_{\lambda \to \infty} \lambda^n \sum\nolimits_1 |I_k| = 0,$$

and therefore

$$\sum\nolimits_1 \int_{I_k} K_\lambda(P-Q)\,d\nu(Q) \to 0$$

as λ tends to infinity.

On the other hand, since $\nu(I_k) = 0$, for every I_k in $\sum_2$ we have

$$\int_{I_k} K_\lambda(P-Q)\,d\nu(Q) = \int_{I_k} [K(P-Q) - K(P-Q_k)]\,d\nu(Q),$$

where Q_k is the center of I_k. Now, on account of the conditions satisfied by Ω and the definition of A_2 it follows, as in Lemma 2 of the preceding chapter, that for every Q in I_k and P in A_2' we have

$$|K(P-Q) - K(P-Q_k)| \leq c\,|P-Q_k|^{-n}\,\omega[c\,|I_k|^{1/n}\,|P-Q_k|^{-1}],$$

and therefore

$$\int_{A'_2} |K(P-Q) - K(P-Q_k)|\,dP \leq \int_{A'_2} c\,|P-Q_k|^{-n}\,\omega[c\,|I_k|^{1/n}\,|P-Q_k|^{-1}]\,dP \leq c,$$

c being a constant. Hence

$$\int_{A'_2} dP \sum \int_{I_k} |K(P-Q) - K(P-Q_k)|\,|d\nu(Q)| \leq c \sum \int_{I_k} |d\nu(Q)| \leq$$
$$\leq 2cy \sum |I_k| = 2cy\,|A_1|,$$

the sum being extended over all intervals in A_1. But this implies that for almost every P in A_2' we have

$$\sum_2 \left| \int_{I_k} K_\lambda(P-Q)\,d\nu(Q) \right| \leq \sum \int_{I_k} |K(P-Q) - K(P-Q_k)|\,|d\nu(Q)| < \infty,$$

and since each of the terms on the left hand side converges as $\lambda \to \infty$ and is majorized by the corresponding term on the right which is independent of λ, it follows that

$$\sum_2 \int_{I_k} K_\lambda(P-Q)\,d\nu(Q)$$

converges as $\lambda \to \infty$.

Thus we have proved that $\tilde{f}_\lambda(P)$ converges almost everywhere to a finite limit.

Finally, the last part of the theorem is an immediate consequence of Theorem 4 of Chapter I and Fatou's lemma.

Remark. Suppose that the mass distribution is differentiable at the point P. Denoting by $\mu'(P)$ the value of the derivative (of course, $\mu'(P)$ is a real-valued function *of the point*) let us consider the mass distribution $\mu_P(Q) = \mu(Q) - \mu'(P)\,\chi(Q)$, where $\chi(Q)$ is the indefinite integral of 1. Due to the properties of the kernel K, the two integrals

$$\int_{E_n} K(P-Q)\,d\mu(Q), \qquad \int_{E_n} K(P-Q)\,d\mu_P(Q)$$

converge or diverge simultaneously, and in the case of convergence their values are the same. Let $\Gamma_\varrho(P)$ denote the sphere with center P and radius ϱ. At almost every point P we have

$$\int_{\Gamma_\varrho(P)} |d\mu_P(Q)| = o(\varrho^n) \qquad (\varrho \to 0),$$

(an analogue of Lebesgue's condition). At every point P at which this condition holds, the convergence of the second integral is to a great extent independent of the shape of the neighborhood excluded around P. For let $D_\varepsilon(P)$ be any set containing $\Gamma_\varepsilon(P)$ and contained in $\Gamma_{M\varepsilon}(P)$, M being a fixed number. Then the difference between the integrals

$$\int_{D'_\varepsilon(P)} K(P-Q)\, d\mu_P(Q), \quad \int_{\Gamma'_\varepsilon(P)} K(P-Q)\, d\mu_P(Q)$$

(where Γ' and D' are the complements of Γ and D) is numerically

$$\leq \int_{\Gamma_{M\varepsilon}(P)-\Gamma_\varepsilon(P)} |K(P-Q)|\, |d\mu_P(Q)| = O(\varepsilon^{-n})\, o(M\varepsilon)^n = o(1),$$

and so tends to 0 with ε. Thus, for almost every P, and for the sets D_ε of the type just described (we might call them *regular* neighborhoods) we have

$$\tilde{f}(P) = \lim_{\varepsilon\to 0} \left\{ \int_{D'_\varepsilon(P)} K(P-Q)\, d\mu(Q) + \mu'(P) \int_{D'_\varepsilon(P)} K(P-Q)\, dQ \right\}$$

almost everywhere. The second integral on the right here exists for every $\varepsilon > 0$. If it tends to 0 with ε (a situation which can occur, due to possible symmetries in the structures of K and of D), then in the last formula we may drop the second term on the right. This is, for example, the case of the kernels (3) and (5) of Introduction, if $D_\varepsilon(P)$ is any square with center P and sides 2ε.

CHAPTER III.

The preceding results can be used to establish differential properties of certain functions. We shall primarily consider the problems of the existence of the first derivatives of the Newtonian potential of a single layer, and of the second derivatives of the logarithmic potential (more general situations we shall consider elsewhere). Thus again we shall be concerned with the kernels of the forms

$$(1) \qquad \frac{x}{(x^2+y^2)^{3/2}}, \quad \frac{x^2-y^2}{(x^2+y^2)^2}, \quad \frac{xy}{(x^2+y^2)^2}.$$

Let us consider a mass distribution μ over the plane, and the Newtonian potential

$$u(x, y, z) = \iint_{E^2} \frac{1}{R} d\mu(s, t), \quad R = [(x-s)^2 + (y-t)^2 + z^2]^{1/2}$$

in the half-space $z > 0$. We assume that the total mass $\int |d\mu|$ is finite. Obviously

$$u_x(x, y, z) = \iint_{E^2} \frac{s-x}{R^3} d\mu(s, t).$$

Suppose that the point (x_0, y_0, z) approaches $(x_0, y_0, 0)$ vertically, and that the mass μ has a density at (x_0, y_0) (by this we mean that μ is differentiable at (x_0, y_0), with respect, say, to concentric circles). Without loss of generality we may assume that $(x_0, y_0) = (0, 0)$ and that the density in question is zero. Let us split the last integral into two, P and Q, the former extended over the circle Γ_z defined by the equation $s^2 + t^2 \leq z^2$, and the latter over the complement Γ_z' of Γ_z. If we set

$$I(r) = \iint_{\Gamma_r} s\, d\mu(s, t)$$

then, as easily seen, $I(r) = o(r^3)$ for $r \to 0$. Hence

$$P = \int_0^z \frac{dI(r)}{(r^2 + z^2)^{3/2}} = \frac{I(z)}{(2z^2)^{3/2}} + 3\int_0^z \frac{rI(r)\,dr}{(r^2 + z^2)^{5/2}} = o(1) + z^{-5}\int_0^z o(r^4)\,dr = o(1),$$

$$Q - \iint_{\Gamma_z'} \frac{s\,d\mu}{(s^2 + t^2)^{3/2}} = \int_z^\infty \left[\frac{1}{(r^2 + z^2)^{3/2}} - \frac{1}{(r^2)^{3/2}}\right] dI(r) =$$

$$= I(z)\left[\frac{1}{(2z^2)^{3/2}} - \frac{1}{(z^2)^{3/2}}\right] + 3\int_z^\infty rI(r)\left[\frac{1}{(r^2 + z^2)^{5/2}} - \frac{1}{r^5}\right] dr =$$

$$= o(z^3)\,O(z^{-3}) + 3\int_z^\infty o(r^4)\,O\left(\frac{z^2}{r^4}\right) dr =$$

$$= o(1) + z^2 \int_z^\infty o(r^{-3})\,dr = o(1).$$

Collecting the results, we see that at every point of differentiability of μ the difference

$$u_x(x, y, z) - \iint_{\Gamma_z'} \frac{x-s}{[(x-s)^2 + (y-t)^2]^{3/2}} d\mu(s, t)$$

converges to 0 as $z \to +0$. In particular, since the integral

$$\iint \frac{x-s}{[(x-s)^2+(y-t)^2]^{3/2}} d\mu(s,t)$$

exists almost everywhere, we see that $\lim u_x(x, y, z)$ exists for almost every (x, y) as $z \to +0$. A slight — and well known — modification of the above argument shows that $\lim u_x(x, y, z)$ exists for almost every point (x, y) as (x, y, z) approaches $(x, y, 0)$ non-tangentially.

Let us now assume that μ is absolutely continuous, that is that $d\mu = f\,ds\,dt$, with f integrable. We shall investigate the problem of the differentiability of the function

$$u(x,y) = \iint_{E^2} \frac{f(s,t)\,ds\,dt}{[(x-s)^2+(y-t)^2]^{1/2}} \tag{2}$$

which is the potential u in the plane $z = 0$.

In what follows, we shall systematically denote by $\Gamma_r(x, y)$ the circle with center (x, y) and radius r. The complementary set will be denoted by $\Gamma_r'(x, y)$. Instead of $\Gamma_r(0, 0)$ and $\Gamma_r'(0, 0)$ we shall simply write Γ_r and Γ_r'.

Theorem 1.

a) *Suppose that f is integrable over the whole plane and that $|f| \log^+ |f|$ is integrable over every finite circle of the plane. Then the integral* (2) *converges over almost every line parallel to the x-axis and represents an absolutely continuous*[1] *function of x. In particular, u_x exists almost everywhere. Moreover*

$$u_x(x,y) = \iint_{E^2} \frac{(s-x)f(s,t)}{[(x-s)^2+(y-t)^2]^{3/2}} ds\,dt$$

almost everywhere.

b) *If f is integrable over the whole plane and belongs to L^q, $q > 2$, over every finite circle, then $u(x, y)$ has a complete differential at almost every point of the plane.*

Proof. It slightly simplifies the argument (though it is not essential for the proof), if we assume that f vanishes outside a sufficiently large circle.

Let us consider

$$u^{(\varepsilon)}(x,y) = \iint_{E^2} f(x-s, y-t) \frac{1}{(s^2+t^2+\varepsilon^2)^{1/2}} ds\,dt, \tag{3}$$

[1] *i.e.* absolutely continuous over every finite interval.

and let us compare

$$u_x^{(\varepsilon)}(x, y) = -\iint_{E^2} f(x-s, y-t) \frac{s}{(s^2+t^2+\varepsilon^2)^{3/2}} \, ds \, dt,$$

with the function

$$\tilde{f}_\varepsilon(x, y) = -\iint_{\Gamma'_\varepsilon} f(x-s, y-t) \frac{s}{(s^2+t^2)^{3/2}} \, ds \, dt.$$

We have

$$-u_x^{(\varepsilon)}(x, y) + \tilde{f}_\varepsilon(x, y) = \iint_{\Gamma_\varepsilon} f(x-s, y-t) \frac{s}{(s^2+t^2+\varepsilon^2)^{3/2}} \, ds \, dt +$$

$$+ \iint_{\Gamma'_\varepsilon} f(x-s, y-t) \left[\frac{s}{(s^2+t^2+\varepsilon^2)^{3/2}} - \frac{s}{(s^2+t^2)^{3/2}} \right] ds \, dt,$$

and if we set

$$N(x, y) = \begin{cases} \dfrac{x}{(x^2+y^2+1)^{3/2}} & \text{for } x^2+y^2 \le 1 \\[2ex] \dfrac{x}{(x^2+y^2+1)^{3/2}} - \dfrac{x}{(x^2+y^2)^{3/2}} & \text{for } x^2+y^2 > 1 \end{cases}$$

it is readily seen that

$$-u_x^{(\varepsilon)}(x, y) + \tilde{f}_\varepsilon(x, y) = \frac{1}{\varepsilon^2} \iint_{E^2} f(s, t) \, N\left(\frac{x-s}{\varepsilon}, \frac{y-t}{\varepsilon}\right) ds \, dt.$$

Now, an application of Lemma 2 in Chapter II shows that, as $\varepsilon \to 0, -u_x^{(\varepsilon)}(x, y) + \tilde{f}_\varepsilon(x, y)$ converges to zero in the mean of order 1, and, according to Theorem 7 of Chapter I, over every set of finite measure $\tilde{f}_\varepsilon(x, y)$ converges in the mean to the function

$$\tilde{f}(x, y) = -\iint_{E^2} f(x-s, y-t) \frac{s}{(s^2+t^2)^{3/2}} \, ds \, dt.$$

Thus, over every set of finite measure, $u_x^{(\varepsilon)}(x, y)$ converges in the mean to $\tilde{f}(x, y)$.

Now we can select a sequence $\varepsilon_n \to 0$ such that, for almost every line $y = y_0$, $u_x^{(\varepsilon)}(x, y_0)$ converge in the mean to $\tilde{f}(x, y_0)$ over sets of finite measure[1], and thus

---

[1] If $f_n(x, y)$ converges in the mean to $f(x, y)$, we have

$$\int dy \int |f_n(x, y) - f(x, y)| \, dx \to 0.$$

Thus the inner integral, as a function of y, converges in the mean to zero, and we can select a sequence n_i such that

$$\int |f_{n_i}(x, y) - f(x, y)| \, dx \to 0$$

for almost every y.

$$\lim_{n\to\infty} [u^{(\varepsilon_n)}(x, y_0) - u^{(\varepsilon_n)}(x_0, y_0)] = \int_{x_0}^{x} \tilde{f}(s, y_0)\, ds.$$

If (as we may assume) $f \geq 0$, the integrand in (3) increases as ε decreases, and this implies that

$$\lim_{\varepsilon\to 0} u^{(\varepsilon)}(x, y) = \iint_{E^2} f(x - s, y - t) \frac{1}{(s^2 + t^2)^{1/2}}\, ds\, dt = u(x, y).$$

Combining the two last results, part a) follows.

Part b) of Theorem 1 asserts that in the neighborhood of almost every point (x_0, y_0), the difference $u(x_0 + h, y_0 + k) - u(x_0, y_0)$ is of the form $Ah + Bk + o(h^2 + k^2)^{1/2}$, where $A = A(x_0, y_0)$, $B = B(x_0, y_0)$. This is more than the mere existence of the partial derivatives $u_x(x_0 y_0)$ and $u_y(x_0, y_0)$ established in a), and it implies, in particular, that u is bounded in the neighborhood of (x_0, y_0). The mere boundedness of u, however, over every finite circle is a direct consequence of Hölder's inequality applied to the integral in (2), since the kernel $(s^2 + t^2)^{-1/2}$ belongs to L^p, $p < 2$, over every finite circle.

Let us now consider any point (x_0, y_0) at which the integral of $|f(x, y) - f(x_0, y_0)|^q$ is differentiable and the derivative is zero. (Generalized Lebesgue condition — it implies ordinary Lebesgue condition with $q = 1$). Let us also assume that both integrals

$$\text{(4)}\qquad \begin{aligned} A &= -\iint_{E^2} f(x - s, y - t) \frac{s}{(s^2 + t^2)^{3/2}}\, ds\, dt, \\ B &= -\iint_{E^2} f(x - s, y - t) \frac{t}{(s^2 + t^2)^{3/2}}\, ds\, dt. \end{aligned}$$

exist. Let $(h^2 + k^2)^{1/2} = \varepsilon$, and let us split the integral defining the difference $\Delta = u(x_0 + h, y_0 + k) - u(x_0, y_0)$ into two, extended respectively over the circle $\Gamma_{2\varepsilon}(x_0, y_0)$ and its complement $\Gamma'_{2\varepsilon}(x_0, y_0)$. Let us denote the integrals so obtained by P and Q respectively. Without loss of generality we may assume that $(x_0, y_0) = (0, 0)$ and that $f(x_0, y_0) = 0$. Then, with $H(x, y) = (x^2 + y^2)^{-1/2}$, we have

$$Q = \iint_{\Gamma'_{2\varepsilon}} f(s, t)\,[H(s - h, t - k) - H(s, t)]\, ds\, dt = h A_{2\varepsilon} + k B_{2\varepsilon} +$$

$$+ \frac{1}{2} \iint_{\Gamma'_{2\varepsilon}} f(s, t)\,[h^2 H_{xx}(s - \theta h, t - \theta k) + 2kh H_{xy}(s - \theta h, t - \theta k) +$$

$$+ k^2 H_{yy}(s - \theta h, t - \theta k)]\, ds\, dt$$

where θ is a function of s and t such that $0 < \theta < 1$, $A_{2\varepsilon}$ and $B_{2\varepsilon}$ are the integrals (4) extended over $\Gamma'_{2\varepsilon}$.

If we replace here $A_{2\varepsilon}$ and $B_{2\varepsilon}$ by A and B, we ultimately commit an error $o(|h|) + o(|k|) = o(h^2 + k^2)^{1/2}$. The first of the remaining three integrals is numerically

$$O(h^2)\iint_{\Gamma'_{2\varepsilon}} |f(s,t)|[(s-\theta h)^2 + (t-\theta k)^2]^{-3/2}\,ds\,dt =$$

$$= O(\varepsilon^2)\iint_{\Gamma'_{2\varepsilon}} |f(s,t)|(s^2+t^2)^{-3/2}\,ds\,dt = O(\varepsilon^2)\int_{2\varepsilon}^{\infty}\frac{dI(r)}{r^3},$$

where $I(r)$ is the integral of $|f|$ extended over the circle Γ_r. Integration by parts and the fact that, by assumption, $I(r) = o(r^2)$, shows that the last expression is $O(\varepsilon^2)\,o(\varepsilon^{-1}) = o(\varepsilon)$.

The same remark applies to the remaining two integrals constituting Q. Hence

$$Q = Ah + Bk + o(\varepsilon). \tag{5}$$

An application of Hölder's inequality to the integrals defining P gives, with $p = \frac{q-1}{q}$.

$$|P| \le \Big[\iint_{\Gamma_{2\varepsilon}} |f|^q\,ds\,dt\Big]^{1/q}\Big\{\Big[\iint_{\Gamma_{2\varepsilon}} H^p\,ds\,dt\Big]^{1/p} + \Big[\iint_{\Gamma_{2\varepsilon}} H(s-h,t-k)^p\,ds\,dt\Big]^{1/p}\Big\} \le$$

$$\le 2\Big[\iint_{\Gamma_{2\varepsilon}} |f|^q\,ds\,dt\Big]^{1/q}\Big[\iint_{\Gamma_{4\varepsilon}} H^p\,ds\,dt\Big]^{1/p} = o(\varepsilon).$$

Hence, using (5), we get $\Delta = P + Q = Ah + Bk + o(\varepsilon)$, and part b) of Theorem 1 is established.

Remarks. Neither part of Theorem 1 admits of much improvement. For, beginning with part a), let us assume that $f(s,t)$ vanishes outside the square S, $0 \le s \le 1$, $0 \le t \le 1$, and that it is constant, equal to $\varphi(s_0)$, along every segment $s = s_0$, $0 \le t \le 1$. Then, integrating with respect to t, one finds that in every smaller square concentric with, and situated similarly to, S the function $u(x,y)$ differs from the logarithmic potential

$$L(x,y) = L(x) = \int_0^1 \varphi(s)\log|x-s|^{-1}\,ds \tag{6}$$

by a bounded function. Let us suppose that φ is non-negative. If $\omega(x)$ is any function tending to ∞ with x, then the integrability of $\omega(f)$ over S is equivalent to the integrability of $\omega(\varphi)$ over $0 \le s \le 1$. If $\omega(x) = x\log^+ x$, then an application of Young's inequality (see [7], p. 64) to the integral shows that the integrability of $\varphi \log^+ \varphi$ implies the boundedness of $L(x, y)$ over S. Suppose, however, that $\omega(x)$ tends to $+\infty$ more slowly than $x \log x$. We can then find a positive function $\varphi(s)$, $0 \leq s \leq 1$, with $\omega[\varphi(s)]$ integrable and such that the integral (6) diverges to $+\infty$ in a set of points x dense in (0, 1). (See [7], p. 99). Hence $u(x, y)$ equals $+\infty$ on a set of segments $s = s_0$, $0 \leq t \leq 1$, dense in S. This shows that at no point interior to S can $u(x, y)$ have a directional derivative in the direction making an angle of $\pm \frac{\pi}{4}$ with the x-axis. Rotating the whole picture by $\frac{\pi}{4}$ we obtain a mass distribution with density $f(s, t)$ such that $\omega(f)$ is integrable and yet the potential $u(x, y)$ has no partial derivative u_x or u_y at any point interior to a square S' obtained by the rotation of S.

It is easily seen that u_x and u_y will be non-existent at almost every point of S', no matter how we modify $u(x, y)$ in any set of measure 0. For, φ being ≥ 0, the function $L(x)$ is lower semicontinuous. Hence given any number $M > 0$, no matter how large, we shall have $L(x) = L(x, y) > M$ in a set of strips parallel to the y-axis ($\varepsilon \leq y \leq 1 - \varepsilon$) and dense in $0 \leq x \leq 1$. Thus no matter how we modify u in a set of measure 0 it will be discontinuous, in the direction $\pm \frac{\pi}{4}$, at almost every point $(x, y) \in S$.

That in part b) we cannot replace the integrability of $|f|^q$, $q > 2$ by the integrability of f^2 (over every finite circle) is even simpler. For the kernel $H(x, y) = (x^2 + y^2)^{1/2}$ is not quadratically integrable near the origin. We can therefore construct a function $f(s, t) \geq 0$ quadratically integrable over every finite circle and such that the convolution u of f and H diverges to $+\infty$ in a set dense over the whole plane. It follows that u remains unbounded in every circle no matter how we change u in a set of measure 0. Thus u cannot have a complete differential at any point, even if we modify u in a set of measure 0.

Of course, we could slightly sharpen part b) by introducing the logarithmic scale of integrability, but this generalization would be of little interest.

It may also be added that, under the assumptions of Theorem 1, the function $u(x, y)$ is absolutely continuous in Tonelli's sense over every finite square I with sides parallel to the axes. This follows from the fact that u is absolutely continuous on almost every line parallel to one of the axes, and that both u_x and u_y are integrable over I.

On account of certain applications we shall state the analogue of Theorem 1 in n dimensions.

Theorem 2. *Suppose that $f(x_1, x_2, \ldots, x_n)$ is integrable over the whole space E^n, and $|f| \log^+ |f|$ is integrable over every finite sphere in the space. Then the potential*

$$u(x_1, \ldots x_n) = u(P) = \int_{E^n} |P - Q|^{-n+1} f(Q)\, dQ$$

converges over almost every line $x_i = \text{const}$, $i = 2, \ldots, n$ and represents an absolutely continuous function of x_1. In particular the partial derivative $\frac{\partial u}{\partial x_1} = u_{x_1}$ exists almost everywhere. Moreover

$$u_{x_1} = (n-1) \int_{E^n} (s_1 - x_1) |P - Q|^{-n-1} f(Q)\, dQ$$

almost everywhere, x_1 and s_1 being the first coordinates of P and Q respectively.

If f is integrable over the whole space E^n and belongs to L^q, $q > n$, then $u(P)$ has a complete differential at almost every point of the space.

The proof follows very closely that of Theorem 1.

Remark. It is not difficult to see that for the most general mass distribution $d\mu$ the potential u has at almost every point $P = (x_1, x_2 \ldots x_n)$ an *approximate differential*, that is $u(x_1 + h_1, \ldots x_n + h_n) - u(x_1 \ldots x_n) = \sum_1^n A_i h_i + o(\sum_1^n |h_i|)$, provided the point $(h_1, h_2 \ldots h_n)$ tends to $(0, 0, \ldots 0)$ through a certain set (depending, in general, on $(x_1 x_2 \ldots x_n)$) having the origin as a point of strong density. For let us make the usual decomposition $d\mu = g + d\nu$, where g is bounded and coincides with $d\mu$ in a perfect set S and equals the average value of μ in certain n-dimensional cubes constituting the complementary open set S'. Correspondingly $u = u_g + u_{d\nu}$. Since g is bounded, u_g has a differential almost everywhere. Lemma 2 of Chapter I easily shows that $u_{d\nu}$ has at almost every point a differential with respect to S. Making S expand, we obtain the result. This argument shows that u has almost everywhere all the approximate partial derivatives u_{x_i}.

We now turn to the logarithmic potential

(8) $$u(x, y) = \iint_{E^2} f(x - s, y - t) \log \frac{1}{(s^2 + t^2)^{1/2}}\, ds\, dt.$$

Since we are only interested in the differential properties of u, we may again assume that f vanishes outside a sufficiently large circle. We shall investigate the existence of the derivatives of the first two orders of u, and the existence of the second differential of u. We shall say that u has a second differential at a point (x_0, y_0) if, for h and k tending to 0,

$$(9)\quad u(x_0 + h, y_0 + k) - u(x_0 y_0) = Ah + Bk + \tfrac{1}{2}(Ch^2 + 2Dhk + Ek^2) + o(h^2 + k^2)$$

where A, B, C, D, E are independent of h and k. The existence of the second differential implies that of the first, and in particular that of $u_x(x_0, y_0) = A$ and $u_y(x_0, y_0) = B$. In general, however, it does not imply the existence of the second partial derivatives in the classical sense. For example, for $k = 0$ the preceding equation reduces to

$$u(x_0 + h, y_0) - u(x_0, y_0) = Ah + \tfrac{1}{2}Ch^2 + o(h^2),$$

which only implies that $u(x, y_0)$ has, for $x = x_0$, a second generalized derivative in the sense of Peano and de la Vallée Poussin (see e.g. [7] p. 257).

Theorem 3. *Suppose that $f(s, t)$ vanishes outside a circle and that $|f| \log^+ |f$ is integrable (in particular $f \in L$). Then*

a) *the integral* (8) *converges absolutely and represents a continuous function $u(x, y)$.*

b) *On almost every line parallel to either axis, $u(x, y)$ is continuously differentiable and the integrals*

$$-\iint_{E^2} f(x-s, y-t)\frac{s}{s^2+t^2}\,ds\,dt; \; -\iint_{E^2} f(x-s, y-t)\frac{t}{s^2+t^2}\,ds\,dt$$

obtained by formal differentiation of the integral (8) *converge and represent $u_x(x, y)$ and $u_y(x, y)$ respectively.*

c) *On almost every line parallel to either axis the derivatives u_x and u_y are absolutely continuous functions. In particular, u_{xx}, u_{yy}, u_{xy}, u_{yx} exist almost everywhere. They are given almost everywhere by the formulae*

$$(9\,a)\quad \begin{aligned} u_{xx}(x, y) &= -\pi f(x, y) + \iint_{E^2} f(x-s, y-t)\frac{s^2-t^2}{(s^2+t^2)^2}\,ds\,dt \\ u_{yy}(x, y) &= -\pi f(x, y) + \iint_{E^2} f(x-s, y-t)\frac{t^2-s^2}{(s^2+t^2)^2}\,ds\,dt \\ u_{xy}(x, y) = u_{yx}(x, y) &= \iint_{E^2} f(x-s, y-t)\frac{2st}{(s^2+t^2)^2}\,ds\,dt; \end{aligned}$$

in particular, $u_{xx} + u_{yy} = -2\pi f$ almost everywhere.

d) *The function $u(x, y)$ is absolutely continuous (i.e. is an integral).*

e) *The function u has almost everywhere a second differential, with C, D and E in (9) equal to u_{xx}, u_{xy} and u_{yy} respectively.*

Proof. That the integral (8) converges uniformly and absolutely follows from the inequality

$$x y \leq x \log^+ x + e^{y-1},$$

(See [7] p. 64) applied to the product $|f| \cdot \frac{1}{2} \log r$.

Whithout loss of generality we may assume that $f \geq 0$. Let us consider the function

$$u^{(\varepsilon)}(x, y) = \frac{1}{2} \iint_{E^2} f(x-s, y-t) \log \frac{1}{s^2 + t^2 + \varepsilon^2} \, ds \, dt, \tag{10}$$

and let us compare $u^{(\varepsilon)}_{xx}$ with the function

$$\tilde{f}_\varepsilon(x, y) = \iint_{\Gamma'_\varepsilon} f(x-s, y-t) \frac{s^2 - t^2}{(s^2 + t^2)^2} \, ds \, dt.$$

We have

$$u^{(\varepsilon)}_{xx} - \tilde{f}_\varepsilon = \iint_{\Gamma_\varepsilon} f(x-s, y-t) \frac{s^2 - t^2 - \varepsilon^2}{(s^2 + t^2 + \varepsilon^2)^2} \, ds \, dt +$$

$$+ \iint_{\Gamma'_\varepsilon} f(x-s, y-t) \left[\frac{s^2 - t^2 - \varepsilon^2}{(s^2 + t^2 + \varepsilon^2)^2} - \frac{s^2 - t^2}{(s^2 + t^2)^2} \right] ds \, dt,$$

and if we set

$$N(x, y) = \begin{cases} \dfrac{x^2 - y^2 - 1}{(x^2 + y^2 + 1)^2} & \text{for } x^2 + y^2 \leq 1, \\[2ex] \dfrac{x^2 - y^2 - 1}{(x^2 + y^2 + 1)^2} - \dfrac{x^2 - y^2}{(x^2 + y^2)^2} & \text{for } x^2 + y^2 > 1, \end{cases}$$

we may write

$$u^{(\varepsilon)}_{xx} - \tilde{f}_\varepsilon = \frac{1}{\varepsilon^2} \iint_{E^2} f(s, t) \, N\left(\frac{x-s}{\varepsilon}, \frac{y-t}{\varepsilon} \right) ds \, dt.$$

Then, by Lemma 2 in Chapter II it follows that, as $\varepsilon \to 0$, $u^{(\varepsilon)}_{xx} - \tilde{f}_\varepsilon$ converges to $-\pi f(x, y)$ in the mean. But according to Theorem 7 in Chapter I, over every set of finite measure $\tilde{f}_\varepsilon(x, y)$ converges in the mean to the function

$$\tilde{f}(x, y) = \iint_{E^2} f(x-s, y-t) \frac{s^2 - t^2}{(s^2 + t^2)^2} \, ds \, dt$$

and thus

$$\text{(11)}\quad \underset{\varepsilon\to 0}{\text{l.i.m.}}\, u_{xx}^{(\varepsilon)} = -\pi f(x,y) + \iint_{E^2} f(x-s,\, y-t)\frac{s^2-t^2}{(s^2+t^2)^2}\,ds\,dt = f_{11}(x\,y).$$

Similarly we get

$$\underset{\varepsilon\to 0}{\text{l.i.m.}}\, u_{xy}^{(\varepsilon)} = \iint_{E^2} f(x-s,\, y-t)\frac{2\,s\,t}{(s^2+t^2)^2}\,ds\,dt = f_{12}(x\,y),$$

(11)

$$\underset{\varepsilon\to 0}{\text{l.i.m.}}\, u_{yy}^{(\varepsilon)} = -\pi f(x,y) + \iint_{E^2} f(x-s,\, y-t)\frac{t^2-s^2}{(s^2+t^2)^2}\,ds\,dt = f_{22}(x\,y).$$

For the first derivative of $u^{(\varepsilon)}$ we have

$$\lim_{\varepsilon\to 0} u_x^{(\varepsilon)}(x\,y) = \lim_{\varepsilon\to 0}\left\{-\iint_{E^2} f(x-s,\, y-t)\frac{s\,ds\,dt}{s^2+t^2+\varepsilon^2}\right\} = -\iint_{E^2} f(x-s,\, y-t)\frac{s\,ds\,dt}{s^2+t^2}$$

at every point where the integral

$$\iint_{E^2} f(x-s,\, y-t)\frac{1}{(s^2+t^2)^{1/2}}\,ds\,dt$$

is finite. But according to Theorem 1 of this chapter, this is in fact so at all points of almost every line $y = y_0$.

Thus for almost every y and every x we have that $u^{(\varepsilon)}(x, y)$ converges and

$$\text{(12)}\quad \lim_{\varepsilon\to 0} u_x^{(\varepsilon)} = -\iint_{E^2} f(x-s,\, y-t)\frac{s}{s^2+t^2}\,ds\,dt = f_1(x, y).$$

An analogous result holds for $u_y^{(\varepsilon)}(x, y)$.

Finally for $u^{(\varepsilon)}(x, y)$ itself we have

$$\lim_{\varepsilon\to 0} u^{(\varepsilon)}(x, y) = u(x, y)$$

everywhere, since the integrand in (10) increases as ε decreases.

We now select a sequence $\varepsilon_n \to 0$ such that, over every set of finite measure of almost every line $y = y_0$, the left-hand sides of (11) converge in the mean to the right hand sides. Let $y = y_0$ be such a line where in addition (12) is satisfied, and take any point (x_0, y_0) on it. Then

$$u^{(\varepsilon_n)}(x, y_0) = \int_{x_0}^{x} (x-s)\, u_{xx}^{(\varepsilon_n)}(s, y_0)\,ds + (x-x_0)\, u_x^{(\varepsilon_n)}(x_0, y_0) + u^{(\varepsilon_n)}(x_0, y_0)$$

$$u^{(\varepsilon_n)}(x, y_0) = \int_{x_0}^{x} u_{xx}^{(\varepsilon_n)}(s, y_0)\,ds + u^{(\varepsilon_n)}(x_0, y_0),$$

and passing to the limit we obtain

$$u(x, y_0) = \int_{x_0}^{x} (x - s) f_{11}(s, y_0)\, ds + (x - x_0) f_1(x_0, y_0) + u(x_0, y_0),$$

$$f_1(x, y_0) = \int_{x_0}^{x} f_{11}(s, y_0)\, ds + f_1(x_0, y_0),$$

for all x. A similar result holds for almost every line $x = x_0$. This proves the absolute continuity of u_x and u_y, and gives the first two formulas (9 a).

Let now (x_0, y_0) be a point such that u be continuously differentiable on $x = x_0$ and on $y = y_0$. Then

$$u^{(\varepsilon)}(x, y) = \int_{x_0}^{x}\int_{y_0}^{y} u_{xy}^{(\varepsilon)}(s, t)\, ds\, dt + u^{(\varepsilon)}(x, y_0) + u^{(\varepsilon)}(x_0, y) - u^{(\varepsilon)}(x_0, y_0)$$

and passing to the limit we have

$$u(x, y) = \int_{x_0}^{x}\int_{y_0}^{y} f_{12}(s, t)\, ds\, dt + u(x, y_0) + u(x_0, y) - u(x_0, y_0).$$

But by a theorem of Tonelli and Fubini [3] for almost every $y = y_0$ the derivative with respect to y of the double integral above exists for all x and is equal to

$$\int_{x_0}^{x} f_{12}(s, y_0)\, ds,$$

and thus is an absolutely continuous function of x whose derivative with respect to x is $f_{12}(x, y_0)$. This completes the proof of part c).

The last formula also shows that $u(x, y)$ is absolutely continuous.

It remains to prove that $u(x, y)$ has almost everywhere a second differential. Let (x_0, y_0) be a point such that

1) the indefinite integral of u_{xy} is differentiable at (x_0, y_0) with respect to regular rectangles and its derivative is $u_{xy}(x_0, y_0)$;

2) $u(x_0 + h, y_0) = u(x_0, y_0) + h\, u_x(x_0, y_0) + \frac{h^2}{2} u_{xx}(x_0, y_0) + o(h^2)$;

3) $u(x_0, y_0 + k) = u(x_0\, y_0) + k\, u_y(x_0\, y_0) + \frac{k^2}{2} u_{yy}(x_0\, y_0) + o(k^2)$.

Since each of these conditions is fulfilled almost everywhere, they will also be satisfied simultaneously almost everywhere. Now

$$u(x_0 + h, y_0 + k) = \int_0^h\int_0^k u_{xy}(x_0 + s, y_0 + t)\, ds\, dt + u(x_0 + h, y_0) +$$
$$+ u(x_0, y_0 + k) - u(x_0, y_0)$$

and on account of 1) the double integral is equal to $h\,k\,u_{xy}(x_0\,y_0) + o\,(h^2 + k^2)$[1]. Thus, taking 2) and 3) into account part e) follows.

We conclude this paper with an extension of Theorem 3 to the potential

$$u(P) = \int_{E^n} f(Q)\,|P - Q|^{-(n-2)}\,dQ \qquad (n > 2) \tag{13}$$

in E^n.

Theorem 4. *Suppose that* $f(P) = f(x_1, x_2, \ldots, x_n)$ *is integrable over* E^n *and that* $|f| \log^+ |f|$ *is integrable over every sphere. Then*

a) *The integral* (13) *converges on almost every two-dimensional plane parallel to a fixed plane and represents a continuous, indeed an absolutely continuous, function there.*

b) *On almost every line parallel to a fixed line,* $u(P)$ *is continuously differentiable and the derivative is absolutely continuous. On almost every line parallel to any coordinate axis all the derivatives* $u_{x_1}, u_{x_2}, \ldots u_{x_n}$ *are absolutely continuous and are given by the formulas*

$$u_{x_i}(P) = \int_{E^n} f(Q)\,\frac{\partial}{\partial x_i}\,|P - Q|^{-(n-2)}\,dQ. \tag{14}$$

In particular, all the second derivatives $u_{x_i x_j}$ *exist almost everywhere. They are given almost everywhere by the formulas*

(15)

$$u_{x_i x_i} = -v_n f(P) + \int_{E^n} f(Q)\,\frac{\partial^2}{\partial x_i^2}\,|P - Q|^{-(n-2)}\,dQ;$$

$$u_{x_i x_j} = \int_{E^n} f(Q)\,\frac{\partial^2}{\partial x_i\,\partial x_j}\,|P - Q|^{-(n-2)}\,dQ \qquad (i \neq j)$$

v_n *denoting the volume of the n-dimensional unit sphere. In particular* $u_{x_1 x_1} + \cdots + u_{x_n x_n} = -n\,v_n f$ *almost everywhere.*

c) *If* $f \in L^q$, $q > \frac{n}{2}$, *then* u *has a second differential almost everywhere.*

Proof. We begin with c). Let P be a point such that

---

[1] Suppose, in fact, that $0 < h < k$. If $\frac{k}{2} < h < k$ our assertion is obviously true. On the other hand, if $h < \frac{k}{2}$ the integral is equal to

$$\int_0^k\int_0^k u_{xy}\,ds\,dt - \int_h^k\int_0^k u_{xy}\,ds\,dt = h\,k\,u_{xy}(x_0, y_0) + o\,(k^2) + o\,[k\,(k - h)].$$

(16)
$$I_\varrho = \int_{\Gamma_\varrho} |f(Q) - f(P)|^q \, dQ = o(\varrho^n);$$

$$J_\varrho = \int_{\Gamma_\varrho} |f(Q) - f(P)| \, dQ = o(\varrho^n)$$ [1]

where Γ_ϱ denotes a sphere of radius ϱ with center at P. Let us suppose in addition that (14) and (15) hold at P. Without any loss of generality we may further assume that $f(P) = 0$.

Let $\bar{e}$ be an arbitrary unit vector with components α_i and consider the expression

(17)
$$u(P + \varrho \bar{e}) - u(P) - \varrho \left(\frac{d u}{d \varrho}\right) - \tfrac{1}{2} \varrho^2 \left(\frac{d^2 u}{d \varrho^2}\right) = \Delta(\varrho)$$

where

$$\frac{d u}{d \varrho} = \sum_i \frac{\partial u}{\partial x_i} \alpha_i; \qquad \frac{d^2 u}{d \varrho^2} = \sum_{ij} \frac{\partial^2 u}{\partial x_i \, \partial x_j} \alpha_i \alpha_j$$

the partial derivatives being taken at the point P.

If we show that $\Delta(\varrho) = o(\varrho^2)$ uniformly in $\bar{e}$, our assertion will be established.

Let us replace in (17) the corresponding integrals. Denoting the complement of Γ_ϱ by Γ'_ϱ, we have

$$\Delta(\varrho) = \int_{\Gamma_{2\varrho}} [|P + \varrho \bar{e} - Q|^{-n+2} - |P - Q|^{-n+2}] f(Q) \, dQ -$$

$$- \varrho \int_{\Gamma_{2\varrho}} \frac{d}{d\varrho} |P - Q|^{-n+2} f(Q) \, dQ - \tfrac{1}{2} \varrho^2 \int_{\Gamma_{2\varrho}} \frac{d^2}{d\varrho^2} |P - Q|^{-n+2} f(Q) \, dQ +$$

$$+ \int_{\Gamma'_{2\varrho}} \Big[|P + \varrho \bar{e} - Q|^{-n+2} - |P - Q|^{-n+2} - \varrho \frac{d}{d\varrho} |P - Q|^{-n+2} -$$

$$- \tfrac{1}{2} \varrho^2 \frac{d^2}{d\varrho^2} |P - Q|^{-n+2} \Big] f(Q) \, dQ = A + B + C + D.$$

First, let us remark that on account of our assumption that $f(P) = 0$ in each of the preceding integrals we may replace $f(Q)$ by $f(Q) - f(P)$.

Then, by Hölder's inequality, we have

$$|A| \le 2 \Big[\int_{\Gamma_{2\varrho}} |f(Q) - f(P)|^q \, dQ \Big]^{1/q} \Big[\int_{\Gamma_{3\varrho}} |P - Q|^{(2-n)p} \, dQ \Big]^{1/p}$$

and on account of (16) we get $|A| = o(\varrho^2)$.

For B we have, again on account of (16),

[1] It is not difficult to show that $I_\varrho = o(\varrho^n)$ implies that $J_\varrho = o(\varrho^n)$. In fact this follows easily by applying Hölder's inequality to J_ϱ.

$$|B| \leq (n-2)\varrho \int_0^{2\varrho} r^{-n+1}\, d J_r = (n-2)\, 2^{-n+1} \varrho^{-n+2} J_{2\varrho} +$$

$$+ (n-2)(n-1)\varrho \int_0^{2\varrho} \frac{J_r}{r^n}\, d r = o(\varrho^2).$$

The integral in C converges to zero and thus it follows that also $|C| = o(\varrho^2)$.

Finally, it is easily seen that the quantity in square brackets in the integral D does not exceed $c\varrho^3 |P - Q|^{-n-1}$. Thus

$$|D| \leq c\varrho^3 \int_{2\varrho}^{\infty} r^{-n-1}\, d J_r = c\, 2^{-n-1} \varrho^{-n+2} J_{2\varrho} + c(n+1)\varrho^3 \int_{2\varrho}^{\infty} \frac{J_r}{r^{n+2}}\, d r = o(\varrho^2),$$

and c) is established.

The continuity of $u(P)$ in almost every plane parallel to a given plane, under the assumptions of the integrability of $|f| \log^+ |f|$, is less trivial here than the continuity of $u(x, y)$ in Theorem 4. It is enough, however, to sketch the proof. We may take the $x_1 x_2$ plane for the fixed plane.

First we consider the function

$$u^{(\varepsilon)}(P) = \int [|P - Q|^2 + \varepsilon^2]^{-\frac{1}{2}(n-2)} f(Q)\, d Q \qquad (f \geq 0)$$

and prove that $u^{(\varepsilon)}$ converges everywhere to $u(P)$, and that its first and second derivatives converge in the mean of order 1 over every set of finite measure. Then we select a subsequence $u^{(\varepsilon_n)}$ in such a way that the derivatives converge in the mean over every set of finite measure on almost every plane parallel to the $x_1 x_2$ plane, and on almost every line parallel to the x_1 or x_2 axis. We may suppose that the $x_1 x_2$ plane is such a plane and the x_1 and x_2 axes are such lines. Then

$$u^{\varepsilon}(x_1, x_2, 0, \ldots, 0) = u^{(\varepsilon)}(0, 0, \ldots, 0) + \int_0^{x_1}\int_0^{x_2} \frac{\partial^2 u^{(\varepsilon)}}{\partial x_1 \partial x_2}\, d x_1\, d x_2 +$$

$$+ \int_0^{x_1} \frac{\partial u^{(\varepsilon)}}{\partial x_1}\, d x_1 + \int_0^{x_2} \frac{\partial u^{(\varepsilon)}}{\partial x_2}\, d x_2,$$

and, passing to the limit,

$$u(x_1, x_2, 0, \ldots, 0) = u(0, 0, \ldots, 0) + \int_0^{x_1}\int_0^{x_2} \lim \frac{\partial^2 u^{(\varepsilon_n)}}{\partial x_1 \partial x_2}\, d x_1\, d x_2 +$$

$$+ \int_0^{x_1} \lim \frac{\partial u^{(\varepsilon_n)}}{\partial x_1}\, d x_1 + \int_0^{x_2} \lim \frac{\partial u^{(\varepsilon_n)}}{\partial x_2}\, d x_2.$$

This shows the existence almost everywhere of the derivatives u_{x_1}, u_{x_2}, $u_{x_1 x_2}$, and the continuity — even absolute continuity — of u in the $x_1 x_2$ plane. The proof of the remaining statements in Theorem 4 is similar.

As in the case of Theorem 2, we may supplement Theorem 4 by the following remark. Let $u(P)$ be the potential of a general mass distribution $d\mu$. Then almost everywhere u has an approximate second differential, in the sense that in the formula (17) the expression $\Delta(\varrho)$ is $o(\varrho^2)$ for almost every P, $\varrho\bar{e}$ tends to 0 through a set of points having 0 as a point of strong density. The proof follows from the same decomposition $d\mu = g + d\nu$ as in the case of Theorem 2. The argument also shows that the approximate second derivatives $u_{x_i x_j}$ exists almost everywhere (being defined as the approximate first derivatives of the ordinary first derivatives) and satisfy the equations (15). In particular, the $u_{x_i x_i}$ satisfy Poisson's equation.

Added in proof, 1. VIII. 52.

1°. When this paper was already accepted for publication, Prof. W. J. Trjitzinsky called our attention to an interesting expository article by S. G. Mikhlin, "Singular integral equations", Uspekhi Matematicheskikh Nauk, No 25 (1948), 29–112, which treats topics similar to those discussed in the present paper and describes the earlier work of Giraud, Tricomi and the author himself. However, only functions of the class L^2 are considered there, and singular integrals are treated in the sense of mean convergence (in the metric L^2). On the other hand, considering vector-functions and matrix-kernels leads the author naturally to the problems of inversion and of the norm preservation of the transform. (For the case $K(z) = 1/z^2$, $f \in L^2$, these problems have also been solved in an unpublished work of Prof. A. Beurling.) Combining those results with the theorems of the present paper, one may present the former in a stronger form, as we hope to show elsewhere.

2°. In Chapters I and II of the present paper we discussed the case of functions f non-periodic and defined over the entire space E^n. Analogous results can be obtained for periodic functions. We shall limit ourselves here to describing only the general idea. Let $\bar{e}_1, \bar{e}_2, \ldots, \bar{e}_n$ be a system of n independent vectors in E^n, which for simplicity we assume to be mutually orthogonal and of length 2π. Let $P_0 = O$, $P_1, P_2, \ldots$ be the sequence of terminal points of the vectors $(P_\nu - O_\nu) = k_1\bar{e}_1 + k_2\bar{e}_2 + \cdots + k_n\bar{e}_n$, where the k_j are arbitrary integers. The series on the right in the formula

$$K^*(P-O) = K(P-O) + \sum_{\nu=1}^{\infty}\{K(P-P_\nu) - K(O-P_\nu)\}$$

converges absolutely and uniformly over any finite sphere in E^n, provided we drop the first few terms. The function K^* is periodic, of period 2π, in each Cartesian coordinate, and the Fourier coefficients of K^*, taken in the principal value sense, are equal to the corresponding values of Fourier transform of K. If $f(P) \sim \Sigma c_{k_1 \ldots k_n} \exp i(k_1 x_1 + \cdots + k_n x_n)$, the function

$$f^*(P) = (2\pi)^{-n} \int_R f(Q) K^*(P-Q)\, dQ,$$

where R stands for the cube $|x_j| \leq \pi$, $j = 1, 2, \ldots, n$, plays a role similar to that of the ordinary conjugate function in E^1. If f^* is integrable, its Fourier coefficients are $c_{k_1 \ldots k_n} \hat{K}_{k_1 \ldots k_n}$, where $\hat{K}_{k_1 \ldots k_n}$ denote the Fourier coefficients of K^*. Familiar results about f^* in E^1 (in which case $K(t) = 1/t$, $K^*(t) = \frac{1}{2} \cot \frac{1}{2} t$) are simple consequences of the theorems established in Chapters I and II and are easily extensible to general n. The simplest cases for $n = 2$ are the kernels $K(z) = z^k / |z|^{k+2}$, $k = \pm 1, \pm 2, \ldots$ The kernel K^* associated with $K(z) = 1/z^2$ is the classical $\wp$ function of Weierstrass.

3°. Let $\bar{e}_1, \bar{e}_2, \ldots, \bar{e}_n$ be any system of independent vectors in E^n and $P_0 = O$, $P_1, P_2, \ldots$ the set of lattice points generated by this system. Let $x_0, x_1, \ldots$ be any sequence of complex numbers such that $\Sigma |x_\mu|^p < \infty$, where $p > 1$ is a fixed number, and let

$$(*) \qquad \tilde{x}_\nu = \sum_{\mu \neq \nu} x_\mu K(P_\nu - P_\mu).$$

Theorem 1 of Chapter I leads to the inequality

$$(\Sigma |\tilde{x}_\nu|^p)^{1/p} \leq A_p (\Sigma |x_\mu|^p)^{1/p}.$$

(For $n = 1$ this remark is due to M. Riesz [9], and the proof in the case of general n follows a similar pattern). The last inequality can be written in the form

$$|\Sigma x_\mu y_\nu K(P_\nu - P_\mu)| \leq A_p (\Sigma |x_\mu|^p)^{1/p} (\Sigma |y_\nu|^q)^{1/q},$$

where $p > 1$, $q > 1$, $1/p + 1/q = 1$. The case $n = 2$, $K(z) = 1/z^2$ is of special interest. The equations (*) can then also be written

$$\tilde{x}_\nu = \sum_{\mu \neq \nu} x_\mu (\nu - \mu)^{-2},$$

where μ and ν now run through all complex integers. This may be considered as the simplest generalization of the Hilbert-Toeplitz linear form to space E^2. The norm of this transformation is the upper bound of the modulus of the function defined by the Fourier series $\Sigma' (k + il)^{-2} \exp i(kx + ly)$.

4°. Theorem 1 of Chapter 1 can be written in the form (which again for $n=1$ was pointed out by M. Riesz)

$$\left| \int_{E^n}\int_{E^n} f(P)\, g(Q)\, K(P-Q)\, dP\, dQ \right| = A_p \|f\|_p \|g\|_q \quad (1/p + 1/q = 1),$$

where the integral on the left is considered as the limit, for $\varepsilon \to 0$, of the integral extended over the portion $|P-Q| \geq \varepsilon$ of the space $E^n \times E^n$.

5°. The case $K(P-O) = (P-O)\,|P-O|^{-n-1}$, $f \in L^2$ is also discussed in a recent unpublished paper of J. Horváth.

References.

[1] G. C. Evans, On potentials of positive masses, Transactions of the American Math. Soc., 37 (1935).

[2] K. O. Friedrichs, A theorem of Lichtenstein, Duke Math. Journal, 14, 67—82 (1947).

[3] G. Fubini and L. Tonelli, Sulla derivata seconda mista di un integrale doppio, Rendiconti Circolo Mat. di Palermo, 40, 295—298 (1915).

[4] B. Jessen, J. Marcinkiewicz and A. Zygmund, Note on the differentiability of multiple integrals, Fundamenta Math., 25, 217—234 (1935).

[5] H. Kober, A note on Hilbert transforms, Journal of the London Math. Soc., 18, 66—71 (1943).

[6] L. Lichtenstein, Über das Poissonsche Integral, Journal für reine und angewandte Mathematik, 141, 12—42 (1912).

[7] A. Zygmund, Trigonometrical Series, Warsaw (1936).

[8] N. Aronszajn, Propriétés de certaines classes hilbertiennes complétées, Comptes rendus de l'Académie des Sciences de Paris, 226, 700—702 (1948).

[9] M. Riesz, Sur les fonctions conjuguées, Math. Zeitschrift, 27, 218—244 (1927).

SOME PROPERTIES OF TRIGONOMETRIC SERIES WHOSE TERMS HAVE RANDOM SIGNS

Dedicated to Professor Hugo Steinhaus for his 65th Birthday

BY

R. SALEM and A. ZYGMUND

Trigonometric series of the type

$$\sum_{1}^{\infty} \varphi_n(t)\,(a_n \cos nx + b_n \sin nx), \tag{0.1}$$

where $\{\varphi_n(t)\}$ denotes the system of Rademacher functions, have been extensively studied in order to discover properties which belong to "almost all" series, that is to say which are true for almost all values of t.[1] We propose here to add some new contributions to the theory.

CHAPTER I

Weighted Means of Ortho-normal Functions

1. Let $\varphi_1, \varphi_2, \ldots, \varphi_n, \ldots$ be a system of functions of x, ortho-normal in an interval (a, b), and let $\gamma_1, \gamma_2, \ldots, \gamma_n, \ldots$ be a sequence of non-negative constants such that

$$S_n = \gamma_1 + \gamma_2 + \cdots + \gamma_n$$

increases indefinitely as n tends to $+\infty$. Under what conditions does the mean

$$R_n(x) = \frac{\gamma_1 \varphi_1(x) + \gamma_2 \varphi_2(x) + \cdots + \gamma_n \varphi_n(x)}{\gamma_1 + \gamma_2 + \cdots + \gamma_n}$$

tend to zero almost everywhere[2] as $n \to \infty$?

[1] Cf., in particular, PALEY and ZYGMUND, *Proc. Cambridge Phil. Soc.*, 26 (1930), pp. 337–357 and 458–474, and 28 (1932), pp. 190–205.

[2] We write briefly $R_n(x) \to 0$ p.p. ("presque partout").

Reprinted from *Acta Math.* 91, 245–301 (1954).

It has been proved[1] that, if $\varphi_n = e^{inx}$, then $R_n(x) \to 0$ p.p., provided $\gamma_n = O(1)$. The proof is applicable without change to any ortho-normal uniformly bounded system. As it was observed in the paper, some condition on the γ_n is indispensable, since, e.g.

$$2^{-n} \sum_{1}^{n} 2^k e^{ikx}$$

does not tend to zero almost everywhere as $n \to \infty$.

More recently, Hill and Kakutani have raised the question whether $R_n(x) \to 0$ p.p. if $\{\varphi_n\}$ is the Rademacher system, the sequence γ_n is monotonically increasing and $\gamma_n = o(S_n)$. The answer has been proved by several authors to be negative.[2]

Here we propose to give a sufficient condition in order that $R_n(x) \to 0$ p.p., when $\{\varphi_n\}$ is any uniformly bounded ortho-normal system in (a, b) and to prove, by the consideration of the trigonometric system, that this condition is the best possible one.

Let us observe first of all that the condition $\gamma_n = o(S_n)$ is trivially necessary in order that $R_n(x) \to 0$ p.p. For $\gamma_n / S_n = \int_a^b R_n \varphi_n \, dx$, and the uniform boundedness of the φ_n implies that $R_n \varphi_n \to 0$ p.p., boundedly, whence $\gamma_n / S_n \to 0$.

As we shall see, the condition $\gamma_n = o(S_n)$ is, in general, not sufficient. Let us note, however, that in the case $\gamma_n = e^{inx}$, if the sequence $\{\gamma_n\}$ is monotonic and $\gamma_n = o(S_n)$, then $R_n(x) \to 0$ everywhere, except for $x \equiv 0$. This follows from an application of summation by parts to the numerator of R_n.

2. (1.2.1) **Theorem.** *Let $\{\varphi_n\}$ be an ortho-normal and uniformly bounded system in (a, b), and let $|\varphi_n| \le M$. Let $\omega(u)$ be a monotonically increasing function of u such that $u/\omega(u)$ increases to $+\infty$ and such that $\sum 1/k\,\omega(k) < \infty$. Then $R_n(x) \to 0$ p.p., provided $\gamma_n = O\{S_n/\omega(S_n)\}$.*

Proof. Let us recall first that, if we set

$$\gamma_n^* = \operatorname*{Max}_m \gamma_m \quad (1 \le m \le n),$$

then also $\gamma_n^* = O\{S_n/\omega(S_n)\}$. For we have $\gamma_n^* = \gamma_p$, where $p = p(n) \le n$ is non-decreasing. Let $Q_n = S_n/\omega(S_n)$. Then

---

[1] Cf. R. Salem, The absolute convergence of trigonometric series, *Duke Math. Journal*, 8 (1941), p. 333.

[2] See Tamotsu Tsuchikura, *Proc. of the Japan Academy*, 27 (1951), pp. 141–145, and the results quoted there, especially Maruyama's result.

$$\frac{\gamma_p}{Q_n} = \frac{\gamma_p}{Q_p} \cdot \frac{Q_p}{Q_n} \leq \frac{\gamma_p}{Q_p},$$

and our assertion follows.

Consequently one also has

$$\frac{\sum_{1}^{N} \gamma_n^2}{S_N^2} = O\left\{\frac{1}{\omega(S_N)}\right\}. \tag{1.2.2}$$

Let us fix a number $\theta > 1$ and let N_j be the first integer such that

$$\theta^j \leq S_{N_j} < \theta^{j+1}.$$

N_j always exists for j large enough. For otherwise there would exist arbitrarily large integers j such that for a suitable m we would have $\theta^j \leq S_m < \theta^{j+1}$ and simultaneously $S_{m-1} < \theta^{j-1}$. This would imply

$$\gamma_m > \theta^j - \theta^{j-1}, \qquad \frac{\gamma_m}{S_m} > \frac{\theta^j - \theta^{j-1}}{\theta^{j+1}} = \frac{\theta - 1}{\theta^2},$$

contradicting the assumption $\gamma_n = o(S_n)$.

Now, by (1.2.2),

$$\sum_j \int_a^b |R_{N_j}|^2 \, dx = \sum_j O\left\{\frac{1}{\omega(\theta^j)}\right\} < \infty$$

by the hypothesis $\Sigma\, 1/k\,\omega(k) < \infty$. Hence $R_{N_j} \to 0$ p.p.

Let now $N_j \leq m < N_{j+1}$. One has

$$R_m = \frac{\sum_{1}^{N_j} \gamma_n \varphi_n}{S_m} + \frac{\sum_{N_j+1}^{m} \gamma_n \varphi_n}{S_m}.$$

The first ratio tends to zero almost everywhere since $S_m \geq S_{N_j}$; the second one has absolute value less than

$$M \frac{\sum_{N_j+1}^{N_{j+1}} \gamma_n}{S_{N_j}} = M \frac{S_{N_{j+1}} - S_{N_j}}{S_{N_j}} < M \frac{\theta^{j+2} - \theta^j}{\theta^j} = M(\theta^2 - 1).$$

It follows that

$$\limsup |R_m| \leq M(\theta^2 - 1)$$

almost everywhere. Since θ can be taken arbitrarily close to 1, the theorem follows.

3. We shall now show that the preceding theorem no longer holds if we allow $\Sigma 1/k\,\omega(k)$ to diverge. This will follow from the following

(1.3.1) **Theorem.** *Given any function $\omega(t)$ increasing to $+\infty$ with t and such that $\int^{\infty} \frac{dt}{t\,\omega(t)} = \infty$, and assuming for the sake of simplicity that $\omega(t)/\log t$ is monotonic, there exists a sequence γ_n such that $\gamma_n = O\{S_n/\omega(S_n)\}$ and that*

$$\limsup \left| \frac{\gamma_1 e^{ix} + \cdots + \gamma_n e^{inx}}{\gamma_1 + \cdots + \gamma_n} \right| = 1$$

almost everywhere.

We shall first prove two lemmas.

(1.3.2) **Lemma.** *Let $\{m_q\}$ be an increasing sequence of integers such that m_q/q is monotonic and $\Sigma 1/m_q = \infty$. Then*

$$\limsup \frac{1}{m_q} \left| \frac{\sin q\, m_q\, x}{\sin q x} \right| = 1$$

almost everywhere.

Proof. By a well-known theorem of Khintchine, the conditions imposed on the sequence $\{m_q\}$ imply that for almost every x there exist infinitely many integers p and q such that

$$\left| \frac{x}{2\pi} - \frac{p}{q} \right| = o\left(\frac{1}{q\, m_q} \right).$$

Hence, fixing an x having this property, one has

$$|q x - 2\pi p| = \varepsilon_q / m_q$$
$$|q\, m_q\, x - 2\pi p\, m_q| = \varepsilon_q$$

for infinitely many p, q, with $\varepsilon_q \to 0$. Hence also, for infinitely many values of q,

$$\frac{1}{m_q} \frac{\sin q\, m_q\, x}{\sin q x} = \frac{1}{m_q} \frac{\sin \varepsilon_q}{\sin (\varepsilon_q / m_q)},$$

and the lemma follows.

As a simple special case we get that $\limsup \left| \frac{\sin q^2 x}{q \sin q x} \right| = 1$ p.p.

Remark. If $\Sigma 1/m_q < \infty$, one has $\lim \frac{\sin q\, m_q\, x}{m_q \sin q x} = 0$ p.p. This follows immediately from the fact that

$$\frac{1}{m_q^2}\int_0^{2\pi}\left(\frac{\sin q\, m_q\, x}{\sin q x}\right)^2 dx = O\left(\frac{1}{m_q}\right).$$

(1.3.3) **Lemma.** *Let $\{m_q\}$ be an increasing sequence of integers such that m_q/q is monotonic and that $\Sigma\, 1/m_q = \infty$. Let us set, for each positive integer q.*

$$h = h(q) = q\, m_q + 1, \quad k = k(q) = (q+1)\, m_q$$

and

$$F_q(x) = e^{h q i x} + e^{(h+1) q i x} + \dots + e^{k q i x}.$$

Then

$$\limsup |F_q(x)|/m_q = 1$$

almost everywhere.

Proof. One has

$$F_q = e^{h q i x}\frac{e^{(k-h+1) q i x} - 1}{e^{q i x} - 1},$$

$$|F_q| = \left|\frac{\sin \frac{1}{2}(k-h+1) q x}{\sin \frac{1}{2} q x}\right| = \left|\frac{\sin \frac{1}{2} q\, m_q\, x}{\sin \frac{1}{2} q x}\right|,$$

and it is enough to apply the preceding lemma.

Proof of Theorem (1.3.1). It will be sufficient to consider the stronger case in which $\omega(t)/\log t$ increases to ∞. We write $\omega(t) = \log t\ \lambda(\log t)$ where $\lambda(u)$ increases to ∞ with u. Observe that

$$\int^{\infty}\frac{du}{u\lambda(u)} = \int^{\infty}\frac{dt}{t\log t\,\lambda(\log t)} = \int^{\infty}\frac{dt}{t\,\omega(t)} = \infty.$$

Hence we can find an increasing sequence of integers m_q such that $m_q/q\,\lambda(q)$ increases to ∞ and that $\Sigma\, 1/m_q = \infty$.

By $\varphi(q)$ we shall denote a function of q increasing to ∞ as slowly as we wish and which we shall determine later on.

Let us now set, for each integer q, as in Lemma (1.3.3),

$$h = h(q) = q\, m_q + 1, \quad k = k(q) = (q+1)\, m_q$$

and let

$$\gamma_n = \frac{e^{q\varphi(q)}}{m_q} = c_q \text{ for } n = \nu q \quad (h \le \nu \le k)$$

$$\gamma_n = 0 \text{ for } n \neq \nu q \quad (hq < n < kq).$$

We note that, F_q being defined as in Lemma (1.3.3), there is no overlapping of terms of F_q and F_{q+1}, since $qk(q) < (q+1)h(q+1)$.

Let us now write

$$\varrho_q = \frac{c_1 F_1 + c_2 F_2 + \cdots + c_q F_q}{c_1 m_1 + c_2 m_2 + \cdots + c_q m_q}.$$

This ratio is equal to $R_n(x)$ for $n = qk(q)$.

One has

$$\varrho_q = \frac{(c_1 F_1 + \cdots + c_{q-1} F_{q-1}) e^{-q\varphi(q)} + F_q/m_q}{(c_1 m_1 + \cdots + c_{q-1} m_{q-1}) e^{-q\varphi(q)} + 1},$$

and since

$$|c_1 F_1 + \cdots + c_{q-1} F_{q-1}| \leq c_1 m_1 + \cdots + c_{q-1} m_{q-1} = \sum_1^{q-1} e^{k\varphi(k)} = o(e^{q\varphi(q)}),$$

one has

$$\limsup |\varrho_q| = \limsup |F_q|/m_q = 1.$$

On the other hand, if e^{nix} occurs in F_q,

$$\gamma_n = \frac{e^{q\varphi(q)}}{m_q} \text{ and } S_n > \sum_1^{q-1} e^{k\varphi(k)} > e^{(q-1)\varphi(q-1)}.$$

Hence

$$\gamma_n/S_n < m_q^{-1}\,[e^{q\varphi(q)-(q-1)\varphi(q-1)}].$$

Since $m_q/q\lambda(q)$ increases to ∞, we can choose $\varphi(q)$ increasing slowly enough in order to have

$$\frac{m_q}{q\varphi(q)\lambda[q\varphi(q)]} > e^{q\varphi(q)-(q-1)\varphi(q-1)}.$$

Therefore,

$$\frac{\gamma_n}{S_n} \leq \frac{1}{q\varphi(q)\lambda[q\varphi(q)]}.$$

But

$$S_n < \sum_1^q e^{k\varphi(k)} \sim e^{q\varphi(q)}.$$

It follows that

$$\frac{\gamma_n}{S_n} = O\left[\frac{1}{\log S_n \, \lambda(\log S_n)}\right] = O\left[\frac{1}{\omega(S_n)}\right],$$

which proves the theorem.

4. As we have mentioned above, the case of Rademacher functions has been completely investigated by Maruyama and Tsuchikura. We give here different proofs of their two results.

(1.4.1) **Theorem.** *Let* $\varphi_1, \varphi_2, \ldots, \varphi_n, \ldots$ *be the system of Rademacher functions. Then* $\gamma_n = o\{S_n/\log\log S_n\}$ *is a sufficient condition in order that* $R_n \to 0$ *p.p.*

We have, as in Section 2,

$$\frac{\sum_1^N \gamma_n^2}{S_N^2} = o\left(\frac{1}{\log\log S_N}\right) = \frac{\varepsilon_N}{\log\log S_N},$$

where $\varepsilon_N \to 0$. Since for the Rademacher functions one has

$$\int_0^1 \left(\sum_1^N \gamma_n \varphi_n\right)^{2k} dx \le k^k \left(\sum_1^N \gamma_n^2\right)^k \quad (k = 1, 2, \ldots) \tag{1.4.2}$$

it follows that, A being a positive constant,

$$\int_0^1 (A R_N)^{2k} dx \le A^{2k} \frac{k^k \left(\sum_1^N \gamma_n^2\right)^k}{S_N^{2k}} = \left(\frac{A^2 k \varepsilon_N}{\log\log S_N}\right)^k.$$

Let us define $k = k(N)$ to be the integral part of $(e A^2 \varepsilon_N)^{-1} \log\log S_N$. We get

$$\int_0^1 (A R_N)^{2k} dx \le \left(\frac{1}{e}\right)^{\frac{\log\log S_N}{e A^2 \varepsilon_N} - 1}.$$

Let us now take $\theta > 1$ and define, as in Section 2, a sequence $\{N_j\}$ such that $\theta^j \le S_{N_j} < \theta^{j+1}$. Let $k_j = k(N_j)$. One has, for j large enough,

$$\frac{1}{e A^2 \varepsilon_{N_j}} > 2.$$

Hence

$$\int_0^1 (A R_{N_j})^{2k_j} dx = O(j^{-2})$$

so that

$$\sum_j (A R_{N_j})^{2k_j} < \infty \quad \text{p.p.}$$

It follows that $\limsup R_{N_j} \leq A^{-1}$ p.p., and so also that $\lim R_{N_j} = 0$ p.p., since A can be taken arbitrarily large. From here we proceed, without change, as in Section 2.

Remark 1. An alternate proof could be given using an extension of the Law of the Iterated Logarithm pertaining to the case of Rademacher functions, but the proof given above is simpler and more direct.

Remark 2. The theorem can obviously be extended to other orthogonal systems for which the inequality of the type (1.4.2) holds, e.g. to certain types of independent functions, and also lacunary trigonometric functions, such as $\cos 2^n x$.

5. (1.5.1) **Theorem.** *Theorem* (1.4.1) *becomes false if we replace in the assumption the order "o" by "O". More precisely, there exists a sequence* $\{\gamma_n\}$ *such that*

$$\gamma_n = O(S_n/\log\log S_n),$$

and such that $\limsup R_n > \alpha$ *p.p.,* α *being any fixed constant less than* 1.

(1.5.2) **Lemma.** *Let* $\varphi_1, \varphi_2, \dots$ *be the system of Rademacher functions and let* $\{m_p\}$ *be an increasing sequence of integers such that*

$$\Delta_p = m_p - m_{p-1} \leq \frac{\log p}{\log 2}.$$

Then, writing

$$F_p = \sum_{m_{p-1}+1}^{m_p} \varphi_k$$

one has

$$\limsup F_p/\Delta_p = 1 \text{ p.p.}$$

The set in which $F_p = \Delta_p$ has measure $2^{-\Delta_p}$. Hence the set where $F_p \neq \Delta_p$ is of measure $1 - 2^{-\Delta_p}$; and the set E_p where $F_p, F_{p+1}, \dots$ are *all* different from Δ_p, $\Delta_{p+1}, \dots$, respectively, is of measure

$$\prod_{q=p}^{\infty} (1 - 2^{-\Delta_q}) = 0$$

by the hypothesis on Δ_q. If, for a given x, $\limsup F_p/\Delta_p < 1$, then, clearly, x belongs to some E_p. Since the sets E_p are all of measure zero, the lemma follows.

We are now able to prove Theorem (1.5.1). Preserving the notation of the lemma, we take a sequence of integers m_p satisfying the conditions

$$\tfrac{1}{3}\log p < m_p - m_{p-1} < \tfrac{1}{2}\log p$$

so that the condition $\Delta_p \le \log p/\log 2$ is satisfied. Let us now take, with $A>1$,

$$\gamma_n = A^p/\Delta_p = c_p, \text{ for } m_{p-1} < n \le m_p,$$

and consider the ratio

$$\varrho_p = \frac{c_1 F_1 + c_2 F_2 + \cdots + c_p F_p}{c_1 \Delta_1 + c_2 \Delta_2 + \cdots + c_p \Delta_p}$$

which is equal to $R_n(x)$ for $n = m_p$. One has

$$\varrho_p = \frac{(c_1 F_1 + \cdots + c_{p-1} F_{p-1}) A^{-p} + F_p \Delta_p^{-1}}{(c_1 \Delta_1 + \cdots + c_{p-1} \Delta_{p-1}) A^{-p} + 1}.$$

Now,

$$|A^{-p}(c_1 F_1 + \cdots + c_{p-1} F_{p-1})| \le A^{-p}(c_1 \Delta_1 + \cdots + c_{p-1} \Delta_{p-1}) = A^{-p} \sum_1^{p-1} A^k.$$

Hence, if A is large enough, using the lemma,

$$\limsup \varrho_p > \alpha \text{ p.p.}$$

On the other hand, if φ_n occurs in F_p, then

$$\gamma_n = A^p \Delta_p^{-1} < 3.A^p/\log p,$$

$$S_{m_{p-1}} < S_n \le S_{m_p},$$

so that $\log\log S_n \sim 2 \log p$, and $\gamma_n = O\{S_n/\log\log S_n\}$.

6. To theorems about the partial sums of divergent series often correspond theorems about the remainders of convergent series, and the results of this chapter admit of such extensions. We shall be satisfied with stating here the following analogue of Theorem (1.2.1), in which the functions $\omega(u)$ and φ_n have the same meaning as there.

(1.6.1) **Theorem.** *If $\gamma_1 + \gamma_2 + \cdots + \gamma_n + \cdots$ is a convergent series with positive terms, and if $\bar{S}_n$ denotes the remainder $\gamma_n + \gamma_{n+1} + \cdots$, then*

$$\bar{R}_n(x) = \frac{\gamma_n \varphi_n + \gamma_{n+1} \varphi_{n+1} + \cdots}{\gamma_n + \gamma_{n+1} + \cdots}$$

tends to 0 p.p., provided $\gamma_n = O\{\bar{S}_n/\omega(1/\bar{S}_n)\}$.

The proof is identical with that of Theorem (1.2.1).

CHAPTER II

The Law of the Iterated Logarithm

1. As proved by Kolmogoroff, the law can be stated as follows. Let

$$z_1, z_2, \dots, z_n, \dots$$

be independent random variables with vanishing mean values and with dispersions $b_1, b_2, \dots, b_n, \dots$, respectively. Let

$$S_N = \sum_1^N z_k, \quad B_N^2 = \sum_1^N b_k .$$

Under the assumptions

$$B_N^2 \to \infty, \quad |z_N| \le m_N = o\left\{\frac{B_N^2}{\log\log B_N}\right\}^{\frac{1}{2}}$$

one has, with probability 1,

$$\limsup \frac{S_N}{(2 B_N^2 \log\log B_N)^{\frac{1}{2}}} = 1. \tag{2.1.1}$$

The result $\limsup \le 1$ has been extended by the authors to the case in which the series of random variables is replaced by a lacunary trigonometric series

$$\Sigma\, (a_k \cos n_k x + b_k \sin n_k x),$$

with $n_{k+1}/n_k > q > 1$.[1] Here we propose to give a theorem equivalent to Kolmogoroff's, valid for almost all trigonometric series of the type (0.1).

(2.1.2) **Theorem.** *Let us consider the series*

$$\sum_1^\infty \varphi_n(t)\,(a_n \cos nx + b_n \sin nx) = \sum_1^\infty \varphi_n(t)\, A_n(x), \tag{2.1.3}$$

where $\{\varphi_n(t)\}$ *is the system of Rademacher functions. Let*

$$c_k^2 = a_k^2 + b_k^2, \quad B_N^2 = \tfrac{1}{2} \sum_1^N c_k^2$$

$$S_N = \sum_1^N \varphi_k(t)\, A_k(x),$$

and let $\omega(p)$ *be a function of* p *increasing to* $+\infty$ *with* p, *such that* $p/\omega(p)$ *increases and that* $\Sigma\, 1/p\,\omega(p) < \infty$. *Then, under the assumptions*

[1] See *Bulletin des Sciences Mathématiques*, 74 (1950).

$$B_N^2 \to \infty, \quad c_N^2 = O\left\{\frac{B_N^2}{\omega(B_N^2)}\right\} \tag{2.1.4}$$

one has, for almost every value of t,

$$\limsup \frac{S_N}{(2B_N^2 \log\log B_N)^{\frac{1}{2}}} = 1$$

almost everywhere in x, that is to say the law of the iterated logarithm is true for almost all series (2.1.3).

The proof will be based on Theorem (1.2.1).

2. For a given x, the law of the iterated logarithm applied to the series (2.1.3) in which t is the variable, gives

$$\limsup \frac{\sum_1^N A_n(x)\,\varphi_n(t)}{\left\{2\sum_1^N A_n^2(x) \log\log \sum_1^N A_n^2(x)\right\}^{\frac{1}{2}}} = 1, \tag{2.2.1}$$

p.p. in t, provided

$$\sum_1^N A_n^2(x) \to \infty, \quad A_n^2(x) = o\left(\frac{\sum_1^N A_n^2(x)}{\log\log \sum_1^N A_n^2(x)}\right). \tag{2.2.2}$$

One has

$$A_n(x) = a_n \cos nx + b_n \sin nx = c_n \cos(nx - \alpha_n),$$

$$A_n^2(x) = \tfrac{1}{2} c_n^2 \{1 + \cos(2nx - 2\alpha_n)\}.$$

Hence, by Theorem (1.2.1) and on account of the condition $c_n^2 = O\{B_n^2/\omega(B_n^2)\}$, one has

$$B_N^{-2} \sum_1^N A_n^2(x) \to 1 \quad \text{p.p. in } x. \tag{2.2.3}$$

On the other hand, condition (2.2.2) is satisfied p.p. in x, because $\Sigma\, 1/p\,\omega(p) < \infty$ and $\omega(p)$ increasing imply $\log p/\omega(p) \to 0$. Hence

$$c_N^2 = o\left\{\frac{B_N^2}{\log\log B_N}\right\},$$

which together with (2.2.3) implies (2.2.2).

Finally, (2.2.1) and (2.2.3) give

$$\limsup \frac{S_N}{(2B_N^2 \log\log B_N)^{\frac{1}{2}}} = 1$$

p.p. in t and x, and so also the theorem as stated.

17 – 533807. *Acta Mathematica.* 91. Imprimé le 30 octobre 1954.

3. Additional remarks. a) We do not know whether the condition

$$c_N^2 = O\left\{\frac{B_N^2}{\omega(B_N^2)}\right\}$$

with $\Sigma\, 1/p\,\omega(p) < \infty$, can be improved or not (this condition is certainly satisfied if $c_n = O(1)$). But the argument used here would break down if $\Sigma\, 1/p\,\omega(p)$ were divergent. This follows from Theorem (1.3.1).

In the rest of this chapter the function $\omega(u)$ will be supposed to have the properties assumed in Theorem (2.1.2).

b) The following is an analogue of Theorem (2.1.2) for power series

(2.3.1) $$\sum_1^\infty \varphi_n(t)\, c_n\, e^{i n x},$$

for which we set

(2.3.2) $$C_N^2 = \sum_1^N |c_k^2|, \quad S_N = \sum_1^N \varphi_k(t)\, c_k\, e^{i k x}.$$

(2.3.3) **Theorem.** *For almost every t we have*

(2.3.4) $$\limsup \frac{|S_N|}{(C_N^2 \log\log C_N)^{\frac{1}{2}}} = 1$$

almost everywhere in x, provided

$$C_N^2 \to \infty, \quad c_N^2 = O\left\{\frac{c_N^2}{\omega(c_N^2)}\right\}.$$

We note that the factor 2 is missing in the denominator in (2.3.4).

Let us set $S_N = U_N + i\, V_N$. From Theorem (2.1.2) it follows that for almost every point (x, t) and every rational α we have

$$\limsup \frac{U_N \cos \alpha\pi + V_N \sin \alpha\pi}{(C_N^2 \log\log C_N)^{\frac{1}{2}}} = 1,$$

and from this we easily deduce that (2.3.4) holds for almost every point (x, t) (compare a similar argument used in Salem and Zygmund, loc. cit.).

c) Kolmogoroff's result quoted at the beginning of this chapter has an analogue for the case in which the series $\Sigma\, b_k$ converges. Writing $R_N = z_N + z_{N+1} + \cdots$, $\beta_N^2 = b_N + b_{N+1} + \cdots$, one has, with probability 1,

$$\limsup R_N / (2\,\beta_N^2 \log\log 1/\beta_N)^{\frac{1}{2}} = 1$$

provided $|z_N| \leq o\{\beta_N^2 / \log\log(1/\beta_N)\}^{\frac{1}{2}}$.

Combining this result with the proof of Theorem (1.2.1), and using Theorem (1.6.1) we get the following

(2.3.5) **Theorem.** *Let us suppose that the sum* $\Sigma (a_k^2 + b_k^2)$ *corresponding to the series* (2.1.3) *is finite, so that the series* (2.1.3) *converges at almost every point* (x, t),[1] *and let us set*

$$c_n^2 = a_n^2 + b_n^2, \quad \beta_N^2 = \tfrac{1}{2} \sum_N^{\infty} c_n^2, \quad R_N = \sum_N^{\infty} \varphi_n(t) A_n(x).$$

Then, for almost every t, *we have*

$$\limsup \frac{R_N}{(2\beta_N^2 \log\log 1/\beta_n)^{\frac{1}{2}}} = 1$$

almost everywhere in x, *provided*

(2.3.6) $$c_N^2 = O\left\{\frac{\beta_N^2}{\omega(1/\beta_N^2)}\right\}.$$

An obvious analogue holds for power series (2.3.1), with $\Sigma |c_k|^2$ finite.

CHAPTER III

The Central Limit Theorem

1. In this Section, $\omega(u)$, c_N, B_N, S_N have the same meaning as in Theorem (2.1.2).

(3.1.1) **Theorem.** *Under the same conditions as in Theorem* (2.1.2), *namely*

$$B_N^2 \to \infty, \quad c_N^2 = O\left\{\frac{B_N^2}{\omega(B_N^2)}\right\},$$

the distribution function of S_N/B_N *tends, for almost every* t, *to the Gaussian distribution with mean value zero and dispersion* 1.

It is easily seen that the assumptions imply also

$$\operatorname*{Max}_{1 \le k \le N} c_k^2 = O\left\{\frac{B_N^2}{\omega(B_N^2)}\right\}.$$

We shall suppose, for the sake of brevity, that the series $\Sigma \varphi_k(t) A_k(x)$ is a purely cosine series (our proof is immediately adaptable to the general case by writing $a_n \cos nx + b_n \sin nx = c_n \cos(nx - \alpha_n)$), and it will be convenient to replace the variable x by $2\pi x$, so that the series becomes $\Sigma \varphi_k(t) a_k \cos 2\pi kx$.

[1] See PALEY and ZYGMUND, loc. cit., or Zygmund, *Trigonometrical Series*, p. 125.

For a given t, let $E_N(y)$ be the set of points x of the interval (0, 1) at which $S_N/B_N \le y$, and let $F_N(y)$ be the measure of $E_N(y)$. Then $F_N(y)$ is the distribution function of S_N/B_N. In order to prove our theorem it will be enough to prove that over every finite range of λ the characteristic function of $F_N(y)$ approaches uniformly that of the Gaussian distribution, for almost every t. This characteristic function is

$$\int_{-\infty}^{+\infty} e^{i\lambda y}\, dF_N(y) = \int_0^1 e^{i\lambda S_N/B_N}\, dx, \tag{3.1.2}$$

and we have to prove that for almost all t the last integral tends to $\exp(-\tfrac{1}{2}\lambda^2)$ uniformly over any finite range of λ.[1]

Let us now fix t. Since

$$e^z = (1+z)\, e^{\frac{1}{2}z^2 + O(|z|^3)} \quad \text{as } z \to 0,$$

one has

$$e^{i\lambda S_N/B_N} = \prod_{k=1}^{N}\left(1 + i\frac{\lambda a_k}{B_N}\varphi_k(t)\cos 2\pi k x\right)\exp\left\{\left(-\tfrac{1}{2}\lambda^2 \sum_1^N \frac{a_k^2}{B_N^2}\cos^2 2\pi k x\right) + o(1)\right\}, \tag{3.1.3}$$

where the term $o(1)$ tends to 0, uniformly in x, as $N \to \infty$, since $\lambda = O(1)$ and

$$\max_{1\le k\le N} a_k^2 = O\left\{\frac{B_N^2}{\omega(B_N^2)}\right\} = o(B_N^2).$$

Observe now that

$$\left|\prod_{k=1}^{N}\left(1 + i\frac{\lambda a_k}{B_N}\varphi_k(t)\cos 2\pi k x\right)\right| \le \prod_{k=1}^{N}\left(1 + \frac{\lambda^2 a_k^2}{B_N^2}\right)^{\frac{1}{2}} \le e^{\lambda^2},$$

and that, writing

$$\sum_1^N \frac{a_k^2}{B_N^2}\cos^2 2\pi k x = 1 + \sum_1^N \frac{a_k^2}{2B_N^2}\cos 4\pi k x = 1 + \xi_N(x),$$

the measure of the set of points at which $|\xi_N(x)| \ge \delta > 0$ is not greater than

$$\delta^{-2}\int_0^1 \xi_N^2\, dx = \tfrac{1}{8}\delta^{-2}(a_1^4 + \cdots + a_N^4)\, B_N^{-4},$$

and that the last quantity tends to 0 as $N \to \infty$.

Hence, with an error tending to zero (uniformly in $\lambda = O(1)$) as $N \to \infty$, the integral (3.1.2) is

$$e^{-\frac{1}{2}\lambda^2}\int_0^1 \prod_1^N \left(1 + i\frac{\lambda a_k}{B_N}\varphi_k(t)\cos 2\pi k x\right) dx,$$

[1] Since the exponential function is continuous, the uniformity of convergence is (as is very well known) not indispensable here.

and in order to prove our theorem we have to show that, p.p. in t,

$$\lim_{N\to\infty}\int_0^1 \prod_1^N (1+i\,\varepsilon_k\,\varphi_k(t)\cos 2\pi kx)\,dx = 1, \tag{3.1.4}$$

where we write $\varepsilon_k=\varepsilon_k(N)=\lambda\, a_k/B_N$.

2. Let us set

$$J_N(t) = \int_0^1 \prod_1^N (1+i\,\varepsilon_k\,\varphi_k(t)\cos 2\pi kx)\,dx - 1$$

$$= \int_0^1 \left\{\prod_1^N (1+i\,\varepsilon_k\,\varphi_k(t)\cos 2\pi kx) - 1\right\} dx.$$

Writing $\prod(x)=\prod_1^N (1+i\,\varepsilon_k\,\varphi_k(t)\cos 2\pi kx)$, one has

$$|J_N(t)|^2 = \int_0^1\int_0^1 \{\textstyle\prod(x)-1\}\{\overline{\textstyle\prod}(y)-1\}\,dx\,dy$$

and

$$\int_0^1 |J_N(t)|^2\,dt = \int_0^1\int_0^1 dx\,dy \int_0^1 \{\textstyle\prod(x)-1\}\{\overline{\textstyle\prod}(y)-1\}\,dt.$$

Now

$$\int_0^1 \textstyle\prod(x)\,dt = \int_0^1 \overline{\textstyle\prod}(y)\,dt = 1$$

and

$$\int_0^1 \textstyle\prod(x)\overline{\textstyle\prod}(y)\,dt = \int_0^1 \prod_1^N \{1+\varepsilon_k^2\cos 2\pi kx\cos 2\pi ky$$

$$+ i\,\varepsilon_k(\cos 2\pi kx - \cos 2\pi ky)\,\varphi_k(t)\}\,dt$$

$$= \prod_1^N (1+\varepsilon_k^2 \cos 2\pi kx\cos 2\pi ky).$$

Hence

$$\int_0^1 |J_N(t)|^2\,dt = \int_0^1\int_0^1 dx\,dy\left\{\prod_1^N (1+\varepsilon_k^2\cos 2\pi kx\cos 2\pi ky) - 1\right\}$$

$$= \int_0^1\int_0^1 \prod_1^N (1+\varepsilon_k^2\cos 2\pi kx\cos 2\pi ky)\,dx\,dy - 1$$

$$< \int_0^1\int_0^1 \left\{\exp \sum_1^N \varepsilon_k^2\cos 2\pi kx\cos 2\pi ky\right\} dx\,dy - 1.$$

Using the fact that $e^u = 1 + u + \frac{1}{2} u^2 e^{\eta u}$, $0 < \eta < 1$, with

$$u = \sum_1^N \varepsilon_k^2 \cos 2\pi k x \cos 2\pi k y,$$

and observing that $|u| \leq \sum_1^N \varepsilon_k^2 = 2\lambda^2$, one has, since $\int_0^1 \int_0^1 u \, dx \, dy = 0$,

$$\int_0^1 |J_N(t)|^2 dt \leq \tfrac{1}{4} e^{2\lambda^2} \int_0^1 \int_0^1 \left(\sum_1^N \varepsilon_k^2 \cos 2\pi k x \cos 2\pi k y \right)^2 dx \, dy$$

$$= \tfrac{1}{8} e^{2\lambda^2} \sum_1^N \varepsilon_k^4 = \tfrac{1}{8} \lambda^4 e^{2\lambda^2} \frac{\sum_1^N a_k^4}{B_N^4},$$

and since

$$\max_{1 \leq k \leq N} a_k^2 = O\left\{ \frac{B_N^2}{\omega(B_N^2)} \right\},$$

one has

$$\int_0^1 |J_N(t)|^2 dt = O\left\{ \frac{1}{\omega(B_N^2)} \right\}.$$

Let us now fix a number $\theta > 1$, and let N_j be the first integer such that $\theta^j \leq B_{N_j}^2 < \theta^{j+1}$. In Chapter I, in connection with the proof of Theorem (1.2.1), we showed that such an integer always exists for j large enough. Thus

$$\int_0^1 |J_{N_j}(t)|^2 dt = O\left\{ \frac{1}{\omega(\theta^j)} \right\}.$$

Since $\Sigma 1/p\,\omega(p) < \infty$, we have $\Sigma 1/\omega(\theta^j) < \infty$, and thus J_{N_j} tends to zero for almost every t. We have therefore shown that

$$\int_0^1 e^{i\lambda S_{N_j}/B_{N_j}} dx \to e^{-\frac{1}{2}\lambda^2},$$

p.p. in t, and uniformly over any finite range of λ.

3. Let us now consider an integer m such that

$$N_j \leq m < N_{j+1}$$

and let

$$\Delta = \int_0^1 e^{i\lambda S_m/B_m} dx - \int_0^1 e^{i\lambda S_{N_j}/B_{N_j}} dx.$$

One has

$$|\Delta| \leq \int_0^1 \left| \lambda \left(\frac{S_m}{B_m} - \frac{S_{N_j}}{B_{N_j}} \right) \right| dx,$$

$$|\Delta|^2 \leq \lambda^2 \int_0^1 \left[\frac{S_m - S_{N_j}}{B_m} + S_{N_j} \left(\frac{1}{B_m} - \frac{1}{B_{N_j}} \right) \right]^2 dx$$

$$\leq \frac{2\lambda^2}{B_m^2} \int_0^1 (S_m - S_{N_j})^2 dx + 2\lambda^2 \frac{(B_{N_{j+1}} - B_{N_j})^2}{B_m^2 B_{N_j}^2} \int_0^1 S_{N_j}^2 dx$$

$$\leq \frac{2\lambda^2}{B_{N_j}^2} \int_0^1 (S_{N_{j+1}} - S_{N_j})^2 dx + 2\lambda^2 \frac{(B_{N_{j+1}} - B_{N_j})^2}{B_{N_j}^4} \int_0^1 S_{N_j}^2 dx$$

$$= 2\lambda^2 \frac{B_{N_{j+1}}^2 - B_{N_j}^2}{B_{N_j}^2} + 2\lambda^2 \frac{(B_{N_{j+1}} - B_{N_j})^2}{B_{N_j}^2}$$

$$\leq 4\lambda^2 \frac{B_{N_{j+1}}^2 - B_{N_j}^2}{B_{N_j}^2} \leq 4\lambda^2 \frac{\theta^{j+2} - \theta^j}{\theta^j} = 4\lambda^2 (\theta^2 - 1).$$

Hence

$$\int_0^1 e^{i\lambda S_m/B_m} dx = \int_0^1 e^{i\lambda S_{N_j}/B_{N_j}} dx + \Delta,$$

where

$$|\Delta| \leq 2|\lambda| (\theta^2 - 1)^{\frac{1}{2}} \text{ and } \lim \int_0^1 e^{i\lambda S_{N_j}/B_{N_j}} dx = e^{-\frac{1}{2}\lambda^2}.$$

Since θ can be taken arbitrarily close to 1, this proves that

$$\int_0^1 e^{i\lambda S_m/B_m} dx \to e^{-\frac{1}{2}\lambda^2} \quad \text{p.p. in } t,$$

uniformly over any finite range of λ, and this completes the proof of Theorem (3.1.1).

Whether the condition $c_N^2 = O\{B_N^2/\omega(B_N^2)\}$ is the best possible one, we are not able to decide.

4. The result that follows is a generalization of Theorem (3.1.1).

(3.4.1) Theorem. *The notation being the same and under the same conditions as in Theorem* (3.1.1), *the distribution function of S_N/B_N on every fixed set G of positive measure tends to the Gaussian distribution, for all values of t with the possible exception* *of a set of measure zero which is independent of the set G. More precisely,* $E_N(y)$ *being the set of points* x *in* (0, 1) *such that* $S_N/B_N \le y$, *and*

$$F_N(y, G) = \frac{\text{meas }[E_N(y) \cdot G]}{\text{meas } G},$$

$F_N(y, G)$ *tends to the Gaussian distribution with mean value zero and dispersion* 1.

We have

$$\int_{-\infty}^{-\infty} e^{i\lambda y}\, dF_N(y, G) = |G|^{-1} \int_G e^{i\lambda S_N/B_N}\, dx,$$

where $|G|$ denotes the measure of G, and we have to prove that the last expression tends to $e^{-\frac{1}{2}\lambda^2}$, uniformly over any finite range of λ, for all values of t except in a set H_t of measure zero, H_t being independent of G.

Our theorem will be established if we prove it in the case when G is an interval with rational end points. For then it would be proved when G is a sum of a finite set of intervals I, whence we would get the result when G is the most general open set in (0, 1). Since every measurable set is contained in an open set of measure differing as little as we please, we would obtain the result in the general case.

Without loss of generality, we may assume that I is an interval of the form $(0, \alpha)$, where α is rational, $0 < \alpha \le 1$. Suppose now we can prove that for a given α,

$$\alpha^{-1} \int_0^{\alpha} e^{i\lambda S_N/B_N}\, dx \to e^{-\frac{1}{2}\lambda^2} \tag{3.4.2}$$

almost everywhere in t, that is to say with the exception of a set $H_t(\alpha)$ of measure zero. Our result will then follow, since the set $H_t = \sum_\alpha H_t(\alpha)$, summation being extended over all rational numbers α, is also of measure zero.

Thus we have to prove that (3.4.2) is, for a given α, true p.p. in t. As in the proof of Theorem (3.1.1), it is enough to show that

$$K_N(t) = \int_0^{\alpha} \prod_1^N (1 + i\varepsilon_k \varphi_k(t) \cos 2\pi k x)\, dx - \alpha$$

tends to zero p.p. in t, where $\varepsilon_k = \lambda a_k / B_N$ and $N \to \infty$.

The proof proceeds exactly in the same way as before until we get

$$\int_0^1 |K_N(t)|^2\, dt = \int_0^{\alpha}\int_0^{\alpha} e^{\sum_1^N \varepsilon_k^2 \cos 2\pi k x \cos 2\pi k y}\, dx\, dy - \alpha^2.$$

Writing now again $e^u = 1 + u + \frac{1}{2} u^2 e^{\eta u}$, $0 < \eta < 1$, we observe that

$$\int_0^\alpha \int_0^\alpha \varepsilon_k^2 \cos 2\pi k x \cos 2\pi k y \, dx \, dy = \frac{\varepsilon_k^2 \sin^2 2\pi k \alpha}{4\pi^2 k^2} < \frac{\varepsilon_k^2}{k^2}$$

and thus

$$\left| \int_0^\alpha \int_0^\alpha u \, dx \, dy \right| \leq 2 \frac{\lambda^2}{B_N^2} \max_{1 \leq k \leq N} a_k^2 = O\left\{ \frac{1}{\omega(B_N^2)} \right\}.$$

Noting also that

$$\int_0^\alpha \int_0^\alpha u^2 \, dx \, dy \leq \int_0^1 \int_0^1 u^2 \, dx \, dy,$$

we get

$$\int_0^1 |K_N(t)|^2 \, dt = O\{1/\omega(B_N^2)\},$$

from which place the proof proceeds as before.

5. Theorems (3.1.1) and (3.4.1) have analogues for power series

$$\sum_1^\infty c_k e^{2\pi i k x} \varphi_k(t), \tag{3.5.1}$$

whose partial sums we shall again denote by $S_N(x)$.

(3.5.2) **Theorem.** *If*

$$C_N^2 = \tfrac{1}{2} \sum_1^N |c_k|^2, \quad c_N^2 = O\{C_N^2/\omega(C_N^2)\},$$

then the two-dimensional distribution function of $S_N(x)/C_N$ tends, for almost every t, to the Gaussian distribution

$$\frac{1}{2\pi} \int_{-\infty}^{\xi} \int_{-\infty}^{\eta} e^{-\frac{1}{2}(\lambda^2 + \mu^2)} \, d\lambda \, d\mu.$$

It is enough to sketch the proof.[1] Let $c_k = |c_k| e^{i\alpha_k}$, and let U_N and V_N denote, respectively, the real and imaginary parts of S_N. Let $F_N(\xi, \eta)$ denote the measure of the set of points x, $0 \leq x < 1$, such that $U_N(x)/C_N \leq \xi$, $V_N(x)/C_N \leq \eta$, simultaneously. The characteristic function of F_N is

[1] See also the authors' notes "On lacunary trigonometric series" part I, *Proc. Nat. Acad.*, 33 (1947), pp. 333–338, esp. p. 337, and part II, *Ibid.* 34 (1948), pp. 54–62.

$$\int_{-\infty}^{+\infty}\int_{-\infty}^{+\infty} e^{i(\lambda\xi+\mu\eta)}\, dF_N(\xi,\eta) = \int_0^1 e^{i(\lambda U_N+\mu V_N)/C_N}\, dx$$

$$= \int_0^1 \exp i C_N^{-1}\left\{\sum_1^N |c_k|\,[\lambda\cos(2\pi kx+\alpha_k)+\mu\sin(2\pi kx+\alpha_k)]\,\varphi_k(t)\right\} dx$$

$$= \int_0^1 \exp\left\{i C_N^{-1}(\lambda^2+\mu^2)^{\frac{1}{2}}\sum_1^N |c_k|\cos(2\pi kx+\alpha_k')\,\varphi_k(t)\right\} dx,$$

where the α_k' now also depend on λ and μ.

To the last integrand we apply a formula analogous to (3.1.3) and we find that for $\lambda^2+\mu^2=O(1)$ our integral is

$$e^{-\frac{1}{2}(\lambda^2+\mu^2)}\int_0^1 \prod_1^N \left\{1+i\,|c_k|\,(\lambda^2+\mu^2)^{\frac{1}{2}}\, C_N^{-1}\,\varphi_k(t)\cos 2\pi kx+\alpha_k')\right\} dx,$$

with an error tending uniformly to zero. The second factor here tends to 1 p.p. in t, since after an obvious change of notation it reduces to the integral in (3.1.4), provided in the latter we replace $\cos 2\pi kx$ by $\cos(2\pi kx+\alpha_k')$, which does not affect the validity of (3.1.4). Hence, p.p. in t, the characteristic function of $F_N(\xi,\eta)$ tends to $e^{-\frac{1}{2}(\lambda^2+\mu^2)}$, which completes the proof of Theorem (3.5.1).

It is clear that the conclusion of the theorem holds if we consider the distribution function of $S_N(x)/C_N$ over any set of positive measure in the interval $0\le x<1$.

This result and Theorem (3.4.2) have analogues in the case when the series are of the class L^2, i.e. when the sum of the squares of the moduli of the coefficients of the series is finite. Then, instead of the normalized partial sums we consider the normalized remainders of the series and show that, under condition (2.3.6), the distribution functions of these expressions tend, p.p. in t, to the Gaussian distribution. The proofs remain unchanged.

6. So far we have been considering only the partial sums or remainders of series. One can easily extend the results to general methods of summability (see, for example, the authors' note cited in the preceding Section, where this is done for lacunary series). We shall, however, confine our attention to the Abel-Poisson method, which is interesting in view of its function-theoretic aspect.

(3.6.1) **Theorem.** *Suppose that* $\Sigma(a_k^2+b_k^2)=\infty$, *and let*

$$c_k^2=a_k^2+b_k^2,\quad B^2(r)=\tfrac{1}{2}\Sigma c_k^2 r^{2k},\quad 0\le r<1.$$

Then, as $r \to 1$, *the distribution function of*

$$f_r(x) = \Sigma (a_k \cos 2\pi k x + b_k \sin 2\pi k x) \varphi_k(t) r^k$$

tends, p.p. in t, to the Gaussian distribution with mean value zero and dispersion 1, *provided*

$$\max_{1 \le k < \infty} (c_k^2 r^{2k}) = O\{B^2(r)/\omega(B^2(r))\},$$

in particular, provided $c_k = O(1)$.

The proof is the same as that of Theorem (3.1.1). Extensions to power series, sets of positive measure and series of the class L^2 are straightforward.

Chapter IV

On the Maximum of Trigonometric Polynomials whose Coefficients have Random Signs

1. In this chapter we shall consider series of the form

$$\sum_1^\infty r_m \varphi_m(t) \cos mx, \tag{4.1.1}$$

where $\{\varphi_m(t)\}$ is the Rademacher system, and where we consider purely cosine series only to simplify writing, there being no difficulty in extending the results to the series of the form $\Sigma r_m \varphi_m(t) \cos(mx - \alpha_m)$.

Writing

$$P_n = P_n(x, t) = \sum_1^n r_m \varphi_m(t) \cos mx,$$

we consider

$$M_n = M_n(t) = \max_x |P_n(x, t)|,$$

and our main problem will be to find, under fairly general conditions, the order of magnitude of M_n for almost every t; more exactly, to determine a function of n, say, $\Omega(n)$, such that

$$c \le \liminf \frac{M_n(t)}{\Omega(n)} \le \limsup \frac{M_n(t)}{\Omega(n)} \le C \tag{4.1.2}$$

almost everywhere in t, $c > 0$ and C being constants.

Analogous results will be given for power series of the form

$$\Sigma r_m e^{i(mx + 2\pi\alpha_m)}, \tag{4.1.3}$$

where the phases α_m are variable. The Rademacher functions are replaced here by the Steinhaus functions $e^{2\pi i \alpha_m}$, which are functions of a single variable t, $0 \leq t \leq 1$, and are obtained from mapping this interval onto the unit cube $0 \leq \alpha_m \leq 1$ of infinitely many dimensions (see Steinhaus, *Studia Math.*, 2 (1930), pp. 21–40).

Part 1. Rademacher Functions

2. We begin by proving a number of lemmas which we shall need later on.

(4.2.1) **Lemma.** *Let* $f_n = \sum_1^n c_m \varphi_m(t)$, *where* $\{\varphi_m\}$ *is the Rademacher system and the* c_m *are real constants. Let* $C_n = \sum_1^n c_m^2$, $D_n = \sum_1^n c_m^4$ *and let* λ *be any real number. Then*

$$e^{\frac{1}{2}\lambda^2 C_n - \lambda^4 D_n} \leq \int_0^1 e^{\lambda f_n} dt \leq e^{\frac{1}{2}\lambda^2 C_n}.$$

Proof. The φ_m being independent functions,

$$\int_0^1 e^{\lambda f_n} dt = \prod_1^n \int_0^1 e^{\lambda c_m \varphi_m} dt = \prod_1^n \frac{e^{\lambda c_m} + e^{-\lambda c_m}}{2}$$
$$= \prod_1^n \left(1 + \frac{\lambda^2 c_m^2}{2!} + \frac{\lambda^4 c_m^4}{4!} + \cdots\right).$$

Since $(2p)! \geq 2^p p!$, one has

$$\int_0^1 e^{\lambda f_n} dt \leq \prod_{m=1}^n \left(\sum_{p=0}^\infty \frac{\lambda^{2p} c_m^{2p}}{2^p p!}\right) = \prod_1^n e^{\frac{1}{2}\lambda^2 c_m^2} = e^{\frac{1}{2}\lambda^2 C_n}.$$

In the opposite direction,

$$\int_0^1 e^{\lambda f_n} dt \geq \prod_1^n \left(1 + \frac{\lambda^2 c_m^2}{2}\right).$$

Using the fact that for $u > 0$ one has $1 + u > e^{u - \frac{1}{2}u^2}$, one gets

$$\int_0^1 e^{\lambda f_n} dt \geq \prod_1^n e^{\frac{1}{2}\lambda^2 c_m^2 - \frac{1}{8}\lambda^4 c_m^4} \geq e^{\frac{1}{2}\lambda^2 C_m - \lambda^4 D_m},$$

which completes the proof.[1]

[1] Later on we shall need the lemma in the case when $n = \infty$ and $\Sigma c_m^2 < \infty$. It is clear that the inequalities of the lemma hold in this case, since $\Sigma c_m \varphi_m$ converges almost everywhere.

The lemma which follows is well known but we state it in order to avoid constant repetitions. It is stated for functions of a single variable but it clearly holds for functions of several variables.

(4.2.2) **Lemma.** *Let $g(x)$, $a \le x \le b$, be a bounded real function. Suppose that*

$$|g(x)| \le A, \quad (b-a)^{-1}\int_a^b g^2(x)\,dx = B.$$

Then, for any positive number μ,

$$(b-a)^{-1}\int_a^b e^{\mu g(x)}\,dx \le 1 + \mu\sqrt{B} + \frac{B}{A^2}e^{\mu A}.$$

In the case when $\int_a^b g(x)\,dx = 0$, this inequality can be replaced by

$$(b-a)^{-1}\int_a^b e^{\mu g}\,dx \le 1 + \frac{B}{A^2}e^{\mu A}.$$

Proof. One has

$$e^{\mu g} = 1 + \frac{\mu g}{11} + \cdots + \frac{\mu^p g^p}{p!} + \cdots.$$

Now

$$(b-a)^{-1}\int_a^b |g|\,dx \le \sqrt{B}, \quad (b-a)^{-1}\int_a^b |g|^p\,dx \le A^{p-2}B = \frac{B}{A^2}A^p.$$

Hence

$$(b-a)^{-1}\int_a^b e^{\mu g}\,dx \le 1 + \mu\sqrt{B} + BA^{-2}\sum_2^\infty \frac{\mu^p A^p}{p!} \le 1 + \mu\sqrt{B} + BA^{-2}e^{\mu A}.$$

The second inequality, if $g(x)$ has mean value zero, is obvious.

(4.2.3) **Lemma.** *Let x be real and $P(x) = \sum_0^n (\alpha_m \cos mx + \beta_m \sin mx)$ be a trigonometric polynomial of order n, with real or imaginary coefficients. Let M denote the maximum of $|P|$ and let θ be a positive number less than 1. There exists then an interval of length not less than $(1-\theta)/n$ in which $|P| \ge \theta M$.*

Proof. Let x_0 be a point at which $|P| = M$, and let x_1 be the first point to the right of x_0 at which $|P| = \theta M$ (if such a point does not exist there is nothing to prove). One has

$$M(1-\theta)=|P(x_0)|-|P(x_1)|$$
$$\leq |P(x_0)-P(x_1)|\leq (x_1-x_0).\ \mathrm{Max}\ |P'|.$$

But, by Bernstein's theorem, max $|P'|\leq nM$. Hence

$$x_1-x_0\geq (1-\theta)/n,$$

as stated.

(4.2.4) **Lemma.** *Let $\varphi(x)\geq 0$, and suppose that*

$$\int_0^1 \varphi\, dx\geq A>0,\quad \int_0^1 \varphi^2\, dx\leq B$$

(clearly, $A^2\leq B$). *Let δ be a positive number less than* 1 *and let $|E|$ denote the measure of the set E in which $\varphi\geq \delta A$. Then $|E|\geq (1-\delta)^2\dfrac{A^2}{B}$.*

If CE denotes the set complementary to E, then $\int_{CE}\varphi\, dx\leq \delta A$ and

$$\int_E \varphi\, dx=\int_0^1 \varphi\, dx-\int_{CE}\varphi\, dx\geq A(1-\delta).$$

But

$$\int_E \varphi\, dx\leq |E|^{\frac{1}{2}}\left(\int_0^1 \varphi^2\, dx\right)^{\frac{1}{2}}\leq |E|^{\frac{1}{2}} B^{\frac{1}{2}}$$

$$A(1-\delta)\leq |E|^{\frac{1}{2}} B^{\frac{1}{2}},$$

so that

$$|E|\geq (1-\delta)^2\frac{A^2}{B}.$$

(4.2.5) **Lemma.** *Let $f_k=\sum_1^k c_m\varphi_m(t)$, where $\{\varphi_m\}$ is the Rademacher system and the c_m are real constants. Let $n(t)$ be a measurable function taking only positive integral values and suppose that $1\leq n(t)\leq n$. Write $C_n=\sum_1^n c_m^2$ and denote by λ a positive number. Then*

$$\int_0^1 e^{\lambda |f_{n(t)}(t)|}\, dt\leq 16\, e^{\frac{1}{2}\lambda^2 C_n}.$$

The proof of this well-known result is included for the convenience of the reader.

Proof. Let k be an integer, $1 \le k \le n$. Then, if (α, β) is any dyadic interval of length 2^{-k} $(\alpha = p\,2^{-k},\ \beta = (p+1)\,2^{-k},\ p$ an integer), one has

$$f_k(t) = (\beta - a)^{-1} \int_\alpha^\beta f_n(u)\,du, \quad \alpha \le t \le \beta.$$

Thus, for all t,

$$|f_k(t)| \le (\beta - \alpha)^{-1} \int_\alpha^\beta |f_n(u)|\,du \le \sup_\theta \frac{1}{\theta - t} \int_t^\theta |f_n(u)|\,du \qquad (0 \le \theta \le 1).$$

Denoting the last member of the inequality by $f_n^*(t)$, we see that

$$|f_{n(t)}(t)| \le f_n^*(t).$$

By the well known inequality of Hardy and Littlewood,

$$\int_0^1 [f_n^*(t)]^q\,dt \le 2\left(\frac{q}{q-1}\right)^q \int_0^1 |f_n(t)|^q\,dt \qquad (q > 1).$$

Hence, if $q \ge 2$,

$$\int_0^1 |f_{n(t)}(t)|^q\,dt \le 8 \int_0^1 |f_n(t)|^q\,dt.$$

Now,

$$\int_0^1 e^{\lambda |f_{n(t)}(t)|}\,dt < \int_0^1 \{e^{\lambda f_{n(t)}(t)} + e^{-\lambda f_{n(t)}(t)}\}\,dt$$

$$= 2 \int_0^1 \sum_0^\infty \frac{\lambda^{2p}\,[f_{n(t)}(t)]^{2p}}{(2p)!}\,dt$$

$$\le 16 \int_0^1 \sum_0^\infty \frac{\lambda^{2p}\,[f_n(t)]^{2p}}{(2p)!}\,dt$$

$$= 8 \int_0^1 e^{\lambda f_n(t)}\,dt + 8 \int_0^1 e^{-\lambda f_n(t)}\,dt$$

$$\le 16\, e^{\frac{1}{2}\lambda^2 C_n},$$

by an application of Lemma (4.2.1).

3. We now pass to the proof of our first theorem.

(4.3.1) **Theorem.** *Consider the series* (4.1.1), *denote by* $P_n = P_n(x, t)$ *the polynomial* $\sum_1^n r_m \varphi_m(t) \cos mx$, *and write* $M_n = M_n(t) = \max_x |P_n(x, t)|$, $R_n = \sum_1^n r_m^2$. *There exists an absolute constant* A *such that*

$$\limsup_{n=\infty} \frac{M_n(t)}{\sqrt{R_n \log n}} \leq A,$$

almost everywhere in t.

Proof. One has, by Lemma (4.2.1)[1],

$$\int_0^1 e^{\lambda P_n} dt \leq e^{\frac{1}{2}\lambda^2 \sum_1^n r_m^2 \cos^2 mx} \leq e^{\frac{1}{2}\lambda^2 R_n}$$

so that

$$\int_0^1 e^{\lambda |P_n|} dt \leq \int_0^1 (e^{\lambda P_n} + e^{-\lambda P_n})\, dt \leq 2\, e^{\frac{1}{2}\lambda^2 R_n}.$$

Hence

$$\int_0^1 dt \int_0^{2\pi} e^{\lambda |P_n|} dx \leq 4\pi\, e^{\frac{1}{2}\lambda^2 R_n}.$$

By Lemma (4.2.1), denoting by θ a fixed positive number less than 1, one has

$$\int_0^{2\pi} e^{\lambda |P_n|} dx > \frac{1-\theta}{n} e^{\theta \lambda M_n}.$$

Hence

$$\int_0^1 e^{\theta \lambda M_n(t)} dt < \frac{4\pi n}{1-\theta} e^{\frac{1}{2}\lambda^2 R_n} = \frac{4\pi}{1-\theta} e^{\frac{1}{2}\lambda^2 R_n + \log n}.$$

Take $\lambda = (\beta R_n^{-1} \log n)^{\frac{1}{2}}$, β being a positive constant to be determined later on. We get

$$\int_0^1 e^{\theta \lambda M_n(t)} dt < \frac{4\pi}{1-\theta} e^{(\frac{1}{2}\beta + 1) \log n}.$$

Since, at present, we are not concerned with improving the value of the constant A, we shall now use rather crude estimates. We have

[1] In what follows λ is always positive.

$$\int_0^1 e^{\theta\lambda M_n - (\frac{1}{2}\beta+2+\eta)\log n}\, dt < \frac{4\pi}{1-\theta} e^{-(1+\eta)\log n},$$

η being a positive number. Hence

$$\sum_1^\infty e^{\theta\lambda M_n - (\frac{1}{2}\beta+2+\eta)\log n} < \infty$$

for almost all t. Hence, for $n \geq n_0 = n_0(t)$,

$$\theta\lambda M_n < (\tfrac{1}{2}\beta + 2 + \eta)\log n$$

and, replacing λ by its value,

$$M_n < \theta^{-1}\beta^{-\frac{1}{2}}(\tfrac{1}{2}\beta + 2 + \eta)(R_n \log n)^{\frac{1}{2}}.$$

This means that, for almost all t,

$$\limsup \frac{M_n(t)}{(R_n \log n)^{\frac{1}{2}}} \leq \frac{\frac{1}{2}\beta + 2 + \eta}{\theta\beta^{\frac{1}{2}}}.$$

Since θ is arbitrarily close to 1, and η as small as we please,

$$\limsup \frac{M_n(t)}{(R_n \log n)^{\frac{1}{2}}} \leq \frac{\beta^{\frac{1}{2}}}{2} + \frac{2}{\beta^{\frac{1}{2}}}.$$

Taking now $\beta = 4$,

$$\limsup \frac{M_n(t)}{(R_n \log n)^{\frac{1}{2}}} \leq 2, \tag{4.3.2}$$

which proves the theorem and shows that the best value of A is ≤ 2. We shall show later on (see Theorem (4.6.1) below) that under certain conditions the best value of A is ≤ 1.

Remarks on Theorem (4.3.1). As we shall see later, the order of magnitude obtained for M_n in Theorem (4.3.1) is not always the best possible one, and additional hypotheses will be required to prove the first inequality (4.1.2) with

$$\Omega(n) = (R_n \log n)^{\frac{1}{2}}.$$

An almost immediate corollary of (4.3.2) is a known result (see Paley and Zygmund, loc. cit.). *If Σr_m^2 is finite, then for almost all t the partial sums*

$$s_n = \sum_1^n r_m \varphi_m(t) \cos mx$$

of the series (4.1.1) *are* $o\,(\log n)^{\frac{1}{2}}$, *uniformly in* x. That these partial sums are $O\,(\log n)^{\frac{1}{2}}$ is obvious. By dropping the first few terms of the series (4.1.1) so as to make the R_n uniformly small, we improve the 'O' to 'o'.

We shall see later on (Section 8) that more precise information can be obtained about the order of s_n for some particular series with $\sum r_m^2 < \infty$.

4. In order to get further results we must now prove another lemma.

(4.4.1) **Lemma.** *Let us again consider the series* (4.1.1) *and keep the notation of Theorem* (4.3.1). *Let* $\{n_j\}$ *be an increasing sequence of positive integers and let* $\mathfrak{M}_j^* = \mathfrak{M}_j^*(t)$ *be the maximum, with respect to* n, *when* $n_j < n \le n_{j+1}$ *of*

$$H_n(t) = \max_x |P_n(x,t) - P_{n_j}(x,t)|.$$

Then, for almost all t,

$$\limsup_j \frac{\mathfrak{M}_j^*(t)}{\{(R_{n_{j+1}} - R_{n_j}) \log n_{j+1}\}^{\frac{1}{2}}} \le 2.$$

Proof. Let $n(t)$ be any measurable function of t taking integral values only and such that $n_j < n(t) \le n_{j+1}$. By Lemma (4.2.5),

$$\int_0^1 e^{\lambda |P_{n(t)} - P_{n_j}|} dt \le 16\, e^{\frac{1}{2}\lambda^2 (R_{n_{j+1}} - R_{n_j})}.$$

Hence

$$\int_0^1 dt \int_0^{2\pi} e^{\lambda |P_{n(t)} - P_{n_j}|} dx \le 32\pi\, e^{\frac{1}{2}\lambda^2 (R_{n_{j+1}} - R_{n_j})}.$$

By Lemma (4.2.3), we have, with $0 < \theta < 1$,

$$\int_0^{2\pi} e^{\lambda |P_{n(t)} - P_{n_j}|} dx > \frac{1-\theta}{n(t)} e^{\theta\lambda H_{n(t)}(t)} \ge \frac{1-\theta}{n_{j+1}} e^{\theta\lambda H_{n(t)}(t)}.$$

Hence

$$\int_0^1 e^{\theta\lambda H_{n(t)}} dt < \frac{n_{j+1}}{1-\theta}\, 32\pi\, e^{\frac{1}{2}\lambda^2 (R_{n_{j+1}} - R_{n_j})},$$

and so

$$\int_0^1 e^{\theta\lambda \mathfrak{M}_j^*(t)} dt \le \frac{32\pi}{1-\theta} e^{\frac{1}{2}\lambda^2 (R_{n_{j+1}} - R_{n_j}) + \log n_{j+1}}.$$

Les us take, as in the proof of Theorem (4.2.1), $\lambda = 2\left(\dfrac{\log n_{j+1}}{R_{n_{j+1}} - R_{n_j}}\right)^{\frac{1}{2}}$ and let $\eta > 0$. Then

$$\int_0^1 e^{\theta\lambda \mathfrak{M}_j^*(t) - (4+\eta)\log n_{j+1}}\, dt \le \frac{32\pi}{1-\theta} e^{-(1+\eta)\log n_{j+1}}$$

and so, exactly as in the proof of Theorem (4.2.1),

$$\limsup_j \frac{\mathfrak{M}_j^*(t)}{\{(R_{n_{j+1}} - R_{n_j}) \log n_{j+1}\}^{\frac{1}{2}}} \le \frac{4+\eta}{2\theta},$$

and the lemma follows by taking η arbitrarily small and θ arbitrarily close to 1.

5. We proceed now to prove, under certain conditions, the first part of inequality (4.1.2), with $\Omega(n) = (R_n \log n)^{\frac{1}{2}}$.

(4.5.1) **Theorem.** *Let us consider again the series* (4.1.1), *the expressions* P_n, M_n, R_n *having the same meaning as in Theorem* (4.2.1). *Let* $T_n = \sum_1^n r_m^4$.

We make the following assumptions:

a) $T_n / R_n^2 = O(n^{-\gamma})$ *for some positive* γ (clearly, $\gamma \le 1$), *this assumption obviously implying that* $\sum r_m^2 = \infty$.

b) $R_{n_1}/R_{n_2} \to 1$ *if* n_1 *and* n_2 *increase indefinitely in such a way that* $n_1/n_2 \to 1$.

Under these assumptions, one has, almost everywhere in t,

$$\liminf_n \frac{M_n(t)}{(R_n \log n)^{\frac{1}{2}}} \ge c(\gamma),$$

$c(\gamma)$ *being a positive constant depending on* γ *only.*

Proof. Let us set $I_n(t) = \dfrac{1}{2\pi}\displaystyle\int_0^{2\pi} e^{\lambda P_n}\, dx$, so that, by Lemma (4.2.1),

$$\int_0^1 I_n(t)\, dt = \frac{1}{2\pi}\int_0^{2\pi} dx \int_0^1 e^{\lambda P_n}\, dt \ge \frac{1}{2\pi}\int_0^{2\pi} e^{\frac{1}{2}\lambda^2 \sum_0^n r_m^2 \cos^2 mx - \lambda^4 T_n}\, dx$$

$$= e^{\frac{1}{4}\lambda^2 R_n - \lambda^4 T_n} \frac{1}{2\pi}\int_0^{2\pi} e^{\frac{1}{4}\lambda^2 \sum_1^n r_m^2 \cos 2mx}\, dx.$$

Now, since

$$e^{\frac{1}{4}\lambda^2 \sum_1^n r_m^2 \cos 2mx} > 1 + \frac{1}{4}\lambda^2 \sum_1^n r_m^2 \cos 2mx,$$

the mean value of the exponential exceeds 1, so that

$$\int_0^1 I_n(t)\,dt > e^{\frac{1}{4}\lambda^2 R_n - \lambda^4 T_n}. \tag{4.5.2}$$

We proceed now to obtain an inequality in the opposite direction for $\int_0^1 I_n^2(t)\,dt$, which, together with (4.5.2), will enable us to apply the result of Lemma (4.2.4).

We have

$$I_n^2(t) = \frac{1}{4\pi^2}\int_0^{2\pi}\int_0^{2\pi} e^{\lambda[P_n(x,t)+P_n(y,t)]}\,dx\,dy,$$

so that

$$\int_0^1 I_n^2(t)\,dt \leq \frac{1}{4\pi^2}\int_0^{2\pi}\int_0^{2\pi} dx\,dy \int_0^1 e^{\lambda[P_n(x,t)+P_n(y,t)]}\,dt.$$

Thus, by Lemma (4.2.1),

$$\int_0^1 I_n^2(t)\,dt \leq \frac{1}{4\pi^2}\int_0^{2\pi}\int_0^{2\pi} e^{\frac{1}{2}\lambda^2 \sum_1^n r_m^2 (\cos mx + \cos my)^2}\,dx\,dy.$$

Writing

$$(\cos mx + \cos my)^2 = 1 + \tfrac{1}{2}\cos 2mx + \tfrac{1}{2}\cos 2my + 2\cos mx \cos my$$

one has, putting

$$S_n = S_n(x, y) = \sum_1^n \{\tfrac{1}{2} r_m^2 \cos 2mx + \tfrac{1}{2} r_m^2 \cos 2my + 2 r_m^2 \cos mx \cos my\},$$

the inequality

$$\int_0^1 I_n^2(t)\,dt \leq e^{\frac{1}{2}\lambda^2 R_n} \cdot \frac{1}{4\pi^2}\int_0^{2\pi}\int_0^{2\pi} e^{\frac{1}{2}\lambda^2 S_n}\,dx\,dy.$$

We now use Lemma (4.2.2) for the function $S_n(x, y)$ of two variables. We observe that $\int_0^{2\pi}\int_0^{2\pi} S_n\,dx\,dy = 0$ and that the system of $3n$ functions $\cos 2mx$, $\cos 2my$, $\cos mx \cos my$ $(m = 1, 2, \dots, n)$ is orthogonal over the square of integration. Therefore,

$$\frac{1}{4\pi^2}\int_0^{2\pi}\int_0^{2\pi} S_n^2\,dx\,dy=\left(\frac{1}{8}+\frac{1}{8}+1\right)T_n=\frac{5}{4}T_n.$$

Since $|S_n|\leq 3R_n$, an application of the second inequality of Lemma (4.2.2) gives

$$\frac{1}{4\pi^2}\int_0^{2\pi}\int_0^{2\pi} e^{\frac{1}{2}\lambda^2 S_n}\,dx\,dy\leq 1+\frac{5}{36}\frac{T_n}{R_n^2}e^{\frac{3}{2}\lambda^2 R_n}$$

so that, by the hypothesis a) of the theorem, we have finally, a denoting a positive constant,

$$\int_0^1 I_n^2(t)\,dt\leq e^{\frac{1}{2}\lambda^2 R_n}\left(1+\frac{a}{n^\gamma}e^{\frac{3}{2}\lambda^2 R_n}\right). \tag{4.5.3}$$

Using the inequalities (4.5.2) and (4.5.3), let us apply Lemma (4.2.4) to determine a lower bound for the measure $|E_n|$ of the set E_n of points t such that

$$I_n(t)\geq n^{-\eta}e^{\frac{1}{4}\lambda^2 R_n-\lambda^4 T_n}, \tag{4.5.4}$$

where the number $\eta>0$ is to be determined later and the factor $n^{-\eta}$ plays role of the δ of Lemma (4.2.4). The lemma gives immediately

$$|E_n|\geq(1-n^{-\eta})^2\frac{e^{\frac{1}{2}\lambda^2 R_n-2\lambda^4 T_n}}{e^{\frac{1}{2}\lambda^2 R_n}\left(1+\frac{a}{n^\gamma}e^{\frac{3}{2}\lambda^2 R_n}\right)},$$

that is to say,

$$|E_n|\geq(1-2n^{-\eta})\cdot e^{-2\lambda^4 T_n}\cdot\left(1-\frac{a}{n^\gamma}e^{\frac{3}{2}\lambda^2 R_n}\right).$$

Let us now fix a number θ, $0<\theta<1$, and let $\lambda=\theta\left(\frac{2}{3}\gamma\frac{\log n}{R_n}\right)^{\frac{1}{2}}$, so that

$$e^{\frac{3}{2}\lambda^2 R_n-\gamma\log n}=e^{(\theta^2-1)\gamma\log n}.$$

One has

$$2\lambda^4 T_n=\theta^4\frac{8}{9}\gamma^2\frac{\log^2 n}{R_n^2}T_n<\frac{b\log^2 n}{n^\gamma},$$

b denoting a positive constant, so that

$$e^{-2\lambda^4 T_n}>1-\frac{b\log^2 n}{n^\gamma},$$

and

$$|E_n| \geq \left(1-\frac{2}{n^\eta}\right)\left(1-\frac{b\log^2 n}{n^\gamma}\right)\left(1-\frac{a}{n^{\gamma(1-\theta^2)}}\right).$$

Choosing now $\eta=\gamma(1-\theta^2)$, we have

$$|E_n| \geq \left(1-\frac{B}{n^{\gamma(1-\theta^2)}}\right),$$

B denoting a constant.

In the set E_n one has, by (4.5.4),

$$e^{\lambda M_n(t)} > I_n(t) > e^{\frac{1}{4}\lambda^2 R_n - \lambda^3 T_n - \eta \log n},$$

so that

$$M_n(t) > \frac{\lambda}{4} R_n - \lambda^3 T_n - \frac{\eta}{\lambda}\log n = \theta\frac{\sqrt{\gamma}}{2\sqrt{6}}(R_n \log n)^{\frac{1}{2}} - \lambda^3 T_n - \frac{\eta}{\lambda}\log n.$$

Now

$$\lambda^3 T_n = O\left\{\left(\frac{\log n}{R_n}\right)^{\frac{3}{2}} T_n\right\} = O\left\{\log n \cdot \frac{T_n}{R_n^2}(R_n \log n)^{\frac{1}{2}}\right\}$$

$$= O\left\{\frac{\log n}{n^\gamma}(R_n \log n)^{\frac{1}{2}}\right\} = o(R_n \log n)^{\frac{1}{2}}$$

and

$$\frac{\eta}{\lambda}\log n = \frac{\gamma(1-\theta^2)}{\theta}\sqrt{\frac{3}{2\gamma}}(R_n \log n)^{\frac{1}{2}} = 6(\theta^{-2}-1)\frac{\theta\sqrt{\gamma}}{2\sqrt{6}}(R_n \log n)^{\frac{1}{2}}$$

so that, writing $\varepsilon_\theta = 6(\theta^{-2}-1)$ and fixing θ close enough to 1 to have ε_θ, say, less than $\frac{1}{2}$, we have for $t \in E_n$,

$$M_n(t) > \frac{\theta\sqrt{\gamma}}{2\sqrt{6}}(1-\varepsilon_\theta - o(1))(R_n \log n)^{\frac{1}{2}}.$$

Let us now take an integer s such that

$$s\gamma(1-\theta^2) > 1.$$

Then the series $\Sigma^* n^{-\gamma(1-\theta^2)}$ extended only over the integers $n=m^s$ $(m=1, 2, \ldots)$ is convergent. Hence, by the lower bound found for $|E_n|$,

$$\sum_{m=1}^{\infty}(1-|E_{m^s}|) < \infty,$$

and thus, for almost all t,

$$\liminf_{n=m^s\to\infty}\frac{M_n(t)}{(R_n \log n)^{\frac{1}{2}}} \geq \frac{\theta\sqrt{\gamma}}{2\sqrt{6}}(1-\varepsilon_\theta). \tag{4.5.5}$$

We must now, in order to complete the proof of the theorem, extend this inequality to the case when n tends to ∞ through *all* integral values. For this purpose we shall use Lemma (4.4.1) and assumption b) of our theorem.

Denoting by $\mathfrak{M}_m^*(t)$ the maximum with respect to n, for $m^s < n \leq (m+1)^s$, of

$$\max_x |P_n(x,t) - P_{m^s}(x,t)|$$

we have, by Lemma (4.4.1),

$$\limsup_m \frac{\mathfrak{M}_m^*(t)}{\{(R_{(m+1)^s} - R_{m^s}) \log (m+1)^s\}^{\frac{1}{2}}} \leq 2$$

so that, since $R_{(m+1)^s}/R_{m^s} \to 1$ by assumption b),

$$\limsup_m \frac{\mathfrak{M}_m^*(t)}{(R_{m^s} \log m^s)^{\frac{1}{2}}} = 0. \tag{4.5.6}$$

Now, since for $m^s < n \leq (m+1)^s$ we have

$$M_n(t) \geq M_{m^s}(t) - \mathfrak{M}_m^*(t),$$

it follows that for almost all t,

$$\liminf_n \frac{M_n(t)}{(R_n \log n)^{\frac{1}{2}}} = \liminf_n \frac{M_n(t)}{(R_{m^s} \log m^s)^{\frac{1}{2}}}$$

$$\geq \liminf \frac{M_{m^s}(t)}{(R_{m^s} \log m^s)^{\frac{1}{2}}} - \limsup \frac{\mathfrak{M}_m^*(t)}{(R_{m^s} \log m^s)^{\frac{1}{2}}}$$

$$\geq \frac{\theta \sqrt{\gamma}}{2\sqrt{6}} (1 - \varepsilon_\theta).$$

It remains now to observe that the last inequality being true for all θ, $0 < \theta < 1$, we can take θ arbitrarily close to 1 and ε_θ arbitrarily small, so that

$$\liminf_n \frac{M_n(t)}{(R_n \log n)^{\frac{1}{2}}} \geq \frac{\sqrt{\gamma}}{2\sqrt{6}},$$

for almost all t, which proves our theorem.

Remarks on Theorem (4.5.1). Let us observe that the preceding argument shows that, with assumption a) alone, one has

$$\limsup_n \frac{M_n(t)}{(R_n \log n)^{\frac{1}{2}}} \geq \frac{\sqrt{\gamma}}{2\sqrt{6}},$$

so that, comparing this with Theorem (4.2.1) we have at least the true order of magnitude for the *superior* limit.

We may also add that we do not use in the proof the full force of assumption b), which is needed only when n_1 and n_2 are of the form m^s and $(m+1)^s$.

6. The case $\gamma = 1$ deserves special attention. We have then $T_n/R_n^2 = O(1/n)$ and this condition is certainly satisfied if the r_m are bounded both above and below (i.e. away from zero). In this case assumption b) of Theorem (4.5.1) is automatically true, if n_1 and n_2 are m^s and $(m+1)^s$ respectively. We shall show that in this case the value of the constant A in Theorem (4.2.1) may be reduced from 2 to 1.

(4.6.1) **Theorem.** *Let us consider the series (4.1.1) and let P_n, M_n, R_n, T_n have the same meaning as before. Then, under the sole assumption $T_n/R_n^2 = O(1/n)$ we have, almost everywhere in t,*

$$\frac{1}{2\sqrt{6}} \leq \liminf \frac{M_n(t)}{\sqrt{R_n \log n}} \leq \limsup \frac{M_n(t)}{\sqrt{R_n \log n}} \leq 1. \tag{4.6.2}$$

This is true, in particular, for the series $\sum \varphi_m(t) \cos mx$.

Proof. From the remark just made it follows that the first inequality (4.6.2) will be proved if we show that, for any integer s, $R_{(m+1)^s}/R_{m^s} \to 1$. Now,

$$[R_{(m+1)^s} - R_{m^s}]^2 \leq [(m+1)^s - m^s]\, T_{(m+1)^s},$$

by Schwarz's inequality, so that

$$\left[\frac{R_{(m+1)^s} - R_{m^s}}{R_{(m+1)^s}}\right]^2 \leq \frac{[(m+1)^s - m^s]\, T_{(m+1)^s}}{R^2_{(m+1)^s}} = O\left\{\frac{(m+1)^s - m^s}{(m+1)^s}\right\} = o(1),$$

as stated.

We now prove the part of the theorem concerning lim sup. We begin as in the proof of Theorem (4.3.1),

$$\int_0^1 e^{\lambda |P_n|}\, dt < 2\, e^{\frac{1}{2}\lambda^2 \sum_1^n r_m^2 \cos^2 mx},$$

but write

$$\sum_1^n r_m^2 \cos^2 mx = \tfrac{1}{2} R_n + \tfrac{1}{2} \sum_1^n r_m^2 \cos 2mx,$$

so that

$$\int_0^1 dt \int_0^{2\pi} e^{\lambda |P_n|}\, dx < 2\, e^{\frac{1}{4}\lambda^2 R_n} \int_0^{2\pi} e^{\frac{1}{4}\lambda^2 \sum_1^n r_m^2 \cos 2mx}\, dx.$$

Applying now Lemma (4.2.2), we have

$$\frac{1}{2\pi}\int_0^{2\pi} e^{\frac{1}{4}\lambda^2 \sum_1^n r_m^2 \cos 2mx}\, dx < 1 + \tfrac{1}{2}\frac{T_n}{R_n^2} e^{\frac{1}{4}\lambda^2 R_n}$$

$$< 1 + \frac{a}{n} e^{\frac{1}{4}\lambda^2 R_n},$$

a being a positive constant, so that

$$\int_0^1 dt \int_0^{2\pi} e^{\lambda |P_n|}\, dx < 4\pi e^{\frac{1}{4}\lambda^2 R_n}\left(1 + \frac{a}{n} e^{\frac{1}{4}\lambda^2 R_n}\right).$$

Taking $\lambda = 2(R_n^{-1}\log n)^{\frac{1}{2}}$ we get

$$\int_0^1 dt \int_0^{2\pi} e^{\lambda |P_n|}\, dx < 4\pi(1+a)\, e^{\log n}.$$

By Lemma (4.2.3), taking a positive number θ less than 1, we get as in the proof of Theorem (4.3.1),

$$\int_0^1 e^{\theta\lambda M_n}\, dt < \frac{4\pi n}{1-\theta}(1+a)\, e^{\log n} = \frac{4\pi(1+a)}{1-\theta} e^{2\log n}.$$

Fixing an $\eta > 0$ we have

$$\int_0^1 e^{\theta\lambda M_n - (2+\eta)\log n}\, dt < \frac{4\pi(1+a)}{1-\theta} n^{-\eta}.$$

Let s be an integer such that $s\eta > 1$. Then $\Sigma\, m^{-s\eta} < \infty$ so that

$$\sum_{n=m^s}^{\infty} e^{\theta\lambda M_n - (2+\eta)\log n} < \infty,$$

for almost all t, with $\lambda = \lambda_n = 2(R_n^{-1}\log n)^{\frac{1}{2}}$. From this we deduce, as in the proof of Theorem (4.3.1), that

$$\limsup_{n=m^s\to\infty} \frac{M_n(t)}{(R_n \log n)^{\frac{1}{2}}} \le \theta^{-1}(1 + \tfrac{1}{2}\eta),$$

for almost all t, and since θ may be taken arbitrarily close to 1,

$$\limsup_{n=m^s\to\infty} \frac{M_n(t)}{(R_n \log n)^{\frac{1}{2}}} \le 1 + \tfrac{1}{2}\eta.$$

To pass to *all* values of n, we use Lemma (4.4.1), just as in the proof of Theorem (4.5.1). We have, with the same notation, $M_n \leq M_{m^s} + \mathfrak{M}_m^*$, for $m^s < n \leq (m+1)^s$, and it follows immediately that

$$\limsup \frac{M_n(t)}{(R_n \log n)^{\frac{1}{2}}} \leq 1 + \tfrac{1}{2}\eta,$$

for almost all t. And since η is arbitrarily small, our result follows.

Remarks to Theorem (4.6.1). It is not excluded that for series having coefficients of sufficiently regular behavior, in particular for the series $\Sigma\, \varphi_m(t) \cos mx$, the expression $M_n/(R_n \log n)^{\frac{1}{2}}$ (for the particular series, $M_n/(n \log n)^{\frac{1}{2}}$) tends to a limit, p.p. in t, as $n \to \infty$. We have not, however, been able to prove a result of this kind, or even to narrow the gap between the constants of Theorem (4.6.1).

We shall see that the gap between the constants is reduced if we replace Rademacher's functions by Steinhaus'.

7. We now proceed to generalize the preceding result.

(4.7.1) **Theorem.** *Let* (α, β) *be a fixed interval contained in* $(0, 2\pi)$ *and let* $M_n(\alpha, \beta)$ *denote the maximum of* $|P_n|$ *for* $\alpha \leq x \leq \beta$. *Then, under the same assumptions as in Theorem* (4.5.1), *and almost everywhere in* t,

$$\liminf \frac{M_n(\alpha, \beta)}{(R_n \log n)^{\frac{1}{2}}} \geq c(\gamma),$$

$c(\gamma)$ *being a positive constant depending on* γ, *which is at least equal to the value* $\sqrt{\gamma}/2\sqrt{6}$ *found for the constant of Theorem* (4.5.1).

Proof. The proof follows the pattern of the proof of Theorem (4.5.1), and we only sketch it briefly to indicate the differences. Writing

$$J_n = J_n(\alpha, \beta, t) = \frac{1}{\beta - \alpha} \int_\alpha^\beta e^{\lambda P_n}\, dx,$$

one has

$$\int_0^1 J_n\, dt \geq e^{\frac{1}{4}\lambda^2 R_n - \lambda^4 T_n} \cdot \frac{1}{\beta - \alpha} \int_\alpha^\beta e^{\frac{1}{4}\lambda^2 \sum_1^n r_m^2 \cos 2mx}\, dx.$$

Now,

$$\left\{\frac{1}{\beta - \alpha} \int_\alpha^\beta \left|\sum_1^n r_m^2 \cos 2mx\right| dx\right\}^2 \leq \frac{1}{\beta - \alpha} \int_\alpha^\beta \left(\sum_1^n r_m^2 \cos 2mx\right)^2 dx \leq \frac{\pi T_n}{\beta - \alpha},$$

and since

$$e^{\frac{1}{4}\lambda^2 \sum_1^n r_m^2 \cos 2mx} \geq 1 + \tfrac{1}{4}\lambda^2 \sum_1^n r_m^2 \cos 2mx,$$

one has

$$\frac{1}{\beta-\alpha}\int_\alpha^\beta e^{\frac{1}{4}\lambda^2 \sum_1^n r_m^2 \cos 2mx}\,dx \geq 1 - \frac{1}{4}\lambda^2\left(\frac{\pi T_n}{\beta-\alpha}\right)^{\frac{1}{2}} \geq 1 - \frac{\lambda^2 T_n^{\frac{1}{2}}}{(\beta-\alpha)^{\frac{1}{2}}},$$

so that

$$\int_0^1 J_n\,dt \geq e^{\frac{1}{4}\lambda^2 R_n - \lambda^4 T_n}\left(1 - \frac{\lambda^2 T_n^{\frac{1}{2}}}{(\beta-\alpha)^{\frac{1}{2}}}\right). \tag{4.7.2}$$

We now find an upper bound for $\int_\alpha^\beta J_n^2\,dt$ by using, as in the proof of Theorem (4.5.1), a double integral and find, with the notation of that theorem,

$$\int_0^1 J_n^2\,dt \leq e^{\frac{1}{2}\lambda^2 R_n}\frac{1}{(\beta-\alpha)^2}\int_\alpha^\beta\int_\alpha^\beta e^{\frac{1}{2}\lambda^2 S_n}\,dx\,dy.$$

As in Theorem (4.5.1), we have $|S_n| \leq 3R_n$. Also

$$\frac{1}{(\beta-\alpha)^2}\int_\alpha^\beta\int_\alpha^\beta S_n^2\,dx\,dy \leq \frac{1}{(\beta-\alpha)^2}\int_0^{2\pi}\int_0^{2\pi} S_n^2\,dx\,dy \leq \frac{4\pi^2}{(\beta-\alpha)^2}\cdot\frac{5}{4}T_n = 5\pi^2\cdot\frac{T_n}{(\beta-\alpha)^2},$$

so that an application of the first inequality of the Lemma (4.2.2) gives

$$\begin{aligned}\frac{1}{(\beta-\alpha)^2}\int_\alpha^\beta\int_\alpha^\beta e^{\frac{1}{2}\lambda^2 S_n}\,dx\,dy &\leq 1 + \frac{1}{2}\lambda^2\frac{\pi\sqrt{5}\sqrt{T_n}}{\beta-\alpha} + \frac{5\pi^2 T_n}{9(\beta-\alpha)^2 R_n^2}e^{\frac{3}{2}\lambda^2 R_n}\\ &\leq 1 + \frac{a\lambda^2\sqrt{T_n}}{\beta-\alpha} + \frac{a T_n}{(\beta-\alpha)^2 R_n^2}e^{\frac{3}{2}\lambda^2 R_n},\end{aligned}$$

a being a positive absolute constant. Hence

$$\int_0^1 J_n^2\,dt \leq e^{\frac{1}{2}\lambda^2 R_n}\left[1 + \frac{a\lambda^2\sqrt{T_n}}{\beta-\alpha} + \frac{a T_n}{(\beta-\alpha)^2 R_n^2}e^{\frac{3}{2}\lambda^2 R_n}\right]. \tag{4.7.3}$$

From here the proof proceeds as in the case of Theorem (4.5.1), using the inequalities (4.7.2) and (4.7.3) instead of (4.5.2) and (4.5.3), and taking into account the hypothesis $T_n/R_n^2 = O(n^{-\gamma})$. One has only to observe that, since $\lambda = O(R_n^{-1} \log n)^{\frac{1}{2}}$, one has

$$\lambda^2 T_n^{\frac{1}{2}} = O\left\{\frac{\log n}{R_n} \cdot T_n^{\frac{1}{2}}\right\} = O(n^{-\frac{1}{2}\gamma} \log n).$$

Once the theorem has been established for a sequence of integers $n = m^s$ one proves it for all n by using again Lemma (4.4.1). The constant $c(\gamma)$ can be taken equal to $\sqrt{\gamma}/2\sqrt{6}$.

The comments on Theorems (4.5.1) and (4.6.1) are applicable here without change.

By taking the end points α, β rational, one sees immediately that if we exclude a certain set of values of t of measure zero then

$$\liminf_n \frac{M_n(\alpha, \beta)}{(R_n \log n)^{\frac{1}{2}}} \geq \frac{\sqrt{\gamma}}{2\sqrt{6}}$$

for *any* fixed interval (α, β).

It is also easily seen that the theorem holds for the intervals (α_n, β_n) whose length and position vary with n, provided that $\beta_n - \alpha_n > n^{-\sigma}$, where σ is a sufficiently small number which can be determined if γ is given. One finds $\sigma < \frac{1}{4}\gamma$, but the constant of the theorem depends then on σ. The details are left to the reader. Finally, Theorem (4.7.1) holds if the interval (α, β) is replaced by a set E of positive measure; it is enough to replace in the proof $\beta - \alpha$ by $|E|$.

8. Some results for the case in which $\sum r_m^2$ is slowly divergent or is convergent.

If the series $\sum r_m^2$ diverges slowly, or is convergent, the assumption a) of Theorem (4.5.1) is not satisfied. In order to show what the situation is in that case, we shall consider examples of series with regularly decreasing coefficients.

(i) The series

$$\sum_1^\infty m^{-(\frac{1}{2}-\varepsilon)} \varphi_m(t) \cos mx$$

presents no difficulty since here T_n/R_n^2 is $O(n^{-4\varepsilon})$ if $\varepsilon < \frac{1}{4}$, is $O(n^{-1} \log n)$ if $\varepsilon = \frac{1}{4}$ and is $O(n^{-1})$ if $\varepsilon > \frac{1}{4}$, so that condition a) of Theorem (4.5.1) is satisfied. Since condition b) of that theorem is also satisfied, the exact order of $M_n(t)$ is

$$(R_n \log n)^{\frac{1}{2}} \sim n^{\varepsilon} (\log n)^{\frac{1}{2}}.$$

(ii) The series

$$\sum_1^\infty m^{-\frac{1}{2}} \varphi_m(t) \cos mx$$

does not satisfy condition a) of Theorem (4.5.1). By Theorem (4.3.1) we have $M_n(t) = O(\log n)$, p.p. in t, and we are going to show that this is the exact order.

We consider the sequence of polynomials

$$Q_n = P_n - P_p = \sum_{p+1}^{n} m^{-\frac{1}{2}} \varphi_m(t) \cos mx,$$

where $p = p(n)$ is a function of n. We take $p = [n^\gamma]$, where γ is a positive number less than 1, to be determined later. We write $M(p, n) = M(p, n, t) = \operatorname*{Max}_x |Q_n|$, the other notations remaining the same as in Theorem (4.5.1).

We observe that for the polynomials Q_n we have

$$\frac{T_n - T_p}{(R_n - R_p)^2} = O\left[\frac{p^{-1} - n^{-1}}{(\log n - \log p)^2}\right] = O(n^{-\gamma} \log^{-2} n)$$

so that condition a) of Theorem (4.5.1) holds; it is easily seen that in the proof of the inequality (4.5.5) of Theorem (4.5.1) the fact that the polynomials are partial sums of the form $\sum_1^n$ is irrelevant, so that the proof of the inequality (4.5.5) applies to the sequence Q_n since condition a) is satisfied [condition b) is not required for the proof of that inequality]. We have to replace T_n by $T_n - T_p$, R_n by $R_n - R_p$, M_n by $M(p, n)$. Hence, for $s\gamma(1-\theta^2) > 1$ and for almost all t,

$$\liminf_{n=m^s} \frac{M(p, n)}{\{(R_n - R_p) \log n\}^{\frac{1}{2}}} \geq \frac{\theta \sqrt{\gamma}}{2\sqrt{6}} (1 - \varepsilon_\theta),$$

θ and ε_θ having the same meaning as in Theorem (4.5.1). Hence

$$\liminf_{n=m^s} \frac{M(p, n)}{\log n} \geq \frac{\theta\sqrt{\gamma}}{2\sqrt{6}} (1 - \varepsilon_\theta) \sqrt{1-\gamma}.$$

Now, by Theorem (4.3.1) we have, for almost all t,

$$\limsup \frac{M_p}{(R_p \log p)^{\frac{1}{2}}} \leq 2, \text{ i.e. } \limsup \frac{M_p}{\log p} \leq 2$$

where, as usual, $M_p = \max_x |P_p|$. In other words,

$$\limsup_p M_p / \log n \leq 2\gamma,$$

so that, if $M_n = \max_x |P_n|$,

$$\liminf_{n=m^s} \frac{M_n}{\log n} \geq \liminf_{n=m^s} \frac{M(p, n)}{\log n} - \limsup \frac{M_p}{\log n} \geq \frac{\theta\sqrt{\gamma}}{2\sqrt{6}} (1-\varepsilon_\theta)\sqrt{1-\gamma} - 2\gamma$$

which is a positive quantity if γ is small enough.

We now pass to the sequence of *all* n, like in Theorem (4.5.1), by using Lemma (4.4.1) which can be applied to the proof since plainly for the series $\Sigma\, m^{-\frac{1}{2}}\, \varphi_n(t) \cos mx$ one has $R_{n_1}/R_{n_2} \to 1$ whenever $n_1/n_2 \to 1$.

Hence, combining our results and observing that θ is as close to 1 as we wish, we have, for almost all t

$$\frac{1}{2\sqrt{6}}\sqrt{\gamma(1-\gamma)} - 2\gamma \le \liminf M_n/\log n \le \limsup M_n/\log n \le 2,$$

where we can take, e.g., $\gamma = 1/100$.

The argument could be applied to show that the exact order of magnitude of $M_n(t)$ for almost all t is again $(R_n \log n)^{\frac{1}{2}}$ for $r_m = m^{-\frac{1}{2}} (\log m)^{-\alpha}$, if $0 < \alpha < \frac{1}{2}$, but it breaks down in the case $\alpha = \frac{1}{2}$ which we are now going to consider.

(iii). *The series* $\Sigma\,(m \log m)^{-\frac{1}{2}}\, \varphi_m(t) \cos mx$. In this case the function $(R_n \log n)^{\frac{1}{2}} \sim$ $\sim (\log n \cdot \log\log n)^{\frac{1}{2}}$ does not give the right order of magnitude for M_n, for we are going to show that in the present case we have $M_n = O(\sqrt{\log n})$ p.p. in t.

In fact, setting

$$\sigma_n = \sum_2^n (m \log m)^{-\frac{1}{2}}\, \varphi_m(t) \cos mx, \qquad P_n = \sum_2^n m^{-\frac{1}{2}}\, \varphi_m(t) \cos mx$$

we have, by Abel's transformation,

$$\sigma_n = \sum_2^{n-1} \left\{ \frac{1}{\sqrt{\log m}} - \frac{1}{\sqrt{\log (m+1)}} \right\} P_m + \frac{1}{\sqrt{\log n}} P_n,$$

and since $\max_x |P_m| = O(\log m)$ p.p. in t we have, again p.p. in t,

$$\max_x |\sigma_n| \le \sum_2^{n-1} O\left\{ \frac{1}{m (\log m)^{\frac{3}{2}}} \log m \right\} + O(\log n)^{\frac{1}{2}} = O(\log n)^{\frac{1}{2}}.$$

It can also be seen that $M_n = \max_x |\sigma_n|$ is $o(\log n)^{\frac{1}{2}}$ for almost *no* t. For suppose that $M_n = o(\log n)^{\frac{1}{2}}$ for t belonging to a set E of positive measure. Then

$$P_n = \sum_2^{n-1} (\sqrt{\log m} - \sqrt{\log (m+1)})\, \sigma_m + (\log n)^{\frac{1}{2}}\, \sigma_n$$

and so, in E, we would have

$$\max_x |P_n| \le \sum_2^{n-1} \left\{ O\left(\frac{1}{m \sqrt{\log m}} \right) o(\sqrt{\log m}) \right\} + o(\log n) = o(\log n),$$

which we know not to be true.

The same argument can be applied to the more general series

$$\sum m^{-\frac{1}{2}} (\log m)^{-\alpha}\, \varphi_m(t) \cos mx$$

if $\frac{1}{2}<\alpha<1$, to show that $M_n=O(\log n)^{1-\alpha}$ but not $o(\log n)^{1-\alpha}$, p.p. in t (see also, the end of Chapter V, Section 5).

Part II. Steinhaus Functions

9. The problem of the series $\sum_1^\infty r_m e^{i(mx+2\pi\alpha_m)}$ $(r_m\geq 0)$ is not essentially different from the problem treated in Part I, and we shall only indicate the relevant modifications of the argument. They lead to better values for the constants. As mentioned in Section 1, we map the hypercube $0\leq\alpha_m\leq 1$ $(m=1, 2,\dots)$ onto the interval $0\leq t\leq 1$ using the classical method.

The following result is an analogue of Lemma (4.2.1).

(4.9.1) **Lemma.** *Let* $f_n=\sum_1^n c_m e^{2\pi i\alpha_m}$, *where the* c_m *are complex constants. Then,* λ *being a real number, one has*

$$\int_0^1 |e^{\lambda f_n}|\,dt=\prod_1^n J_0(i\lambda|c_m|), \tag{4.9.2}$$

where J_0 *in the Bessel function of order zero:*

$$J_0(iz)=\sum_0^\infty\frac{(z^2/4)^k}{(k!)^2}=\int_0^1 e^{z\cos 2\pi u}\,du, \tag{4.9.3}$$

and writing $C_n=\sum_1^n|c_m|^2$, $D_m=\sum_1^n|c_m|^4$, *one has*

$$e^{\frac{1}{4}\lambda^2C_n-\lambda^4D_n}\int_0^1|e^{\lambda f_n}|\,dt\leq e^{\frac{1}{4}\lambda^2C_n}. \tag{4.9.4}$$

Proof. One has, if $c_m=|c_m|\,e^{2\pi i\varphi_m}$,

$$\int_0^1|e^{\lambda f_n}|\,dt=\prod_1^n\int_0^1\left|e^{\lambda|c_m|e^{2\pi i(\alpha_m+\varphi_m)}}\right|d\alpha_m$$

$$=\prod_1^n\int_0^1 e^{\lambda|c_m|\cos 2\pi(\alpha_m+\varphi_m)}\,d\alpha_m$$

$$=\prod_1^n\int_0^1 e^{\lambda|c_m|\cos 2\pi u}\,du=\prod_1^n J_0(i\lambda|c_m|).$$

Now, if z is real,

$$J_0(iz) \leq \sum_0^\infty \frac{1}{k!}\left(\frac{z^2}{4}\right)^k = e^{\frac{1}{4}z^2}$$

and

$$J_0(iz) \geq 1 + \frac{1}{4}z^2 \geq e^{\frac{1}{4}z^2 - z^4}.$$

Hence

$$e^{\frac{1}{4}\lambda^2 C_n - \lambda^4 D_n} \leq \prod_1^n J_0(i\lambda c_m) \leq e^{\frac{1}{4}\lambda^2 C_n},$$

which proves (4.9.4).

(4.9.5) **Lemma.** *The notation being the same as in Lemma* (4.9.1), *one has*

$$\int_0^1 e^{\lambda |f_n|} dt \leq e^{\frac{1}{4}\lambda^2 C_n (1+\varepsilon_n)} \tag{4.9.6}$$

where $\varepsilon_n \to 0$ *if* $C_n \to \infty$.

Proof. If k is a positive integer,

$$\int_0^1 |f_n|^{2k} dt = \int_0^1 \cdots \int_0^1 |f_n|^{2k} d\alpha_1 \dots d\alpha_n = \sum \left(\frac{k!}{k_1! \dots k_n!}\right)^2 |c_1|^{2k_1} \cdots |c_n|^{2k_n},$$

the summation being extended over all combinations such that $k_j \geq 0$, $\sum k_j = k$. Thus

$$\int_0^1 |f_n|^{2k} dt \leq k! \sum \frac{k!}{k_1! \cdots k_n!} |c_1|^{2k_1} \cdots |c_n|^{2k_n} = k!\, C_n^k.$$

Hence

$$\int_0^1 e^{\lambda |f_n|} dt \leq \int_0^1 e^{\lambda |f_n|} dt + \int_0^1 e^{-\lambda |f_n|} dt = 2\sum_0^\infty \frac{\lambda^{2k}}{(2k)!} \int_0^1 |f_n|^{2k} dt \leq 2 \sum_0^\infty (\lambda^2 C_n)^k \frac{k!}{(2k)!}.$$

By Stirling's formula, asymptotically,

$$\frac{k!}{(2k)!} \sim \frac{\sqrt{\pi k}}{4^k \cdot k!},$$

so that the general term of the series is

$$2\lambda^{2k} C_n^k \frac{k!}{(2k)!} \sim \frac{2(\frac{1}{4}\lambda^2)^k C_n^k}{k!} \sqrt{\pi k},$$

asymptotically, and it follows that

$$\int_0^1 e^{\lambda |f_n|} dt \leq e^{\frac{1}{4}\lambda^2 C_n (1+\varepsilon_n)}$$

where ε_n can be taken as small as we please if C_n is large enough.

Let now

$$P_n = \sum_1^n r_m e^{i(mx+2\pi\alpha_m)}, \quad M_n = \operatorname*{Max}_x |P_n|, \quad R_n = \sum_1^n r_m^2, \quad T_n = \sum_1^n r_m^4.$$

Theorem (4.3.1) is now replaced by a corresponding theorem for the series $\sum r_m e^{i(mx+2\pi\alpha_m)}$, where the constant 2 can be replaced by $\sqrt{2}$, on account of the inequality (4.9.6).

For the proof of the result corresponding to Theorem (4.5.1) we use the integral

$$H_n(t) = \frac{1}{2\pi} \int_0^{2\pi} |e^{\lambda P_n}| \, dx,$$

so that Lemma (4.9.1) leads immediately to

$$\int_0^1 H_n(t)\, dt = \frac{1}{2\pi} \int_0^{2\pi} dx \int_0^1 |e^{\lambda P_n}| \, dt \geq e^{\frac{1}{4}\lambda^2 R_n - \lambda^4 T_n}, \tag{4.9.7}$$

by inequality (4.9.4).

On the other hand, as in the proof of Theorem (4.5.1),

$$\int_0^1 H_n^2(t)\, dt = \frac{1}{4\pi^2} \int_0^{2\pi}\int_0^{2\pi} dx\, dy \int_0^1 |e^{\lambda [P_n(x,t) + P_n(y,t)]}| \, dt.$$

Since

$$P_n(x,t) + P_n(y,t) = \sum_1^n r_m (e^{imx} + e^{imy}) e^{2\pi i \alpha_m}$$

and

$$|e^{imx} + e^{imy}|^2 = |1 + e^{im(x-y)}|^2 = 2(1 + \cos m(x-y)),$$

one has, by (4.9.4),

$$\int_0^1 |e^{\lambda [P_n(x,t)+P_n(y,t)]}| \, dt = e^{\frac{1}{2}\lambda^2 R_n + \frac{1}{2}\lambda^2 \sum_1^n r_m^2 \cos m(x-y)}.$$

Thus

$$\int_0^1 H_n^2(t)\,dt \le e^{\frac{1}{2}\lambda^2 R_n}\cdot\frac{1}{4\pi^2}\int_0^{2\pi}\int_0^{2\pi} e^{\frac{1}{2}\lambda^2\sum_1^n r_m^2\cos m(x-y)}\,dx\,dy$$

$$= e^{\frac{1}{2}\lambda^2 R_n}\cdot\frac{1}{2\pi}\int_0^{2\pi} e^{\frac{1}{2}\lambda^2\sum_1^n r_m^2\cos mx}\,dx$$

so that, by Lemma (4.2.2),

$$\int_0^1 H_n^2(t)\,dt \le e^{\frac{1}{2}\lambda^2 R_n}\left[1+\frac{T_n}{2R_n^2}\,e^{\frac{1}{2}\lambda^2 R_n}\right]. \tag{4.9.8}$$

The inequalities (4.9.7) and (4.9.8) now lead to the proof of the analogue of Theorem (4.5.1) and, due to the disappearance of the factor 3 in the exponential in the brackets of (4.9.8), the constant $c(\gamma)$ can be taken equal to $\sqrt{\gamma}/2\sqrt{2}$.

The analogue of Theorem (4.6.1) is as follows.

(4.9.9) **Theorem.** *Considering the series* $\sum_1^\infty r_m e^{i(mx+2\pi\alpha_m)}$, *the expressions* P_n, M_n, R_n, T_n *being the same as above, under the sole assumption* $T_n/R_n^2 = O(1/n)$, *we have, for almost all series,*

$$2^{-\frac{3}{2}} \le \liminf M_n/(R_n \log n)^{\frac{1}{2}} \le \limsup M_n/(R_n \log n)^{\frac{1}{2}} \le 1.$$

This applies, in particular, to the series $\sum e^{i(mx+2\pi\alpha_m)}$.

Remarks. In the proofs of the analogues of Theorems (4.5.1) and (4.6.1) we need, of course, an analogue, for Steinhaus' functions, of Lemma (4.2.5). The proof of the latter, though a little troublesome, follows the same pattern and we sketch it briefly here, the notation being the same as that of Lemma (4.9.1).

Let k be an integer, $n \ge k \ge 1$, and let (α, β) be a dyadic interval, $\alpha = p2^{-q}$, $\beta = (p+1)2^{-q}$, where $q = \frac{1}{2}k(k+1)$. Then it is known (see e.g. Kaczmarz and Steinhaus, *Orthogonalreihen*, pp. 137–138) that

$$\int_\alpha^\beta f_k(u)\,du = \int_\alpha^\beta f_n(u)\,du,$$

and, if both u and t are in the interior of (α, β), then

$$|f_k(u) - f_k(t)| \le 2\pi(|c_1|2^{-k} + |c_2|2^{-(k-1)} + \cdots + |c_k|2^{-1})$$

so that

$$\left|\frac{1}{\beta-\alpha}\int_{\alpha}^{\beta} f_k(u)\,du - f_k(t)\right| \leq a\,C_k^{\frac{1}{2}} \leq a\,C_n^{\frac{1}{2}}$$

and

$$|f_k(t)| \leq \frac{1}{\beta-\alpha}\int_{\alpha}^{\beta} |f_n(u)|\,du + a\,C_n^{\frac{1}{2}} \leq \operatorname*{Sup}_{\theta}\frac{1}{\theta-t}\int_{t}^{\theta} |f_n(u)|\,du + a\,C_n^{\frac{1}{2}},$$

$$|f_{n(t)}(t)| \leq f_n^*(t) + a\,C_n^{\frac{1}{2}},$$

$f_{n(t)}(t)$ and $f_n^*(t)$ having the same meaning as in Lemma (4.2.5), and a being an absolute constant.

For $c \geq 2$,

$$|f_{n(t)}(t)|^q \leq 2^{q-1}\{|f_n(t)|^q + a^q\,C_n^{\frac{1}{2}q}\}$$

$$\int_0^1 |f_{n(t)}(t)|^q\,dt \leq 2^{q-1}\cdot 2\left(\frac{q}{q-1}\right)^q \int_0^1 |f_n(t)|^q\,dt + 2^{q-1}a^q\,C_n^{\frac{1}{2}q}$$

$$\leq 2^{q+2}\int_0^1 |f_n(t)|^q\,dt + 2^{q-1}a^q\int_0^1 |f_n(t)|^q\,dt \leq A^q\int_0^1 |f_n|^q\,dt,$$

where A is an absolute constant. Hence, by the same sequence of inequalities which led to Lemma (4.2.5),

$$\int_0^1 e^{\lambda|f_{n(t)}(t)|}\,dt \leq 2\,e^{\frac{1}{2}\lambda^2A^2C_n}.$$

The introduction of the constant A in the exponent will lead to the replacement of 2 by $2A$ in the inequality of Lemma (4.4.1), but clearly will have no effect upon the inequality analogous to (4.5.6).

Chapter V

Continuity of Trigonometric Series whose Terms have Random Signs

1. Given a trigonometric series $\sum_1^{\infty} r_m\,\varphi_m(t)\cos(mx-\alpha_m)$, where $\{\varphi_n(t)\}$ is the Rademacher system and $\Sigma\,r_m^2<\infty$, we shall say, briefly, that the series is "randomly continuous" if it represents a continuous function for almost every value of t. We propose to give here some new contributions to the theory of such series (which have already been studied).[1] Without impairing generality we shall simplify writing

[1] See Paley and Zygmund, loc. cit., and R. Salem, *Comptes Rendus*, 197 (1933), pp. 113–115 and *Essais sur les séries trigonométriques*, Paris (Hermann), 1940.

by considering purely cosine series

$$\sum_{1}^{\infty} r_m \varphi_n(t) \cos mx. \tag{5.1.1}$$

In what follows, A will denote an absolute constant, not necessarily the same at every occurrence.

(5.1.2) **Lemma.** *Let $Q = Q(t, x)$ denote the polynomial $\sum_{p+1}^{n} r_m \varphi_m(t) \cos mx$, and let $M = M(t) = \max_x |Q(t, x)|$. Let $R = \sum_{p+1}^{n} r_m^2$. Then $\int_0^1 M\,dt \le A(R \log n)^{\frac{1}{2}}$.*

Proof. By the argument of Theorem (4.3.1),

$$\int_0^1 e^{\theta\lambda M}\,dt \le \frac{4\pi}{1-\theta} e^{\frac{1}{2}\lambda^2 R_n + \log n} \qquad (0 < \theta < 1).$$

Hence

$$e^{\theta\lambda\int_0^1 M\,dt} \le e^{\frac{1}{2}\lambda^2 R + \log n + \log\frac{4\pi}{1-\theta}}.$$

Taking $\lambda = (2R^{-1} \log n)^{\frac{1}{2}}$, we get

$$\int_0^1 M\,dt \le \theta^{-1}(2R\log n)^{\frac{1}{2}} + \theta^{-1}\log\left(\frac{4\pi}{1-\theta}\right)\cdot\left(\frac{R}{2\log n}\right)^{\frac{1}{2}}$$

$$= \theta^{-1}(2R\log n)^{\frac{1}{2}}\left[1 + \log\left(\frac{4\pi}{1-\theta}\right)\cdot\frac{1}{2\log n}\right]$$

$$\le A(R\log n)^{\frac{1}{2}},$$

which proves the theorem. We see that we could take A as close to $\sqrt{2}$ as we wish if n is large enough, but this is irrelevant for our purposes.

We could also, by writing

$$\cos mx = \cos px \cos(m-p)x - \sin px \sin(m-p)x$$

prove easily that

$$\int_0^1 M\,dt \le A\{R\log(n-p)\}^{\frac{1}{2}},$$

but we shall not make use of this slightly stronger inequality.

(5.1.3) **Lemma.** *Let*

$$Q_n = \sum_{p+1}^{n} r_m \varphi_m(t) \cos mx, \qquad M_n = \max_x |Q_n|,$$

and let, for fixed m,

$$M^*(t) = \max_n M_n(t) \qquad \text{when} \qquad p+1 < n \leq m.$$

Then, with $R = \sum_{p-1}^{m} r_k^2$, *we have*

$$\int_0^1 M^* \, dt \leq A (R \log m)^{\frac{1}{2}}. \tag{5.1.4}$$

Proof. By the argument of Lemma (4.4.1) we have

$$\int_0^1 e^{\theta \lambda M^*} dt \leq \frac{32\pi}{1-\theta} e^{\frac{1}{2}\lambda^2 R + \log n}, \qquad 0 < \theta < 1,$$

which leads, exactly as in the proof of the preceding lemma, to (5.1.4).

(5.1.5) **Theorem.** *Let* R_n *denote*[1] *the remainder* $\sum_{n+1}^{\infty} r_m^2$ *of the convergent series* $\sum r_m^2$. *If* $\sum n^{-1} (\log n)^{-\frac{1}{2}} \sqrt{R_n} < \infty$, *the series* (5.1.1) *is randomly continuous. Moreover, for almost every t that series converges uniformly in x.*

Proof. Let us divide the series into blocks $Q_0, Q_1, \ldots$ such that

$$Q_k = \sum_{n_k+1}^{n_{k+1}} r_m \varphi_m(t) \cos mx \qquad (n_0 = 0, \; n_k = 2^{2^k}).$$

Let $M_k = \max_x |Q_k|$. By lemma (5.1.2),

$$\int_0^1 M_k \, dt \leq A (2^k R_{2^{2^k}})^{\frac{1}{2}}.$$

By Cauchy's theorem, the convergence of $\Sigma n^{-1} (\log n)^{-\frac{1}{2}} \sqrt{R_n}$ implies that of $\Sigma k^{-\frac{1}{2}} \sqrt{R_{2^k}}$ which, in turn, implies the convergence of $\Sigma (2^k R_{2^{2^k}})^{\frac{1}{2}}$. Hence $\Sigma \int_0^1 M_k \, dt < \infty$, which shows that ΣM_k converges for almost every t, i.e. for almost every t the series ΣQ_k converges uniformly in x and the first part of the theorem has been established.

---

[1] Attention of the reader is called to the fact that R_n, and later on T_n, has not the same meaning here as in the preceding chapter.

To prove the second part of the theorem it is enough to observe that, writing

$$M_k^*(t) = \max_{n_k+1<m\leq n_{k+1}} \{\max_x |\sum_{n_k+1}^{m} r_n \varphi_n(t) \cos nx|\}$$

one has, by Lemma (5.1.3),

$$\int_0^1 M_k^* \, dt \leq A(2^k R_{2^k})^{\frac{1}{2}},$$

so that, for almost all t, $\Sigma M_k^* < \infty$, and so also $M_k^* \to 0$.

2. Remarks on Theorem (5.1.5). a) The condition $\Sigma n^{-1} (\log n)^{-\frac{1}{2}} \sqrt{R_n} < \infty$ is merely sufficient, but not necessary, for the random continuity of the series (5.1.1). It is enough to consider the series $\Sigma p^{-2} \varphi_{2^{2^p}}(t) \cos 2^{2^p} x$.

b) On the other hand, the condition $\Sigma n^{-1} (\log n)^{-\frac{1}{2}} \sqrt{R_n} < \infty$ is the best possible of its kind. In other words, there exist series (5.1.1) which represent a continuous function for no value of t and such that

$$\Sigma \frac{\sqrt{R_n}}{n (\log n)^{\frac{1}{2}} \omega(n)} < \infty,$$

$\omega(n)$ being a given function, increasing to ∞ with n, as slowly as we please.

To see this, let us consider the series $\Sigma \frac{1}{p\psi(p)} \varphi_{2^p}(t) \cos 2^p x$, where $\psi(p)$ increases to ∞ with p but $\Sigma 1/p\psi(p) = \infty$. The series being lacunary, it cannot represent a continuous function for any value of t. But

$$R_{2^p} = \sum_{p+1}^{\infty} \frac{1}{k^2 \psi^2(k)} < \frac{1}{p\psi^2(p)},$$

so that

$$\left(\frac{R_{2^k}}{k}\right)^{\frac{1}{2}} \cdot \frac{1}{\Omega(k)} < \frac{1}{k\psi(k)\Omega(k)}.$$

Now, no matter how slowly $\Omega(k)$ increases, we can find $\psi(k)$ such that

$$\Sigma 1/k\psi(k) = \infty, \quad \Sigma 1/k\psi(k)\Omega(k) < \infty,$$

and this proves the statement, if we set $\omega(2^k) = \Omega(k)$ and apply Cauchy's theorem.

c) A *necessary* condition for random continuity of the series (5.1.1) is known (see Paley and Zygmund, loc. cit.). Let us divide the series into blocks

(5.2.1) $$P_k = \sum_{2^k+1}^{2^{k+1}} r_m \varphi_m(t) \cos mx,$$

and let

(5.2.2) $$\Delta_k = \sum_{2^k+1}^{2^{k+1}} r_m^2.$$

The condition $\Sigma \sqrt{\Delta_k} < \infty$ is *necessary*, for random continuity.

But the condition is not sufficient, as seen on the following example.

It will be slightly simpler to deal with a series of exponentials $\Sigma\, r_m \varphi_m(t) e^{imx}$. We shall construct a series of this type which for no value of t does represent a continuous function, although the series $\Sigma\, \Delta_k^{\frac{1}{2}}$ will be convergent.

We shall make use of the familiar fact, namely, that if $\psi_1, \psi_2, \ldots, \psi_N$ are any distinct Rademacher functions, then

$$\max_x \left| \sum_1^N \psi_k(t) e^{i 2^k x} \right| \geq A\, N,$$

no matter what value we give to t. Let us now determine the coefficients r_m in each polynomial P_k as follows (compare (5.2.1) and (5.2.2)):

$$r_m = k^{-\frac{3}{4}} \qquad \text{for } m = 2^k + 2^s \quad (s = 1, 2, \ldots, k)$$

$$r_m = 0 \qquad \text{for other values of } m.$$

Then, by the remark just made, we have for all t,

$$\max_x |P_k| > A\, k \cdot k^{-\frac{3}{4}} = A\, k^{\frac{1}{4}}$$

while

$$\Delta_k = k \cdot k^{-\frac{3}{2}} = k^{-\frac{1}{2}},$$

so that $\Delta_k^{\frac{1}{2}} = k^{-\frac{1}{4}}$. Let us now consider an increasing sequence of integers n_q such that $\Sigma\, n_q^{-\frac{1}{4}} < \infty$ (e.g. $n_q = q^5$) and construct the series

$$P_{n_1} + P_{n_2} + \cdots + P_{n_q} + \cdots$$

with the polynomials just defined. The series, having infinitely many Hadamard gaps, must, if it represents a bounded function, have its partial sums of order corresponding to the beginning or end of the gaps uniformly bounded.[1] In particular, P_{n_q} must be uniformly bounded. Since $\max |P_{n_q}| > A\, n_q^{\frac{1}{4}}$, this is impossible, no matter what value we give to t. And yet, for this series, $\Sigma\, \Delta_k^{\frac{1}{2}} = \Sigma\, n_q^{-\frac{1}{4}} < \infty$.

[1] See Zygmund, *Trigonometrical Series*, p. 251.

We shall return to the problem of necessary and sufficient conditions for random continuity at the end of this chapter.

3. (5.3.1) **Theorem.** *Suppose that the series (5.1.1.) is randomly continuous. Denote by $\{n_q\}$ any lacunary sequence of positive integers (i.e. such that $n_{q+1}/n_q \geq \lambda > 1$). Write $S_n = \sum_1^n r_m \varphi_m(t) \cos mx$. Then, for almost every t, the partial sums S_{n_q}, of order n_q, converge uniformly in x.*

Proof. Observe first that, t_0 being a fixed number—not a dyadic rational—the series

$$\Sigma\, r_m\, \varphi_m(t_0)\, \varphi_m(t) \cos mx \tag{5.3.2}$$

is randomly continuous, if (5.1.1) is. For let E be the set of measure 1 such that, when $t \in E$, (5.1.1) represents a continuous function. Let $t \in E$ and define t' by

$$\varphi_m(t')\, \varphi_m(t_0) = \varphi_m(t).$$

It is easy to see (e.g. by the consideration of dyadic intervals) that the set of t' corresponding to the $t \in E$ is also of measure 1.

Let us now divide the series (5.1.1) into blocks $P_q = \sum_{n_q+1}^{n_{q-1}} r_m \varphi_m(t) \cos mx$ and consider the two series

$$P_0 + P_1 + P_2 + P_3 + \cdots$$

$$P_0 - P_1 + P_2 - P_3 + \cdots$$

It follows from our remarks that the series $P_0 + P_2 + P_4 + \cdots$ and $P_1 + P_3 + P_5 + \cdots$ are both randomly continuous. But both series are series with Hadamards gaps, so that the partial sums S_{n_q} of order n_q of the series (5.1.1) converge uniformly in x, for almost every t.

Remark on Theorem (5.3.1). Let us consider alongside (5.1.1) the series (5.3.2), where now $\varphi_m(t_0) = 1$ for $m = n_q$ $(q = 1, 2, \ldots)$ and $\varphi_m(t_0) = -1$ for $m \neq n_q$ $(n_{q+1}/n_q \geq \lambda > 1)$. An application of the preceding argument leads to the conlusion that the random continuity of (5.1.1) implies the random continuity of $\Sigma\, r_{n_q} \varphi_{n_q}(t) \cos n_q x$. The sequence $\{n_q\}$ being lacunary this implies that $\Sigma\, |r_{n_q}| < \infty$. In other words, *if (5.1.1) is randomly continuous, the moduli of any lacunary subsequence of its coefficients have a finite sum.* This is of course a consequence of the necessary condition discussed in Section 2, c), but the proof given here is much simpler.

4. (5.4.1) **Theorem.** *Suppose that the series* (5.1.1) *is randomly continuous. Then, writing*

$$R_n = \sum_{n+1}^{\infty} r_m^2, \quad T_n = \sum_{n+1}^{\infty} r_m^4$$

we necessarily have $R_n \log T_n \to 0$ *as* $n \to \infty$.

Proof. Let us observe first that the series (5.1.1) converges almost everywhere in x for almost every t (see e.g. Zygmund, *Trigonometrical Series*, p. 125). Let $Q_n = \sum_{n+1}^{\infty} r_m \varphi_m(t) \cos mx$. By lemma (4.2.1),

$$\int_0^1 \int_0^{2\pi} e^{\lambda \sum_{n+1}^{N} r_m \varphi_m(t) \cos mx} \, dt\, dx \leq 2\pi e^{\frac{1}{2}\lambda^2 R_n}$$

so that, by the theorems of Fatou and Fubini, $(2\pi)^{-1} \int_0^{2\pi} e^{\lambda Q_n} dx = I_n(t)$ exists for almost all t and is integrable.

The argument used in the proof of Theorem (4.5.1) can then be applied without change, though Q_n is not a polynomial here but an infinite series (see footnote to Lemma (4.2.1)), and we get

$$\int_0^1 I_n \, dt \geq e^{\frac{1}{4}\lambda^2 R_n - \lambda^4 T_n},$$

$$\int_0^1 I_n^2 \, dt \leq e^{\frac{1}{2}\lambda^2 R_n} \left[1 + A \frac{T_n}{R_n^2} e^{\frac{3}{2}\lambda^2 R_n} \right].$$

Taking $\lambda = \{\frac{2}{3} R_n^{-1} \log (R_n^2/T_n)\}^{\frac{1}{2}}$, we have

$$\int_0^1 I_n(t)\, dt \geq e^{\frac{1}{4}\lambda^2 R_n - \lambda^4 T_n}.$$

$$\int_0^1 I_n^2(t)\, dt \leq (1+A)\, e^{\frac{1}{2}\lambda^2 R_n},$$

so that, by Lemma (4.2.4), if we denote by E_n the set of points t such that

$$I_n \geq \tfrac{1}{2} e^{\frac{1}{4}\lambda^2 R_n - \lambda^4 T_n},$$

we have

$$|E_n| \geq [4(A+1)]^{-1} e^{-2\lambda^4 T_n}.$$

Now,

$$2\lambda^4 T_n = \frac{8}{9} R_n^{-2} T_n \log^2 (R_n^2/T_n) \le u^{-1} \log^2 u,$$

where $u = R_n^2/T_n \ge 1$, so that $2\lambda^4 T_n \le B$, B being an absolute constant, and

$$|E_n| \ge [4(A+1)e^B]^{-1} = \delta,$$

δ being a positive absolute constant.

Now, writing $M_n = M_n(t) = \max_x |Q_n(t, x)|$ for a given t for which $I_n(t)$ exists (M_n can be $+\infty$) one has, if $t \in E_n$,

$$e^{\lambda M_n} \ge I_n(t) \ge e^{\frac{1}{4}\lambda^2 R_n - \lambda^4 T_n - \log 2},$$

that is to say,

$$M_n \ge \frac{\lambda}{4} R_n - \lambda^3 T_n - \frac{\log 2}{\lambda}.$$

Now,

$$\frac{\lambda}{4} R_n = \frac{1}{2\sqrt{6}} \{R_n \log (R_n^2/T_n)\}^{\frac{1}{2}} \tag{5.4.2}$$

$$\lambda^3 T_n < T_n R_n^{-\frac{3}{2}} \log^{\frac{3}{2}} (R_n^2/T_n) = \frac{\log (R_n^2/T_n)}{R_n^2/T_n} \{R_n \log (R_n^2/T_n)\}^{\frac{1}{2}} \tag{5.4.3}$$

and

$$\lambda^{-1} \log 2 = \log 2 \cdot (\tfrac{3}{2})^{\frac{1}{2}} R_n^{\frac{1}{2}} \log^{-\frac{1}{2}} (R_n^2/T_n) = (\tfrac{3}{2})^{\frac{1}{2}} \cdot \log 2 \cdot \frac{\{R_n \log (R_n^2/T_n)\}^{\frac{1}{2}}}{\log (R_n^2/T_n)}. \tag{5.4.4}$$

Suppose now that $R_n \log R_n^2/T_n$ does not tend to zero. Since $R_n \to 0$, we see that R^2/T_n is unbounded. Hence we can find a sequence $\{n_q\}$ of integers with the following properties:

a) $R_{n_q}^2/T_{n_q}$ increases to $+\infty$

b) $R_{n_q}(\log R_{n_q}^2/T_{n_q}) \ge c > 0$

c) $n_{q+1}/n_q \ge 2$ for all q.

It follows then from (5.4.2), (5.4.3), (5.4.4) that $M_{n_q}(t) > \sqrt{c}/10$ in a set E_{n_q} of measure $\ge \delta$. But this is impossible if (5.1.1) is randomly continuous. In fact, $\{n_q\}$ being lacunary, Q_{n_q} must tend then to zero for almost every t uniformly in x, by the preceding theorem. Now, consider the set $\mathcal{E}$ of points t for which $Q_{n_q} \to 0$ uniformly in x. Every $t \in \mathcal{E}$ must belong to all the complementary sets $C E_{n_q}$ after a certain rank. Hence

$$\mathcal{E} = \prod_1^\infty C E_{n_q} + \prod_2^\infty C E_{n_q} + \cdots$$

If we denote the products on the right by $F_1, F_2, \cdots$, respectively, then

$$F_1 \subset F_2 \subset \cdots \subset F_k \subset \cdots, \quad |F_k| \le |C E_{n_k}| \le 1 - \delta.$$

Thus $|\mathcal{E}| \le 1-\delta$, so that, since obviously $\mathcal{E}$ must be of measure 0 or 1, we have $|\mathcal{E}|=0$. In other words, if $R_n \log R_n^2/T_n$ is not $o(1)$, almost no series (5.1.1) represents a continuous function.

The proof of the theorem is completed by observing that $R_n \to 0$ implies $R_n \log R_n \to 0$ so that the condition $R_n \log (R_n^2/T_n) \to 0$ is equivalent to $R_n \log T_n \to 0$.

Corollary. If $\{r_m\}$ is a decreasing sequence, the condition $\Sigma r_m^2 < \infty$ implies $m r_m^2 \to 0$ so that, for n large enough,

$$T_n < (n+1)^{-2} + (n+2)^{-2} + \cdots < 1/n,$$

and so $R_n \log n \to 0$ is a necessary condition for random continuity.[1] This is of course true, more generally, if $\sum_{n+1}^{\infty} r_m^4 = O(n^{-\varepsilon})$ for some $\varepsilon > 0$.

5. We shall now indicate a case of "regularity" in which the convergence of $\Sigma n^{-1} (\log n)^{-\frac{1}{2}} \sqrt{R_n}$ is both necessary and sufficient for the random continuity of (5.11.1).

(5.5.1) **Theorem.** *If the sequence $\{r_m\}$ is decreasing and if there exists a $p > 1$ such that $R_n (\log n)^p$ is increasing, then the convergence of $\Sigma n^{-1} (\log n)^{-\frac{1}{2}} \sqrt{R_n}$ is both necessary and sufficient for the random continuity of* (5.1.1).

In view of Theorem (5.1.5) it is sufficient to prove the necessity of the condition.

The hypothesis is better understood if we observe that the boundedness of $R_n (\log n)^p$ for some $p > 1$ implies $\Sigma n^{-1} (\log n)^{-\frac{1}{2}} \sqrt{R_n} < \infty$. Thus we have to assume that $R_n (\log n)^p$ is unbounded; our "regularity" condition consists in assuming the monotonicity of the latter expression for some $p > 1$.

(5.5.2) **Lemma.** *If* (5.5.1) *is randomly continuous and if two following conditions are satisfied*

$$\frac{R_{2^k} - R_{2^{k+1}}}{R_{2^k}} = O\left(\frac{1}{k}\right), \qquad k R_{2^k} = O(1),$$

then $\Sigma n^{-1} (\log n)^{-\frac{1}{2}} \sqrt{R_n} < \infty$.

Proof of the Lemma. Using the notation of Section 3,

$$\Delta_k = R_{2^k} - R_{2^{k+1}}$$

we know (by the result of Paley and Zygmund quoted there) that if (5.1.1) is randomly continuous, then $\Sigma \Delta_n^{\frac{1}{2}} < \infty$. Now

[1] In particular, the series, $\Sigma m^{-\frac{1}{2}} (\log m)^{-1} \varphi_m(t) \cos mx$, for which $R_n \log n$ is bounded but does not tend to zero, is not randomly continuous.

$$\sum_{1}^{n} k^{-\frac{1}{2}} R_{2k} = \sum_{1}^{n-1} O(k^{\frac{1}{2}}) [\sqrt{R_{2k}} - \sqrt{R_{2k+1}}] + O(n R_{2n})^{\frac{1}{2}}$$
$$= \sum_{1}^{n-1} O\left(k^{\frac{1}{2}} \frac{\Delta_k}{\sqrt{R_{2k}}}\right) + O(n R_{2n})^{\frac{1}{2}}$$
$$= \sum_{1}^{n-1} O(\Delta_k^{\frac{1}{2}}) + O(1),$$

so that $\sum k^{-\frac{1}{2}} \sqrt{R_{2k}} < \infty$ and the lemma follows by an application of Cauchy's theorem.

Proof of Theorem (5.5.1). The sequence $\{r_m\}$ being decreasing, the random continuity of (5.1.1) implies $R_n \log n \to 0$ (see the Corollary of Theorem (5.4.1)). In particular, $k R_{2k} \to 0$.

Moreover, since $R_n (\log n)^p$ increases,

$$R_{2k} k^p < R_{2k+1} (k+1)^p$$
$$R_{2k+1}/R_{2k} > \left(\frac{k}{k+1}\right)^p > 1 - \frac{A}{k}.$$

Hence $1 - (R_{2k+1}/R_{2k}) \leq A/k$, and the theorem follows from the lemma.

6. It is clear that the results of this chapter hold when the Rademacher functions are replaced by those of Steinhaus, viz. for the series of the type $\sum_{1}^{\infty} r_m e^{i(mx+2\pi\alpha_m)}$. In particular if $r_m > 0$ is decreasing, $R_m \log m = o(1)$ is necessary for random continuity. It might be interesting to recall in this connection that, if the sequence $\{1/r_m\}$ is monotone and concave, *no matter how slow is the convergence of* $\sum r_m^2$, there always exists a particular sequence $\{\alpha_m\}$ such that the series $\sum_{1}^{\infty} r_m e^{i(mx+2\pi\alpha_m)}$ converges uniformly (see Salem, *Comptes Rendus*, 201 (1935), p. 470, and *Essais sur les séries trigonométriques*, Paris (Hermann), 1940), although the series need not be randomly continuous, e.g. if $R_n \log n \neq o(1)$.

The problem whether an analogous result holds for the series of the type $\sum_{1}^{\infty} r_m \cos mx\, \varphi_m(t)$, where $\{\varphi_m\}$ is the sequence of Rademacher functions, is open.

CHAPTER VI

The Case of Power Polynomials

1. Let us consider a power series $\sum_{0}^{\infty} \alpha_k x^k$ of radius of convergence 1 and let us also consider the power series $\sum^{\infty} \alpha_k \varphi_k(t) x^k$ and its partial sums

$$P_n = \sum_0^n \alpha_k \varphi_k(t) x^k,$$

where $\varphi_0, \varphi_1, \varphi_2, \dots$ is the sequence of Rademacher functions. We may consider the problem of the order of magnitude, for almost all t, of

$$M_n(t) = \max_{-1 \le x \le +1} |P_n|,$$

assuming, for the sake of simplicity, that the coefficients α_k are real.

From Theorem (4.3.1), using the principle of maximum, we see at once that

$$M_n(t) = O(R_n \log n)^{\frac{1}{2}}$$

almost everywhere in t, with $R_n = \sum_0^n \alpha_k^2$. We shall see, however, that better estimates than that can be found and that the problem has some curious features distinguishing it from the corresponding problem for trigonometric polynomials.

(6.1.1) **Theorem.** *If* $R_n \to \infty$ *and*

$$\alpha_k^2 = o\{R_k/\log \log R_k\}$$

then

(6.1.2) $$\limsup M_n(t)/\{2R_n \log \log R_n\}^{\frac{1}{2}} = 1$$

for almost every t. *On the other hand,*

(6.1.3) $$\liminf M_n(t)/R_n^{\frac{1}{2}} = O(1),$$

almost everywhere in t.

Thus, unlike in the theorems of Chapter IV, even in the simplest cases (e.g. for $\alpha_0 = \alpha_1 = \dots = 1$) the maximum $M_n(t)$ has no *definite* order of magnitude p.p. in t.

Proof. The inequality (6.1.2) is a rather simple consequence of the Law of the Iterated Logarithm.

For let $M_n'(t)$ and $M_n''(t)$ denote the maximum of $|P_n|$ on the intervals $0 \le x \le 1$ and $-1 \le x \le 0$ respectively. It is enough to prove (6.1.2) with $M_n(t)$ replaced by $M_n'(t)$. For then the inequality will follow for $M_n''(t)$ (since it reduces to the preceding case if we replace α_k by $(-1)^k \alpha_k$), and so also for $M_n(t) = \max\{M_n'(t), M_n''(t)\}$.

Let us set

$$S_m(t) = \sum_0^m \alpha_k \varphi_k(t), \quad S_n^*(t) = \max_{1 \le m \le n} |S_m(t)|.$$

Since

$$P_n = \sum_0^n \alpha_k \varphi_k(t) x^k = \sum_0^{n-1} S_k (x^k - x^{k+1}) + x^n S_n$$

we immediately obtain

(6.1.4) $$M_n(t) \leq S_n^*(t).$$

On the other hand,

$$S_n^*(t) = |S_m(t)| \qquad \text{for some } m = m(n) \leq n,$$

so that

$$\limsup_n M_n(t)/(2R_n \log\log R_n)^{\frac{1}{2}} \leq \limsup_m |S_m(t)|/(2R_n \log\log R_n)^{\frac{1}{2}} \leq 1$$

by the Law of the Iterated Logarithm, and this gives (6.1.2) with '=' replaced by '≤'.

The opposite inequality follows from the fact that $M_n(t) \geq |S_n(t)|$ and that $\limsup |S_n(t)|/(2R_n \log\log R_n)^{\frac{1}{2}} = 1$ p.p. in t.

As regards (6.1.3), it is enough to prove it with M_n replaced by S_n^*, on account of (6.1.4). By Lemma (4.2.5),

$$\int_0^1 e^{\lambda S_n^*} dt \leq 16\, e^{\frac{1}{2}\lambda^2 R_n}.$$

Let us consider any function $\omega(n)$ increasing to $+\infty$ with n. In the inequality

$$I_n = \int_0^1 e^{\lambda S_n^* - \omega(n)} dt \leq 16\, e^{\frac{1}{2}\lambda^2 R_n - \omega(n)}$$

we set $\lambda = R_n^{-\frac{1}{2}} \omega^{\frac{1}{2}}(n)$. Then $I_n \leq \exp\{-\frac{1}{2}\omega(n)\}$. Thus, if $\{n_j\}$ increases fast enough, we have $\Sigma I_{n_j} < \infty$ so that, for almost all t and for $n = n_j$ large enough, we shall have $\lambda S_n^* \leq \omega(n)$, that is

(6.1.5) $$\liminf_n S_n^*(t)/\{R_n \omega(n)\}^{\frac{1}{2}} \leq 1, \qquad \text{p.p. in } t.$$

From this it is easy to deduce the validity of (6.1.3), with M_n replaced by S_n^*, for almost every t. For suppose that (6.1.3) does not hold in a set E of positive measure. Then $S_n^*(t)/R_n^{\frac{1}{2}}$ tends to infinity in E. Using the theorem of Egoroff, we may assume that this convergence to ∞ is uniform in E. We can then find a function $\omega(n)$ monotonically increasing to ∞ and such that $S_n^*(t)/\{R_n \omega(n)\}^{\frac{1}{2}}$ still tends to ∞ in E, and with this function $\omega(n)$ the inequality (6.1.5) is certainly false. This completes the proof of the theorem.

The argument leading to (6.1.3) is obviously crude and there is no reason to expect that it gives the best possible result. It is included here only to show that under very general conditions the maximum $M_n(t)$ has no definite order of magnitude for almost every t. Under more restrictive conditions, involving third moments, Chung has shown (see his paper in the *Transactions of the American Mathematical Soc.*,

64 (1948), pp. 205–232)[1] that

$$\liminf_{n\to\infty} S_n^*(t) \Big/ \left(\frac{R_n}{\log\log R_n}\right)^{\frac{1}{2}} = \frac{\pi}{\sqrt{8}} \tag{6.1.6}$$

almost everywhere. (This equality holds, in particular for $\alpha_1 = \alpha_2 = \cdots = 1$.) Owing to (6.1.4) this leads, under Chung's conditions, to

$$\liminf M_n(t) \Big/ \left(\frac{R_n}{\log\log R_n}\right)^{\frac{1}{2}} \leq \frac{\pi}{\sqrt{8}}, \tag{6.1.7}$$

an inequality stronger than (6.1.3). Unfortunately, we know nothing about the inequality opposite to (6.1.7)[2].

[1] We are grateful to Dr. Erdös for calling our attention to Chung's paper. It may be added that (6.1.6) generalizes an earlier result of Erdös who showed that in the case $\alpha_1 = \alpha_2 = \cdots = 1$ the left side of (6.1.6) is almost everywhere contained between two positive absolute constants.

[2] (*Added in proof.*) Dr. Erdös has communicated us that in the case $a_1 = a_2 = \cdots = 1$ he can prove that, for every $\varepsilon > 0$,

$$\liminf \frac{M_n(t)}{n^{\frac{1}{2}-\varepsilon}} > 0$$

almost everywhere, and even a somewhat stronger result.

EXTRAIT DE STUDIA MATHEMATICA, T. XIV. (1954)

Singular integrals and periodic functions

by

A. P. CALDERÓN (Columbus, Ohio) and A. ZYGMUND *) (Chicago, Illinois)

1. The purpose of this note is to extend to periodic functions some of the results about singular integrals known for the non-periodic case. We shall be more specific later and begin by recalling basic facts.

Let $x=(\xi_1,\ldots,\xi_k)$, $y=(\eta_1,\ldots,\eta_k),\ldots$ denote points in the k-dimensional Euclidean space E^k. By x we shall also denote the vector joining the origin $O=(0,\ldots,0)$ with the point x. The length of the vector x will be denoted by $|x|$. If $x\neq 0$, by x' we shall mean the projection of x onto the unit sphere Σ having O for centre. Thus

$$x'=\frac{x}{|x|}, \qquad |x'|=1.$$

We shall consider kernels $K(x)$ of the form

$$K(x)=\frac{\Omega(x')}{|x|^k}=\frac{\Omega(x')}{r^k} \qquad (r=|x|), \tag{1.1}$$

where Ω is a scalar (real or complex) function defined on Σ and satisfying the following conditions:

1° $\Omega(x')$ is continuous on Σ and its modulus of continuity $\omega(\delta)$ satisfies the Dini condition

$$\int_0^1 \frac{\omega(\delta)}{\delta}\,d\delta<\infty;$$

2° The integral of $\Omega(x')$ extended over Σ is zero.

Condition 1° is certainly satisfied if the function Ω satisfies a Lipschitz condition of positive order. It could be considerably relaxed in very important special cases, but for the problems discussed in this paper it is the most suitable one. On the other hand, condition 2° is absolutely essential, as explained in [2][1].

*) Fellow of the John Simon Guggenheim Foundation.

[1]) Numbers in brackets refer to the bibliography at the end of the paper.

Reprinted from *SM* 14, 249–271 (1954).

We shall denote by $K_\varepsilon(x)$, and call it the ***truncated kernel***, the expression defined by the equations

$$K_\varepsilon(x)=\begin{cases} K(x) & \text{for } |x|\geqslant\varepsilon, \\ 0 & \text{elsewhere.}\end{cases}$$

Let $f(x)\in L^p$, $p\geqslant 1$. The convolution

$$\tilde{f}_\varepsilon(x)=\int_{E^k} f(y)K_\varepsilon(x-y)\,dy=\int_{E^k} K_\varepsilon(y)f(x-y)\,dy$$

exists as an absolutely convergent integral for all x. We write here dy for $d\eta_1\ldots d\eta_k$. The limit of $\tilde{f}_\varepsilon(x)$ as $\varepsilon\to 0$, it it exists, will be denoted by $\tilde{f}(x)$ and called the ***Hilbert transform***, associated with the kernel K, of $f(x)$. The classical Hilbert transform corresponds, except for a numerical constant, to $k=1$ and $K(x)=1/x$.

We state without proof a number of known results concerning these transforms [2]).

THEOREM A. *The function $\tilde{f}(x)$ exists almost everywhere if $f\in L^p$, $p\geqslant 1$. More generally, if F is a totally additive function of set in E^k, the limit*

(1.1a) $$\lim_{\varepsilon\to 0}\int_{E^k} K_\varepsilon(x-y)F(dy)$$

exists almost everywhere.

THEOREM B. *If $f(x)$ belongs to L^p, $1<p<\infty$, so does $\tilde{f}(x)$. Moreover,*

(1.2) $$\|\tilde{f}\|_p\leqslant A_p\|f\|_p.$$

Here and hereafter constants like A_p are positive and depend not only on the parameters shown explicitly but also on the kernel K (*i. e.* on the function Ω). By A (or $B, C, \ldots$) we shall denote constants depending at most on the kernel K, and we may use identical notation for different constants. By $\|f\|_p$ we mean the norm

$$\Big(\int_{E^k}|f|^p\,dx\Big)^{1/p}.$$

Theorem B breaks down for $p=1$ and $p=\infty$. If $f\in L$, the function $\tilde{f}$ need not be integrable, even locally. Nor does the boundedness of f imply that of $\tilde{f}$ (even if we assume the existence of $\tilde{f}$, which is not guaranteed here by Theorem A). For these extreme cases we have the following results:

---

[2]) Theorems A-F, as well as part 1° of Theorem G, are proved in [2]. Cases 2° and 3° of Theorem G are established by Mr Cotlar [3].

THEOREM C. *If* $|f|(1+\log^+|f|)$ *is integrable over* E^k, *then* $\tilde{f}$ *is integrable over every set* S *of finite measure and*

(1.3) $$\int_S |\tilde{f}|\,dx \leqslant A_S \int_{E^k} |f|(1+\log^+|f|)\,dx + B_S.$$

THEOREM D. *If* $f \in L$, *then* $|\tilde{f}|^a$ *is integrable over every set* S *of finite measure if* $0<a<1$. *Moreover,*

(1.4) $$\int_S |\tilde{f}|^a dx \leqslant \frac{A}{1-a}\,|S|^{1-a}\Big(\int_{E^k} |f|\,dx\Big)^a.$$

THEOREM E. *If* $f \in L$, *then the measure of the set* E_y *in which we have* $|\tilde{f}(x)| \geqslant y > 0$, *satisfies the inequality*

(1.5) $$|E_y| \leqslant \frac{A}{y}\,\|f\|_1.$$

THEOREM F. *Suppose that* $|f(x)| \leqslant 1$ *for all* x, *and let*

(1.6) $$g(x) = \lim_{\varepsilon\to+0} \int_{E^k} \{K_\varepsilon(x-y) - K_1(x-y)\} f(y)\,dy.$$

Then, for any bounded set S,

(1.7) $$\int_S \exp \lambda |g(x)|\,dx \leqslant A_S,$$

provided $0<\lambda\leqslant\lambda_0=\lambda_0(S,K)$. *If* f *is, in addition continuous, the last integral is finite for every* $\lambda>0$.

THEOREM G. *Given any* f, *let*

$$\Phi(x) = \sup_{\varepsilon>0} \Big| \int_{E^k} K_\varepsilon(x-y) f(y)\,dy \Big|.$$

Then in the three cases

(i) $$f \in L^p, \qquad p>1;$$

(ii) $$|f|(1+\log^+|f|) \in L;$$

(iii) $$f \in L,$$

the inequalities (1.2), (1.3), (1.4) *hold, respectively, with* $\tilde{f}$ *replaced by* Φ.

2. Let us now consider a system $e_1, e_2, \ldots, e_k$ of independent vectors in E^k and let $x_0=0, x_1, x_2, \ldots$ be the sequence of all lattice points generated by these vectors:

(2.1) $$x_\nu = m_1 e_1 + m_2 e_2 + \ldots + m_k x_k,$$

where $m_1, m_2, \ldots, m_k$ are arbitrary integers. If we define a function $K^*(x)$ by the formula

(2.2) $$K^*(x) = K(x) + \sum_{\nu=1}^{\infty} \{K(x-x_\nu) - K(-x_\nu)\},$$

the series on the right converges absolutely and uniformly over any bounded set provided we discard the first few terms. (Of course we can also replace the terms in curly brackets by $K(x+x_\nu)-K(x_\nu)$). For if $x\in S$ and ν is large enough, then

$$|K(x-x_\nu)-K(-x_\nu)|\leqslant \frac{1}{|x_\nu|^k}\,\omega\left(\frac{A}{|x_\nu|}\right) \tag{2.3}$$

(see [2], p. 95), and condition 1° imposed on the kernel K implies that the terms on the right here form a convergent series.

It is also easy to see that the function K^* has periods $e_1, e_2, \dots, e_k$. For let us assign to the kernel $K(x)$ the value 0 at the origin. We may then write the series (2.2) in the form $\sum_{\nu=0}^{\infty}\{K(x-x_\nu)-K(-x_\nu)\}$. Hence

$$K^*(x)-K^*(x+e_1)=\sum_{\nu=0}^{\infty}\{K(x-x_\nu)-K(x-x_\nu^*)\},$$

where $x_\nu^*=x_\nu-e_1$. The last series converges absolutely. Let us fix the values of $m_2,\dots,m_k$ and consider the sum of the terms with variable m_1. This sum is zero due to the cancellation of "adjoining" terms and the fact that $K(x)$ tends to zero as $|x|$ tends to infinity. Hence the sum of the whole last series is zero. It follows that $K^*(x)$ has the period e_1, and so also periods $e_2,\dots,e_k$.

Of course, the passage from $K(x)$ to $K^*(x)$ is a clasical one. If $k=1$, $e_1=2\pi$, $K(x)=1/x$, then

$$K^*(x)=\frac{1}{2}\cot\frac{1}{2}x.$$

If $k=2$, $K(z)=1/z^2$, then $K^*(z)$ is the classical $\wp$ function of Weierstrass, etc.

Let us now suppose for the sake of simplicity that the vectors $e_1, e_2, \dots, e_k$ are all mutually orthogonal and of length 1. We may assume that they are situated on coordinate axes. Let $f(x)=f(\xi_1,\xi_2,\dots,\xi_k)$ be a function of period 1 in each ξ_j and integrable over every bounded set. We shall consider the convolution

$$f^*(x)=\int_R f(y)K^*(x-y)\,dy=\int_R K^*(y)f(x-y)\,dy, \tag{2.4}$$

where R is the "fundamental" cube

$$|\xi_j|\leqslant\frac{1}{2} \qquad (j=1,2,\dots,k) \tag{R}$$

and the integral is taken in the principal value sense. For such functions f^* we easily obtain results analogous to Theorems A-G.

In what follows, $f(x)=f(\xi_1,\xi_2,\dots,\xi_k)$ will systematically denote a function of period 1 in each variable (for simplicity, we shall use the abbreviation "periodic function"), integrable over R. By $\mathfrak{M}_p[f]$ we shall denote the norm

$$\Big(\int_R |f|^p\,dx\Big)^{1/p}.$$

THEOREM 1. *The integral* (2.4) *exists for almost every value of* x. *The same holds for the integral*

$$\int_R K^*(x-y)F(dy) \tag{2.5}$$

for any totally additive function of set F.

THEOREM 2. *If* $f\in L^p$, $1<p<\infty$, *then* f^* *also belongs to* L^p *and*

$$\mathfrak{M}_p[f^*]\leqslant A_p\mathfrak{M}_p[f]. \tag{2.6}$$

THEOREM 3. *If* $|f|\log^+|f|$ *is integrable over* R, *then*

$$\int_R |f^*|\,dx\leqslant A\int_R |f|\log^+|f|\,dx+B. \tag{2.7}$$

THEOREM 4. *If* f *is merely integrable over* R, *then for every* $0<a<1$ *and every set* $S\subset R$,

$$\int_S |f^*|^a\,dx\leqslant |S|^{1-a}\frac{A}{1-a}\Big(\int_R |f|\,dx\Big)^a. \tag{2.8}$$

THEOREM 5. *If* f *is integrable over* R, *then the measure of the set* E_y *of points* $x\in R$ *at which* $|f^*|\geqslant y>0$, *satisfies the inequality*

$$|E_y|\leqslant\frac{A}{y}\,\mathfrak{M}_1[f]. \tag{2.9}$$

THEOREM 6. *If* $|f|\leqslant 1$, *then*

$$\int_R \exp\lambda|f^*|\,dx\leqslant A, \tag{2.10}$$

provided λ *is small enough*, $0<\lambda\leqslant\lambda_0(K^*)$. *If* f *is also continuous, then the last integral is finite for every* $\lambda>0$.

Given a function f, let

$$\varphi(x)=\sup_{0<\varepsilon<1/2}\Big|\int_R f(x-y)K^*_\varepsilon(y)\,dy\Big|,$$

where $K^*_\varepsilon(x)$ is the function equal to $K^*(x)$ except in the ε-neighbourhoods of the lattice points x_ν, in which it is equal to zero.

THEOREM 7. *According as*

(i) $$f \in L^p, \qquad p>1,$$

(ii) $$|f| \log^+ |f| \in L,$$

(iii) $$f \in L,$$

we have respectively the inequalities (2.6), (2.7), (2.8), *with* f^* *replaced by* φ.

3. These results can be easily obtained from Theorems A-G. For let R_1 denote the cube

$$(\mathrm{R}_1) \qquad |\xi_j| \leqslant 1 \qquad (j=1,2,\ldots,k),$$

and let $f_1(x)$ be the function equal to $f(x)$ in R_1 and to zero elsewhere. The difference K^*-K being bounded in R,

$$|K^*(x)-K(x)|<B \quad \text{for} \quad x \in R,$$

and the function $f(x-y)$ being identical with $f_1(x-y)$ for x and y in R, we see immediately that the second integral (2.4) converges for $x \in R$ if and only if the integral

$$(3.1) \qquad \int_R K(y) f_1(x-y)\, dy \qquad (x \in R)$$

does, and that the difference between these two integrals is numerically less than

$$B \int_R |f(x-y)|\, dy = B \int_R |f(y)|\, dy.$$

On the other hand, if we replace R in (3.1) by E^k we obtain the function $\tilde{f}_1(x)$, and the error committed will not exceed

$$C \int_{E^k-R} |f_1(x-y)|\, dy \leqslant C \int_R |f_1(y)|\, dy = C \int_R |f(y)|\, dy.$$

Collecting results we see that, for x in R, the integral $f^*(x)$ exists if and only if $\tilde{f}_1(x)$ does, and that

$$(3.2) \qquad |f^*(x)-\tilde{f}_1(x)| \leqslant A\mathfrak{M}_1[f] \qquad (x \in R).$$

In particular, $f^*(x)$ exists almost everywhere in R. A similar argument applies to the integral (2.5) and Theorem 1 is established.

Passing to Theorem 2 we observe that (3.2) leads to

$$\mathfrak{M}_p[f] \leqslant \mathfrak{M}_p[\tilde{f}_1] + A\mathfrak{M}_1[f] \leqslant \|\tilde{f}_1\|_p + A\mathfrak{M}_p[f] \leqslant A_p \|f_1\|_p + A\mathfrak{M}_p[f] \leqslant A_p \mathfrak{M}_p[f],$$

which completes the proof.

If $|f|\log^+|f|$ is integrable, (3.2) and Theorem C, with $S=R$, give

$$\int_R |f^*|\,dx \leqslant \int_R |\tilde{f}_1|\,dx + A\mathfrak{M}_1[f]$$
$$\leqslant A\int_{R_1} |f_1|\log^+|f_1|\,dx + A + A\mathfrak{M}_1[f] \leqslant A\int_R |f|\log^+|f|\,dx + B,$$

and (2.7) is established.

Passing to Theorem 4, let us integrate the inequality

$$|f^*|^a \leqslant |\tilde{f}_1|^a + A^a\mathfrak{M}_1^a[f],$$

which follows from (3.2), over S. We get

$$\int_S |f^*|^a dx \leqslant \int_S |\tilde{f}_1|^a dx + |S|\,A^a\mathfrak{M}_1^a[f]$$
$$\leqslant \frac{A}{1-a}\,|S|^{1-a}\Big(\int_{E^k} |f_1|\,dx\Big)^a + |S|\,A^a\mathfrak{M}_1^a[f]$$
$$\leqslant \frac{A}{1-a}\,|S|^{1-a}\Big(\int_R |f|\,dx\Big)^a + |S|\,A^a\mathfrak{M}_1^a[f] \leqslant \frac{A}{1-a}\,|S|^{1-a}\mathfrak{M}_1^a[f],$$

since $|S|\leqslant 1$. This gives (2.8).

For the proof of Theorem 5 we may suppose that $\mathfrak{M}_1[f]=1$. Let us denote the constants A in (1.5) and (3.2) by A' and A'' respectively. Let us also temporarily assume that $y/2 \geqslant A''$. From (3.2) we see that the set of points $x\in R$ at which $|f^*|\geqslant y$ is contained in the set of points x at which $|\tilde{f}|\geqslant y/2$ and so, by Theorem E, has a measure not exceeding

$$A'\left(\frac{1}{2}\,y\right)^{-1}\int_{E^k} |f_1|\,dx = A'2^{k+1}y^{-1}.$$

This gives (2.9), with $A=2^{k+1}A'$, provided $y\geqslant 2A''$. Increasing the constant A so that A/y exceeds 1 for $0<y\leqslant 2A''$ we shall have the inequality (2.9) trivially satisfied for such y's. Thus (2.9) holds for all positive y.

Suppose now that $|f|\leqslant 1$. The inequality (3.2) shows that $|f^*-\tilde{f}_1|\leqslant A$ in R. If we denote by $g_1(x)$ the function (1.6) corresponding to f_1, then clearly $|\tilde{f}_1-g_1|\leqslant A$ in R. Hence, by (3.2),

$$|f^*(x)-g_1(x)|\leqslant A \quad \text{for} \quad x\in R.$$

Using this and (1.7) for $S=R$, we obtain

$$\int_R \exp\lambda|f^*(x)|\,dx \leqslant \int_R e^{\lambda A}e^{\lambda|g_1(x)|}\,dx \leqslant A \quad \text{for} \quad \lambda\leqslant\lambda_0,$$

which is (2.10). Similarly we obtain the result concerning continuous functions f.

The proof of Theorem 7 is analogous to the proofs of the preceding results provided we use Theorem G and the inequality

$$|\tilde{f}_\varepsilon(x) - \tilde{f}_\varepsilon^*(x)| \leqslant A\mathfrak{M}_1[f],$$

analogous to (3.2) and established similarly.

Remarks. 1° The best possible constant A_p in (1.2) satisfies inequalities

$$(3.3) \qquad A_p \leqslant \frac{A}{p-1} \quad \text{for} \quad 1<p\leqslant 2, \qquad A_p \leqslant Ap \quad \text{for} \quad p \geqslant 2.$$

Moreover, if $\bar{A}_p$ is the constant corresponding to the kernel $K(-x)$, then

$$\bar{A}_p = A_{p'} \quad \text{for} \quad p' = \frac{p}{p-1} \qquad (1<p<\infty)$$

(see [2]). It follows from the proof of Theorem 2 that the same results hold for the constant A_p in (2.6).

That the constant A_p in part (i) of Theorem G satisfies inequality (3.3) is proved in [3]. Therefore this inequality also holds in part 1° of Theorem 7.

2° Using the second inequality (3.3) we can deduce (2.10) from (2.6) by a familiar argument. For, if $|f| \leqslant 1$, then

$$\int_R \exp\lambda|f^*|\,dx < 2\int_R \cosh\lambda|f^*|\,dx = 2\left(1 + \sum_{\nu=1}^{\infty}\frac{\lambda^{2\nu}}{(2\nu)!}\int_R |f^*|^{2\nu}dx\right)$$

$$\leqslant 2 + 2\sum_{\nu=1}^{\infty}\frac{(2A\lambda\nu)^{2\nu}}{(2\nu)!}\int_R |f|^{2\nu}dx \leqslant 2 + 2\sum_{\nu=1}^{\infty}\frac{(2A\lambda\nu)^{2\nu}}{(2\nu)!},$$

and the last series is finite provided $\lambda A < e^{-1}$.

Since the second estimate (3.3) holds if we replace f^* by φ in (2.6), it follows that also in (2.10) we can replace f^* by φ.

3° Theorem D has an analogue for the functions (1.1a), with $\|\tilde{f}\|_1$ in (1.4) replaced by the total variation of F over E^k (see [2]). A corresponding result holds for the function (2.5).

4. We now pass to a different group of theorems. Let us consider a periodic function $f(x) = f(\xi_1, \xi_2, \ldots, \xi_k)$ and its Fourier series

$$(4.1) \qquad f(x) \sim \sum c_{\mu_1,\mu_2,\ldots,\mu_k} e^{2\pi i(\mu_1\xi_1 + \ldots + \mu_k\xi_k)} = \sum c_m e^{2\pi i(m,x)},$$

where $m = (\mu_1, \ldots, \mu_k)$, and (m, x) is the scalar product $\mu_1\xi_1 + \ldots + \mu_k\xi_k$ of m and x.

If $f \in L^p$, $p>1$, or if only $|f|\log^+|f|$ is integrable, then the periodic function $f^*(x)$ is also integrable and we may consider its Fourier series

$$f^*(x) \sim \sum c_m^* e^{2\pi i(m,x)}. \tag{4.2}$$

Our next problem will be to consider relations between the series (4.1) and (4.2). For $k=1$ and $K(x)=1/x$, the series (4.2) is the conjugate of (4.1), and in the general case we shall also call (4.2) the *conjugate* of (4.1), corresponding to the kernel K.

We first compute the Fourier coefficients γ_m of $K^*(x)$,

$$\gamma_m = \int_R K^*(y) e^{-2\pi i(m,y)} y, \tag{4.2 a}$$

where the integral is taken in the principal value sense. That these coefficients exist, is clear if we observe that in the neighbourhood of the origin $K^*(x)$ differs from $K(x)$ by a bounded function, that for $|x|$ small

$$K(x)e^{-2\pi i(m,x)} = K(x)\{1+O(|x|)\} = K(x)+O(|x|^{-k+1}),$$

and that the integral of K over R, in the principal value sense, exists owing to condition 2^0 imposed on K.

Next, we show that for $m \neq (0,0,\ldots,0)$ we have

$$\gamma_m = \int_{E^k} K(y) e^{-2\pi i(m,y)} dy = \lim_{\varepsilon\to 0} \int_{E^k} K_\varepsilon(y) e^{-2\pi i(m,y)} dy, \tag{4.3}$$

i. e. the Fourier coefficient of K^* is equal to the corresponding Fourier transform of K.

One remark is indispensable here. The function K_ε, being in L^2, has by the theorem of Plancherel a Fourier transform, also in L^2. This transform is, however, defined almost everywhere only, while here we insist on its existence at the lattice points m. We must therefore show that under the conditions imposed on K the last integral in (4.3), defined as

$$\lim_{\varrho\to\infty} \int_{|y|\leqslant\varrho} K_\varepsilon(y) e^{-2\pi i(m,y)} dy = \lim_{\varrho\to\infty} \int_{\varepsilon\leqslant|y|\leqslant\varrho} K(y) e^{-2\pi i(m,y)} dy, \tag{4.4}$$

exists for each m. This will follow, as we are going to show, from the existence — already established — of the γ_m. We may assume that $m \neq (0,0,\ldots,0)$, since in the remaining case the limit (4.4) clearly exists and is zero.

Let R_ν be the cube with centre x_ν and congruent to R; thus $R=R_0$. Let $\Gamma(\varepsilon)$ denote the sphere with centre at the origin and radius ε, $0<\varepsilon\leqslant 1/2$. Using (2.2) we have

$$\begin{aligned}(4.5)\qquad &\int_{R-\Gamma(\varepsilon)} K^*(y) e^{-2\pi i(m,y)} dy \\ &= \int_{R-\Gamma(\varepsilon)} K(y) e^{-2\pi i(m,y)} dy + \sum_{\nu=1}^{\infty} \int_{R-\Gamma(\varepsilon)} \{K(y-x_\nu)-K(-x_\nu)\} e^{-2\pi i(m,y)} dy.\end{aligned}$$

Owing to the convergence of the series of the right sides in (2.3) we see that if in the last sum in (4.5) we replace the domain of integration $R-\Gamma(\varepsilon)$ by R, we commit an error $O(\varepsilon)$. Suppose that in the new series we only retain terms with $|x_\nu|\leqslant\varrho$. The contribution of the omitted terms tends to zero as $\varrho\to\infty$, and the sum retained is

$$\sum_{0<|x_\nu|\leqslant\varrho}\int_R\{K(y-x_\nu)-K(-x_\nu)\}e^{-2\pi i(m,y)}dy=\sum_{0<|x_\nu|\leqslant\varrho}\int_{R_\nu}K(y)e^{-2\pi i(m,y)}dy.$$

Observing that the measure of the union of the sets R_ν with $|x_\nu|\leqslant\varrho, \nu\neq 0$, differs from the measure of the sphere $|y|\leqslant\varrho$ by $O(\varrho^{k-1})$, and that $K(y)=O(|y|^{-k})$ for $|y|\to\infty$, we can write (4.5) in the form

$$\int_{R-\Gamma(\varepsilon)}K^*(y)e^{-2\pi i(m,y)}dy=\int_{2\leqslant|y|\leqslant\varrho}K(y)e^{-2\pi i(m,y)}dy+O(\varepsilon)+o_\varrho(1),$$

where $o_\varrho(1)$ is a quantity tending to zero as $\varrho\to\infty$. This formula not only proves the existence of (4.4) but also the equations (4.3), provided $m\neq(0,0,\ldots,0)$.

If $m=(0,0,\ldots,0)$, the preceding argument shows that

$$\gamma_0=-\lim_{\varrho\to\infty}\sum_{0<|x_\nu|\leqslant\varrho}K(-x_\nu)=-\lim_{\varrho\to\infty}\sum_{0<|x_\nu|\leqslant\varrho}K(x_\nu).$$

The latter quantity need not be zero and the formula (4.3) fails then. However, γ_0 is zero, for example in the case when the sum of the $K(x_\nu)$ extended over the x_ν situated in the circle $|x_\nu|\leqslant\varrho$ is zero. It is clearly so in the classical case $n=1$, $K(x)=1/x$. If $k=2$ and

$$K(z)=\frac{e^{im\varphi}}{|z|^2}=\frac{z^m}{|z|^{m+2}}\qquad(\varphi=\arg z,\ m=\pm1,\pm2,\ldots),$$

we have $\gamma_0=0$ if m is not divisible by 4. In particular, $\gamma_0=0$ for $K(z)=1/z^2$.

Remarks. 1° An argument similar to the proof of (4.3) shows that if the integral of f over R is zero, *i. e.* if $c_0=0$, then

$$f^*(x)=\int_{E^k}f(y)K(x-y)\,dy=\lim_{\varepsilon\to0}\left\{\lim_{\varrho\to\infty}\int_{\varepsilon\leqslant|y|\leqslant\varrho}K(y)f(x-y)\,dy\right\},$$

the inner limit existing for all x and the outer one if and only if the integral (2.4) converges. Also

$$f^*(x)=\int_{E^k}f(y)K(x-y)\,dy=\lim_{\varepsilon\to0}\left\{\lim_{\varrho\to\infty}\int_{\varepsilon\leqslant|y|\leqslant\varrho}f(y)K(x-y)\,dy\right\}.$$

2° We might slightly modify the definition of K^* so that all the previous results hold and (4.3) is also valid for $m=(0,0,\ldots,0)$. For if we set

$$I_\nu=\int_{R_\nu} f(y)\,dy,$$

$$K^*(x)=K(x)+\sum_{\nu=1}^{\infty}\{K(x+x_\nu)-I_\nu\},$$

the new kernel K^* differs from the old one by a constant only and we now have $\gamma_0=0$.

5. THEOREM 8. *If $f^*(x)$ is integrable (in particular if $f\in L^p$, $p>1$) then the Fourier series* (4.2) *of f^* has coefficients*

$$c_m^*=c_m\gamma_m, \tag{5.1}$$

where γ_m is given by (4.2a) *or, for $m\neq(0,0,\ldots,0)$, by* (4.3).

Given any trigonometric series $\Sigma c_m e^{2\pi i(m,x)}$, we may call

$$\sum c_m\gamma_m e^{2\pi i(m,x)}$$

the *conjugate* of the former, corresponding to the kernel K^*. Thus the result may be stated that if the conjugate function f^* is integrable, the Fourier series of f^* is the conjugate of the Fourier series of f.

This theorem is very well known in the one-dimensional case (see [8], pp. 153, 163). It is obtained there either through complex methods or by considering a certain definition of integral more general than that of Lebesgue. While the first approach fails for general k, the second is applicable straightforwardly.

Let $f(x)=f(\xi_1,\ldots,\xi_k)$ be measurable and periodic. Let us consider any partition P of the cube R into a finite number of parallelpipedes with edges parallel to the axes. These parallelpipeds will be denoted by $\Delta_1,\Delta_2,\ldots,\Delta_N$ and their measures by $|\Delta_1|,\ldots,|\Delta_N|$. Let $x_j=(\xi_1^j,\xi_2^j,\ldots,\xi_k^j)$ be any point of Δ_j and let $t=(\tau_1,\tau_2,\ldots,\tau_k)$ be any vector with $0\leqslant\tau_j\leqslant 1$. Let us consider the sum

$$\sum_{j=1}^{N} f(x_j+t)\,|\Delta_j|. \tag{5.2}$$

It will be denoted by S or $S(t)$, $S[f]$, $S(f,t)$; of course it also depends on P and the points x_j.

If S converges in measure to a limit I as the norm of the partition P (*i. e.* the largest diameter of the Δ_j) tends to 0, we shall say that the function $f(x)$ is *integrable B over R* and that

$$(B)\int_R f(x)\,dx=I.$$

This is an immediate extension of the definition familiar in the case $k=1$ (see [8], p. 151) and the same proof as there shows that, *if f is integrable L over R it is also integrable B and both integrals have the same value*. Therefore, Theorem 8 will be a corollary of the following result:

THEOREM 9. *Suppose that $f(x)$ is periodic and L integrable. Then the function $f^*(x)$, as well as the functions $f(x)e^{-2\pi i(m,x)}$, are B-integrable over R, and the Fourier coefficients of f^*, in the B-sense, satisfy the equation* (5.1).

We first observe that, if f is a trigonometric monomial $e^{2\pi i(m,x)}$, then the conjugate function is (see (2.4))

$$\int_R K^*(y)e^{2\pi i(m,x-y)}dy = e^{2\pi i(m,x)}\int_R K^*(y)e^{-2\pi i(m,y)}dy = \gamma_m e^{2\pi i(m,x)}.$$

Hence, for any trigonometric polynomial $\sum c_m e^{2\pi i(m,x)}$ the conjugate function is $\sum c_m\gamma_m e^{2\pi i(m,x)}$.

We shall now show that for any $f\epsilon L$ and for any lattice point m the function $f(x)e^{-2\pi i(m,x)}$ is integrable B over R and the integral is $c_m\gamma_m$. In view of the preceding remark we may assume that $c_m=0$. Let us set $f=f'+f''$, where f' is a trigonometric polynomial and $\mathfrak{M}_1[f'']$ is small. Without loss of generality we may assume that the coefficient of $e^{2\pi i(m,x)}$ in f' is zero. Thus $f^*=f'^*+f''^*$ and

$$S(f^*e^{-2\pi i(m,x)},t)=S(f'^*e^{-2\pi i(m,x)},t)+S(f''^*e^{-2\pi i(m,x)},t). \tag{5.3}$$

The first term on the right tends to zero as the norm of the partition approaches zero. The absolute value of the second term is

$$\Big|\sum_{j=1}^N f''^*(x_j+t)e^{-2\pi i(m,x_j)}|\Delta_j|\Big|,$$

and the sum between the signs of absolute value is the conjugate g^* of the function

$$g(t)=\sum f''(x_j+t)e^{-2\pi i(m,x_j)}|\Delta_j|.$$

By Theorem 4, with $a=1/2$ and $S=R$, the quantity $\mathfrak{M}_{1/2}[g^*]$ does not exceed a fixed multiple of

$$\mathfrak{M}_1[g(t)]\leqslant\sum_{j=1}^N|\Delta_j|\int_R|f''(x_j+t)|\,dt=\int_R|f''(t)|\,dt,$$

and so is small. It follows that g^*, and so also the last term in (5.3), is small except for t's belonging to a set of small measure, no matter what is the partition P. Therefore the left side of (5.3) is small except for t's in a small set, provided the norm of P is small enough. This shows that $f^*(x)e^{-2\pi i(m,x)}$ is integrable B over R and the integral is zero. Thus the proof of Theorem 9 is completed.

6. We shall now give a few illustrations for the results obtained.

Suppose that the function Ω in the numerator of the kernel K is a spherical harmonic Y_n of order n. In other words, $Y_n(x')|x|^n$ is a homogeneous polynomial P of degree n in the variables $\xi_1, \xi_2, \ldots, \xi_k$, satisfying in these variables Laplace's equation $\Delta P = 0$. Thus

$$(6.1) \qquad K(x) = \frac{Y_n(x')}{|x|^k} \qquad (n=1,2,\ldots).$$

In the simplest case $k=2$, we have $Y_n = e^{in\varphi}$.

It is well known that the Fourier transform of K in (6.1) is a numerical multiple of Y_n; more precisely,

$$(6.2) \qquad (2\pi)^{-k/2} \int_{E^k} \frac{Y_n(y')}{|y|^k} e^{-2\pi i(x,y)} dy = i^n Y_n(x') 2^{-k/2} \frac{\Gamma\left(\frac{1}{2}n\right)}{\Gamma\left(\frac{1}{2}n + \frac{1}{2}k\right)} \text{ [3]}.$$

Combining this with previous results we have the following theorem:

THEOREM 10. *Let $P(x) = P_n(x)$ be a homogeneous polynomial of degree n in the variables $\xi_1, \xi_2, \ldots, \xi_k$, satisfying Laplace's equation $\Delta P = 0$. Given any L-integrable function*

$$(6.3) \qquad f(x) \sim \sum c_m e^{2\pi i(m,x)},$$

consider the series

$$(6.4) \qquad \sum_{m\neq 0} c_m P(m') e^{2\pi i(m,x)} = \sum_{m\neq 0} c_m \frac{P(m)}{|m|^n} e^{2\pi i(m,x)}.$$

Then,

(i) *If $f \in L^p$, $p > 1$, the series* (6.4) *is the Fourier series of a function f^* of the class L^p, and f^* satisfies* (2.6);

(ii) *If $|f| \log^+ |f|$ is integrable,* (6.4) *is the Fourier series of an $f^* \in L$ and satisfying* (2.7);

---

[3]) See Bochner [1]. A different and independent proof for the case $k=3$ was obtained at about the same time by Prof. Szegö, but never published. In the case $k=2$, $y_n = e^{in\varphi}$, $n > 0$, the proof of the formula (6.2) is very simple and the formula itself apparently much older though we cannot give any reference. See also Giraud [4]. It may be added that Bochner sums the integral (6.2) near $y = \infty$ by Abel's method but since the integral converges the sum in both cases must be the same.

Developping the function Ω into a series of spherical harmonics, $\Omega(y') \sim \Sigma Y_n(y')$, and using the formulas (6.2) we formally obtain the Fourier transform of the kernel $K = \Omega/r^n$. It can be shown that this argument is justified under very general conditions on Ω. We shall return to this problem elsewhere.

(iii) *If f is merely integrable,* (6.4) *is the Fourier series, in the B-sense, of a function f^* satisfying* (2.8);

(iv) *If $|f|\leqslant 1$ the function f^* satisfies* (2.10).

The simplest cases here are

$$P(x)=P_1(x)=\xi_j, \tag{6.5}$$

or

$$P(x)=P_2(x)=\xi_i\,\xi_j \qquad (i\neq j). \tag{6.6}$$

In these cases part (i) of Theorem had been proved by Marcinkiewicz [5]. His method does not yield the remaining parts of Theorem 9 since it consists of repeated application of a result from the case $k=1$ and somewhat looses strength as k increases. His proof works for more general cases than (6.5) or (6.6). Of course, also the argument given above applies to general series (5.1a), provided we know that the multipliers γ_m are the Fourier coefficients of a suitable kernel K of our type. To the problem what properties of the γ_m guarantee that assertion we shall return elsewhere.

7. We shall now prove results concerning the behaviour of the conjugate series in the case the function f satisfies a Lipschitz (Hölder) condition

$$|f(x+h)-f(x)|\leqslant C\,|h|^{\alpha},$$

with C independent of x and h. If this condition is satisfied we shall write

$$f\in\Lambda_\alpha.$$

If we disregard constant functions, only the case $0<\alpha\leqslant 1$ need be considered.

In the one-dimensional case there is a familiar result, due to Privalov, asserting that if f is in Λ_α, $0<\alpha<1$, so is f^* (see *e. g.* [8], p. 156). The result is false for $\alpha=1$ (see below). The theorem which follows is an extension of Privalov's result to the k-dimensional case:

THEOREM 11. *Suppose that $f(x)$ is periodic and of the class Λ_α, $0<\alpha<1$, and that $\Omega\in\Lambda_\beta$, $\beta>\alpha$, on Σ. Then $f^*\in\Lambda_\alpha$.*

Proof. Suppose that $|h|$ is small. We may assume that the integral of K^* over R is zero. Then, denoting by $\Gamma(x,r)$ the sphere with centre x and radius r, we may write f^* in the form

$$\begin{aligned} f^*(x)&=\int_R [f(x-t)-f(x)]K^*(t)\,dt \\ &=\int_{R-\Gamma(0,3|h|)} [f(x-t)-f(x)]K^*(t)\,dt+O(|h|^{\alpha}), \end{aligned} \tag{7.1}$$

since the integrand here is $O(|t|^\alpha)\cdot O(|t|^{-k})$, and the integral of the latter function over $\Gamma(0,3|h|)$ is $O(|h|^\alpha)$. Thus

$$
\begin{aligned}
f^*(x+h) &= \int_{R-\Gamma(0,3|h|)} [f(x+h-t)-f(x)]K^*(t)\,dt + O(|h|^\alpha) \\
&= \int_{R(-h)-\Gamma(-h,3|h|)} [f(x-t)-f(x)]K^*(t+h)\,dt + O(|h|^\alpha),
\end{aligned}
\tag{7.2}
$$

where $R(-h)$ denotes the cube R translated by $-h$. A simple calculation shows that if we replace in the last integral the domain of integration by $R-\Gamma(0,3|h|)$ we commit an error $O(|h|)+O(|h|^\alpha)=O(|h|)^\alpha$; for in the neighbourhood of the boundary of R the integrand is $O(1)$, and in the shell $|h|\leqslant|t|\leqslant 3|h|$ the integrand is $O(|h|^\alpha)\cdot O(|h|^{-k})$.

Thus

$$
f^*(x+h)-f^*(x)=O(|h|^\alpha)+\int_{R-\Gamma(0,3|h|)} [f(x-t)-f(x)][K^*(t+h)-K^*(t)]\,dt. \tag{7.3}
$$

The first factor in the integrand here is $O(|t|^\alpha)$. In estimating the second factor, in which $|h|\leqslant|t|/3$, we use the series (2.2) and the formula

$$
\frac{\Omega[(t+h)']}{|t+h|^k}-\frac{\Omega(t')}{|t|^k}=\frac{\Omega[(t+h)']-\Omega(t')}{|t+h|^k}+\Omega(t')\left[\frac{1}{|t+h|^k}-\frac{1}{|t|^k}\right].
$$

The contribution of the term $K(t)$ on the right is

$$
O\left(\frac{|h|^\beta}{|t|^{k+\beta}}\right)+O\left(\frac{|h|}{|t|^{k+1}}\right)=O\left(\frac{|h|^\beta}{|t|^{k+\beta}}\right),
$$

and the contribution of the remaining terms is

$$
O(|h|^\beta)+O(|h|)=O(|h|^\beta).
$$

Thus, collecting results, we see that the last integral does not exceed

$$
O(|h|^\beta)\int_{|t|\geqslant 3|h|}\frac{dt}{|t|^{k+\beta-\alpha}}+O(|h|^\beta)\int_R O(1)\,dt=O(|h|^\alpha)+O(|h|^\beta)=O(|h|^\alpha),
$$

which shows that $f^*\in\Lambda_\alpha$.

The following result is an obvious corollary of Theorem 11:

THEOREM 12. *Suppose that the function $f(x)$ given by* (6.3) *is of the class $\Lambda_\alpha, 0<\alpha<1$, and that $P(x)$ is the same as in Theorem* 10. *Then the series* (6.4) *is the Fourier series of a function in Λ_α.*

Let now $f(x)$ denote any, not necessarily periodic, function defined in a domain $D\subset E^k$. We shall say that the *function $f(x)$ satisfies condition Λ_**, and write

$$
f\in\Lambda_*,
$$

if $f(x)$ is continuous and if

$$|f(x+h)+f(x-h)-2f(x)|\leqslant C|h|, \tag{7.4}$$

for any x and h such that $x, x\pm h$ are in D, with C independent of x and h.

Clearly, $\Lambda_1 \subset \Lambda_*$. If the left side in (7.4) is $o(|h|)$ as $|h|\to 0$, uniformly in $x\in D$, we shall write $f\in\lambda_*$.

In a number of problems the class Λ_* seems to be a more natural one to consider than Λ_1. For example, the theorem of Privalov quoted above does not hold for $a=1$, but it can be shown that if $f(x)\sim\sum c_m e^{2\pi imx}$ is in Λ_* (in particular, if $f\in\Lambda_1$) then the conjugate function

$$\sum c_m(-i\operatorname{sign} m)\,e^{2\pi imx}$$

is also in Λ_* (see [7]). An analogous result could be established for the k-dimensional case, but the proof then is decidedly more difficult and unlike the proof of Theorem 11 is not an imitation of the argument in the one-dimensional case. For this reason we shall confine our attention here to a rather special result which is of interest on account of certain applications.

Let us suppose that f is periodic and of the class Λ_1 and that Ω is merely bounded. We then still have (7.1) and (7.2). For $f^*(x+h)+f^*(x-h)-2f^*(x)$ we get an expression analogous to the right side of (7.3) with $K^*(t+h)-K^*(t)$ replaced by $K^*(t+h)+K^*(t-h)-2K^*(t)$. Suppose that for $|t|\geqslant 2|h|$ we have an inequality

$$|K(t+h)+K(t-h)-2K(t)|\leqslant A\frac{|h|^\gamma}{|t|^{k+\gamma}} \qquad (\gamma>1) \tag{7.5}$$

valid for some γ and A independent of t and h. Then, using formula (2.2), we find, as before, that

$$K^*(t+h)+K^*(t-h)-2K^*(t)=O\left(\frac{|h|^\gamma}{|t|^{k+\gamma}}\right)+O(|h|^\gamma)$$

and

$$f^*(x+h)+f^*(x-h)-2f^*(x)$$

$$=O(|h|)+O(|h|^\gamma)\int\limits_{|t|\geqslant 3|h|}\frac{dt}{|t|^{k+\gamma-1}}+O(|h|^\gamma)\int\limits_R O(1)\,dt=O(h),$$

so that $f^*\in\Lambda_*$.

A similar argument shows that if f is *continuously differentiable* (by this we mean that all derivatives of f of order 1 exist and are continuous) then $f^*\in\lambda_*$. For we may write $f=f_1+f_2$, where f_1 is a finite polynomial and $|f_2(x+t)-f_2(x)|\leqslant\varepsilon|t|$ for $|t|$ small enough. Then clearly,

$$|f_2^*(x+h)+f_2^*(x-h)-2f_2^*(x)|\leqslant C\varepsilon|h|,$$

for $|h|$ small enough, and since f_1^* is a trigonometric polynomial and so satisfies condition λ_*, the function $f^* = f_1^* + f_2^*$ also satisfies condition λ_*.

Inequality (7.5) is certainly valid, with $\gamma=2$, if Ω has bounded second derivatives with respect to the spherical coordinates. For assuming, as we may, that Ω is real-valued, by the Mean-Value Theorem

$$K(t+h)+K(t-h)-2K(t)=|h|^2K''(t+\theta h)=O(|h|^2)$$

for $|t|\geqslant 2|h|$, K'' denoting here the second directional derivative and $-1<\theta<1$. Thus

THEOREM 13. *Suppose that K satisfies (7.5). Then the assumption $f\in\Lambda_1$ implies $f^*\in\Lambda_*$, and if f is continuously differentiable, f^* satisfies condition λ_*. The conclusions hold, in particular, if f^* is defined by the series (6.4) and $P(m)$ is the same as in Theorem 10.*

In the concluding section of this paper we shall give a few observations about functions $f\in\lambda_*$.

8. In this section we prove a few results about discrete analogous of the Hilbert transform.

Let $X=(\ldots,x_{-1},x_0,x_1,\ldots,x_n,\ldots)$ be any two-way infinite sequence of real or complex numbers. For any $p>0$ we shall denote by $\|X\|_p$ the p-th norm of X:

$$\|X\|_p=\left(\sum|x_n|^p\right)^{1/p}.$$

The class of sequences X with $\|X\|_p$ finite will be denoted by l^p.

Let $\tilde{X}=(\ldots,\tilde{x}_{-1},\tilde{x}_0,\tilde{x}_1,\ldots)$ denote the sequence

$$\tilde{x}_m=\sum_n{}' \frac{x_n}{m-n}, \tag{8.1}$$

the prime indicating that the term $m=n$ is omitted in summation. This is a discrete analogue of the Hilbert transform

$$\tilde{f}(x)=\int_{-\infty}^{+\infty}\frac{f(y)}{x-y}\,dy, \tag{8.2}$$

and it is very well known that

$$\|\tilde{X}\|_2\leqslant A\|X\|_2,$$

a result which was extended by Riesz [6] to

$$\|\tilde{X}\|_p\leqslant A_p\|X\|_p,\qquad p>1. \tag{8.3}$$

The best value of A_2 is π; for other p's the best value of A_p is unknown.

Riesz [6] deduced (8.3) from the inequality $\|\tilde{f}\|_p \leqslant A_p \|f\|_p$, valid for the function (8.2). His argument is applicable to a more general class of discrete transforms which we are going to introduce now.

Let us again consider the space E^k, and a kernel $K(x) = \Omega(x')/|x|^k$ with properties described in section 1. Let $e_1, e_2, \ldots, e_k$ be a system of k linearly independent vectors in E^k; thus. in particular, all the e_j are different from zero. Let $p_0 = 0, p_1, p_2, \ldots$ be the sequence of all lattice points in E^k generated by this system, *i. e.* the p's are of the form $\mu_1 e_1 + \mu_2 e_2 + \ldots + \mu_k e_k$, where the coefficients μ_j are arbitrary real integers. For any sequence $X = (x_0, x_1, x_2, \ldots)$ of real or complex numbers we define the transform $\tilde{X} = (\tilde{x}_0, \tilde{x}_1, \tilde{x}_2, \ldots)$ by the formulae

$$\tilde{x}_m = \sum_n{}' x_n K(p_m - p_n). \tag{8.4}$$

For such sequences $\tilde{X}$ we have the following result generalizing (8.3):

THEOREM 14. *If X is in l^p, $p > 1$, so is $\tilde{X}$, and*

$$\|\tilde{X}\|_p \leqslant A_p \|X\|_p,$$

where A_p depends only on p and the kernel K.

The series on the right are all absolutely convergent if $X \epsilon l^p, p \geqslant 1$. For $p=1$ this is immediate, and for $p>1$ follows by an application of Hölder's inequality; for since $K(x) = O(|x|^{-k})$ as $|x| \to \infty$, the series $\sum_n{}' |K(p_m - p_n)|^q$ is finite for every $q > 1$.

For the sake of simplicity we assume that the vectors $e_1, e_2, \ldots, e_k$ are all mutually orthogonal, of length 1, and situated on the coordinate axes. The proof in the general case remains essentially the same. Let R_m denote the cube with centre p_m and edges of length 1, parallel to the axes. By R'_m we shall denote the concentric and similarly situated cube with edges 1/2. Given a sequence $X = (x_0, x_1, \ldots, x_m, \ldots)$ let $f(x)$ denote the function taking the value x_m at the points of R'_m $(m = 0, 1, 2, \ldots)$ and equal to zero elsewhere in E^k. The function f is in L^p if and only if X is in l^p, and the ratio $\|X\|_p / \|f\|_p$ depends on k and p only. Hence, on account of Theorem B of Section 1,

$$\sum_m \int_{R'_m} |\tilde{f}(x)|^p dx = \int_{\cup R'_m} |\tilde{f}(x)|^p dx \leqslant \|\tilde{f}\|_p^p \leqslant A_p^p \|f\|_p^p \leqslant A_p^p \sum_m |x_m|^p. \tag{8.6}$$

For $x \epsilon R'_m$ we may write

$$\tilde{f}(x) = \sum_{n \neq m} x_n \int_{R'_n} K(x-y)\,dy + x_m \int_{R'_m} K(x-y)\,dy. \tag{8.7}$$

Let $\omega(\delta)$ be the modulus of continuity of Ω on Σ. Without loss of generality we may assume that $\omega(\delta) \geqslant \delta$, since otherwise in the inequalities that follow we replace $\omega(\delta)$ by $\omega_1(\delta)=\operatorname{Max}(\delta,\omega(\delta))$. We easily verify that, for $x\in R'_m$ and $y\in R'_n$

$$|K(x-y)-K(p_m-p_n)|\leqslant A\,\frac{\omega\left(\dfrac{1}{|p_m-p_n|}\right)}{|p_m-p_n|^k},$$

and (8.7) may be written

$$\tilde{f}(x)=2^{-k}\sum_{n\neq m}x_nK(p_m-p_n)+x_m\int_{R'_m}K(x-y)\,dy+O\left\{\sum_{n\neq m}\frac{\omega\left(\dfrac{1}{|p_n-p_m|}\right)}{|p_n-p_m|^k}\right\}$$

$$=2^{-k}\tilde{x}_m+x_m\tilde{\varphi}_m(x)+O\left\{\sum_{n\neq m}|x_n|\,|p_n-p_m|^{-k}\omega\left(\frac{1}{|p_m-p_n|}\right)\right\},$$

where φ_m is the characteristic function of the cube R'_m. Thus $|\tilde{x}_m|^p$ does not exceed a fixed multiple of the sum of the three expressions

$$|\tilde{f}(x)|^p,\qquad |x_m|^p|\tilde{\varphi}_m(x)|^p,\qquad \sum_{n\neq m}|x_n|\,|p_m-p_n|^{-k}\omega\left(\frac{1}{|p_m-p_n|}\right).$$

Let us integrate these expressions over the cube R'_m and sum the results over all m. It is enough to show that all the three sums are majorized by a fixed multiple of $\Sigma|x_n|^p$.

This is certainly the case for the sum involving $|\tilde{f}|^p$ (see (8.6)). Since

$$\int_{R'_m}|\tilde{\varphi}_m(x)|^p\,dx\leqslant\|\tilde{\varphi}_m\|_p^p\leqslant A_p^p\|\varphi_m\|_p^p\leqslant A_p^p,$$

also the second sum satisfies the condition. Finally, setting

$$a_m=|p_m|^{-k}\omega\left(\frac{1}{|p_m|}\right)\qquad\text{for}\quad m>0,\ a_0=0,$$

and observing that the Dini condition imposed on ω implies that $\Sigma a_m=a$ is finite, we may write the following inequalities, in which p' is the exponent conjugate to p:

$$\sum_m\{\sum_n|x_n|\,a_{m-n}\}^p=\sum_m\{\sum_n|x_n|\,a_{m-n}^{1/p}a_{m-n}^{1/p'}\}^p$$

$$\leqslant\sum_m\{\sum_n|x_n|^pa_{m-n}\}\{\sum_n a_{m-n}\}^{p/p'}=a^{p/p'}\sum_m\{\sum_n|x_n|^pa_{m-n}\}$$

$$=a^{1+p/p'}\sum_n|x_n|^p=a^p\sum_n|x_n|^p.$$

This completes the proof of Theorem 14.

Of course, Theorem 14 can be restated in the language of bilinear (or quadratic) forms:

$$\left|\sum x_m y_n K(p_m - p_n)\right| \leqslant A_p \|X\|_p \|Y\|_{p'}, \qquad p>1.$$

9. Let us now change our notation slightly and let m, and similarly n, denote the general lattice point in E^k, *i. e.* $m=\mu_1 e_1+\ldots+\mu_k e_k$, where the μ's are arbitrary integers and $e_1,\ldots,e_k$ are unit vectors mutually orthogonal.

THEOREM 15. *The series*

$$\sum_m{}' K(m) e^{2\pi i(m,x)} \tag{9.1}$$

is the Fourier series of a bounded function $\chi(x)$. *The number*

$$M = \text{ess sup} |\chi(x)|$$

is the norm of the linear transformation

$$\tilde{x}_m = \sum_n{}' K(m-n) x_n, \tag{9.2}$$

considered as a transformation from l^2 *into* l^2.

We already know that the transformation (9.2) is bounded and from this fact we shall be able to deduce the boundedness of the function (9.1). Since $\Sigma|K(m)|^2$ is finite, (9.1) is in any case the Fourier series of a function $\chi \epsilon L^2$. Similarly, if $\sum |x_m|^2$ converges,

$$\sum x_m e^{2\pi i(m,x)} \sim \psi(x) \epsilon L^2.$$

By (9.2), $\sum \tilde{x}_m e^{2\pi i(m,x)}$ is the Fourier series of the integrable function

$$\varphi(x) = \psi(x)\chi(x). \tag{9.3}$$

Since $\sum |\tilde{x}_m|^2$ is finite, the function φ is even quadratically integrable and we can write

$$\sum |\tilde{x}_m|^2 = \int_R |\varphi|^2 dx = \int_R |\psi|^2 |\chi|^2 dx \leqslant M^2 \int_R |\psi|^2 dx = M^2 \sum |x_m|^2,$$

so that the number $M(\leqslant\infty)$ is not less than the norm of the transformation (9.2). Moreover, one immediately sees that M is actually equal to the norm of the transformation. Since the transformation is bounded, the theorem follows.

An interesting illustration is provided in the two-dimensional case by the transformation

$$\tilde{x} = \sum_n{}' \frac{x_n}{(m-n)^2}, \tag{9.4}$$

where m and n denote complex integers. This seems to be the most natural extension of the classical Hilbert transformation (8.1) to the two-dimensional case. The norm of (9.4) is the upper bound of the modulus of the function given by the Fourier series

$$\sum_{\mu,\nu}' \frac{e^{2\pi i(\mu x+\nu y)}}{(\mu+i\nu)^2}.$$

The latter function occurs already in the work of Kronecker on elliptic functions and is expressible in terms of elliptic theta functions.

10. We conclude by a few remarks concerning *smooth functions*.

Suppose a function $f(x)$ is determined in the neighbourhood of a point $x_0 \in E^k$. We say that f is *smooth* at x_0, if

(10.1) $$f(x_0+h)+f(x_0-h)-2f(x_0)=o(h) \quad \text{as} \quad |h|\to 0.$$

If f is smooth *and continuous* at every point of an open set D, f will be called *smooth in D*. The latter notion has close connection with condition λ_* introduced in Section 7, the only difference being that the notion of smoothness in D does not presuppose the uniformity of the "o" in (10.1) with respect to $x_0 \in D$.

Clearly, if f is differentiable at x_0 (*i. e.*, if it has a total differential at x_0), then f is smooth at x_0, but the converse need not be true. Thus smooth functions may be considered as a generalization of differentiable functions. Similarly, functions of the class λ_* may be considered as a generalization of continuously differentiable functions (*i. e.*, functions with continuous first partial derivatives).

The notion of smoothness of functions is familiar in the simplest case $k=1$ (see [7]), and the definition (10.1) seems to be a natural extension of that special case to general k. In what follows we shall prove a few simple results concerning smooth functions. We shall not presuppose any longer that the functions f considered are periodic, and the results themselves will have little connection with the previous discussion.

(a) *If f is smooth in D and real-valued, and if f has a maximum (or minimum) at x_0, then f is differentiable at x_0 and the partial derivatives of f at x_0 with respect to the coordinates are zero.*

This is immediate since (10.1) can be written

$$\{f(x_0+h)-f(x_0)\}+\{f(x_0-h)-f(x_0)\}=o(|h|),$$

and since for $|h|$ small enough both terms in curly brackets are of the same sign, we get $f(x_0+h)-f(x_0)=o(|h|)$, which is the desired result.

(b) *If f is smooth in D and real-valued, then the set S of the points of differentiability of f is dense in D; indeed, it is of the power of the continuum in every sphere K totally contained in D.*

We may suppose that the closure of K is in D. Let $g(x)$ be a real-valued and continuously differentiable function vanishing on the boundary of K, positive inside K and taking a large value at the centre of K. Then the sum $h=f+g$ certainly has a maximum at a point $x_0 \epsilon K$, and so is differentiable at x_0. Hence, also f is differentiable at x_0, which shows that S is dense in D.

Let now $l(x)=a_1\xi_1+a_2\xi_2+\ldots+a_k\xi_k$ be a real-valued linear function with coefficients $a_1, a_2, \ldots, a_k$ numerically small but otherwise quite arbitrary. The function $h(x)=f(x)+g(x)+l(x)$ will then still have a maximum at a point $x_0(\xi_1, \ldots, \xi_k)\epsilon K$, and so will be differentiable at that point. Moreover, the first partial derivatives of $h(x)$ at $x_0(\xi_1, \ldots, \xi_k)$ will be zero, and so the first partial derivatives of $f(x)+g(x)$ at that point will be $-a_1, \ldots, -a_k$. Thus the point $x_0(a_1, \ldots, a_k)$ varies with the system $(a_1, \ldots, a_k)$. It follows that the set of points $x_0(a_1, \ldots, a_k)$ is of the power of the continuum, $h=f+g+l$ is differentiable in a subset of K of the power of the continuum, and the same holds for f.

(c) *The partial derivatives* $u_{\xi_i}(\xi_1, \ldots, \xi_k)$ *of the potential*

$$u(\xi_1, \ldots, \xi_k)=u(x)=\int_{E^k} f(y)\frac{dy}{|x-y|^{k-2}} \qquad (k>2) \tag{10.2}$$

corresponding to a continuous density f, *satisfy condition* λ_* *in every finite sphere.*

Without loss of generality we may suppose that $f(y)$ vanishes for $|y|$ large. It is a classical fact that under the assumption of continuity of f the partial derivatives u_{ξ_i} exist everywhere, are continuous and given by the formulae

$$u_{\xi_i}(x)=-(k-2)\int_{E^k} f(y)\frac{\xi_i-\eta_i}{|x-y|^k}\,dy. \tag{10.3}$$

It is also very well known that the second partial derivatives of u need not exist at individual points, and statement (c) is a substitute for the existence of these derivatives.

It is enough to give a sketch of proof since the whole argument follows familiar lines. On the right of the last formula we have a convolution of f with the kernel $K(x)=-(k-2)\xi_i/|x|^k$. In the integrals

$$\int_{E^k} f(y)K(x-y)\,dy, \qquad \int_{E^k} f(y)K(x\pm h-y)\,dy,$$

we consider separately the parts extended over the sphere $|y-x|\leqslant 2|h|$ and over the remainder of the space E^k. Since f is bounded, and

$K(z) = O(|z|^{-k+1})$ for small $|z|$, the parts extended over the sphere are all $O(|h|)$. Since

$$\Big|\int_{|x-y|\geqslant 2|h|} f(y)\{K(x+h-y)+K(x-h-y)-2K(x-y)\}\,dy\Big|$$

$$< O(|h|^2)\int_{|x-y|\geqslant 2|h|} |f(y)|\,|x-y|^{-k-1}dy = O(|h|),$$

collecting results we see that $u_{\xi_t}(x+h)+u_{\xi_t}(x-h)-2u_{\xi_t}(x) = O(|h|)$.

So far we have only used the boundedness of f and showed that then $u_{\xi_t}(x)$ satisfies condition Λ_* in every sphere. Since in the formula (10.3) we may replace $f(y)$ on the right by $f(y)-f(x)$, the condition Λ_* refines to λ_* if f is continuous.

Of course, (c) also holds for $k=2$ if we replace (10.2) by the logarithmic potential. The result in this case was pointed to us by W. H. Oliver, and clearly the proof for $k>2$ is essentially the same.

Bibliography

[1] S. Bochner, *Theta relations with spherical harmonics*, Proc. Nat. Academy of Sciences 37 (1951), p. 804-808.

[2] A. P. Calderón and A. Zygmund, *On the existence of certain integrals*, Acta Math. 88 (1952), p. 85-139.

[3] M. Cotlar, *On Hilbert's transform*, Ph. D. thesis presented at the University of Chicago (to appear soon).

[4] G. Giraud, *Sur une classe générale d'équations à intégrales principales*, Comptes Rendus 202 (1936), p. 2124-2125.

[5] J. Marcinkiewicz, *Sur les multiplicateurs des séries de Fourier multiples*, Studia Math. 8 (1938), p. 78-91.

[6] M. Riesz, *Sur les fonctions conjuguées*, Math. Zeitschrift 27 (1927), p. 218-244.

[7] A. Zygmund, *Smooth functions*, Duke Math. Journal 12 (1945), p. 47-76.

[8] — *Trigonometrical series*, Monografie Matematyczne, Warszawa 1935.

(Reçu par la Rédaction le 15. 11. 1953)

THÉORIE DES ENSEMBLES. — *Sur un théorème de Piatetçki-Shapiro.* Note (*) de MM. **Raphaël Salem** et **Antoni Zygmund**, présentée par M. Joseph Pérès.

Par ensemble U nous désignerons un ensemble d'unicité pour le développement trigonométrique. Soit C la classe des entiers algébriques positifs θ dont tous les conjugués (autres que θ lui-même) sont, en module, strictement inférieurs à 1. Dans tout ce qui suit, $\theta \in C$, $\theta > 2$, et n désignera le degré de θ. Par E nous désignerons l'ensemble du type de Cantor dont les points sont donnés par la formule $(\theta - 1)(\varepsilon_1\theta^{-1} + \varepsilon_2\theta^{-2} + \ldots)$ où les ε sont égaux à 0 ou 1. Tout récemment Piatetçki-Shapiro a démontré que si $\theta > 2^n$, E est un ensemble U. Le but de cette Note est de modifier la démonstration de Piatetçki-Shapiro de manière à s'affranchir de l'hypothèse $\theta > 2^n$ et de montrer que E est un ensemble U dès que $\theta > 2$ appartient à C (¹).

Soit F l'ensemble homothétique de E dont les points x sont donnés par $(\varepsilon_1\theta^{-1} + \varepsilon_2\theta^{-2} + \ldots)$. Soient $\alpha_1, \alpha_2, \ldots, \alpha_{n-1}$ les conjugués de θ, $P(z)$ le polynome irréductible à coefficients entiers ayant les racines $\theta, \alpha_1, \ldots, \alpha_{n-1}$; Q le polynome réciproque de P; $R(z)$ un polynome quelconque de degré $n-1$ à coefficients entiers, $T(z)$ son réciproque; $P'(z)$ la dérivée de P. On a

$$\frac{R(z)}{Q(z)} = \sum_0^\infty c_m z^m = \frac{\lambda}{1-\theta z} + \sum_1^{n-1} \frac{\mu_i}{1-\alpha_i z},$$

où les c_m sont entiers, $\lambda = T(\theta)/P'(\theta)$, $\mu_i = T(\alpha_i)/P'(\alpha_i)$, $\lambda\theta^m = c_m + \delta_m$, avec $\delta_m = -\sum_1^{n-1} \mu_i \alpha_i^m \to 0$ pour $m \to \infty$. On peut supposer $\lambda > 0$. On a

$$(1) \qquad \sum_0^\infty |\delta_m| \leq \sum_1^{n-1} \frac{|\mu_i|}{1-|\alpha_i|}.$$

(*) Séance du 16 mai 1955.

(¹) *Cf.* R. Salem, *Trans. Amer. Math. Soc.*, 54, 1943, p. 218-228; 63, 1948, p. 595-598; Piatetçki-Shapiro, *Uspekhi Matematicheskikh Nauk*, 8, 1953, p. 167-170; Uçenye Zapiski *de l'Université de Moscou*, année 1954.

Reprinted from *CRAS* 240, 2040–2042 (1955).

Soit N un entier positif fixe. Si $x \in F$ et m est un entier $\geqq 0$, on a, *modulo* 1,

$$(2) \qquad \lambda\theta^m x \equiv \lambda\left(\frac{\varepsilon_{m+1}}{\theta} + \ldots + \frac{\varepsilon_{m+N}}{\theta^N}\right) + \lambda\left(\frac{\varepsilon_{m+N+1}}{\theta^{N+1}} + \ldots\right) + (\varepsilon_m\delta_0 + \ldots + \varepsilon_1\delta_{m-1})$$

Lemme. — *Soit V_k le vecteur de l'espace euclidien R^n ayant pour coordonnées $c_{k+1}, c_{k+2}, \ldots, c_{k+n}$. La suite V_k est « normale » au sens de Piatetçki-Shapiro, c'est-à-dire que $a_1, \ldots, a_n$ étant des entiers fixes non tous nuls,*

$$|a_1 c_{k+1} + \ldots + a_n c_{k+n}| \to \infty \quad \textit{quand} \quad k \to \infty.$$

En effet

$$a_1 c_{k+1} + \ldots + a_n c_{k+n} = a_1\lambda\theta^{k+1} + \ldots + a_n\lambda\theta^{k+n} + a_1\delta_{k+1} + \ldots + a_n\delta_{k+n}.$$

Or, pour $k \to \infty$, $a_j\delta_{k+j} \to 0$ et $\lambda\theta^{k+1}|a_1 + \ldots a_n\theta^{n-1}| \to \infty$ puisque θ est de degré n.

Démontrons maintenant que E est un ensemble U. Les n coefficients entiers de $T(z)$ peuvent (théorème de Minkowski) être déterminés de manière que

$$(3) \qquad \left|\frac{T(0)}{P'(0)}\right| \frac{1}{\theta^N(\theta-1)} \leq \frac{1}{8n\,2^{\frac{N}{n}}}, \qquad \left|\frac{T(\alpha_i)}{P'(\alpha_i)}\right| \frac{1}{1-|\alpha_i|} \leq \frac{1}{8n\,2^{\frac{N}{n}}},$$

à condition que $(8n2^{N/n})^{-n} > |\Delta|\theta^{-N}$, où $\Delta = \Delta(\theta)$ est un déterminant ne dépendant que de θ, et non de N. Puisque $\theta > 2$ ceci est possible en choisissant un entier N qui restera fixe, tel que $(\theta/2)^N > (8n)^n|\Delta|$. On a alors, d'après (1), (2), (3), pour tout m, *modulo* 1,

$$(4) \qquad \left|\lambda\theta^m x - \lambda\left(\frac{\varepsilon_{m+1}}{\theta} + \ldots + \frac{\varepsilon_{m+N}}{\theta^N}\right)\right| \leq \frac{1}{8\,2^{\frac{N}{n}}}.$$

Soit g_m la partie fractionnaire de $\lambda(\varepsilon_{m+1}\theta^{-1} + \ldots + \varepsilon_{m+N}\theta^{-N})$. Soit, dans R^n, O_k le point de coordonnées $(g_{k+1}, \ldots, g_{k+n})$. Quelles que soient les valeurs de k et des ε, le nombre de points O_k distincts ne dépasse pas 2^{N+n-1}. En effet g_{k+1} prend au plus 2^N valeurs distinctes; g_{k+1} étant fixé, g_{k+2} ne peut prendre que deux valeurs différentes; g_{k+1} et g_{k+2} étant fixés, g_{k+3} ne peut prendre que deux valeurs différentes, et ainsi de suite. Soit maintenant $x \in F$ et P_k le point de R^n dont les coordonnées sont les parties fractionnaires de $\lambda\theta^{k+1}x, \ldots, \lambda\theta^{k+n}x$ respectivement.

D'après (4) P_k est à l'intérieur d'un cube de côté $1/2^{(N/n)+2}$ ayant pour centre, soit O_k, soit un point dont certaines coordonnées diffèrent de celles de O_k d'une unité. Comme il y a au plus 2^{N+2n-1} cubes, leur volume total ne dépasse pas

$$2^{N+2n-1}\left(2^{-\frac{N}{n}-2}\right)^n = \frac{1}{2}.$$

Il existe donc dans l'« hypercube unité » de R^n un cube fixe ne contenant aucun point P_k. Ceci reste vrai (pour $k > k_0$) pour les points dont les coordonnées sont les parties fractionnaires de $c_{k+1}x, \ldots, c_{k+n}x$ respectivement.

L'ensemble F est donc du type dénommé $H^{(n)}$ par Piatetçki-Shapiro à qui l'on doit d'avoir démontré que tout ensemble de ce type est un ensemble U. F est donc un ensemble U, et E un ensemble U par homothétie.

ON A PROBLEM OF MIHLIN

BY

A. P. CALDERÓN AND A. ZYGMUND

1. Let $x=(\xi_1, \xi_2, \cdots, \xi_k)$, $y=(\eta_1, \eta_2, \cdots, \eta_k)$, $z=(\zeta_1, \zeta_2, \cdots, \zeta_k)$, $\cdots$ denote points of the k-dimensional Euclidean space E^k. Here $k\geqq 1$ but only the case $k\geqq 2$ will be of interest. The space may also be treated as a vector space by identifying x with the vector joining the origin $0=(0, \cdots, 0)$ with the point x. The rules for addition of vectors and for multiplying them by scalars are the usual ones, and the norm is defined by the formula

$$|x| = (\xi_1^2 + \xi_2^2 + \cdots + \xi_k^2)^{1/2}.$$

By $x'=(\xi_1', \xi_2', \cdots, \xi_k')$ we shall systematically denote the point of intersection of the ray $0x$ $(x\neq 0)$ with the unit sphere $\Sigma=\Sigma_{k-1}$ defined by the equation $|x|=1$. Thus,

$$x' = x/|x|, \qquad |x'| = 1.$$

In this note we shall consider the problem of the existence of the integral

$$(1.1) \qquad \int_{E^k} K(x, y)f(y)dy$$

where $dy=d\eta_1 d\eta_2 \cdots d\eta_k$, f is a function of the class L^2 over E^k, and $K(x, y)$ is a singular kernel satisfying certain conditions. In general, we shall have

$$(1.2) \qquad K(x, y) = \frac{\Omega(x, z')}{|z|^k}$$

where, systematically, $z=x-y$. Thus $K(x, y)$ depends on the point x and on the direction from x to y.

In a special but important case, K may depend on z only. We then have

$$K(x, y) = \frac{\Omega(z')}{|z|^k} \qquad (z = x - y)$$

i.e.

$$(1.3) \qquad K(x, y) = K(x - y) \quad \textit{with} \quad K(x) = \frac{\Omega(x')}{|x|^k}.$$

It is well known (see [2]) that if $K(x)$ satisfies certain regularity conditions and the indispensable condition

Presented to the Society, September 5, 1953; received by the editors September 15, 1953.

Reprinted from *TAMS* 78, 209–224 (1955).

$$\int_{\Sigma} \Omega(x')dx' = 0,$$

and if $f \in L^p$, $p \geqq 1$, then the integral

$$\int_{E^k} K(x-y)f(y)dy \tag{1.4}$$

exists in the principal value sense for almost every x. (By the principal value of the integral (1.4) we mean the limit, for $\epsilon \to 0$, of the integral extended over the exterior of the sphere with center x and radius ϵ.) Moreover, the value $\tilde{f}(x)$ of (1.4) has many properties similar to those of the Hilbert transform

$$\int_{-\infty}^{+\infty} \frac{f(y)dy}{x-y}$$

in E^1.

The purpose of the present note is to prove some results about the more general case (1.2). We fix our kernel $K(x, y)$ once for all and use the notation

$$\tilde{f}_\epsilon(x) = \int_{|z| \geqq \epsilon} K(x, y)f(y)dy \qquad (z = x - y). \tag{1.5}$$

By $\tilde{f}(x)$ we shall mean the limit of $\tilde{f}_\epsilon(x)$ as $\epsilon \to 0$. This limit may be considered pointwise or in some norm. In this note we shall be concerned exclusively with convergence in norm. We shall systematically use the notation

$$\|f\|_p = \left(\int_{E^k} |f(y)|^p dy\right)^{1/p},$$

but in the case $p = 2$ we shall simply write $\|f\|$ for $\|f\|_2$.

In what follows, by A with various subscripts we shall mean a constant depending on the kernel K and on the parameters displayed in the subscripts. In particular, by A without any subscript we shall mean constants depending on K at most. The constants need not be the same at every occurrence.

We shall first state the main theorem of this note. Comments and generalizations are postponed to a later section.

THEOREM 1. *Suppose that the kernel K defined by* (1.2) *satisfies for each x the following two conditions*:

$$\int_{\Sigma} \Omega(x, z')dz' = 0, \tag{1.6}$$

$$\int_{\Sigma} |\Omega(x, z')|^2 dz' \leqq A, \tag{1.7}$$

with A independent of x. Let $f \in L^2$. Then for each x and $\epsilon > 0$ the integral (1.5) *converges absolutely and the function* $\tilde{f}_\epsilon(x)$ *tends to a limit* $\tilde{f}(x)$ *in norm* L^2. *Moreover*,

$$\|\tilde{f}_\epsilon\| \leqq A\|f\| \qquad (\epsilon > 0), \tag{1.8}$$

$$\|\tilde{f}\| \leqq A\|f\|. \tag{1.9}$$

This theorem was stated as a problem by Mihlin in [6] (see also [5]). He settled the case $k=2$ only, in a somewhat weaker form since he defines $\tilde{f}(x)$ not as a limit of $\tilde{f}_\epsilon(x)$ but as a linear operator which for sufficiently smooth (say, differentiable) functions f is defined directly by the then everywhere convergent integral (1.4) and is subsequently extended by continuity to all functions $f\in L^2$. For $k>2$ he has to replace condition (1.7) by much stronger conditions involving partial derivatives of Ω.

2. Let us fix x and develop $\Omega(x, z')$ into a series of spherical harmonics

$$\Omega(x, z') \sim \sum_{n=1}^{\infty} a_n(x)Y_n(z') \tag{2.1}$$

where $Y_n(z')$ is an (ultra) spherical function of order n, i.e. is the value on Σ of a homogeneous polynomial $P(\zeta_1, \cdots, \zeta_k)$ satisfying Laplace's equation $\Delta P=0$. The development begins with $n=1$ since, on account of (1.6), the term $n=0$ of the development is zero. If $k=2$ we may also write (2.1) in the form

$$\sum_{-\infty}^{+\infty}{}' a_n(x)e^{inz}.$$

We may always normalize the Y_n and assume that

$$\|Y_n\| = \left(\frac{1}{|\Sigma|}\int_{\Sigma} |Y_n(z')|^2dz'\right)^{1/2} = 1,$$

$|\Sigma|$ denoting the $(k-1)$-dimensional measure of Σ. No misunderstanding will occur if we use the same notation for the norm in two different cases, those of the whole space E^k and of the sphere Σ.

The functions $Y_n(z')$ form an orthonormal system on Σ and Bessel's inequality combined with (1.6) gives

$$\left(\sum_{n=1}^{\infty} |a_n(x)|^2\right)^{1/2} \leqq A. \tag{2.2}$$

It will be convenient to modify the definition (1.5) by inserting the factor $(2\pi)^{-k/2}$ in the integral. Thus

$$\tilde{f}_\epsilon(x) = (2\pi)^{-k/2}\int_{|z|\geqq\epsilon} f(y)\,\frac{\Omega(x, z')}{|z|^k}\,dy \qquad (z = x - y). \tag{2.3}$$

This integral converges absolutely for each x since $f \in L^2$ and $\Omega(x, z')|z|^{-k}$ is, on account of (1.7), quadratically integrable over the set $|z| \geqq \epsilon$.

Our first step will be to replace the function Ω in (2.3) by the development (2.2) and prove the equation

$$\tilde{f}_\epsilon(x) = \sum_{n=1}^{\infty} a_n(x)\tilde{f}_{n,\epsilon}(x), \tag{2.4}$$

where

$$\tilde{f}_{n,\epsilon}(x) = (2\pi)^{-k/2} \int_{|z| \geqq \epsilon} f(y) \frac{Y_n(z')}{|z|^k} dy. \tag{2.5}$$

Let us denote the Nth partial sum of the series (2.1) by $S_N(x, z')$. Since, for fixed x, $S_N(x, z')$ converges to $\Omega(x, z')$ over Σ, in norm L^2, the product $S_N(x, z')|z|^{-k}$ converges, in the same norm, to $\Omega(x, z')|z|^{-k}$ over the set $|z| \geqq \epsilon$. Hence, by Schwarz's inequality,

$$\int_{|z| \geqq \epsilon} f(y)S_N(x, z')|z|^{-k}dy \to \int_{|z| \geqq \epsilon} f(y)\Omega(x, z')|z|^{-k}dy,$$

which is (2.4).

It may be added that the series in (2.4) converges absolutely. On account of (2.2) it is enough to prove the convergence of $\sum |\tilde{f}_{n,\epsilon}(x)|^2$. For this purpose we observe that the integrals (2.5) are the Fourier coefficients of $f(y) = f(x-z)$ with respect to the functions equal to $(2\pi)^{-k/2}Y_n(z')|z|^{-k}$ for $|z| \geqq \epsilon$ and equal to zero elsewhere. The latter functions form an orthogonal system in E^k, on account of the orthogonality of the $Y_n(z')$ over Σ. The norms of these functions are not 1 but are bounded away from zero so that the convergence of $\sum |\tilde{f}_{n,\epsilon}(x)|^2$ is a consequence of Bessel's inequality.

Let us now integrate the inequality (see (2.4))

$$|\tilde{f}_\epsilon(x)|^2 \leqq \left(\sum_1^\infty |a_n(x)|^2\right)\left(\sum_1^\infty |\tilde{f}_{n,\epsilon}(x)|^2\right) \leqq A \sum_1^\infty |\tilde{f}_{n,\epsilon}(x)|^2 \tag{2.6}$$

over E^k. We get

$$\|\tilde{f}_\epsilon\|^2 \leqq A \sum_1^\infty \|\tilde{f}_{n,\epsilon}(x)\|^2. \tag{2.7}$$

Suppose we can prove that

$$\|\tilde{f}_{n,\epsilon}\| \leqq \frac{A}{n}\|f\| \qquad (n = 1, 2, \cdots) \tag{2.8}$$

with A independent of n and ϵ. The inequality (1.8) will then follow.

From (2.8) will also follow that

$$\|\tilde{f}_\epsilon - \tilde{f}_{\epsilon'}\| \to 0 \qquad (\epsilon, \epsilon' \to 0)$$

since it is known that for each particular value of n we have

$$\|\tilde{f}_{n,\epsilon} - \tilde{f}_{n,\epsilon'}\| \to 0 \qquad \text{as } \epsilon, \epsilon' \to 0$$

(see [2, p. 89]). Hence there will exist a function $\tilde{f} \in L^2$ such that

$$\|\tilde{f} - \tilde{f}_\epsilon\| \to 0,$$

which in conjunction with (1.8) implies (1.9).

Thus our main problem now is to prove (2.8).

Let us revert to (2.5) and let us denote the Fourier transform of any function $\hat{f}$ by f. In other words,

$$\hat{f}(x) = (2\pi)^{-k/2} \int_{E^k} e^{-i(x,y)} f(y) dy$$

where (x, y) denotes the scalar product $\xi_1\eta_1 + \xi_2\eta_2 + \cdots + \xi_k\eta_k$ of x and y. The equation (2.5) tells us that $\tilde{f}_\epsilon(x)$ is the convolution (with normalizing factor $(2\pi)^{-k/2}$) of f and of the function $g_{n,\epsilon}(y)$ equal to $Y_n(y')|y|^{-k}$ for $|y| \geqq \epsilon$ and to zero otherwise. Suppose that f is both in L^2 and L. Then $\tilde{f}_{n,\epsilon}$ is of the class L^2 and its Fourier transform is the product of the transforms of f and $g_{n,\epsilon}$:

$$\widehat{\tilde{f}_{n,\epsilon}} = \hat{f}\hat{g}_{n,\epsilon}.$$

Moreover, by the Plancherel-Parseval theorem,

$$\|\tilde{f}_{n,\epsilon}\| = \|\widehat{\tilde{f}_{n,\epsilon}}\| = \|\hat{f}\,\hat{g}_{n,\epsilon}\| \leqq \operatorname*{Sup}_x |\hat{g}_{n,\epsilon}(x)|\,\|\hat{f}\| = \operatorname*{Sup}_x |\hat{g}_{n,\epsilon}(x)|\,\|f\|,$$

and (2.8) will follow, at least for f simultaneously in L and L^2, if we show that

$$(2.9) \qquad |\hat{g}_{n,\epsilon}(x)| \leqq \frac{A}{n}.$$

That this will, in turn, imply (2.8) for general f quadratically integrable is immediate. For we may first apply (2.8) to the function $f_R(x)$ which coincides with $f(x)$ for $|x| \leqq R$ and is zero elsewhere, and then making R tend to infinity and applying Fatou's lemma we obtain (2.8) in the general case.

Thus our problem has been reduced to (2.9). The proof of the latter inequality requires two lemmas.

3. LEMMA 1. *Let $Y_n(y')$ be a spherical function of order n and suppose that $\|Y_n(y')\| = 1$. Then*

$$(3.1) \qquad |Y_n(y')| \leqq An^{(k-2)/2}.$$

The case $k=2$ being trivial, we may assume that $k \geqq 3$. We shall systematically use the notation

$$\lambda = (k-2)/2 \tag{3.2}$$

and shall denote by P_n^λ the ultraspherical (Gegenbauer) polynomials defined by the equation

$$(1 - 2w\cos\gamma + w^2)^{-\lambda} = \sum_{n=0}^{\infty} w^n P_n^\lambda(\cos\gamma). \tag{3.3}$$

Then, for any x with $|x| = 1$ we have

$$Y_n(x) = \frac{\Gamma(\lambda)(n+\lambda)}{2\pi^{\lambda+1}} \int_\Sigma P_n^\lambda(\cos\gamma) Y_n(y')dy' \tag{3.4}$$

where γ is the angle between the vectors x and y' and the integral is zero if we replace $Y_n(y')$ by $Y_m(y')$, with $m \neq n$ (see [4]). By Schwarz's inequality,

$$\begin{aligned} |Y_n(x)| &\leq An \left\{\int_\Sigma [P_n^\lambda(\cos\gamma)]^2 dy'\right\}^{1/2} \left\{\int_\Sigma |Y_n(y')|^2 dy'\right\}^{1/2} \\ &\leq An \left\{\int_\Sigma [P_n^\lambda(\cos\gamma)]^2 dy'\right\}^{1/2}. \end{aligned} \tag{3.5}$$

The value of the last integral may be obtained if for Y_n we take the spherical function P_n^λ $(\cos\delta)$, where δ is the angle of y' with a fixed axis through the origin. Then for x on that axis, $Y_n(x) = P_n^\lambda(1)$, and (3.4) reduces to

$$P_n^\lambda(1) = \frac{\Gamma(\lambda)(n+\lambda)}{2\pi^{\lambda+1}} \int_\Sigma [P_n^\lambda(\cos\gamma)]^2 dy'. \tag{3.6}$$

On the other hand, (3.3) gives

$$\sum_0^\infty P_n^\lambda(1) w^n = (1-w)^{-(k-2)},$$

which shows that $P_n^\lambda(1)$ is exactly of the order n^{k-3} and this in conjunction with (3.5) and (3.6) proves the lemma.

Let us now consider the Fourier transform of $g_{n,\epsilon}$. We have

$$\begin{aligned} \hat{g}_{n,\epsilon}(x) &= (2\pi)^{-k/2} \int_{|y|\geq\epsilon} e^{i(x,y)} \frac{Y_n(y')}{|y|^k} dy \\ &= (2\pi)^{-k/2} \int_{|y|\geq\epsilon} e^{ir\rho\cos\gamma} \frac{Y_n(y')}{|y|^k} dy, \end{aligned} \tag{3.7}$$

where

$$r = |x|, \qquad \rho = |y|, \qquad r\rho\cos\gamma = (x, y).$$

The last integral is defined as the limit for $R\to\infty$ of

$$(3.8)\qquad \int_{\epsilon\leq\rho\leq R} e^{ir\rho\cos\gamma}\frac{Y_n(y')}{|y|^k}dy = \int_\epsilon^R \frac{d\rho}{\rho}\int_\Sigma e^{ir\rho\cos\gamma}Y_n(y')dy'$$
$$= \int_{\epsilon r}^{Rr}\frac{d\rho}{\rho}\int_\Sigma e^{i\rho\cos\gamma}Y_n(y')dy'.$$

Let us first consider the case $k\geqq 3$ and use the expansion

$$e^{i\rho\cos\gamma} = 2^\lambda\Gamma(\lambda)\sum_{m=0}^\infty (m+\lambda)i^m\frac{J_{m+\lambda}(\rho)}{\rho^\lambda}P_m^\lambda(\cos\gamma)$$

(see [9, p. 368], J_k is here Bessel's function of order k) which converges absolutely and uniformly for ρ remaining within any finite interval. On account of (3.4), the last integral (3.8) reduces, except for a multiplicative constant depending on λ only, to

$$Y_n(x)\int_{\epsilon r}^{\epsilon R}\frac{J_{n+\lambda}(\rho)}{\rho^{1+\lambda}}d\rho \to Y_n(x)\int_{\epsilon r}^\infty \frac{J_{n+\lambda}(\rho)}{\rho^{1+\lambda}}d\rho \qquad (as\ R\to\infty).$$

Since, by Lemma 1, $|Y_n|\leqq An^\lambda$, the inequality (2.9) will follow if we prove the following

LEMMA 2.

$$(3.9)\qquad \left|\int_h^\infty \frac{J_{n+\lambda}(\rho)}{\rho^{1+\lambda}}d\rho\right| \leqq \frac{A}{n^{1+\lambda}}, \qquad for\ 0<h<\infty;\ n=1,2,\cdots.$$

Let us revert for a moment to the case $k=2$. We may then set

$$x = re^{i\theta},\qquad y=\rho e^{i\phi},\qquad Y_n(y') = \alpha e^{in\phi}+\bar\alpha e^{-in\phi}.$$

The inner integral on the right of (3.8) reduces then to

$$\int_0^{2\pi} e^{ir\rho\cos(\phi-\theta)}(\alpha e^{in\phi}+\bar\alpha e^{-in\phi})d\phi = 2\pi J_n(r\rho)i^n(\alpha e^{in\theta}+\bar\alpha e^{-in\theta}),$$

and we are again led to the inequality (3.9) with $\lambda=0$, as it should be.

We could not find Lemma 2 in literature—though the formula for the integral (3.9) in the case $h=0$ is classical (see [9, p. 391] or [8, p. 182]). We are therefore forced to give here a proof of it which is a straightforward adaptation of a proof communicated to us by Professor Szegö for the case of $\lambda=0$. This proof is long and possibly could be simplified. We postpone it to the last section of the paper. Taking Lemma 2 temporarily for granted we may consider Theorem 1 as proved.

In view of the complicated character of the proof of Lemma 2 the following remark may be of some interest. Suppose that the function f is differentiable sufficiently many times and vanishes outside a compact set. Then the integral defining $\tilde{f}(x)$ is obviously convergent at every point x. Moreover the relation (2.4) will hold for $\epsilon=0$. For such functions f we have the inequality

$$\|\tilde{f}\| \leqq A\|f\|, \tag{3.10}$$

provided we have (3.9) with $h=0$, a result which, as we have already observed, is well known. Thus the operator $\tilde{f}$ can be defined directly for functions f forming a set dense in L^2 and for these functions we have (3.10) with A independent of f. Such an operator can be extended to the whole L^2 with preservation of (3.10). This kind of extension is systematically used by Mihlin [5; 6]. Of course using this argument we lose the fact that $\tilde{f}_\epsilon$ converges to $\tilde{f}$ in the metric L^2 for every f in L^2, as $\epsilon\to 0$.

Let us finally make two remarks of a chronological type. 1° Developments of the numerator Ω into series of spherical harmonics was already considered by Giraud [3] (see also Mihlin, loc. cit.). 2° The computation of the Fourier transform of the function $\tilde{g}_{n,\epsilon}$ contains implicitly the important fact that the Fourier transform of $Y_n(y')|y|^{-k}$ is, apart from a numerical factor, the harmonics $Y_n(y')$ itself. This fact was not unknown in the case $k=2$ but for higher values of k seems to have been first published by Bochner [1] whose argument is used in our proof. (For $k=3$ the result was also obtained—independently and almost simultaneously—by Professor G. Szegö, who communicated it to us in a letter. His proof was not published.) Bochner sums the Fourier transforms by the method of Abel, but since we know that the transform actually converges [2, p. 89], this point is irrelevant.

4. The convergence of the series in (2.7) follows from the inequality (2.8) and the convergence of the series $\sum n^{-2}$. The latter fact is rather crude and we have here considerable leeway which might conceivably be used to strengthen Theorem 1. Using instead of (2.6) the inequality

$$|\tilde{f}_\epsilon(x)|^2 \leqq \left(\sum_{n=1}^{\infty} |a_n(x)|^2 n^{-1+\delta}\right)\left(\sum_{n=1}^{\infty} |\tilde{f}_{n,\epsilon}(x)|^2 n^{1-\delta}\right),$$

and applying (2.8) we see that the conclusion of Theorem 1 holds if instead of the boundedness of the function $\sum|a_n(x)|^2$—which is equivalent to condition (1.7)—we assume the boundedness of $\sum|a_n(x)|^2 n^{-1+\delta}$. This will lead us to the following result.

THEOREM 2. *The conclusion of Theorem* 1 *holds if condition* (1.7) *is replaced by*

$$\int_\Sigma |\Omega(x, z')|^p dz' < A \tag{4.1}$$

for any

$$(4.2) \qquad p > 2\frac{k-1}{k}\cdot$$

Thus for $k=2$ any $p>1$ will do. However, as k increases indefinitely the expression $2(k-1)/k$ tends to 2 and the exponent 2 of Theorem 1 is the best one valid for all k. In the last section of this paper (see Theorem 3 below) we shall show that for no k can we replace the exponent $2(k-1)/k$ of Theorem 2 by a smaller one. On the other hand, we do know (see [2, p. 91]) that in the special case of $\Omega=\Omega(z')$ condition (4.1) can be replaced by a weaker one, namely

$$(4.3) \qquad \int_{\Sigma} |\Omega(z')| \log^{+} |\Omega(z')| dz' < \infty$$

for all k. We shall also show later that the proofs of Theorems 1 and 2 give slightly more than actually stated in the theorems.

Let us recall some familiar facts about the polynomials P_n^λ (for all this see, for example, [7, pp. 80 sqq.]). We have

$$(4.4) \quad (1-x^2)^{\lambda-1/2}P_n^\lambda(x) = \frac{(-2)^n}{n!}\frac{\Gamma(n+\lambda)\Gamma(n+2\lambda)}{\Gamma(\lambda)\Gamma(2n+2\lambda)}\frac{d^n}{dx^n}(1-x^2)^{n+\lambda-1/2}.$$

Functions $g(x)$ defined on the segment $-1\leqq x\leqq 1$ can be developed into Fourier series

$$g(x) \sim \sum_0^\infty c_n P_n^\lambda(x)$$

where

$$(4.5) \quad c_n = \frac{2^{2\lambda-1}}{\pi}\Gamma^2(\lambda)\frac{(n+\lambda)\Gamma(n+1)}{\Gamma(n+2\lambda)}\int_{-1}^{+1} g(x)(1-x^2)^{\lambda-1/2}P_n^\lambda(x)dx.$$

We shall, in particular, consider the Fourier series of the function

$$(4.6) \qquad g(x) = (1-x)^{-\alpha}$$

where the number α is positive and will be fixed presently. Clearly, we must have

$$\alpha < \lambda + 1/2.$$

Using the formulas (4.5) and (4.4) and applying repeated integration by parts we find (the computation is simple and is omitted here) that c_n is exactly of the order $n^{2\alpha-2\lambda}$.

Let us now consider on Σ the function

$$(4.7) \qquad G = (1-\cos\theta)^{-\alpha}$$

where θ is the angle with the polar axis. The function G belongs to the class L^q on Σ if and only if the integral

$$\int_0^\pi \frac{\sin^{k-2}\theta}{(1-\cos\theta)^{\alpha q}}\,d\theta$$

is finite, that is, for

$$q < (k-1)/2\alpha. \tag{4.8}$$

The development of G into spherical harmonics is obtained by replacing x in the Fourier series $\sum_0^\infty c_n P_n^\lambda(x)$ by $\cos\theta$, i.e. is

$$\sum c_n P_n^\lambda(\cos\theta), \quad \text{with} \quad |c_n| \sim n^{2\alpha-2\lambda}.$$

Let us now consider any function $F(x')$ integrable on Σ and an arbitrary function $G(x')$ also integrable on Σ which is however a function of the polar angle θ only. The integral

$$H(x') = \frac{1}{|\Sigma|}\int_\Sigma F(y')G(\cos\gamma)dy', \tag{4.9}$$

where γ denotes the angle between the vectors x' and y', is a sort of convolution of the functions F and G; we may call it the *spherical convolution*. It has many properties of the usual convolution in E^k. In particular, the familiar inequality of W. H. Young

$$\|H\|_t \leqq \|F\|_p \|G\|_q \qquad \left(p, q, t \geqq 1; \frac{1}{t} = \frac{1}{p} + \frac{1}{q} - 1\right)$$

remains valid here, with proof unchanged.

If the developments of F and G into series of spherical harmonics are

$$\sum a_n Y_n(x') \quad \text{and} \quad \sum c_n P_n^\lambda(\cos\theta), \tag{4.10}$$

respectively, then the spherical development of H is

$$\frac{2\pi^{\lambda+1}}{|\Sigma|\,\Gamma(\lambda)} \sum_n a_n c_n \frac{1}{n+\lambda} Y_n(x'). \tag{4.11}$$

The formal proof immediately follows from (4.9) and (3.4). This formal proof is perfectly rigorous if one at least of the developments (4.10) converges absolutely and uniformly. In particular, if we denote by $F_r(x')$ the Abel-Poisson means of the spherical development of $F(x')$ $(0 \leqq r < 1)$, the spherical composition of F_r and G is given by the series (4.11) with a_n replaced by $a_n r^n$. But as is easily seen from the representation of F_r as a Poisson integral of F, we have, on Σ,

$$\|F - F_r\|_1 \to 0 \quad \text{as} \quad r \to 1,$$

and so, by Young's inequality with $p=q=1$, the spherical composition of F_r and G tends in norm L to the spherical composition of F and G, which immediately proves that (4.11) is the development of H into spherical harmonics.

Suppose now that for our G we take the function (4.7). Then the cofactor of $a_n Y_n$ in (4.11) is exactly of the order $n^{2\alpha-2\lambda-1}$. If $2\alpha-2\lambda-1$ exceeds $-1/2$, i.e. if

$$2\alpha > k - 3/2, \tag{4.12}$$

and if the function H is quadratically integrable over Σ, then the series

$$\sum |a_n|^2 n^{-1+\delta} \tag{4.13}$$

converges for some $\delta>0$. Thus the last series will converge provided $F\in L^p$ and provided we can find a $q>1$ and a number $\alpha>0$ such that the conditions (4.8) and $1/2 \geq p^{-1}+q^{-1}-1$ are satisfied. It is easily seen that the last two conditions can be satisfied if we have (4.2).

Summarizing, for any function $F(z') \sim \sum a_n Y_n(z')$ on Σ and of the class L^p, with p satisfying (4.2), the series (4.13) converges for some $\delta>0$. If for $F(z')$ we take $\Omega(x, z')$, with x fixed, and if we consider the development (2.1), then the assumptions (4.1) and (4.2) will imply the uniform boundedness of $\sum |a_n(x)|^2 n^{-1+\delta}$ for some $\delta>0$ and the proof of Theorem 2 is completed.

In the above proof we implicitly assumed that $k>2$. For $k=2$ the proof is analogous if we take $g=(1-e^{i\theta})^{-\alpha}$. In this case we could also appeal to a well known result of Hardy and Littlewood asserting that if $F\in L^p$, then the fractional integral of order β belongs to the class L^t with t defined by the equation $1/t=1/p-\beta$, so that again we would have the convergence of (4.13) for some $\delta>0$ provided $F\in L^p$, $p>1$. The proof given previously is of course more elementary.

5. **Remarks.** 1° We have mentioned above that from the proofs of Theorems 1 and 2 we can deduce slightly more than actually stated. For let $\epsilon_1, \epsilon_2, \cdots$ be a sequence of positive variables and let us replace on the right the factors $\tilde{f}_{n,\epsilon}$ by $\tilde{f}_{n,\epsilon_n}$. Let the resulting sum be denoted by $\tilde{f}_{\epsilon_1\epsilon_2\cdots\epsilon_n\cdots}$ so that

$$\tilde{f}_{\epsilon_1\epsilon_2\cdots\epsilon_n\cdots} = \sum_1^\infty a_n(x)\tilde{f}_{n,\epsilon_n}(x).$$

Then the proof of Theorem 1 shows that under its assumption we have

$$\|\tilde{f}_{\epsilon_1\epsilon_2\cdots\epsilon_n\cdots}\| \leqq A \tag{5.1}$$

and

$$\|\tilde{f}_{\epsilon_1\epsilon_2\cdots\epsilon_n\cdots} - \tilde{f}\| \to 0 \tag{5.2}$$

if each individual ϵ_n tends to zero.

2° Let us consider for a moment the case $k=2$ and suppose that the function Ω depends on z' only, i.e. is a function of an angle θ, $0\leqq\theta\leqq 2\pi$. In this case the equation (2.1) takes the form

$$\Omega(\theta) \sim \sum_{-\infty}^{+\infty}{}' a_n e^{in\theta},$$

where the a_n are constants independent of x. The inequality (2.8) becomes

$$\|\tilde{f}_{n,\epsilon}\| \leqq \frac{A}{|n|}\|f\| \qquad (n = \pm 1, \pm 2, \cdots).$$

Let us now assume that the function $\Omega(\theta)\log^+|\Omega(\theta)|$ is integrable over $(0, 2\pi)$. It is known that then the series $\sum'|a_n n^{-1}|$ converges absolutely and

$$\text{(5.3)} \qquad \sum_{-\infty}^{+\infty}{}' \left|\frac{a_n}{n}\right| \leqq A\int_0^{2\pi} |\Omega| \log^+ |\Omega|\, d\theta + A$$

(see [10, p. 235, Ex. 5]) and from 2.4—or rather its analogue in the case $k=2$—we obtain the inequalities (1.8) and (1.9).

We obtained these inequalities under the assumption that $\Omega\log^+|\Omega|$ is integrable over $0\leqq\theta\leqq 2\pi$. Of course, the result is not new, but the present argument shows that the generalizations (5.1) and (5.2) are valid in the case $k=2$ if $\Omega\log^+|\Omega|$ is integrable. The argument is not extensible to higher values of k since, though the inequality (5.3) is extensible to general uniformly bounded orthonormal systems (loc. cit.), the condition of boundedness is essential here and the orthonormal systems $\{Y_n(z')\}$ we come across when $k\geqq 3$ are no longer uniformly bounded.

6. We shall now prove Lemma 2. As we have observed, the inequality (3.9) is certainly true if $h=0$, so that Lemma 2 is equivalent to the inequality

$$\text{(6.1)} \qquad \left|\int_0^h \frac{J_{n+\lambda}(\rho)}{\rho^{1+\lambda}}\, d\rho\right| \leqq \frac{A}{n^{1+\lambda}} \qquad (n = 1, 2, \cdots; h \geqq 0).$$

Let us write $\nu=n+\lambda$. We shall consider four special cases, namely,

1° $0 \leqq h \leqq \nu/2$, 2° $\nu/2 \leqq h \leqq \nu$, 3° $\nu \leqq h \leqq 2\nu$, 4° $h \geqq 2\nu$.

In case 1°, the classical formula (see e.g. [9, p. 48])

$$|J_\nu(\rho)| \leqq \left|\frac{(\rho/2)^\nu}{\Gamma(\nu + 1/2)\Gamma(1/2)}\int_{-1}^{+1} (1 - t^2)^{\nu-1/2} e^{i\nu t}dt\right| \leqq \frac{A(\rho/2)^\nu}{(\nu - 1)!}$$

coupled with Stirling's formula for $(\nu-1)!$ shows that the integral in (6.1) is uniformly $\leqq Aq^\nu$, where q is a positive number less than 1, and (6.1) is surely true.

In case 2° we use the formula [9, p. 257, (6)]

$$\int_0^{\nu} J_\nu(\rho)d\rho \simeq \frac{1}{3},$$

and since $J_\nu(\rho)$ is positive for $0 \leqq \gamma \leqq \rho$, we have

$$\int_{\nu/2}^{h} \frac{J_\nu(\rho)}{\rho^{\lambda+1}} d\rho \leqq A\nu^{-\lambda-1} \int_{\nu/2}^{\nu} J_\nu(\rho)d\rho \leqq A\nu^{-\lambda-1} \leqq An^{-\lambda-1},$$

which in conjunction with case 1° again gives (6.1).

In case 4° we use the differential equation of J_ν which may be written

$$\frac{J_\nu(\rho)}{\rho^{\lambda+1}} = -\frac{J_\nu'(\rho)}{\rho^{\lambda}(\rho^2-\nu^2)} - \frac{J_\nu''(\rho)}{\rho^{\lambda-1}(\rho^2-\nu^2)}.$$

Let us integrate this over $h \leqq \rho < \infty$. Since $|J_\nu'(\rho)| \leqq 1$, the first term on the right is, numerically, $\leqq A\rho^{-2-\lambda}$ and its integral is $\leqq Ah^{-\lambda-1} \leqq A\nu^{-\lambda-1}$. To the integral of the second term we apply the second mean-value theorem and remove the decreasing factor $[\rho^{\lambda-1}(\rho^2-\nu^2)]^{-1}$, which shows again that the integral is $\leqq A\nu^{-\lambda-1}$. This proves (3.9) in case 4°.

It remains to prove (6.1) in case 3°. Let $\nu \leqq h \leqq 2\nu$. The second mean-value theorem gives

$$\int_{\nu}^{h} \frac{J_\nu(\rho)}{\rho^{\lambda+1}} d\rho = \nu^{-\lambda-1} \int_{\nu}^{h'} J_\nu(\rho)d\rho \qquad (\nu < h' < 2\nu)$$

which indicates that it is enough to prove the boundedness of the last integral. We write this condition in the form

$$\int_1^{\xi} J_\nu(\nu\rho)d\rho = O\left(\frac{1}{\nu}\right) \qquad (1 < \xi < 2). \tag{6.2}$$

This part of the argument is the least simple. We set $\rho = \sec\beta$, and use Watson's formula [9, p. 252, (5)] valid in the "transitional region":

$$\begin{aligned} J_\nu(\nu\sec\beta) &= (1/3)\tan\beta\cdot\cos\nu B[J_{-1/3}(t) + J_{1/3}(t)] \\ &\quad + 3^{-1/2}\tan\beta\cdot\sin\nu B[J_{-1/3}(t) - J_{1/3}(t)] + O(1/\nu), \end{aligned}$$

where

$$B = \tan\beta - (1/3)\tan^3\beta - \beta, \qquad t = (1/3)\nu\tan^3\beta,$$

and the "O" is an absolute one, provided $0 \leqq \beta \leqq \beta_0$, β_0 being any fixed constant (in our case $\sec\beta_0 = 2$).

If we drop the term $O(1/\nu)$, we obtain an approximate formula for J_ν and the error committed in the integral (6.2) will be $O(1/\nu)$, a quantity unimportant for our purposes.

First let us suppose that $t\leqq 1$ in the interval $1\leqq\rho\leqq\xi$. Then $J_{-1/3}\pm J_{1/3} = O(t^{-1/3})$ and the left side of (6.2) is, numerically,

$$\leqq \int_{t\leqq 1} |J_\nu(\nu\sec\beta)| \frac{\sin\beta}{\cos^2\beta} d\beta \leqq A\int_{t\leqq 1} \tan^2\beta t^{-1/3} d\beta \leqq A\nu^{-1/3}\int_{t\leqq 1}\tan\beta d\beta$$

$$\leqq A\nu^{-1/3}\int_{t\leqq 1}\beta d\beta \leqq A\nu^{-1/3}\cdot\nu^{-2/3} = A\nu^{-1}.$$

Second, let $t>1$. Then

$$J_{\mp 1/3}(t) = \left(\frac{2}{\pi t}\right)^{1/2}\cos\left(t\pm\frac{\pi}{6}-\frac{\pi}{4}\right)+O(t^{-3/2}),$$

$$J_{-1/3}+J_{1/3} = 3^{1/2}\left(\frac{2}{\pi t}\right)^{1/2}\cos\left(t-\frac{\pi}{4}\right)+O(t^{-3/2}),$$

$$J_{-1/3}-J_{1/3} = -\left(\frac{2}{\pi t}\right)^{1/2}\sin\left(t-\frac{\pi}{4}\right)+O(t^{-3/2}),$$

$$J_\nu(\nu\sec\beta) = 3^{-1/2}\tan\beta\left(\frac{2}{\pi t}\right)^{1/2}\cos\left(\nu B+t-\frac{\pi}{4}\right)+O(\tan\beta\cdot t^{-3/2})$$

$$= 3^{-1/2}\tan\beta\left(\frac{2}{\pi t}\right)^{1/2}\cos\left[\nu(\tan\beta-\beta)-\frac{\pi}{4}\right]+O(\beta t^{-3/2}).$$

The contribution of the "O" term here to that part of the integral (6.2) which corresponds to $t\geqq 1$ does not exceed

$$A\int_{t\geqq 1,\,\beta\leqq\beta_0}\beta^2 t^{-3/2}dt \leqq A\nu^{-3/2}\int_{t\geqq 1}\beta^{-5/2}d\beta \leqq A\nu^{-3/2}\cdot A\nu^{1/2} = A\nu^{-1},$$

and it remains to consider the integral (in which $\beta_1\leqq\beta_0$)

$$\int_{t\geqq 1,\,\beta\leqq\beta_1}\tan\beta\cdot t^{-1/2}\cdot\cos\left[\nu(\tan\beta-\beta)-\frac{\pi}{4}\right]\tan\beta\sec\beta d\beta$$

$$= \int A\nu^{-1/2}\tan^{1/2}\beta\sec\beta\cos\left[\nu(\tan\beta-\beta)-\frac{\pi}{4}\right]d\beta.$$

Let us now introduce a new variable $x=\tan\beta-\beta$ which is an increasing function of β. Clearly, for $\beta\to 0$ we have

$$x\simeq\frac{1}{3}\beta^3,\qquad \tan^{1/2}\beta\sec\beta\simeq Ax^{1/6}\qquad \frac{d\beta}{dx}\simeq Ax^{-2/3}$$

and the ratios of both sides in the last two equivalences are monotone functions. Therefore, applying the second mean-value theorem we see that the last integral is

$$A\nu^{-1/2}\int x^{-1/2}\cos\left(\nu x-\frac{\pi}{4}\right)dx$$

where the integral is extended over some interval of the positive real axis. Making the substitution $\nu x=y$ and observing that $\left|\int y^{-1/2}\cos(y-\pi/4)dy\right| \leqq A$, we see that (6.4) is absolutely $A\nu^{-1}$, and the proof of Lemma 2 is complete.

7. THEOREM 3. *If in the inequality* (4.1) *we take* $p=2(k-1)/k$, *the transform* $\tilde{f}$ *of an* $f\in L^2$ *need not be in* L^2.

For let us take for $f(y)$ the function equal to 1 for $|y|\leqq 1$ and equal to zero elsewhere. It belongs to L^2 (to any L^q, $q>0$). Let p be any positive number.

Let us assume that $\Omega(x, z')=0$ for $|x|\leqq 2$. For $|x|>2$ we define $\Omega(x, z')$ as follows:

1° It is equal to $|x|^{(k-1)/r}$ for the points y of each ray from x intersecting Σ;

2° It is equal to $-|x|^{(k-1)/r}$ for the points y of the rays opposite to those in 1°;

3° It is equal to zero on all other rays from x.

Then

$$\int_\Sigma \Omega(x, z')dz' = 0, \qquad \int_\Sigma |\Omega(x, z')|^p dz' < A,$$

It is not difficult to see that for the function f just defined,

$$|\tilde{f}(x)| = \left|\int_\Sigma \Omega(x, z')|z|^{-k}f(y)dy\right| \simeq \frac{C}{|x|^\alpha} \qquad \text{as } |x|\to\infty,$$

where C is a constant and $\alpha=k-(k-1)/p$. If we want $\tilde{f}(x)$ to be in L^2 we must assume that $\alpha>k/2$, or, what is the same thing, $p>2(k-1)/k$. This completes the proof.

If we assume something more about the symmetrical structure of the kernel K, the conclusion of Theorem 2 may be considerably strengthened. To this we shall return in another paper.

Added in Proof. A new and simpler proof of Lemma 2 was found by L. Lorch and P. Szegö and is to appear in Duke Math. J.

REFERENCES

1. S. Bochner, *Theta relations with spherical harmonics*, Proc. Nat. Acad. Sci. U.S.A. vol. 37 (1951) pp. 804–808.

2. A. P. Calderón and A. Zygmund, *On the existence of certain integrals*, Acta Math. vol. 88 (1952) pp. 85–139.

3. G. Giraud, *Sur une classe générale d'équations à intégrales principales*, C. R. Acad. Sci. Paris vol. 202 (1936) pp. 2124–2125.

4. E. Heine, *Handbuch der Kugelfunktionen*, vols. I and II, 2d ed., Berlin, 1878, 1888.
5. S. Mihlin, *Singular integral equations*, Uspehi Mat. Nauk. vol. 3 (1948) pp. 29–112.
6. ———, *Concerning a theorem on the boundedness of a singular integral operator*, Uspehi Mat. Nauk vol. 8 (1953) pp. 213–217.
7. G. Szegö, *Orthogonal polynomials*, Amer. Math. Soc. Colloquium Publications, vol. 23, 1939.
8. E. C. Titchmarsh, *Introduction to the theory of Fourier's integral*, Oxford University Press, 1948.
9. G. N. Watson, *A treatise on the theory of Bessel functions*, Cambridge University Press, 1922.
10. A. Zygmund, *Trigonometrical series*, Warsaw, 1935.

Ohio State University,
Columbus, Ohio.
University of Chicago,
Chicago, Ill.

Reprinted from the
TRANSACTIONS OF THE AMERICAN MATHEMATICAL SOCIETY
Vol. 84, No. 2, pp. 559-560
March, 1957

ADDENDA TO THE PAPER *ON A PROBLEM OF MIHLIN*(1)

BY
A. P. CALDERÓN AND A. ZYGMUND

The proof of the formula (2.8) in the paper quoted is incomplete. Though meanwhile we have proved and generalized Theorem 1 by a different method [1], the initial approach still seems to be indispensable in establishing Theorem 2. The needed addition is not difficult but perhaps not obvious and so we give it here.

In (2.1) the functions $Y_n(z')$ depend on x, that is $Y_n(z') = Y_n(x, z')$ but the proof of (2.8) as it stands in the paper applies only to the case when $Y_n(x, z')$ is independent of x. We indicate now how to complete the argument. We write $Y_n(x, z') = \sum_m \alpha_m(x) Y_{nm}(z')$ where $Y_{nm}(z')$ is a complete orthonormal system of (fixed) spherical harmonics of degree n and $\sum |\alpha_m(x)|^2 = 1$. Then $\tilde{f}_{n\epsilon}(x) = \sum_m \alpha_m(x) \tilde{f}_{nm\epsilon}(x)$ where $\tilde{f}_{nm\epsilon}$ is given by (2.5) with $Y_{nm}(z')$ in the integral instead of $Y_n(z')$, and by Schwarz's inequality it follows that $|\tilde{f}_{n\epsilon}(x)|^2 \leq \sum_m |\tilde{f}_{nm\epsilon}(x)|^2$. Integrating and applying Plancherel's theorem to the righthand side we obtain

$$\|\tilde{f}_{n,\epsilon}\|^2 \leq \int \sum_m |\hat{f}_{nm\epsilon}(x)|^2 dx, \tag{1}$$

where $\hat{f}_{nm\epsilon}$ is the Fourier transform of $\tilde{f}_{nm\epsilon}$. Now $\hat{f}_{nm\epsilon}$ is the product of the Fourier transform $\hat{f}(x)$ of $f(x)$ and the Fourier transform of the kernel equal to

$$(2\pi)^{-k/2} Y_{n,m}(x') \, |x|^{-k}$$

for $|x| > \epsilon$ and zero elsewhere. By Lemma 2 the latter transform is in absolute value less than $An^{-1-\lambda} |Y_{n,m}(x')|$, A being an absolute constant. Thus from (1) we obtain

$$\begin{aligned} \|\tilde{f}_{n,\epsilon}\|^2 &\leq A^2 n^{-2-2\lambda} \int |\hat{f}(x)|^2 \sum_m Y_{nm}(x')^2 dx \\ &\leq A^2 n^{-2-2\lambda} \|\hat{f}\|^2 \max_{x'} \left[\sum_m Y_{nm}(x')^2 \right]. \end{aligned} \tag{2}$$

Now

(1) Transactions of the American Mathematical Society, vol. 78 (1955) pp. 209–224. The reexamination of our paper was prompted by a criticism by Mihlin (*On the theory of multidimensional singular equations*, Bull. of the Leningrad Univ., no. 1, (1956), Series on Math., Mechanics and Astronomy). Although the dependence of $Y_m(z')$ on x was implicitly assumed in the paper it was overlooked in the proof of (2.8).

$$\max_{x'}\left[\sum_m Y_{nm}(x')^2\right] = \max_{\beta,x'}\left[\sum_m \beta_m Y_{nm}(x')\right]^2,$$

where $\sum \beta_m^2 = 1$. Since $\sum_m \beta_m Y_{nm}(x')$ is a normalized spherical harmonic of degree n, according to lemma 1 its absolute value is dominated by Bn^λ, where B is again an absolute constant. Finally since $\|\hat{f}\| = \|f\|$ from (2) and the last estimate we obtain (2.8).

We conclude with one more remark. It is clear from the context of the paper that in the case of Theorem 2 the function $\tilde{f}_\epsilon$ was understood to be defined by (2.4). Actually one can prove that the integral in (1.5) is almost everywhere absolutely convergent and therefore can be used to define $\tilde{f}_\epsilon$ directly. We will return to this and related problems on another occasion.

Reference

1. A. P. Calderón and A. Zygmund, *On singular integrals*, Amer. J. Math. vol. 78 (1956) pp. 289–309.

A NOTE ON THE INTERPOLATION OF SUBLINEAR OPERATIONS.*

By A. P. CALDERÓN and A. ZYGMUND.[1]

The purpose of this note is to give an extension of M. Riesz' interpolation theorem for linear operations to certain non-linear ones.

Let R be a measure space. This means that we have a set function $\mu(E)$, non-negative and countably additive, defined for some ('measurable') subsets E of R. For any measurable (with respect to μ) function f defined on R we write

$$\left(\int_R |f|^r d\mu\right)^{1/r} = \| f \|_{r,\mu} \qquad (0 < r < \infty),$$

and denote by $\| f \|_{\infty,\mu}$ the essential (with respect to μ) upper bound of $|f|$. The set of functions f such that $\| f \|_{r,\mu}$ is finite $(0 < r \leqq \infty)$ is denoted by $L^{r,\mu}$. If no confusion arises, we write $\| f \|_r$, L^r for $\| f \|_{r,\mu}$, $L^{r,\mu}$.

Let R_1 and R_2 be two measure spaces with measures μ and ν respectively. Let $h = Tf$ be a transformation of functions $f = f(u)$ defined (almost everywhere) on R_1 into functions $h = h(v)$ defined on R_2. The most important special case is when T is a *linear operation*. This means that if Tf_1 and Tf_2 are defined, and if α_1, α_2 are complex numbers, then $T(\alpha_1 f_1 + \alpha_2 f_2)$ is defined and

$$T(\alpha_1 f_1 + \alpha_2 f_2) = \alpha_1 Tf_1 + \alpha_2 Tf_2.$$

Let $r > 0$, $s > 0$. A linear operation $h = Tf$ will be said to be *of type* (r, s) if it is defined for each $f \in L^{r,\mu}$ and if

$$\| Tf \|_{s,\nu} \leqq M \| f \|_{r,\mu}, \tag{1}$$

where M is independent of f. The least value of M is called the (r, s) *norm* of the operation.

Denote by (α, β) points of the square

$$0 \leqq \alpha \leqq 1, \qquad 0 \leqq \beta \leqq 1. \tag{Q}$$

* Received September 9, 1955.

[1] The research resulting in this paper was supported in part by the office of Scientific Research of the Air Force under contract AF 18(600)-1111.

Reprinted from *Amer. J. Math.* 78, 282–288 (1956).

The Riesz interpolation theorem (in the form generalized by Thorin (see [1]-[6] of the References at the end of the note) asserts that if a linear operation $h = Tf$ is simultaneously of types $(1/\alpha_1, 1/\beta_1)$ and $(1/\alpha_2, 1/\beta_2)$, with norms M_1 and M_2 respectively, and if

$$\alpha = (1-t)\alpha_1 + t\alpha_2, \quad \beta = (1-t)\beta_1 + t\beta_2, \qquad (0 < t < 1) \tag{2}$$

then T is also of type $(1/\alpha, 1/\beta)$, with norm

$$M \leqq M_1^{1-t} M_2^{t}. \tag{3}$$

The significance of this theorem is by now widely recognized, and its applications are many. Riesz himself deduced the result, through appropriate passages to limits, from a theorem about bilinear forms, and in this argument the linearity of T plays an important role. The same can be said of other proofs. There are however a number of interesting operations which are not linear and to which therefore the theorem cannot be applied. For the sake of illustration we mention one of them, first considered by Littlewood and Paley (see [7]), which has important application in Fourier series.

Given any $f \varepsilon L(0, 2\pi)$, we consider the function $F(z)$ regular for $|z| < 1$, whose real part is the Poisson integral of f, and imaginary part is zero at the origin. The Littlewood-Paley function is

$$g(\theta) = \{ \int_0^1 (1-\rho) | F'(\rho e^{i\theta}) |^2 d\rho \}^{\frac{1}{2}}.$$

The operation $g = Tf$ is clearly not linear. It satisfies, however, the following relations

$$| T(f_1 + f_2) | \leqq | Tf_1 | + | Tf_2 |, \tag{4}$$

$$| T(kf) | = | k | \, | Tf |, \tag{5}$$

for any constant k.

There are other interesting non-linear operations which have the same properties and it may be of interest to study the problem of interpolation of such operations. This is the object of this note.

We begin with general definitions.

We call an operation $h = Tf$ *sublinear*, if the following conditions are satisfied:

(i) Tf is defined (uniquely) if $f = f_1 + f_2$, and Tf_1 and Tf_2 are defined;

(ii) For any constant k, $T(kf)$ is defined if Tf is defined;

(iii) Conditions (4) and (5) hold.

In view of (5) we may, as in the linear case, consider inequalities (1) and introduce the notions of the *type* and *norm* of a sublinear operation. In what follows, the functions f will be defined (almost everywhere) on a space R_1 with measure μ, and the $h = Tf$ on a space R_2 with measure ν.

THEOREM. *Let (α_1, β_1) and (α_2, β_2) be any two points of the square Q. Suppose that a sublinear operation $h = Tf$ is simultaneously of types $(1/\alpha_1, 1/\beta_1)$ and $(1/\alpha_2, 1/\beta_2)$ with norms M_1 and M_2 respectively. Let (α, β) be given by* (2). *Then T is also of type $(1/\alpha, 1/\beta)$, with norm M satisfying* (3).

Proof. We easily deduce from conditions (i), (ii), (iii) that, if Tf_1, $Tf_2, \cdots, Tf_n$ are defined so is $T\{n^{-1}(f_1 + \cdots + f_n)\}$ and

$$(6) \qquad | T\{(f_1 + f_2 + \cdots + f_n)/n\} | \leqq n^{-1}(| Tf_1 | + \cdots + | Tf_n |).$$

We may suppose that $\alpha_1 \leqq \alpha_2$. Thus

$$(7) \qquad \alpha_1 \leqq \alpha \leqq \alpha_2.$$

Consider any f in $L^{1/\alpha, \mu}$ and write $f = f_1 + f_2$, where f_1 equals f at the points at which $|f| \leqq 1$, and equals 0 elsewhere. By (7),

$$| f_1 |^{1/\alpha_1} \leqq | f_1 |^{1/\alpha} \leqq | f |^{1/\alpha},$$

so that $f_1 \varepsilon L^{1/\alpha_1}$ and Tf_1 is defined, by hypothesis. Similarly $f_2 \varepsilon L^{1/\alpha_2}$ and Tf_2 is defined. It follows from (i) that $Tf = T(f_1 + f_2)$ is defined. Our task is to show that the $(1/\alpha, 1/\beta)$ norm M of T is finite and satisfies (3).

We assume for the time being that $\alpha > 0$, $\beta < 1$. Take any $f \varepsilon L^{1/\alpha}$. Without loss of generality we may suppose that

$$\| f \|_{1/\alpha} = 1.$$

Clearly

$$(8) \qquad \| Tf \|_{1/\beta} = \sup_g \int_{R_2} | Tf | \cdot g d\nu,$$

where g is non-negative and satisfies $\| g \|_{1/(1-\beta)} = 1$. We may confine our attention to functions g which are simple (a function is called *simple* if it takes only a finite number of values and is distinct from 0 on a set of finite measure; simple functions are dense in every L^s, $0 < s < \infty$; and in L^∞, if the space has finite measure). We make one more assumption, of which we shall free ourselves later, namely that f is also simple. We fix f and g and consider the integral

$$(9) \qquad I = \int_{R_2} | Tf | \cdot g d\nu.$$

Let $c_1, c_2, \cdots, c_m$ be the distinct from 0 (and different from each other) values of f. Let E_k be the set in which $f = c_k$, and let $\chi_k = \chi_k(u)$ be the characteristic function of E_k. Similarly let $c'_1, c'_2, \cdots, c'_n$ be the different from 0 values of g, E'_l the set where $g = c'_l$, and $\chi_l = \chi_l(v)$ the characteristic function of E'_l. Hence $f = \sum | c_k | \epsilon_k \chi_k$, $g = \sum c'_l \chi'_l$, where $| \epsilon_k | = 1$ and $c'_l > 0$.

Let $\alpha(z)$ and $\beta(z)$ be the right sides of (2), with z for t. Consider the non-negative function

$$(10) \qquad \Phi(z) = \int_{R_2} | T(| f |^{\alpha(z)/\alpha} \operatorname{sign} f) | \, g^{(1-\beta(z))/(1-\beta)} \, d\nu \qquad (z = x + iy),$$

which reduces to I for $z = t$. We show that $\Phi(z)$ is continuous and $\log \Phi(z)$ is subharmonic, in the whole plane.

Since, for each z, $| f |^{\alpha(z)/\alpha} \operatorname{sign} f$ is simple, and so is in L^{1/α_1}, the function $T(| f |^{\alpha(t)/\alpha} \operatorname{sign} f)$ is in L^{1/β_1}, and in particular is integrable over the set where $g > 0$. Hence $\Phi(z)$ exists for each z.

We have

$$\begin{aligned}(11) \qquad \Phi(z) &= \int_{R_2} | T\{\sum | c_k |^{\alpha(z)/\alpha} \epsilon_k \chi_k\} | \, \{\sum c'_l{}^{(1-\beta(z))(1-\beta)} \chi_l\} \, d\nu \\ &= \sum \int_{E'_l} c'_l{}^{(1-\beta(z))/(1-\beta)} \, | T\{ \sum | c_k |^{\alpha(z)/\alpha} \epsilon_k \chi_k\} | \, d\nu \\ &= \sum \int_{E'_l} T\{c'_l{}^{(1-\beta(z))/(1-\beta)} \sum | c_k |^{\alpha(z)/\alpha} \epsilon_k \chi_k\} | \, d\nu,\end{aligned}$$

and it is enough to show that each integral of the last sum is continuous and its logarithm is subharmonic. In proving this we shall make repeated use of the inequality $| \, | Tf_1 | - | Tf_2 | \, | \leqq | T(f_1 - f_2) |$, which is a consequence of (4).

We therefore fix l and write

$$\psi_z = \sum_k c'_l{}^{(1-\beta(z))/(1-\beta)} \, | c_k |^{\alpha(z)/\alpha} \epsilon_k \chi_k, \qquad \Psi(z) = \int_{E'_l} | T\psi_z | \, d\nu.$$

Clearly

$$\begin{aligned}| \Psi(z + \Delta z) - \Psi(z) | &\leqq \int_{E'_l} | T(\psi_{z+\Delta z} - \psi_z) | \, d\nu \\ &\leqq \| T(\psi_{z+\Delta z} - \psi_z) \|_{1/\beta_1} \{\nu(E'_l)\}^{1-\beta_1} \\ &\leqq M_1 \| \psi_{z+\Delta z} - \psi_z \|_{1/\alpha_1} \{\nu(E'_l)\}^{1-\beta_1},\end{aligned}$$

and since $\psi_{z+\Delta z} - \psi_z$ is zero outside $\cup E_k$ and tends to 0, uniformly in u, as $\Delta z \to 0$, the norm $\| \psi_{z+\Delta z} - \psi_z \|_{1/\alpha_1}$ tends to 0, and Ψ is continuous at z. Hence Φ is continuous.

It is very well known that $\log \Psi(z)$ is subharmonic if and only if $\Psi(z)e^{h(z)}$ is subharmonic for every harmonic $h(z)$. We fix a harmonic function $h(z)$, and denote by $H(z)$ the analytic function whose real part is $h(z)$. Since the problem is local, we may consider h and H in a given circle. Write

$$\psi^*_z = \psi_z e^{H(z)}, \qquad \Psi^*(z) = \Psi(z)e^{h(z)} = \int_{E'_l} | T\psi^*_z | \, d\nu.$$

We fix z, take a $\rho > 0$, and denote by $z_1, z_2, \cdots, z_p$ a system of points equally spaced over the circumference of the circle with center z and radius ρ. We have

$$\psi^*_z(u) = \lim_{p\to\infty} 1/p \sum_{j=1}^{p} \psi^*_{z_j}(u),$$

uniformly in u. Since

$$\int_{E'_l} | T(\psi^*_z - 1/p \sum_1^p \psi_{z_j}) | \, d\nu \leqq \| T(\psi^*_z - 1/p \sum_1^p \psi^*_{z_j} \|_{1/\beta_1} \{\nu(E'_l)\}^{1-\beta_1}$$

$$\leqq M_1 \| \psi^*_z - 1/p \sum_1^p \psi^*_{z_j} \|_{1/\alpha_1} \{\nu(E'_l\}^{1-\beta_1}$$

the left side tends to 0 as $p \to \infty$. In particular, as $p \to \infty$ we have

$$\delta_p = \int_{E'_l} | \, | T\psi^*_z | - | T(1/p \sum_1^p \psi^*_{z_j}) | \, | \, d\nu \to 0,$$

$$\int_{E'_l} | T\psi^*_z | \, d\nu \leqq \delta_p + 1/p \sum_1^p \int_{E'_l} | T\psi^*_{z_j} | \, d\nu,$$

$$\Psi^*(z) \leqq \lim_{p\to\infty} 1/p \sum_1^p \Psi^*(z_j) = 1/(2\pi) \int_0^{2\pi} \Psi^*(z + \rho e^{it}) dt.$$

Hence $\Psi^*(z)$ is subharmonic.

We have therefore proved that $\Phi(z)$ is continuous in the whole plane and its logarithm is subharmonic. Moreover $\Phi(z)$ is bounded in every vertical strip of the plane, since from (11), (4) and (5) we deduce that

$$\Phi(z) \leqq \sum_{k,l} | c_k |^{\alpha(z)/\alpha} c'_l{}^{(1-\beta(x))/(1-\beta)} \int_{E'_l} | T(\chi_k) | \, d\nu.$$

Next we show that $\Phi \leqq M_1$ on the line $x = 0$, and $\Phi \leqq M_2$ on $x = 1$. It is enough to prove the first inequality. If $x = 0$, then

$$\Phi(z) \leqq \| T(| f |^{\alpha(z)/\alpha} \operatorname{sign} f) \|_{1/\beta_1} \| g^{(1-\beta_1)/(1-\beta)} \|_{1/(1-\beta_1)}$$

$$\leqq M_1 \| \, | f |^{\alpha(z)/\alpha} \operatorname{sign} f \|_{1/\alpha_1} \leqq M_1 \| \, | f |^{\alpha_1/\alpha} \|_{1/\alpha_1} = M_1.$$

Since $\log \Phi(z)$ is bounded above and subharmonic in the strip $0 \leqq x \leqq 1$, and does not exceed $\log M_1$ and $\log M_2$ on the lines $x=0$ and $x=1$ respectively, an application of the Three-Line Theorem for subharmonic functions shows that

$$\log \Phi(z) \leqq (1-t)\log M_1 + t \log M_2$$

on the line $x=t$ and, in particular, $I=\Phi(t) \leqq M_1^{1-t}M_2^{t}$.

Summarizing results, we have proved that

$$\| Tf \|_{1/\beta} \leqq M_1^{1-t} M_2^{t} \| f \|_{1/\alpha} \tag{12}$$

for each simple f. We show that this holds for every $f \varepsilon L^{1/\alpha}$.

We fix such an f, and for each $m=1,2,\cdots$ consider the decomposition $f=f'_m+f''_m$, in which $f'_m=f$ wherever $|f|\leqq m$, and $f'_m=0$ elsewhere; hence $|f''_m|$ is either 0 or else greater than m. Let f_m be a simple function equal to 0 wherever $f'_m=0$ and such that $|f_m-f'_m|<1/m$ everywhere. Then

$$||Tf|-|Tf_m|| \leqq |T(f-f_m)| \leqq |T(f'_m-f_m)| + |Tf''_m|. \tag{13}$$

If we show that each term on the right tends to 0 almost everywhere as m tends to $+\infty$ through a sequence of values, then the inequality (12) with f_m for f, will lead, by Fatou's lemma, to the inequality for f.

Now f'_m-f_m is in $L^{1/\alpha}$, and so also in L^{1/α_1}, since $|f'_m-f_m|<1/m$. It follows that

$$\| T(f'_m-f_m) \|_{1/\beta_1} \leqq M_1 \| f'_m-f_m \|_{1/\alpha_1} \leqq M_1 \| f'_m-f_m \|_{1/\alpha}^{\alpha_1/\alpha} \rightarrow 0. \tag{14}$$

as $m\rightarrow\infty$. Similarly

$$\| Tf''_m \|_{1/\beta_2} \leqq M_2 \| f''_m \|_{1/\alpha_2} \leqq M_2 \| f''_m \|_{1/\alpha}^{\alpha_2/\alpha} \rightarrow 0. \tag{15}$$

The inequalities (14) and (15) imply that there is a sequence of m tending to $+\infty$ and such that $|T(f'_m-f_m)|$ and $|Tf''_m|$ tend to 0 almost everywhere. This completes the proof of the theorem.

It remains however to consider the two extreme cases $\alpha=0$ and $\beta=1$, which we previously put aside. These two cases cannot occur simultaneously. If $\beta=1$, we replace the right side of (8) by $\int_{R_2} |Tf|\, d\nu$ and the function $\Phi(z)$ of (10) by

$$\int_{R_2} | T(|f|^{\alpha(z)/\alpha} \operatorname{sign} f) |\, d\nu.$$

After that the proof proceeds as before. If $\alpha=0$, then also $\alpha_1=\alpha_2=0$;

but it is immediately seen that whenever $\alpha_1 = \alpha_2$ the theorem is a corollary of Hölder's inequality.

Massachusetts Institute of Technology and
The University of Chicago.

REFERENCES.

[1] M. Riesz, "Sur les maxima des formes billinéaires et sur les fonctionnelles linéaires," *Acta Mathematica*, vol. 49 (1926), pp. 465-497.

[2] G. O. Thorin, "An extension of a convexity theorem due to M. Riesz," *Kungl. Fysiografiska Saellskapets i Lund Forhaendliger*, 8 (1939), Nr. 14.

[3] ———, *Convexity theorems*, Uppsala, 1948, pp. 1-57.

[4] R. Salem, "Convexity theorems," *Bulletin of the American Mathematical Society*, vol. 55 (1949), pp. 851-860.

[5] J. D. Tamarkin and A. Zygmund, "Proof of a theorem of Thorin," *Bulletin of the American Mathematical Society*, vol. 50 (1944), pp. 279-282.

[6] A. P. Calderón and A. Zygmund, "On the theorem of Hausdorff-Young," *Annals of Mathematical Studies*, vol. 25 (1950), pp. 166-188.

[7] J. E. Littlewood and R. E. A. C. Paley, "Theorems on Fourier series and power series, I," *Journal of the London Mathematical Society*, vol. 6 (1931), pp. 230-233.

ON SINGULAR INTEGRALS.*

By A. P. Calderón and A. Zygmund.[1]

1. Introduction. In earlier work [1] we considered certain singular integrals arising in various problems of Analysis and studied some of their properties. Here we present a new approach to such integrals. Unlike the method used in [1] it is based on the theory of Hilbert transforms of functions of one variable, but otherwise it is simpler and yields most results obtained previously, under far less restrictive assumptions. Unfortunately some important cases ($f \varepsilon L$ for instance) seem to be beyond its scope. We have been unable to decide whether the corresponding theorems as presented in [1] can be likewise strengthened.

Our present results can be summarized in the theorems presented below.

Let $x, y, z, \cdots$ denote vectors in n-dimensional Euclidean space E_n, $|x|$ the length of x and $x' = x|x|^{-1}$. Consider the integral

1.1 $$\tilde{f}_\epsilon(x) = \int_{|x-y|>\epsilon} K(x,y) f(y)\,dy,$$

where dy denotes the element of volume in E_n.

Theorem 1. *If $K(x,y) = N(x-y)$, where $N(x)$ is a homogeneous function of degree $-n$, i.e. such that $N(\lambda x) = \lambda^{-n} N(x)$ for every x and $\lambda > 0$, and if $N(x)$ has in addition the following properties*

i) *$N(x)$ is integrable over the sphere $|x| = 1$ and its integral is zero,*

ii) *$N(x) + N(-x)$ belongs to $L \log^+ L$ on $|x| = 1$,*

then, if $f(x) \varepsilon L^p$, $1 < p < \infty$, $\tilde{f}_\epsilon(x)$ as defined in 1.1 converges to a limit $\tilde{f}(x)$ in the mean of order p, and pointwise almost everywhere as $\epsilon \to 0$. Furthermore $\bar{f}(x) = \sup_\epsilon |\tilde{f}_\epsilon(x)|$ belongs to L^p and $\|\bar{f}\|_p \leqq A \|f\|_p$, where A is a constant depending on p and K, and $\|f\|_p$ is the L^p norm of f.

The condition that $N(x) + N(-x)$ be in $L \log^+ L$ on $|x| = 1$ cannot be relaxed. For given a function $\phi(t)$ such that $\phi(t)/t \log t \to 0$ as $t \to \infty$

* Received September 9, 1955.

[1] The research resulting in this paper was supported in part by the office of Scientific Research of the Air Force under contract AF 18(600)-1111.

Reprinted from *Amer. J. Math.* 78, 289–309 (1956).

there exists a function satisfying i) such that $\phi[|N(x)+N(-x)|]$ is integrable on $|x|=1$ but whose Fourier transform is unbounded, so that even if the pointwise limit of $\tilde{f}_\epsilon(x)$ exists (as is the case of $f(x)$ continuously differentiable and vanishing outside a bounded set), no relationship of the form $\|\tilde{f}\|_2 \leqq A\|f\|_2$ holds.

The Fourier transform $M(x)$ of $N(x)$ is a homogeneous function of degree zero and can easily be shown to be given by the formula

$$M(x)=\int N(y')[i\tfrac{1}{2}\pi\,\mathrm{sg}\cos\theta+\log|\cos\theta|]d\sigma,$$

where θ is the angle between the unit vectors x' and y', and $d\sigma$ is the element of "area" of the sphere $|x|=1$ over which the integral is extended. It is the presence of the term $\log|\cos\theta|$ which makes the class $L\log^+ L$ the best possible. Since we merely want to indicate this fact we omit further details.

THEOREM 2. *If*

$$K(x,y)=N(x,x-y)$$

where $N(x,y)$ is homogeneous of degree $-n$ in y and

i) *for every x, $N(x,y)$ is integrable over the sphere $|y|=1$ and its integral is zero,*

ii) *for a $q>1$ and every x, $|N(x,y)|^q$ is integrable over the sphere $|y|=1$ and its integral is bounded,*

then the same conclusions as in Theorem 1 hold about $\tilde{f}_\epsilon(x)$, provided that $f\,\varepsilon\,L^p$ with $q/(q-1)\leqq p<\infty$.

The condition that $p\geqq q/(q-1)$ is essential. We shall show by means of an example that if $p<q/(q-1)$, then $\tilde{f}_\epsilon(x)$ need no longer be in L^p.

A third type of integrals suggested by the theory of spherical summability of Fourier integrals is the object of the next two theorems.

THEOREM 3. *If*

$$K(x,y)=N(x,x-y)\psi(|x-y|)$$

where $\psi(t)$ is a Fourier-Stieltjes transform, $N(x,y)$ is homogeneous of degree $-n$ in y and

i) *$|N(x,y)|\leqq F(y)$ where $F(y)$ is a homogeneous function of degree $-n$ integrable over $|y|=1$,*

ii) $\psi(t)$ *is an even function and* $N(x,y)$ *is odd in* y, *i.e.* $N(x,y) = -N(x,-y)$, *or* $\psi(t)$ *is odd and* $N(x,y)$ *is even in* y,

then the same conclusions as in Theorem 1 *hold about* $\bar{f}_\epsilon(x)$.

THEOREM 4. *If* $K(x,y)$ *is the same as in the previous theorem with condition* i) *replaced by*

i′) *for some* $q > 1$ *and every* x, $|N(x,y)|^q$ *is integrable over the sphere* $|y| = 1$ *and its integral is bounded,*

then the same conclusions about $\bar{f}_\epsilon(x)$ *hold provided that* $f \varepsilon L^p$, $q/(q-1) \leqq p < \infty$.

In the cases of Theorems 2, 3 and 4 we may also consider the transposed integral 1.1, that is $\int_{|x-y|>\epsilon} K(y,x)f(y)dy$. The convergence in the mean of this integral in an immediate consequence of those theorems. The pointwise convergence does not follow readily though, and at individual points the integral may actually diverge even if f is continuously differentiable and vanishes outside a bounded set.

A straightforward application of Theorem 3 will yield the following statement about spherical summability of Fourier integrals.

THEOREM 5. *If the number* n *of variables of* f *is odd and* $f \varepsilon L^p$, $1 < p \leqq 2$, *then the spherical means of order* $\frac{1}{2}(n-1)$ *of the Fourier integral representation of* f *converge to* f *in the mean of order* p.

Whether this theorem remains valid for n even is an open question.

Finally, we might also mention two generalizations of the maximal theorem of Hardy and Littlewood which are obtained using the same ideas. These extensions are needed in the proofs of Theorems 1 and 2.

THEOREM 6. *Let* $K_\epsilon(x,y) = \epsilon^{-n} N(x-y)\psi(\epsilon^{-1}|x-y|)$ *where* $N(x)$ *is a non-negative homogeneous function of degree zero, integrable over* $|x| = 1$, *and* $\psi(t)$ *is a non increasing function of the real variable* t *such that* $\psi(|x|)$ *is integrable in* E_n. *Then if* $f \varepsilon L^p$, $1 < p \leqq \infty$, *and*

$$f^*(x) = \sup_\epsilon \left| \int K_\epsilon(x,y)f(y)dy \right|,$$

f^* *belongs to* L^p *and*

$$\| f^* \|_p \leqq A \| f \|_p,$$

where A *is a constant depending on* N, p *and* ψ.

The case when $N(x) \equiv 1$ and $x(t)$ is the characteristic function of the interval $(0, 1)$ is well known.

THEOREM 7. *If* $K_\epsilon(x, y) = \epsilon^{-n} N(x, x-y) \psi(\epsilon^{-1} | x-y |)$ *where* $N(x, y)$ *is homogeneous of degree zero in* y, $| N(x, y) |^q$, $q > 1$, *is integrable over the sphere* $| y | = 1$ *and its integral is bounded, and* $\psi(t)$ *is the same as in the previous theorem, then* f^* *as defined in Theorem 6 is in the same* L^p *class as* f *and* $\| f^* \|_p \leqq A \| f \|_p$, *provided that* $q/(q-1) \leqq p < \infty$.

2. We start by showing that the integral 1.1 is meaningful. In the cases of Theorems 1 and 3 this is not quite evident.

In either case we have $| K(x, y) | \leqq F(| x-y |)$, where $F(y)$ is a homogeneous function of degree $-n$, integrable over the sphere $| y | = 1$, and thus it will be sufficient to show that

$$\int_{|x-y|>\epsilon} F(x-y) | f(y) | \, dy$$

is absolutely convergent for almost every x and any $\epsilon > 0$.

Let y' be a unit vector, t a real number and S a full sphere in E_n of diameter d. Then the integral

$$\int_\Sigma F(y') dy' \int_S dx \int_\epsilon^\infty | t^{-1} f(x-ty') | \, dt, \tag{2.1}$$

where dy' is the element of area of the unit sphere Σ, is finite. In fact, the inner integral is less than or equal to

$$A \left[\int_{-\infty}^{+\infty} | f(x-ty') |^p \, dt \right]^{1/p},[2]$$

where A depends on ϵ and p but not on f. Substituting this expression for the inner integral in 2.1 and applying Hölder's inequality to the integral over S we find that the latter is dominated by

$$A \, | S |^{(p-1)/p} \left[\int_S dx \int_{-\infty}^{+\infty} | f(x-ty') |^p \, dt \right]^{1/p}.$$

Now the integral with respect to x can be computed first along lines parallel to y' and then over the space of such lines, rendering evident that its value does not exceed $d \| f \|_p^p$. Thus the integral over S in 2.1 is a bounded function of y' and 2.1 is therefore finite. Hence

$$\int_\Sigma F(y') dy' \int_\epsilon^\infty | t^{-1} f(x-ty') | \, dt$$

[2] Throughout the rest of the paper the letter A will stand for a constant, not necessarily the same in each occurrence.

is finite for almost all x. But the last is nothing but the expression of $\int_{|x-y|>\epsilon} F(x-y)|f(y)|\,dy$ in polar coordinates with origin at x. In other words, for any $\epsilon>0$, 1.1 is absolutely convergent for almost every x.

3. In this section we shall prove Theorems 3 and 4.

Let $g(t)$ be a function of the real variable t belonging to L^p in $-\infty<t<\infty$, $1<p<\infty$, and let $\epsilon(s)$ be an arbitrary positive measurable function in $-\infty<s<\infty$. Then the integral

$$\int_{|s-t|>\epsilon(s)} g(t)/(s-t)\,dt$$

represents a function whose L^p norm does not exceed the L^p norm of g multiplied by a constant A which depends on p but not on g or the function $\epsilon(s)$ (see [1], Chapter II, Theorem 1). More generally, the same holds for

$$e^{irs}\int_{|s-t|>\epsilon(s)} e^{-irt}g(t)/(s-t)\,dt,$$

with the same constant as before. Thus if $\mu(r)$ is a function of bounded variation in $-\infty<r<\infty$ from Minkowski's integral inequality it follows that the L^p norm of the function of s given by

$$\int_{-\infty}^{+\infty} e^{irs}\Big[\int_{|s-t|>\epsilon(s)} e^{-irt}g(t)/(s-t)\,dt\Big]d\mu(r)$$

is not larger than the L^p norm of g multiplied by the constant A above and by the total variation of μ. Now interchanging the order of integration in the expression above (which we may) and observing that the function $\epsilon(s)$ is positive and measurable but otherwise arbitrary we conclude that if

3.1 $$\bar{g}(s)=\sup_{\epsilon}\Big|\int_{|s-t|>\epsilon}\{\psi(s-t)/(s-t)\}g(t)\,dt\Big|,$$

where $\psi(s)=\int_{-\infty}^{+\infty} e^{isr}\,d\mu(r)$, then $\|\bar{g}\|_p\leqq AV(\mu)\|g\|_p$, $V(\mu)$ being the total variation of μ and A being a constant which only depends on p.

Let now $f(x)$ be a given function of L^p, $1<p<\infty$, in E_n, y' a unit vector, and define

3.2 $$\tilde{f}_\epsilon(x,y')=\int_{|t|>\epsilon} t^{-1}f(x-ty')\psi(t)\,dt,$$

3.3 $$\bar{f}(x,y')=\sup_{\epsilon}|\tilde{f}_\epsilon(x,y')|.$$

Clearly $\bar{f}_\epsilon$ exists for almost all (x, y') and is a measurable function of (x, y'). Furthermore, for almost all (x, y') it is a continuous function of ϵ, so that if we restrict ϵ to rational values in 3.3 we obtain the same value for $\bar{f}$ almost everywhere in (x, y'), which shows that $\bar{f}$ is also measurable.

Now it is readily seen that $\bar{f}(x, y')$ restricted to any straight line parallel to y' is precisely the integral in 3.1 of the function $f(x)$ restricted to the same line. Consequently

$$\int_{-\infty}^{+\infty} \bar{f}(x - ty', y')^p \, dt \leqq A^p V(\mu)^p \int_{-\infty}^{+\infty} | f(x - ty') |^p \, dt,$$

and integrating this inequality over the space of lines parallel to y' we obtain

3.4 $$\int \bar{f}(x, y')^p \, dx \leqq A^p V(\mu)^p \int | f(x) |^p \, dx.$$

Define now

3.5 $$f^\dagger(x) = \tfrac{1}{2} \int_\Sigma \bar{f}(x, y') F(y') \, dy',$$

3.6 $$\tilde{f}_\epsilon(x) = \tfrac{1}{2} \int_\Sigma \tilde{f}_\epsilon(x, y') N(x, y') \, dy',$$

where F and N are the functions introduced in Theorem 3. On account of 3.3 it follows that $| \tilde{f}_\epsilon(x) | \leqq f^\dagger(x)$, and Minkowski's integral inequality applied to 3.5, and 3.4 gives

3.7 $$\| f^\dagger \|_p \leqq \tfrac{1}{2} A V(\mu) \int_\Sigma F(y') \, dy' \, \| f \|_p.$$

But the function $\tilde{f}_\epsilon(x)$ defined in 3.6 coincides with the integral 1.1 as specified in Theorem 3. To see this one merely has to substitute $\tilde{f}_\epsilon(x, y')$ for its value in 3.6 and observe that one obtains 1.1 in polar coordinates with origin at x. Interchanging the order of integration is permissible wherever 1.1 is absolutely convergent, that is, almost everywhere. Thus we have proved that under the assumptions of Theorem 3, $\bar{f}(x) = \sup_\epsilon | \tilde{f}_\epsilon(x) |$ belongs to L^p, and that $\| \bar{f} \|_p \leqq A \| f \|_p$. A more explicit estimate of the constant involved appears in the right-hand side of 3.7, where A depends only on p.

We now prove that the same holds under the assumptions of Theorem 4.

We redefine $f^\dagger(x)$ and $\tilde{f}_\epsilon(x)$ by means of the formulas

$$f^\dagger(x) = \tfrac{1}{2} \int_\Sigma \bar{f}(x, y') | N(x, y') | \, dy', \qquad \tilde{f}_\epsilon(x) = \tfrac{1}{2} \int_\Sigma \tilde{f}_\epsilon(x, y') N(x, y') \, dy'.$$

First we observe that the $\tilde{f}_\epsilon(x)$ just introduced coincides with the $\tilde{f}_\epsilon(x)$

in 1.1. For the last integral above is nothing but 1.1 expressed in polar coordinates with origin at x. On account of 3.3 it follows again that $|\bar{f}_\epsilon(x)| \leqq f^\dagger(x)$, and Hölder's inequality and Fubini's theorem yield

$$\int f^\dagger(x)^p\,dx = 2^{-p}\int [\int_\Sigma \bar{f}(x,y')\,|N(x,y')|\,dy']^p\,dx$$
$$\leqq 2^{-p}\int [\int_\Sigma \bar{f}(x,y')^p\,dy'][\int_\Sigma |N(x,y')|^{p'}\,dy']^{p/p'}\,dx$$
$$\leqq 2^{-p}\int_\Sigma dy'[\int \bar{f}(x,y')^p[\int_\Sigma |N(x,y')|^{p'}\,dy']^{p/p'}\,dx],$$

where $p' = p/(p-1)$. Now $p' \leqq q$, so that condition i) in Theorem 4 implies that the innermost integral in the last expression above is bounded. Hence if $B^{p'}$ is an upper bound for this integral and ω is the "area" of the unit sphere in E_n, 3.4 yields $\| f^\dagger \|_p \leqq \frac{1}{2}AV(\mu)\omega^{1/p}B \, \| f \|_p$. Thus we find again that $\| \bar{f}(x) \|_p \leqq A \, \| f \|_p$.

Now we can prove that $\bar{f}_\epsilon(x)$ converges in the mean and pointwise almost everywhere. The argument clearly covers both Theorem 3 and Theorem 4.

Let $\rho(t)$ be an even and continuously differentiable function equal to 1 for $t = 0$ and vanishing outside the interval $(-1, 1)$. It is readily seen that the Fourier transform of $\psi(t)\rho(t)$ is bounded and integrable. Consider now the function equal to t^{-1} for $|t| > \epsilon$ and zero otherwise. An easy computation shows that its Fourier transform is bounded uniformly in ϵ and converges pointwise as $\epsilon \to 0$. Consequently it follows from Parseval's formula that

3.8 $$\int_{|t|>\epsilon} \psi(t)\rho(t)/t\,dt$$

converges as $\epsilon \to 0$. Thus under the hypotheses of either Theorem 3 or Theorem 4, the integral

$$\int_{|x-y|>\epsilon} K(x,y)\rho(|x-y|)\,dy$$

converges as $\epsilon \to 0$. To see this one merely has to compute this integral in polar coordinates and use the fact pointed out above that 3.8 converges.

Let now $f(x)$ be continuously differentiable and vanish outside a bounded set. Then

$$\bar{f}_\epsilon(x) = \int_{|x-y|>\epsilon} K(x,y)f(y)\,dy = \int_{|x-y|>\epsilon} K(x,y)\,[f(y) - f(x)\rho(|x-y|)]\,dy$$
$$+ f(x)\int_{|x-y|>\epsilon} K(x,y)\rho(|x-y|)\,dy.$$

The integrand in the first integral on the right is absolutely integrable over E_n, and the second integral converges as $\epsilon \to 0$. Consequently $\tilde{f}_\epsilon(x)$ converges.

In the general case, given $f(x)$ in L^p and $\delta > 0$ there exists a continuously differentiable g vanishing outside a bounded set such that $f = g + h$ and $\| h \|_p < \delta$. Since

$$\tilde{f}_\epsilon(x) = \tilde{g}_\epsilon(x) + \tilde{h}_\epsilon(x), \quad |\tilde{h}_\epsilon(x)| \leqq \bar{h}(x)$$

and $\| \bar{h} \|_p \leqq A \| h \|_p < A\delta$, and since $\tilde{g}_\epsilon(x)$ converges everywhere,

$$\overline{\lim}\, \tilde{f}_\epsilon(x) - \underline{\lim}\, \tilde{f}_\epsilon(x) \leqq 2\bar{h}(x),$$

and this implies that $\tilde{f}_\epsilon(x)$ converges almost everywhere because $\bar{h}(x)$ has arbitrarily small L^p norm. Finally, since $\tilde{f}_\epsilon(x) \to \tilde{f}(x)$ almost everywhere and $|\tilde{f}_\epsilon(x)| \leqq \bar{f}(x)$, the theorem on dominated convergence yields $\| \tilde{f}_\epsilon - \tilde{f} \|_p \to 0$ as $\epsilon \to 0$.

Theorems 3 and 4 are thus established.

4. The proof of Theorems 6 and 7 is based on the same technique we used in the preceding section.

Let $f(t) \geqq 0$ be a function defined in $-\infty < t < \infty$, and $\phi(t) = \psi(t) t^{n-1}$, where $\psi(t)$ is the function introduced in Theorem 6. Then $\phi(t)$ is integrable in $(0, \infty)$. Set

$$f^*(s) = \sup_\epsilon \epsilon^{-1} \int_0^\infty f(s+t)\phi(t\epsilon^{-1})\,dt, \qquad \epsilon > 0,$$

$$F_s(t) = t^{-1} \int_0^t f(s+t)\,dt, \qquad t > 0,$$

and $G(s) = \sup_t F_s(t)$. Then integration by parts gives

$$\epsilon^{-1} \int_0^\infty f(s+t)\phi(t\epsilon^{-1})dt = -\epsilon^{-1} \int_0^\infty tF_s(t)d\phi(t\epsilon^{-1}) \leqq -\epsilon^{-1}G(s) \int_0^\infty td\phi(t\epsilon^{-1})$$

$$= G(s) \int_0^\infty \phi(t)dt,$$

and consequently $f^*(s) \leqq G(s) \int_0^\infty \phi(t)\,dt$.

Now, a theorem of Hardy and Littlewood (see [3], p. 244), asserts that if $f \in L^p$, $1 < p < \infty$, then $G \in L^p$ and $\| G \|_p \leqq A \| f \|_p$, where A depends on p only. Therefore $\| f^* \|_p \leqq A \| f \|_p$, A now depending on p and ϕ.

Let now $f(x) \geqq 0$ be a function from L^p, $1 < p < \infty$ in E_n and y' a unit vector. Define

$$f^*(x, y') = \sup_\epsilon \epsilon^{-1} \int_0^\infty f(x + ty')\phi(t\epsilon^{-1})\,dt.$$

Then, as we have shown above,

$$\int_{-\infty}^{+\infty} f^*(x+ty',y')^p\,dt \leqq A \int_{-\infty}^{+\infty} f(x+ty')^p\,dt,$$

where A depends only on p and ϕ. Integrating over the space of lines parallel to y' we obtain

$$\text{4.1} \qquad \int f^*(x,y')^p\,dx \leqq A \int f(x)^p\,dx.$$

Under the assumptions of Theorem 6 we have

$$f^*(x) = \sup_\epsilon \left| \int K_\epsilon(x,y)f(y)dy \right| \leqq \sup_\epsilon \epsilon^{-n} \int N(x-y)\psi(|x-y|\epsilon^{-1})f(y)dy$$

$$= \sup_\epsilon \int_\Sigma N(y')[\epsilon^{-1}\int_0^\infty f(x+ty')\phi(t\epsilon^{-1})dt]dy' \leqq \int_\Sigma N(y')f^*(x,y')dy',$$

and an application of Minkowski's integral inequality and 4.1 yield

$$\| f^* \|_p \leqq A \int_\Sigma N(y')\,dy' \,\| f \|_p,$$

where A depends on p and ψ only.

On the other hand, under the assumptions of Theorem 7 we have

$$f^*(x) = \sup_\epsilon \left| \int K_\epsilon(x,y)f(y)dy \right|$$

$$\leqq \sup_\epsilon \epsilon^{-n} \int |N(x,x-y)|\,\psi(|x-y|\epsilon^{-1})f(y)dy$$

$$= \sup_\epsilon \int_\Sigma |N(x,y')|\,[\epsilon^{-1}\int_0^\infty f(x+ty')\phi(t\epsilon^{-1})dt]dy'$$

$$\leqq \int_\Sigma |N(x,y')|\,f^*(x,y')dy'.$$

Hence

$$\int f^*(x)^p\,dx \leqq \int [\int_\Sigma |N(x,y')|\,f^*(x,y')\,dy']^p\,dx$$

$$\leqq \int [\int_\Sigma f^*(x,y')^p\,dy'][\int_\Sigma |N(x,y')^{p'}dy']^{p/p'}\,dx,$$

where $p' = p/(p-1)$. But $p' \leqq q$ so that the integral of $|N(x,y')|^{p'}$ is bounded. If $B^{p'}$ is a bound for this integral and ω is the area of the unit sphere in E_n, interchanging the order of integration and applying 4.1 we obtain finally $\| f^* \|_p \leqq A\omega^{1/p}B \,\| f \|_p$, where A only depends on p and ψ.

The proof of Theorems 6 and 7 is thus completed.

Remark. The methods used so far still yield results under slightly less restrictive assumptions about the type of integrability of $f(x)$. For instance if f vanishes outside a bounded set and $|f|\log^+|f|$ is integrable, one can still prove that under the assumptions of either Theorem 3 or Theorem 4, $\tilde{f}_\epsilon(x)$ converges in the mean order 1 on any bounded set.

This result is needed in the next section but only in the special case when $K(x,y)$ is the kernel of M. Riesz (see below), and in this form it is also contained in Theorem 7, Chapter 1 of [1]. We may thus safely omit further details.

5. The proof of Theorems 1 and 2 in their full generality is more complicated. In special cases they are contained in Theorems 3 and 4 respectively. In fact, if in Theorem 1 the function $N(x)$ is such that $N(x)=-N(-x)$, then Theorem 1 reduces to Theorem 3 with $N(x,y)=N(y)$ and $\psi(t)=1$. Similarly, if $N(x,y)$ in Theorem 2 is such that $N(x,y)=-N(x,-y)$, then Theorem 2 reduces to Theorem 4 with the same N and $\psi(t)=1$.

Since it is always possible to decompose the functions $N(x)$ and $N(x,y)$ in the sum of two,

$$N(x)=N_1(x)+N_2(x), \qquad N(x,y)=N_1(x,y)+N_2(x,y),$$

where

$$N_1(x)=N_1(-x), \qquad N_2(x)=-N_2(-x),$$

and

$$N_1(x,y)=N_1(x,-y), \qquad N_2(x,y)=-N_2(x,-y),$$

we need only treat the cases $N(x)=N(-x)$, $N(x,y)=N(x,-y)$ and for this purpose we shall use the device of representing f as a singular integral with the kernel of M. Riesz. Our original integral 1.1 will then appear as an iterated integral to which we shall be able to apply the preceding results.

In what follows we shall use vector valued functions but we shall introduce no special notation for them. When talking about the L^p norm of a vector valued function we shall be meaning the L^p norm of its absolute value. The inner product of two vectors will be denoted by their symbols with a dot in-between.

The kernel R of M. Riesz is vector valued and odd

$$R(x)=\pi^{-\frac{1}{2}(n+1)}\Gamma(\tfrac{1}{2}n+\tfrac{1}{2})x\,|x|^{-n-1}.$$

If

$$g_\epsilon(x)=-\int_{|x-y|>\epsilon}R(x-y)f(y)dy,$$

and $f \varepsilon L^p$, $1 < p < \infty$, then $g_\epsilon(x)$ is a vector valued function which, as $\epsilon \to 0$, converges in the mean of order p to a function $g(x)$. This follows from Theorem 1 by applying it to each component. On the other hand, if

$$f_\epsilon(x) = \int_{|x-y|>\epsilon} R(x-y)\cdot g(y)\,dy,$$

where the integrand is the inner product of the vectors displayed, then $f_\epsilon(x)$ converges likewise to $f(x)$. That it converges to a function h is again a consequence of Theorem 1. That $h = f$ can be verified for f bounded and vanishing outside a bounded set by taking Fourier transforms (see [2]), whence the general case follows from the continuity in L^p of the linear operation taking f into h.

Let $\phi(t)$ be a continuously differentiable function of the real variable t, $t \geqq 0$, equal to zero in $(0, \frac{1}{4})$ and to 1 in $(\frac{3}{4}, \infty)$, and $F(x)$ a homogeneous function of degree $-n$, such that $F(x) = F(-x)$ and that $|F| \log^+ |F|$ is integrable on the sphere $|x| = 1$. Suppose in addition that the integral of $F(x)$ over $|x| = 1$ is zero and consider

5.1 $$\int_{|x-y|>\epsilon} R(x-y)F(y)\,dy,$$

5.2 $$\int_{|x-y|>\epsilon} R(x-y)F(y)\phi(|y|)\,dy.$$

Since $|R(x)| \leqq A|x|^{-n}$ the second integral is absolutely convergent. The first has a singularity at $y = 0$, but, owing to the fact that $R(x)$ is continuously differentiable if $x \neq 0$, it can be given a natural meaning if $|x| > \epsilon$ by integrating outside a small sphere with center at $y = 0$ and taking the limit of the value obtained as the radius of the small sphere tends to zero.

The properties of the integrals above which we need are summarized in the following

Lemma. *Under the preceding assumptions, as* $\epsilon \to 0$, 5.2 *converges in the mean of order* 1 *on any compact set, and* 5.1 *converges on any compact set not containing the point* $x = 0$. *The corresponding limits,* $F_2(x)$ *and* $F_1(x)$, *are odd functions, i.e.* $F_1(x) = -F_1(-x)$, $F_2(x) = -F_2(-x)$. *The function* $F_1(x)$ *is homogeneous of degree* $-n$, *and, for* $|x| \geqq 1$,

5.3 $$|F_1(x) - F_2(x)| \leqq A \int_\Sigma |F(y')|\,dy'\,|x|^{-n-1}.$$

There exists a homogeneous function $G(x)$ *of degree zero such that for* $|x| \leqq 1$,

5.4 $$|F_2(x)| \leqq G(x),$$

and

5.5
$$\int_\Sigma G(y')\,dy' < \infty.$$

If for some q, $1 < q < \infty$, $\int_\Sigma |F(y')|^q\,dy' < \infty$, *then the inequalities*

5.6
$$\int_\Sigma |F_1(y')|^q\,dy' \leqq A \int_\Sigma |F(y')|^q\,dy',$$

5.7
$$\int_\Sigma G(y')^q\,dy' \leqq A \int_\Sigma |F(y')|^q\,dy'$$

hold, with the constants A *depending on* q *but not on* F, *and the integral in* 5.2 *converges to its limit in the mean of order* q.

That the functions $F_1(x)$ and $F_2(x)$, if existent, are odd is clear. That $F_1(x)$ is homogeneous of degree $-n$ is also clear. To see that 5.1 converges in the mean between two spheres of radii $\rho_1 < \rho_2$ we observe that the contributions to the integral from the sphere $|y| \leqq \frac{1}{2}\rho_1$ and from the exterior of the sphere $|y| = 2\rho_2$ is bounded, and to the integral extended over $\frac{1}{2}\rho_1 \leqq |y| \leqq 2\rho_2$ we may apply the remark of Section 4 and obtain immediately the desired result. On the other hand, an application of Theorem 4 to each component of the vector valued integral 5.1 gives that the integral of $|F_1(x)|^q$ extended to the region between two spheres with center at $x = 0$ is dominated by the q-th power of the right hand of 5.6. Since $F_1(x)$ is homogeneous, 5.6 follows. A similar argument applies to the integral 5.2, except that in this case it will not be necessary to exclude a neighborhood of $x = 0$.

For the difference between $F_1(x)$ and $F_2(x)$ we get the following estimates

$$\begin{aligned}|F_2(x) - F_1(x)| &\leqq \left|\int R(x-y)F(y)[\phi(|y|)-1]\,dy\right| \\ &= \left|\int [R(x-y)-R(x)]F(y)[\phi(|y|)-1]\,dy\right| \\ &\leqq \int_{|y|\leqq 3/4} |F(y)|\,|R(x-y)-R(x)|\,dy.\end{aligned}$$

Now, it is readily seen that for $|x| \geqq 1$ and $|y| \leqq \frac{3}{4}$ we have the inequality

$$|R(x-y) - R(x)| \leqq A\,|x|^{-n-1}\,|y|.$$

Substituting this in the preceding integral we obtain 5.3.

In order to prove 5.4, 5.5 and 5.7 we proceed as follows. First we observe that, owing to the fact that $F(y)\phi(|y|)$ vanishes in $|y| \leqq \frac{1}{4}$, $F_2(x)$

is continuous and bounded in $|x| \leqq \frac{1}{8}$, and that $A \int_{\Sigma} |F(y')|\, dy'$ with an appropriate constant A independent of F is an upper bound for $|F_2(x)|$ in this particular domain $|x| \leqq \frac{1}{8}$.

In $\frac{1}{8} \leqq |x| \leqq 1$ we have

$$|F_2(x)| \leqq \phi(|x|)|F_1(x)| + |F_2(x) - \phi(|x|)F_1(x)|$$

$$\leqq |F_1(x)| + \left| \int R(x-y)[\phi(|y|)F(y) - \phi(|x|)F(y)]dy \right.$$

$$= |F_1(x)| + \left| \int [R(x-y) - \chi(|y|)R(x)]\,[\phi(|y|) - \phi(|x|)]F(y)dy \right|,$$

where χ is the characteristic function of the interval $(0, 1)$.

Now, one verifies easily that if $\frac{1}{8} \leqq |x| \leqq 1$, then

$$|R(x-y) - \chi(|y|)R(x)| \leqq A\, |y|^{\frac{1}{2}} \cdot |x-y|^{-n}.$$

On the other hand, since $\phi(t)$ has a bounded derivative,

$$|\phi(|x|) - \phi(|y|)| \leqq A\, \big|\, |x| - |y| \,\big| \leqq A\, |x-y|.$$

Thus in the last integral, substituting this we get

$$|F_2(x)| \leqq |F_1(x)| + A \int |y|^{\frac{1}{2}} |F(y)|\, |x-y|^{-(n-1)} dy,$$

and since $|x| \geqq \frac{1}{8}$, it follows that

$$|F_2(x)| \leqq A\Big[|x|^n |F_1(x)| + |x|^{n-\frac{3}{2}} \int |y|^{\frac{1}{2}} |F(y)|\, |x-y|^{-(n-1)} dy \Big].$$

The integral on the right represents a homogeneous function of degree $-n + \frac{3}{2}$, and $F_1(x)$ is homogeneous of degree $-n$. Hence the right-hand side of the inequality is a homogeneous function of degree zero. It remains only to show 5.5 and 5.7. The contribution of the term $|x|^n |F_1(x)|$ can be estimated either by the fact that $F_1(x)$ is integrable between two concentric spheres with center at the origin, or by 5.6. The contribution of the other term is estimated by splitting the integral in two, one extended over the set $\frac{1}{16} \leqq |y| \leqq 2$, the other over the complement of this set. The second integral is readily seen to be bounded by $A \int_{\Sigma} |F(y')|\, dy'$ or $A\big[\int_{\Sigma} |F(y')|^q dy'\big]^{1/q}$ and an application of the theorem of Young (see [3], page 71, where the theorem is stated and proved in a special case; but the proof extends obviously to the most general situation) to the first shows that it represents a function of the same class as $F(x)$ in $\frac{1}{16} \leqq |x| \leqq 2$. But the function under consideration is homogeneous of degree zero and, consequently, an estimate for its norm in $\frac{1}{2} \leqq |x| \leqq \frac{3}{2}$ gives an estimate for its norm on the sphere $|x| = 1$. Collecting results, 5.5 and 5.7 follow and the proof of our lemma is complete.

Let now $N(x)$ have the property that $N(x) = N(-x)$ and satisfy the conditions of Theorem 1. Then, by the lemma above,

$$N_1(x) = \lim_{\epsilon \to 0} \int_{|y-x|>\epsilon} R(x-y) N(y)\, dy$$

also satisfies the conditions of Theorem 1 and $N_1(x) = -N_1(-x)$. Furthermore, if

$$N_2(x) = \lim_{\epsilon \to 0} \int_{|x-y|>\epsilon} R(x-y) N(y) \phi(|y|)\, dy,$$

then, for $|x| \geqq 1$,

5.8 $$|N_2(x) - N_1(x)| \leqq A\, |x|^{-n-1},$$

and for $|x| \leqq 1$,

5.9 $$|N_2(x)| \leqq G(x),$$

where $G(x)$ is a homogeneous function of degree zero integrable over the sphere $|x| = 1$.

Consider now the vector valued function $g(x)$ and

$$f(x) = \lim_{\epsilon \to 0} \int_{|x-y|>\epsilon} R(x-y) \cdot g(y)\, dy.$$

As we already know, if $|g(x)| \in L^p$, $1 < p < \infty$, then $f(x) \in L^p$ and $\| f \|_p \leqq A \| g \|_p$. Furthermore the integral above converges in the mean of order p and every function $f(x)$ in L^p can be thus represented.

We shall prove the following identity:

5.10 $$\int N(x-y) \phi(|x-y|\,\epsilon^{-1}) f(y)\, dy = \epsilon^{-n} \int N_2(\epsilon^{-1}(x-y)) \cdot g(y)\, dy.$$

If g is continuously differentiable and vanishes outside a bounded set, then, on account of absolute integrability,

5.11 $$\int N(x-y) \phi(|x-y|\,\epsilon^{-1})\, dy \int_{|y-z|>\delta} R(y-z) \cdot g(z)\, dz$$
$$= \int \Big[\int_{|y-z|>\delta} N(x-y) \phi(|x-y|\,\epsilon^{-1}) R(y-z)\, dy \Big] \cdot g(z)\, dz.$$

Now, as $\delta \to 0$, by the lemma above and by changing variables, the inner integral on the right is seen to converge to $\epsilon^{-n} N_2((x-y)\epsilon^{-1})$ in the mean of order 1 on any compact set. Therefore the right-hand side above converges to

$$\epsilon^{-n}\int N_2((x-y)\epsilon^{-1})\cdot g(y)dy.$$

On the other hand,

$$\int_{|y-z|>\delta} R(y-z)\cdot g(z)dz=\int_{|y-z|>\delta} R(y-z)\cdot[g(z)-g(y)]dz,$$

and on account of the continuous differentiability of g, the right-hand side is readily seen to converge uniformly as $\delta\to 0$ and to be independent of δ for $\delta<1$ and $|y|$ sufficiently large. Therefore, the left-hand side of 5.11 is seen to converge to

$$\int N(x-y)\phi(|x-y|\epsilon^{-1})f(y)dy.$$

Thus 5.10 is proved for g continuously differentiable and vanishing outside a bounded set. In the general case, given $g\in L^p$, we take a sequence of continuously differentiable functions g_k, each vanishing outside a compact set and such that $\|g_k-g\|_p\to 0$ and $\sum_1^\infty \|g_{k+1}-g_k\|_p<\infty$. If

$$f_k(x)=\lim_{\epsilon\to 0}\int_{|x-y|>\epsilon} R(x-y)\cdot g_k(y)dy,$$

then $\|f_k-f\|_p\to 0$ and $\sum_1^\infty \|f_{k+1}-f_k\|_p<\infty$. From the finiteness of the series $\sum\|g_{k+1}-g_k\|_p$ and $\sum\|f_{k+1}-f_k\|_p$ it follows that the functions $\bar g=|g_1|+\sum|g_{k+1}-g_k|$ and $\bar f=|f_1|+\sum|f_{k+1}-f_k|$ are finite almost everywhere and belong to L^p. Thus the sequences

$$g_k(x)=g_1(x)+\sum_1^{k-1}[g_{j+1}(x)-g_j(x)] \text{ and } f_k(x)$$

converge almost everywhere and are dominated in absolute value by $\bar g$ and $\bar f$ respectively. Now the considerations of Section 2 show that the integral

$$\int |N(x-y)|\phi(|x-y|\epsilon^{-1})\bar f(y)dy\leqq\int_{|x-y|>\epsilon/4}|N(x-y)|\bar f(y)dy$$

is finite for almost all x. On the other hand, on account of 5.8 we have

$$\int |N_2((x-y)\epsilon^{-1})|\bar g(y)dy\leqq\int_{|x-y|<\epsilon}|N_2((x-y)\epsilon^{-1})|\bar g(y)dy$$
$$+\int_{|x-y|>\epsilon}|N_1((x-y)\epsilon^{-1})|\bar g(y)dy+A\int_{|x-y|>\epsilon}|x-y|^{-n-1}\bar g(y)dy.$$

The first and last integrals on the right are absolutely convergent for almost all x owing to the absolute integrability of $N_2(x)$ in $|x| \leqq 1$, and the remaining integral, by the considerations of Section 2, is also finite for almost all x. Hence both

$$\int |N(x-y)|\phi(|x-y|\epsilon^{-1})\bar{f}(y)dy$$

and

$$\int |N_2((x-y)\epsilon^{-1})|\bar{g}(y)dy$$

are finite for almost all x. Consequently we can pass to the limit in

$$\int N(x-y)\phi(|x-y|\epsilon^{-1})f_k(y)dy = \epsilon^{-n}\int N_2((x-y)\epsilon^{-1})\cdot g_k(y)dy,$$

and we find that in the general case 5.10 holds for almost all x.

Now the proof of Theorem 1 is nearly completed. We have

$$\begin{aligned}\tilde{f}_\epsilon(x) &= \int_{|x-y|>\epsilon} N(x-y)f(y)dy = \int N(x-y)\phi(|x-y|\epsilon^{-1})f(y)dy \\ &\quad - \int_{|x-y|<\epsilon} N(x-y)\phi(|x-y|\epsilon^{-1})f(y)dy \\ &= \epsilon^{-n}\int N_2((x-y)\epsilon^{-1})\cdot g(y)dy - \int_{|x-y|<\epsilon} N(x-y)\phi(|x-y|)\epsilon^{-1})f(y)dy,\end{aligned}$$

and on account of 5.8 and 5.9 we find that

$$\begin{aligned}|\tilde{f}_\epsilon(x)| \leqq \Big|\int_{|x-y|>\epsilon} N_1(x-y)\cdot g(y)dy\Big| &+ \epsilon^{-n}\int_{|x-y|<\epsilon} G((x-y)/|x-y|)|\,g(y)|\,dy \\ &+ A\epsilon^{-n}\int_{|x-y|>\epsilon} |(x-y)\epsilon^{-1}|^{-n-1}|g(y)|\,dy \\ &+ A\epsilon^{-n}\int |N((x-y)/|x-y|)|\,|f(y)|\,dy,\end{aligned}$$

whence Theorem 1 applied to the first term on the right, and Theorem 6 applied to the remaining ones yield

$$\|\bar{f}\|_p = \|\sup_\epsilon |\tilde{f}_\epsilon|\,\|_p \leqq A\,\|g\|_p + A\,\|f\|_p \leqq A\,\|f\|_p.$$

From this convergence in the mean and almost everywhere of $\tilde{f}_\epsilon(x)$ follows as in the proof of Theorems 3 and 4. Theorem 1 is thus proved.

The proof of Theorem 2 proceeds along similar lines but the differences justify its presentation. Let $K(x,y)$ be as specified in Theorem 2 with the additional property that $K(x,y) = K(x,-y)$ and define

$$K_1(x,y)=\lim_{\epsilon\to 0}\int_{|y-z|>\epsilon} K(x,z)R(y-z)\,dz,$$

$$K_2(x,y)=\lim_{\epsilon\to 0}\int_{|y-z|>\epsilon} K(x,z)\phi(|z|)R(y-z)\,dz.$$

Both K_1 and K_2 are odd functions in y, and K_1 satisfies the conditions of Theorem 2. Furthermore, for $|y|\geqq 1$,

5.12 $$|K_2(x,y)-K_1(x,y)|\leqq A\,|y|^{-n-1},$$

with A independent of x, and for $|y|\leqq 1$

5.13 $$|K_2(x,y)|\leqq G(x,y),$$

where G is homogeneous of degree zero in y and such that

5.14 $$\int_{\Sigma} G(x,y')^q\,dy'$$

is bounded. All this follows from our lemma. Let now $g(x)$ be a vector valued function in L^p, $p\geqq q/(q-1)$ and set

$$f(x)=\lim_{\epsilon\to 0}\int_{|x-y|>\epsilon} R(x-y)\cdot g(y)\,dy.$$

We shall prove the identity

5.15 $$\int K(x,x-y)\phi(|x-y|\epsilon^{-1})f(y)\,dy = \epsilon^{-n}\int K_2(x,(x-y)\epsilon^{-1})\cdot g(y)\,dy.$$

If $g(y)$ is continuously differentiable and vanishes outside a bounded set this identity is proved the same way we proved 5.10. In the general case, given $g\in L^p$ we take a sequence of continuously differentiable functions g_n, each vanishing outside a bounded set, and tending to g in the mean of order p. Since both $K(x,x-y)\phi(|x-y|\epsilon^{-1})$ and $K_2(x,(x-y)\epsilon^{-1})$ as functions of y are of integrable power $p/(p-1)$, and since the functions f_n corresponding to the g_n converge to f in the mean of order p, a passage to the limit under the integral sign yields 5.15 in its full generality.

Now we have

$$\begin{aligned}\tilde f_\epsilon(x)&=\int_{|x-y|>\epsilon}K(x,x-y)f(y)\,dy=\int K(x,x-y)\phi(|x-y|\epsilon^{-1})f(y)\,dy\\&\quad-\int_{|x-y|<\epsilon}K(x,x-y)\phi(|x-y|\epsilon^{-1})f(y)\,dy\\&=\epsilon^{-n}\int K_2(x,(x-y)\epsilon^{-1})\cdot g(y)\,dy-\int_{|x-y|<\epsilon}K(x,x-y)\phi(|x-y|\epsilon^{-1})f(y)\,dy,\end{aligned}$$

and on account of 5.12 and 5.13 we find that

$$| \tilde{f}_\epsilon(x)| \leqq | \int_{|x-y|>\epsilon} K_1(x, x-y) \cdot g(y) dy | + \epsilon^{-n} \int_{|x-y|<\epsilon} G(x, (x-y)/|x-y|)| g(y)| dy$$

$$+ A\epsilon^{-n} \int_{|x-y|>\epsilon} |(x-y)\epsilon^{-1}|^{-n-1} | g(y)| dy + A\epsilon^{-n} \int_{|x-y|<\epsilon} | K(x, (x-y)/|x-y|)| | f(y)| dy,$$

whence Theorem 2 applied to the first term on the right and Theorem 7 applied to the remaining ones yield

$$\| \tilde{f} \|_p = \| \sup_\epsilon | \tilde{f}_\epsilon(x) | \|_p \leqq A (\| g \|_p + \| f \|_p) \leqq A \| f \|_p.$$

The argument can now be completed as before.

We close this section by showing that Theorem 2 ceases to hold for functions in L^p, $p < q/(q-1)$.

Let $p < q/(q-1)$ and take an integer n and $\alpha > 0$ so that

$$1/\alpha \leqq p/(p-1) - q, \qquad n/\alpha = p/(p-1).$$

Define $f(x)$ in E_n as follows: $f(x) = 1$ for $|x| \leqq 1$ and $f(x) = 0$ otherwise. Set

$$K(x, y) = |x|^\alpha |y|^{-n} \text{ for } |x| \geqq 1 \text{ and } |x' + y'| \leqq 1/|x|,$$

$$K(x, y) = -|x|^\alpha |y|^{-n} \text{ for } |x| \geqq 1 \text{ and } |x' - y'| \leqq 1/|x|$$

and $K(x, y) = 0$ otherwise. Then one sees readily that

$$\int_\Sigma | K(x, y')|^q dy' \leqq A |x|^{\alpha q - n + 1} \qquad |x| \geqq 1$$

But

$$\alpha q - n + 1 = \alpha q - \alpha p/(p-1) + 1 = \alpha[q - p/(p-1)] + 1 \leqq 0,$$

and consequently the integral above is bounded. On the other hand,

$$| \tilde{f}(x)|^p \geqq A |x|^{(\alpha - n)p} \text{ as } |x| \to \infty,$$

and

$$(n - \alpha)p = \alpha p/(p-1) = n,$$

so that $\tilde{f} \notin L^p$.

Remark. In order to simplify our presentation as far as possible we have omitted to give explicit estimates for the constant A in the inequalities

$\| \tilde{f} \|_p \leqq A \| f \|_p$. In the paper that follows this, though, we shall need to know more about A. If, in Theorem 1, $N(x)$ is such that

$$\int_\Sigma | N(y') |^q \, dy' < \infty, 1 < q < \infty,$$

then

$$A = A_{pq} [\int_\Sigma | N(y') |^q \, dy']^{1/q},$$

where A_{pq} depends on p and q but *not* on $N(x)$. The reader will have little difficulty in verifying this statement himself, by estimating step by step the constants in the preceding proofs.

6. In this last section we shall prove Theorem 5. We restrict ourselves to the case of three or more variables.

Let $f(x)$ be a function L^p, $1 < p \leqq 2$. Then $f(x)$ has a Fourier transform f given by

6.1 $$\hat{f}(x) = \lim_{r \to \infty} \int_{|y| < r} e^{i(x \cdot y)} f(y) \, dy,$$

the limit being understood as a limit in the mean of order $p/(p-1)$. The spherical means of order $\frac{1}{2}(n-1)$ of the Fourier integral of f are given by

6.2 $$\sigma_r(f, y) = (2\pi)^{-n} \int_{|x| < r} \hat{f}(x) e^{-i(x \cdot y)} (1 - | x |^2 r^{-2})^{\frac{1}{2}(n-1)} \, dx.$$

If we assume that f vanishes outside a compact set we may replace $\hat{f}$ by its value 6.1 and interchange the order of integration obtaining

6.3 $$\sigma_r(f, y) = (2\pi)^{-n} \int [\int_{|x| < r} e^{ix \cdot (z-y)} (1 - | x |^2 \, r^{-2})^{\frac{1}{2}(n-1)} \, dx] f(z) \, dz.$$

If we set

$$F(z) = \int_{|y| \leqq 1} e^{-i(y \cdot z)} (1 - | y |^2)^{\frac{1}{2}(n-1)} \, dy,$$

6.3 becomes

6.4 $$\sigma_r(f, y) = (2\pi)^{-n} r^n \int F[r(y - z)] f(z) \, dz.$$

Now the function $F(z)$ is in L^2 and bounded and consequently it belongs to L^q for every $q \geqq 2$. Therefore 6.4 can be extended to an arbitrary f in L^p by a passage to the limit.

Our next step will be to prove that

$$\| \sigma_r(f, y) \|_p \leqq A \| f \|_p.$$

For this purpose we shall show that

$$F(z) = \psi(|z|)|z|^{-n}$$

where $\psi(t)$ is an odd Fourier-Stieltjes integral. We take an orthogonal coordinate system in E_n whose first coordinate axis coincides with z and denote by t the corresponding coordinate. Then

$$|z|^n F(z) = |z|^n \int_{|y|\leqq 1} e^{-i(y\cdot z)}(1-|y|^2)^{\frac{1}{2}(n-1)}\,dy$$

$$= |z|^n \omega_{n-2} \int_{-1}^{+1} e^{i|z|t}\Big[\int_0^{(1-t^2)^{\frac{1}{2}}} [1-(t^2+s^2)]^{\frac{1}{2}(n-1)} s^{n-2}\,ds\Big]dt,$$

where ω_{n-2} denotes the "area" of the $n-1$ dimensional unit sphere.

Now by setting $s^2 = v(1-t^2)$ the inner integral on the right is readily seen to be equal to

$$\tfrac{1}{2}(1-t^2)^{n-1}\int_0^1 (1-v)^{\frac{1}{2}(n-1)} v^{\frac{1}{2}(n-3)}\,dv,$$

and thus

$$|z|^n F(z) = A|z|^n \int_{-1}^{+1} e^{i|z|t}(1-t^2)^{n-1}\,dt = \psi(|z|),$$

where

$$\psi(t) = At^n \int_{-1}^{+1} e^{ist}(1-s^2)^{n-1}\,ds.$$

Since n is odd, $\psi(t)$ is odd and an n-fold integration by parts shows that $\psi(t)$ is a Fourier-Stieltjes transform.

Thus 6.4 becomes

6.5 $$\sigma_r(f,y) = (2\pi)^{-n}\int \psi(r|y-z|)|y-z|^{-n} f(z)\,dz.$$

Since, for each r, $\psi(rt)$ is the Fourier-Stieltjes transform of a function whose total variation is independent of r, Theorem 3 applied to the preceding integral yields

6.6 $$\|\sigma_r(f,y)\|_p \leqq A\|f\|_p,$$

with A independent of r.

Suppose now that $f(x)$ has continuous derivatives of all orders and vanishes outside a bounded set. Then $\dot{f}$ is absolutely integrable and the inversion theorem for Fourier transforms shows that $\sigma_r(f,y)$ converges to $f(y)$ as $r\to\infty$. Furthermore $\sigma_r(f,y)$ is bounded uniformly in r and an inspection of 6.5 shows that

$$|\sigma_r(f,y)| \leqq A|y|^{-n}$$

for $|y|$ sufficiently large, regardless of the value of r. This makes it clear that

6.7 $$\| \sigma_r(f, y) - f(y) \|_p \to 0$$

as $r \to \infty$.

In the general case, given $f \varepsilon L^p$ and $\epsilon > 0$ there exists a function g with continuous derivatives of all orders and vanishing outside a bounded set such that $\| f - g \|_p < \epsilon$. Then

$$\| \sigma_r(f, y) - f(y) \|_p \leqq \| \sigma_r(g, y) - g(y) \|_p + \| \sigma_r(f - g, y) \|_p + \| f - g \|_p,$$

and according to 6.6 and 6.7 as $r \to \infty$ the right-hand side has an upper limit not exceeding $(A + 1)\epsilon$. Since ϵ is arbitrary, the proof is complete.

In closing this section we point out that if $\sigma_r(f, y)$ is defined directly by means of 6.4, then the theorem holds for $f \varepsilon L^p$ for any p, $1 < p < \infty$. We also remark that the same method of proof can be used to establish the corresponding result for other appropriate methods of spherical summation.

MASSACHUSETTS INSTITUTE OF TECHNOLOGY AND
THE UNIVERSITY OF CHICAGO.

REFERENCES.

[1] A. P. Calderón and A. Zygmund, "On the existence of certain singular integrals," *Acta Mathematica*, vol. 88 (1952), pp. 85-139.

[2] J. Horvàth, "Sur les fonctions conjuguées à plusieurs variables," Koninkligke Nederlandse Akademie van Wetenschappen. Indagationes Mathematicae ex Actis Quibus Titulis. *Proceedings of the Section of Sciences*, vol. 15, No. 1 (1953), pp. 17-29.

[3] A. Zygmund, *Trigonometrical series*, Warsaw (1935).

ALGEBRAS OF CERTAIN SINGULAR OPERATORS.*

By A. P. CALDERÓN and A. ZYGMUND.[1]

1. In this note we study composition of singular integral operators of a type we have considered in earlier work.

Let $x=(\xi_1,\xi_2,\cdots,\xi_n)$ denote either a point of Euclidean n-space of coordinates $\xi_1,\xi_2,\cdots,\xi_n$, or the vector from $0=(0,0,\cdots,0)$ to $(\xi_1,\xi_2,\cdots,\xi_n)$, and $|x|$ its length, that is $(\xi_1^2+\cdots+\xi_n^2)^{\frac{1}{2}}$.

If $K(x)$ is a homogeneous function of degree $-n$, i.e. such that

$$K(\lambda x)=\lambda^{-n}K(x)$$

for every x and every $\lambda>0$, and if

$$(1.1)\qquad \int_\Sigma K(x)\,d\sigma=0 \quad\text{and}\quad \int_\Sigma |K(x)|^p\,d\sigma<\infty$$

for some $p>1$, the integral being taken over the unit sphere Σ, $|x|=1$, and $d\sigma$ denoting the elements of "area" of Σ, then for $f\in L^r$

$$(1.2)\qquad \tilde{f}_\epsilon(x)=\int_{|x-y|>\epsilon} K(x-y)f(y)\,dy$$

converges pointwise almost everywhere and in the mean order r as ϵ tends to zero, and the operation of taking f into the limit $\tilde{f}$ of the integral above is continuous in L^r, and

$$(1.3)\qquad \|\tilde{f}\|_r\leqq A_{r,p}\Big[\int_\Sigma |K(x)|^p\,d\sigma\Big]^{1/p}\|f\|_r,$$

where A_{rp} is a constant depending on p and r only (see [3], remark to Section 5).

This result suggests studying composition of operators of the form

$$(1.4)\qquad \mathcal{K}(f)=\alpha f+\tilde{f},$$

where α is a complex constant.

We shall consider

i) the class $\mathfrak{A}$ of all operators with $K(x)$ in C^∞ in $|x|>0$, that is with $K(x)$ possessing derivatives of all orders if $x\neq 0$;

* Received September 9, 1955.

[1] The research resulting in this paper was supported in part by the office of Scientific Research of the Air Force under contract AF 18(600)-1111.

Reprinted from *Amer. J. Math.* 78, 310–320 (1956).

ii) the class $\mathcal{A}_p$, $p > 1$, of all operators $\mathcal{K}$ for which

$$\| \mathcal{K} \|_p = | \alpha | + [\int_\Sigma | K(x) |^p d\sigma]^{1/p} < \infty. \tag{1.5}$$

Such classes are obviously closed under addition and multiplication by scalars. To show that they are also closed under composition (operator multiplication) is one of the purposes of this note. Moreover, we intend to prove that $\mathcal{A}_p$, when endowed with the norm (1.5), is a commutative semi-simple Banach algebra. This will follow from the inequality

$$\| \mathcal{K}\mathcal{H} \|_p \leqq A_p \| \mathcal{K} \|_p \| \mathcal{H} \|_p, \tag{1.6}$$

where A_p is a constant depending on p only.

Since all operators under consideration are continuous in L^2 and commute with translations, if $\mathcal{F}(g)$ denotes the Fourier transform of g, and $f \varepsilon L^2$, we must have

$$\mathcal{F}[\mathcal{K}(f)] = \mathcal{F}(f)\mathcal{F}(\mathcal{K}),[2] \tag{1.7}$$

where $\mathcal{F}(\mathcal{K})$ is a bounded function which we shall call the *Fourier transform of* $\mathcal{K}$. Further, and according to (1.3) and (1.5), and assuming that the constant $A_{r,p}$ in (1.3) is $\geqq 1$,

$$| \mathcal{F}(\mathcal{K}) | \leqq A_{2,p} \| \mathcal{K} \|_p. \tag{1.8}$$

As we shall see, $\mathcal{F}(\mathcal{K})$ is actually a homogeneous function of degree zero, continuous in $x \neq 0$. If $\mathcal{K} \varepsilon \mathcal{A}$, then $\mathcal{F}(\mathcal{K})$ is in addition of class C^∞ in $x \neq 0$. Conversely, every homogeneous function of degree zero possessing derivatives of all orders in $x \neq 0$ is the Fourier transform of an operator in $\mathcal{A}$.

Finally we shall prove that an operator in $\mathcal{A}$ or $\mathcal{A}_p$ has an inverse in the same class if and only if its Fourier transform does not vanish. Since we may identify homogeneous functions of degree zero with their restrictions to the sphere $| x | = 1$, we can translate the last statement into the language of Banach Algebras and assert that the space of maximal ideals of $\mathcal{A}_p$ is homeomorphic with the sphere $| x | = 1$.

For the convenience of the reader we summarize our results in the following formal statements.

---

[2] In our special case this known general statement also follows from the fact that the Fourier transform of $K_\lambda(x)$, defined to be $K(x)$ for $| x | > \lambda$ and zero otherwise, converges boundedly to a bounded function as $\lambda \to 0$ (see [2], pp. 89-91). For if $f \varepsilon L^2$, the Fourier transform of (1.1) converges in L^2 to the product of a bounded function depending on K only and the Fourier transform of f.

THEOREM 1. *If $\mathfrak{A}$ is the class of all operators defined in* i), *then $\mathfrak{A}$ is closed under addition and operator multiplication. The Fourier transform of an operator in this class is a homogeneous function of degree zero and of C^∞ in $x \neq 0$, and conversely every such homogeneous function is the Fourier transform of an operator in $\mathfrak{A}$. The Fourier transform of the product of two operators is the product of their Fourier transforms, and consequently an operator in $\mathfrak{A}$ has an inverse in $\mathfrak{A}$ if and only if its Fourier transform does not vanish. If $k(x)$ is the Fourier transform of the kernel $K(x)$, and $\beta(k)$ is the least common upper bound for the absolute value of k and of its derivatives up to order $n+1$ evaluated in $|x| \geqq 1$, then for $|x| = 1$ we have*

$$|K(x)| \leqq A\beta(k), \tag{1.9}$$

where A is a constant independent of K.

THEOREM 2. *If $\mathfrak{A}_p$ is the class of all operators defined in* ii) *and endowed with the norm* (1.5), *then $\mathfrak{A}_p$ becomes a semisimple commutative Banach Algebra under operator multplication, and* (1.6) *holds for the norm of the product of two operators. Then Fourier product of an operator in $\mathfrak{A}_p$ is a homogeneous function of degree zero continuous in $|x| \neq 0$, and the Fourier transform of a product is the product of the Fourier transforms of the factors.*

The existence of inverses and the functional calculus of operators in $\mathfrak{A}_p$ is based on the following two theorems.

THEOREM 3. *Let $g(x)$ be a function defined on the sphere Σ ($|x| = 1$) which is locally a restriction to Σ of Fourier transforms of operators in $\mathfrak{A}_p$; that is, every $x_0 \in \Sigma$ is contained in a neighborhood where $g(x)$ coincides with the restriction to Σ of the Fourier transform of an operator in $\mathfrak{A}_p$. Then there exists a single operator $\mathfrak{H}$ in $\mathfrak{A}_p$ whose Fourier transform coincides with g at all points of Σ.*

THEOREM 4. *Let $g(x)$ be a function defined on the sphere Σ, which is locally an analytic function of the restriction $h(x)$ to Σ of the Fourier transform of an operator in $\mathfrak{A}_p$; that is, for every $x_0 \in \Sigma$ there exists a power series $\Sigma a_n Z^n$ with positive radius of convergence such that $g(x) = \Sigma a_n[h(x) - h(x_0)]^n$ for x in some neighborhood of x_0. Then $g(x)$ is a restriction to Σ of the Fourier transform of an operator in $\mathfrak{A}_p$.*

COROLLARY. *An operator in $\mathfrak{A}_p$ has an inverse in $\mathfrak{A}_p$ if and only if its Fourier transform does not vanish. The space of maximal ideals of $\mathfrak{A}_p$ is homeomorphic to the sphere Σ ($|x| = 1$).*

One thing here must be stressed.

If an operator in $\mathfrak{A}_p$ is thought of as acting in the space L^2 of square integrable functions, then the fact that its Fourier transform does not vanish implies immediately that there is a bounded operator in L^2 which is the inverse of the given one. The fact, however, that this operator is in $\mathfrak{A}_p$ is non-trivial, and this fact is, of course, the essence of the preceding corollary. A similar remark applies to Theorems 2, 3 and 4.

The content of Theorem 4 can be described briefly by saying that an analytic function of the Fourier transform $h(x)$ of an operator in $\mathfrak{A}_p$ is again the Fourier transform of an operator in $\mathfrak{A}_p$. This analytic function need not be single valued, and the values of $h(x)$ might even be allowed to go through branch points of the function, provided that the conditions of Theorem 4 are respected at such points x. For example, if the function $h(x)$ has a continuous square root and coincides locally with Fourier transforms of operators in $\mathfrak{A}_p$ at all points where $h(x)$ vanishes then $h(x)$ has a square root in $\mathfrak{A}_p$.

The preceding theorems apply immediately to systems of singular integral operators in $\mathfrak{A}$ or $\mathfrak{A}_p$. Such systems may be thought of as a convolution of a square singular matrix kernel with a vector function plus a numerical matrix applied to the same function. The condition of invertibility then becomes that the matrix of the corresponding Fourier transforms have a nonvanishing determinant.

2. With things organized as we have them here it will be convenient to study first operators in $\mathfrak{A}$. Once the basic facts about such operators are established and the validity of (1.6) is proved in this special case, everything else will be relatively simple.

The following partly standard notation will be sufficient for our purposes. We shall write $f \cdot g$ for the (absolutely convergent) integral of $f(x)\bar{g}(x)$, $f * g$ for the convolution of f and g, $\mathfrak{F}(f)$ for the Fourier transform of f. We shall also write $g^\lambda(x) = \lambda^n g(\lambda x)$, and denote by $g_\lambda(x)$ the function equal to g if $|x| \geqq \lambda$ and to zero otherwise (the latter notation will apply to kernels only and will not conflict with the notation on the left side of (1.2)).

By Γ we shall denote the class of all functions g of C^∞ such that g and all its derivatives are $O(|x|^{-k})$ as $|x| \to \infty$, for each $k > 0$. The Fourier transform of a function in Γ is in Γ; this we easily see by differentiating under the integral sign and integrating by parts.

We shall call a function f *radial* if it only depends on $|x|$. Fourier transforms of radial functions are radial (see [1] page 67). By a *corradial* function on the other hand we shall mean a function which is orthogonal to all radial functions, i.e. such that $f \cdot g = 0$ for all radial g. The fact just quoted clearly implies that Fourier transforms of corradial functions are corradial.

Homogeneous functions satisfying the first condition 1.1 are corradial in an obvious sense, and conversely every corradial homogeneous function satisfies that condition. We shall therefore refer to homogeneous functions satisfying 1.1 as *corradial homogeneous functions.*

The argument which follows is based on a certain representation of homogeneous functions of a given degree.

Suppose that $g(x)$ is a corradial function in Γ. Then $g(0) = 0$ and

$$\int_0^\infty g^\lambda(x)\lambda^{-n-1+r}d\lambda \tag{2.1}$$

converges absolutely for $r > -1$. Moreover it represents a corradial homogeneous function of degree $-r$. Differentiation under the integral sign shows that this function is of C^∞ in $x \neq 0$.

Conversely, every corradial homogeneous function $K(x)$ of degree $-r$ can thus be represented by setting $g(x) = K(x)\rho(|x|)$ where $\rho(t)$ has continuous derivatives of all orders, vanishes in a neighborhood of 0 and ∞ nad such that

$$\int_0^\infty \lambda^{-1}\rho(\lambda)d\lambda = 1. \tag{2.2}$$

Let $K(x)$ be corradial homogeneous of degree $-n$ and of C^∞ in $x \neq 0$. Then, if $x \neq 0$, $K_\lambda(x)$ converges to $K(x)$ as $\lambda \to 0$, and it is not difficult to prove (see [2], pp. 89-91) that also $\mathfrak{F}(K_\lambda)$ converges pointwise and boundedly to a limit which we shall denote by $\mathfrak{F}(K)$. Consequently, if $f \varepsilon L^2$, $K_\lambda * f$ converges in mean of order 2, and the Fourier transform of its limit is $\mathfrak{F}(f)\mathfrak{F}(K)$. Thus the Fourier transform $\mathfrak{F}(\mathcal{K})$ of the operator in (1.4) is precisely $\alpha + \mathfrak{F}(K)$.

We shall prove presently that this function is homogeneous of degree zero and of C^∞ in $x \neq 0$, and that conversely every function with such properties is of this form. This will imply immediately that $\mathfrak{A}$ is closed under operator multiplication, as we stated in Theorem 1.

Let $\rho(x)$ be a radial function of C^∞ such that $\rho(0) = 1$, $\rho(x) = 0$ for $|x| \geqq 1$, and let $f(x)$ be any function of C^∞ vanishing outside a bounded set. Then $K_\lambda \cdot \rho = 0$ and

$$\begin{aligned} \mathfrak{F}(f)\cdot\mathfrak{F}(K) &= \lim_{\lambda\to 0}\mathfrak{F}(f)\cdot\mathfrak{F}(K_\lambda) = \lim_{\lambda\to 0}(f\cdot K_\lambda) \\ &= \lim_{\lambda\to 0}[f - f(0)\rho]\cdot K_\lambda = [f - f(0)\rho]\cdot K, \end{aligned} \tag{2.3}$$

the last integral being absolutely convergent since $f(x)-f(0)\rho(x)$ vanishes at 0. We now represent $K(x)$ by the formula (2.1) with $r=n$ and obtain

$$[f-f(0)\rho]\cdot K=[f-f(0)\rho]\cdot\int_0^\infty \lambda^{-1}g^\lambda d\lambda=\int_0^\infty \lambda^{-1}[f-f(0)\rho]\cdot g^\lambda d\lambda,$$

the change of the order of integration being justified by absolute convergence.

Now since g^λ is corradial and ρ radial we have $g^\lambda\cdot\rho=0$, and since

$$\mathfrak{F}(g^\lambda)=\lambda^n\mathfrak{F}(g)^{\lambda^{-1}},$$

as seen by changing variables in the Fourier integral of g^λ, setting $\lambda^{-1}=\mu$ we may further write

$$\begin{aligned}(2.4)\quad &\int_0^\infty \lambda^{-1}[f-f(0)\rho]\cdot g^\lambda d\lambda=\int_0^\infty \lambda^{-1}(f\cdot g^\lambda)d\lambda=\int_0^\infty \mathfrak{F}(f)\cdot\mathfrak{F}(g^\lambda)\lambda^{-1}d\lambda\\ &=\int_0^\infty \mathfrak{F}(f)\cdot\mathfrak{F}(g)^\mu\mu^{-n-1}d\mu=\mathfrak{F}(f)\cdot\int_0^\infty \mathfrak{F}(g)^\mu\mu^{-n-1}d\mu\end{aligned}$$

changes of the order of integration being again justified by the absolute convergence of the integrals involved.

From the equality of the left side of (2.3) and the right side of (2.4) we conclude that if

$$(2.5)\qquad K(x)=\int_0^\infty \lambda^{-1}g^\lambda(x)\,d\lambda,$$

then

$$(2.6)\qquad \mathfrak{F}(K)=\int_0^\infty \mathfrak{F}(g)^\lambda\lambda^{-n-1}d\lambda.$$

Since these integrals represent the most general corradial homogeneous functions of degrees $-n$ and 0 respectively, of C^∞ in $x\neq 0$, we have proved *that* $\mathfrak{F}(\mathcal{K})$ is corradial homogeneous of degree zero and of C^∞ in $x\neq 0$, and that conversely every function with these properties is an $\mathfrak{F}(\mathcal{K})$.

We now pass to the proof of (1.9). For this purpose we assume that $K(x)$ is represented as in (2.5), and that $|x|=1$. Then

$$\begin{aligned}|K(x)|&=\left|\int_0^\infty g(\lambda x)\lambda^{n-1}d\lambda\right|\\ &\leq \sup|g(y)|\int_0^1 \lambda^{n-1}d\lambda+\sup|g(y)|\,|y|^{n+1}\int_1^\infty \lambda^{-2}d\lambda,\end{aligned}$$

and we only have to estimate $\sup|g(y)|$ and $\sup|g(y)|\,|y|^{n+1}$ in terms of $\mathfrak{F}(K)$.

Let $\hat{g}$ denote the Fourier transform of g, and let $\xi_1,\xi_2,\cdots,\xi_n$ be the coordinates of x, and $\eta_1,\cdots,\eta_n$ those of y. Then

$$g(x)=\int e^{2\pi i(x\cdot y)}\hat{g}(y)dy,$$

$$\xi_k^{n+1}g(x)=(2\pi i)^{-n-1}\int e^{2\pi i(x\cdot y)}(\partial^{n+1}/\partial\eta_k^{n+1})\hat{g}(y)dy.$$

By Hölder's inequality $|x|^{n+1}\leqq n^{\frac{1}{2}(n-1)}\sum|\xi_i|^{n+1}$, and thus

$$|g(x)||x|^{n+1}\leqq(2\pi)^{-n-1}n^{\frac{1}{2}(n-1)}\sum_k\int|(\partial^{n+1}/\partial\eta_k^{n+1})\hat{g}(y)|\,dy,$$

$|g(x)|\leqq\int|\hat{g}(y)|\,dy$, and it only remains to estimate the integral on the right in terms of $\mathfrak{F}(K)$.

For this purpose we set, as we may, $\hat{g}(y)=\mathfrak{F}(K)\rho(|y|)$, where $\rho(\lambda)$ is of C^∞, vanishes outside $1\leqq\lambda\leqq 2$ and satisfies (2.2). This function we choose once for all independently of the particular kernel K under consideration. This makes it clear that $\hat{g}(y)$ and its derivatives can be estimated in terms of the derivatives of $\mathfrak{F}(K)$ and $\mathfrak{F}(K)$ itself. Furthermore $\hat{g}(y)$ vanishes outside $1\leqq|y|\leqq 2$, and this fact makes it possible to estimate the integral above in terms of $\beta[\mathfrak{F}(K)]$. Collecting estimates we obtain (1.9).

Finally, from (1.9) we readily obtain

$$\|\mathcal{K}\|_p\leqq A\beta[\mathfrak{F}(\mathcal{K})] \tag{2.8}$$

for any operator $\mathcal{K}$ in $\mathcal{A}$, A being a constant independent of p and $\mathcal{K}$, but not necessarily the same as in (1.9).

3. In this section we prove (1.6) for operators in $\mathcal{A}$.

Let K and H be two corradial homogeneous functions of degree $-n$ of C^∞ in $x\neq 0$, and consider

$$K*H_\lambda=\lim_{\mu\to 0}K_\mu*H_\lambda. \tag{3.1}$$

As $\lambda\to 0$, this function converges pointwise for $x\neq 0$, and its limit J is a homogeneous function of degree $-n$. On the other hand, the Fourier transform of (3.1) converges to $\mathfrak{F}(K)\mathfrak{F}(H)$. We will show that J is corradial and that

$$\mathfrak{F}(K)\mathfrak{F}(H)=\alpha+\mathfrak{F}(J), \tag{3.2}$$

where α is a constant, and that

$$\|\mathcal{J}\|_p\leqq A_p\|\mathcal{K}\|_p\|\mathcal{H}\|_p, \tag{3.3}$$

$$|\alpha|\leqq A_p\|\mathcal{K}\|_p\|\mathcal{H}\|_p, \tag{3.4}$$

$\mathcal{H}$, $\mathcal{J}$, $\mathcal{K}$ being the convolution operators with kernels H, J, K, respectively.

First we easily see that $H_\lambda = (H_1)^{\lambda^{-1}}$; $K * H_\lambda = (K * H_1)^{\lambda^{-1}}$. Thus $K * H_\lambda - J_\lambda = (K * H_1 - J_1)^{\lambda^{-1}}$. Since H_1 is in L^2 the same holds for $K * H_1$. It follows that $K * H_1 - J_1$ is integrable over bounded sets. On the other hand it is not difficult to see that $K * H_1 - J_1$ is of order $|x|^{-n-1}$ as $|x| \to \infty$. Hence $K * H_1 - J_1$ is absolutely integrable.

Let now f be a function in Γ. A change of variables gives

$$(K * H_\lambda - J_\lambda) \cdot f = (K * H_1 - J_1)^{\lambda^{-1}} \cdot f = (K * H_1 - J_1) \cdot \lambda^{-n} f^\lambda$$

As $\lambda \to 0$, $\lambda^{-n} f^\lambda$ tends to $f(0)$ while remaining bounded, and this implies that $(K * H_\lambda - J_\lambda) \cdot f$ converges as $\lambda \to 0$. Now

$$K * H_\lambda \cdot f = \mathfrak{F}(K)\, \mathfrak{F}(H_\lambda) \cdot \mathfrak{F}(f)$$

also converges as $\lambda \to 0$, and consequently the same holds for $J_\lambda \cdot f$. But if $f(0) \neq 0$, $J_\lambda \cdot f$ cannot converge unless J is corradial. Hence J is corradial and $\mathfrak{F}(J_\lambda)$ converges boundedly to a limit $\mathfrak{F}(J)$.

Suppose now that $f(0) = 0$. Then

$$\begin{aligned}[\mathfrak{F}(K)\mathfrak{F}(H) - \mathfrak{F}(J)] \cdot \mathfrak{F}(f) &= \lim_{\lambda \to 0} [\mathfrak{F}(K)\mathfrak{F}(H_\lambda) - \mathfrak{F}(J_\lambda)] \cdot \mathfrak{F}(f) \\ &= \lim_{\lambda \to 0} (K * H_\lambda - J_\lambda) \cdot f.\end{aligned}$$

Since $f(0) = 0$, the last limit is zero, as we pointed out above. Consequently, if $g = \mathfrak{F}(f)$ we have

$$[\mathfrak{F}(K)\mathfrak{F}(H) - \mathfrak{F}(J)] \cdot g = 0$$

for any $g \in \Gamma$ with vanishing integral, and this is possible only if $\mathfrak{F}(K)\mathfrak{F}(H) - \mathfrak{F}(J)$ is a constant.

Next we estimate $\| \mathcal{J} \|_p$. First we note that, on account of homogeneity,

$$\left[\int | J_\lambda(x) |^p \, dx \right]^{1/p} = [n(p-1)\lambda^{(p-1)n}]^{-1/p} \| \mathcal{J} \|_p, \tag{3.5}$$

and similarly for H and K. Next, for $|x| \geqq 2$ we have

$$\begin{aligned}J_2(x) &= [(K - K_1) + K_1] * [(H - H_1) + H_1] \\ &= (K - K_1) * (H - H_1) + (K - K_1) * H_1 + (H - H_1) * K_1 + H_1 * K_1,\end{aligned}$$

and since the first term in the last sum vanishes for $|x| \geqq 2$, we see that

$$J_2(x) = K * H_1 + H * K_1 - H_1 * K_1$$

for $|x| \geqq 2$. Now (3.5) and (1.3) applied to this inequality yield (3.3). And from (1.8), (3.3) and (3.2) we easily derive (3.4).

It is clear that (3.3) and (3.4) imply (1.6).

4. The extension of (1.6) to operators in $\mathcal{A}_p$ is straightforward.

Given two operators $\mathcal{H}$ and $\mathcal{K}$ in $\mathcal{A}_p$, we take two sequences of operators $\mathcal{H}_n$, $\mathcal{K}_n$ in $\mathcal{A}$ such that

$$\| \mathcal{H}_n - \mathcal{H} \|_p \to 0; \qquad \| \mathcal{K}_n - \mathcal{K} \|_p \to 0.$$

Then from the validity of (1.6) for operators in $\mathcal{A}$ it follows that $\mathcal{H}_n \mathcal{K}_n$ is a Cauchy sequence in $\mathcal{A}_p$, and therefore converges to a limit $\mathcal{J}$ in $\mathcal{A}_p$ for which the inequality $\| \mathcal{J} \|_p \leqq A_p \| \mathcal{H} \|_p \| \mathcal{K} \|_p$ holds. Consequently, if we show that $\mathcal{J} = \mathcal{H}\mathcal{K}$ we will have shown that $\mathcal{A}_p$ is closed under multiplication (composition) and that (1.6) holds for the product.

Consider (1.3) and (1.5). Assuming, as we may, that $A_{r,p} \geqq 1$, we see that $\mathcal{K}(f) = \alpha f + \tilde{f}$ satisfies $\| \mathcal{K}(f) \|_r \leqq A_{r,p} \| \mathcal{K} \|_p \| f \|_r$. Consequently the operator norm of $\mathcal{K}$ as an operator in L^r, which is defined as

$$\sup_f \| \mathcal{K}(f) \|_r / \| f \|_r,$$

is dominated by $A_{r,p} \| \mathcal{K} \|_p$. Since $\mathcal{H}_n \to \mathcal{H}$ and $\mathcal{K}_n \to \mathcal{K}$ in $\mathcal{A}_p$, the same holds in the operator topology, and consequently $\mathcal{H}_n \mathcal{K}_n \to \mathcal{H}\mathcal{K}$ in the operator topology. On the other hand, $\mathcal{H}_n \mathcal{K}_n \to \mathcal{J}$ in $\mathcal{A}_p$, and consequently the same holds in the operator topology. Hence $\mathcal{J} = \mathcal{H}\mathcal{K}$ and the proof is completed.

This also completes the proof of Theorem 2 since the fact that the Fourier transforms of operators in $\mathcal{A}_p$ are continuous homogeneous functions of degree zero follows readily from (1.8) and the fact that $\mathcal{A}$ is dense in $\mathcal{A}_p$ and its elements have continuous Fourier transforms.

5. We now proceed to prove Theorems 3 and 4.

We might observe here that if we knew already that every maximal ideal in $\mathcal{A}_p$ is the set of all operators whose Fourier transforms vanish at a point of the unit sphere Σ ($|x| = 1$), then Theorems 3 and 4 would merely be standard facts from Banach Algebras. In our present setup though we can prove Theorems 3 and 4 directly with comparatively little additional effort and obtain the structure of the maximal ideals as a consequence.

For simplicity of notation we shall denote the Fourier transform of an operator $\mathcal{H}$ in $\mathcal{A}_p$ by h. The symbol $\| h \|_p$ will now stand for $\| \mathcal{H} \|_p$, and $\beta(h)$ will denote, as in Theorem 1 or Section 2, the least upper bound for the absolute value of h and its derivatives of order $n+1$, evaluated in $|x| \geqq 1$. Occasionally, instead of working with homogeneous functions of degree zero we shall work with their restrictions to the unit sphere Σ.

Let $g(x)$ be a function on Σ, and suppose that for each x_0 there is a neighborhood N_{x_0} of x_0 and an operator in $\mathfrak{H}_{x_0}$ whose Fourier transform h_{x_0} (restricted to Σ) coincides with $g(x)$ in N_{x_0}. Let N_{x_i}, $i=1,2,\cdots$ be a finite collection of such neighborhoods covering Σ. Let further $k_i \geqq 0$ be functions in C^∞, each vanishing outside N_{x_i} and such that $\Sigma\, k_i(x) > 0$. Then

$$k'_i(x) = k_i(x)\,[\sum_j k_j(x)]^{-1}$$

is also in C^∞ and vanishes outside N_{x_i}. Furthermore $\Sigma\, k'_i(x) = 1$, and consequently

$$g(x) = \Sigma\, g(x) k'_i(x) = \Sigma\, h_{x_i}(x) k'_i(x),$$

since $h_{x_i}(x) = g(x)$ wherever $k'_i(x) \neq 0$. Since $k'_i(x)$ is a restriction to Σ of a homogeneous function of degree zero of C^∞ in $x \neq 0$, which is in turn the Fourier transform of an operator $\mathcal{K}_i$ in $\mathfrak{A}$, the last expression on the right is precisely the restriction to Σ of the Fourier transform of $\Sigma\, \mathfrak{H}_{x_i}\mathcal{K}_i$, and Theorem 3 is thus established.

To prove Theorem 4 we begin by observing that, as an easy computation shows, if $\alpha > |h(x)|$ and $h(x)$ is in C^∞, then $\beta(h^k) = O(\alpha^k)$ as $k \to \infty$. Consequently it follows from (2.8) that $\| h^k \|_p = O(\alpha^k)$.

We now extend this result to the Fourier transform $h(x)$ of an arbitrary operator $\mathfrak{H}$ in $\mathfrak{A}_p$. Given such an $h(x)$ and $\alpha > |h(x)|$, we take α_0 so that $\alpha > \alpha_0 > |h(x)|$, and $h_0(x)$ in C^∞ so that $A_p \| h - h_0 \|_p + \alpha_0 < \alpha$ and $|h_0(x)| < \alpha_0$. Then, by (1.6),

$$\| h^k \|_p = \| [(h - h_0) + h_0]^k \|_p \leqq \sum_{i=0}^{k} \binom{k}{i} \| (h - h_0)^i h_0^{k-i} \|_p$$

$$= O[\sum_{i=0}^{k} A_p^i \binom{k}{i} \| h - h_0 \|_p^i \alpha_0^{k-i}] = O[(A_p \| h - h_0 \|_p + \alpha_0)^k] = O(\alpha^k).$$

Let now $F(z) = \Sigma\, a_k (z - z_0)^k$ be analytic in $|z - z_0| < 2\epsilon$, $h(x)$ the Fourier transform of an operator in $\mathfrak{A}_p$, and $h(x_0) = z_0$. If we show that $g(x) = F[h(x)]$ coincides with the Fourier transform of an operator in $\mathfrak{A}_p$ in some neighborhood of x_0, a repeated application of this result and Theorem 3 will yield Theorem 4. For this purpose we take $0 \leqq k(x) \leqq 1$ homogeneous of degree zero, equal to 1 in a neighborhood of x_0 and vanishing wherever $|h(x) - h(x_0)| \geqq \epsilon$, and define

$$h'(x) = h(x_0) + k(x)\,[h(x) - h(x_0)].$$

Clearly we have $|h'(x) - h(x_0)| < \epsilon$ and $h'(x) = h(x)$ in a neighborhood of x_0. The series

$$F[h'(x)] = \Sigma\, a_k [h'(x) - h(x_0)]^k$$

coincides with $F[h(x)]$ in a neighborhood of x_0. But since

$$\|[h'(x) - h(x_0)]^k\|_p = O(\epsilon^k),$$

the corresponding series of operators converges in $\mathfrak{A}_p$, and the Fourier transform of its sum is precisely $F[h'(x)]$. Theorem 4 is thus established.

Regarding the Corollary to Theorem 4 we observe that if $h(x)$ does not vanish then $h^{-1}(x)$ satisfies the conditions of Theorem 4 and consequently it is the Fourier transform of an operator in $\mathfrak{A}_p$.

To determine the structure of the maximal ideals in $\mathfrak{A}_p$ we observe that if $h_i = \mathcal{F}(\mathfrak{H}_i)$ and the $\mathfrak{H}_i$ belong to a proper ideal I in $\mathfrak{A}_p$, the $h_i(x)$ must be necessity have a common zero. For otherwise there would exist a finite number of such $h_i(x)$ without common zero, and the function $h(x) = \Sigma h_i\bar{h}_i > 0$ would be the Fourier transform of an invertible operator in I, and I would not be a proper ideal. Consequently a maximal ideal in $\mathfrak{A}_p$ consists of all operators whose Fourier transform vanish at a point of Σ, and conversely.

MASSACHUSETTS INSTITUTE OF TECHNOLOGY AND
THE UNIVERSITY OF CHICAGO.

REFERENCES.

[1] S. Bochner and K. Chandrasekharan, "Fourier transforms," *Annals of Mathematics Studies*, 19 (1949).

[2] A. P. Calderón and A. Zygmund, "On the existence of certain singular integrals," *Acta Mathematica*, vol. 88 (1952), 85-139.

[3] ———, "On singular integrals," *American Journal of Mathematics*, vol. 78 (1956), pp. 289-309.

On a theorem of Marcinkiewicz concerning interpolation of operations;

PAR A. ZYGMUND
(Université de Chicago),

1. INTRODUCTION. — Some 30 years ago M. Riesz proved a theorem about linear operations which not only simplified the proofs of existing results but supplied a general and unifying point of view in a number of problems seemingly without connection (*see* [6]; numbers in square brackets refer to the bibliography at the end of the paper). The gist of Riesz's theorem is that linear operations defined for certain classes of functions can be " interpolated ", that is defined for some "intermediate" classes of functions. A precise formulation of Riesz's theorem is given below.

There are important cases to which Riesz's theorem does not apply. In 1939, in a brief Note in the *Comptes rendus*, Marcinkiewicz stated [4] without proof a result about the interpolation of (not necessarily linear) operations which in some cases succeeds where Riesz's theorem fails. Neither theorem however is more general than the other.

The Note of Marcinkiewicz seems to have escaped attention and one does not find allusion to it in the existing literature. The purpose of this paper is to supply a proof of the theorem of Marcinkiewicz, present the theorem in a more general form, and find some new applications for it ([1]).

([1]) *See* footnote ([2]).

Reprinted from *J. Math. Pures Appl.* (9) 35, 223–248 (1956).

2. THE RIESZ-THORIN INTERPOLATION THEOREM. — To make the meaning and significance of the theorem of Marcinkiewicz clear we begin by giving a short account of the theorem of Riesz in the form generalized by Thorin (*see* [10]; also [10], [9], [2], [7], [8]).

Denote by $a, b, a_1, b_1, \ldots$ numbers of the closed interval $1 \leqq a \leqq \infty$, and by $\alpha = \frac{1}{a}$, $\beta = \frac{1}{b}$, $\alpha_1 = \frac{1}{\alpha_1}, \cdots$ their reciprocals. Thus α, β, $\alpha_1, \ldots$ are points of the closed interval $(0, 1)$.

Consider two spaces R and S —for simplicity Euclidean spaces— with non-negative and completely additive measures μ and ν respectively.

We say that

$$h = Tf$$

is a linear operation *of type* (a, b), if :

1. Tf is defined for each $f \in L^a_\mu(R)$, that is for each f measurable with respect to μ and such that

$$\|f\|_{a,\mu} = \left(\int_R |f|^a d\mu\right)^{\frac{1}{a}}$$

is finite, the right side being interpreted as the essential upper bound (with respect to μ) of $|f|$ if $a = \infty$;

2. for every $f \in L^a_\mu(R)$, $h = Tf$ is in $L^b_\nu(S)$ and

$$\|h\|_{b,\nu} \leqq M \|f\|_{a,\mu}, \tag{2.1}$$

where M is independent of f;

3. Tf is additive, i. e.

$$T(k_1 f_1 + k_2 f_2) = k_1 Tf_1 + k_2 Tf_2 \tag{2.2}$$

for all scalars k_1, k_2.

The least M in (2.1) is *the* (a, b) *norm* of the operation T.

Suppose now that $h = Tf$ is simultaneously of types $\left(\frac{1}{\alpha_1}, \frac{1}{\beta_1}\right)$ and $\left(\frac{1}{\alpha_2}, \frac{1}{\beta_2}\right)$, i. e. that

$$\|h\|_{\frac{1}{\beta_k},\nu} \leqq M_k \|f\|_{\frac{1}{\alpha_k},\mu}$$

for $k=1, 2$, and denote by (α, β) any point of the segment

$$\text{(L)} \qquad \alpha=(1-t)\alpha_1+t\alpha_2, \qquad \beta=(1-t)\beta_1+t\beta_2 \qquad (0\leq t\leq 1).$$

The Riesz-Thorin theorem asserts that we can then extend the operation T, in a unique way, to all functions f in $L^{\frac{1}{\alpha}}(R)$ so that T becomes of type $\left(\frac{1}{\alpha}, \frac{1}{\beta}\right)$, and moreover that the least M in (2.1) satisfies the inequality

$$M\leq M_1^{1-t}M_2^t$$

(i. e. log M is a convex function of the point on the segment L).

The most interesting application of this result was given by M. Riesz himself who showed that the Hausdorff-Young-F. Riesz theorem about Fourier coefficients is an immediate consequence of it. A number of other applications have been found since.

There are, however, situations when an operation T is of type $\left(\frac{1}{\alpha}, \frac{1}{\beta}\right)$ for each point (α, β) interior to the segment L but not for both endpoints. Two cases illustrate this possibility.

a. The operation

$$\tilde{f}=Tf,$$

where f is an integrable function of period 2π, and

$$\tilde{f}(x)=-\frac{1}{\pi}\int_{-\pi}^{\pi} f(x+t)\frac{1}{2}\operatorname{cotg}\frac{1}{2}t\,dt$$

is the conjugate function of f, is known to be of type $\left(\frac{1}{\alpha}, \frac{1}{\alpha}\right)$ for each $0<\alpha<1$. In other words,

$$(2.3) \qquad \left(\int_0^{2\pi} |\tilde{f}|^p dx\right)^{\frac{1}{p}}\leq A_p\left(\int_0^{2\pi} |f|^p dx\right)^{\frac{1}{p}} \qquad (1<p<\infty).$$

The operation is however neither of type (1, 1) nor of type (∞, ∞), since the conjugate of an integrable function need not be integrable, and the conjugate of a bounded function need not be bounded. It follows that the Riesz interpolation theorem cannot be applied to interpolate between $p=1$ and $p=\infty$.

b. Suppose that $\varphi_1(x)$, $\varphi_2(x)$, ... is an orthonormal system defined on $0 \leq x \leq 1$, and that

$$|\varphi_n(x)| \leq M$$

for all n. By the Paley generalization of a theorem of Hardy and Littlewood (*see* e. g. [10], p. 202, sqq.), the Fourier coefficients

$$c_n = \int_0^1 f(x)\varphi_n(x)\,dx$$

satisfy an inequality

$$(2.4) \qquad \left(\sum_{n=1}^{\infty} |c_n|^p n^{p-2}\right)^{\frac{1}{p}} \leq A_p M^{\frac{2-p}{p}} \left(\int_0^1 |f|^p dx\right)^{\frac{1}{p}},$$

provided

$$1 < p \leq 2.$$

For $p = 2$, (2.4) is obviously a consequence of Bessel's inequality, but for $p = 1$ the inequality (2.4) fails : the left side may be infinite if f is merely integrable. A simple example is provided by

$$f(x) \sim \sum_{n=2}^{\infty} \frac{\cos 2\pi nx}{\log n}.$$

The left side of (2.4) can be written

$$(2.5) \qquad \left(\sum |nc_n|^p n^{-2}\right)^{\frac{1}{p}} = \|h\|_{p,\nu},$$

if $h(x)$ denotes a function which takes the value nc_n at the point $x = n (n = 1, 2, \ldots)$, and is arbitrary elsewhere; and the measure ν of the set consisting of a single point n is $\frac{1}{n^2}$, and is zero for any set which contains no point n. Hence, by (2.4), the operation

$$(2.6) \qquad h = \{nc_n\} = Tf$$

is of type $\left(\frac{1}{\alpha}, \frac{1}{\alpha}\right)$ for each $\frac{1}{2} \leq \alpha < 1$, but not for $\alpha = 1$. It follows that (2.4) cannot be deduced from Riesz's theorem by interpolating between $\alpha = 1$ and $\alpha = \frac{1}{2}$.

We shall see in Sections 4 and 5 below that both a and b can be treated by means of the interpolation theorem of Marcinkiewicz.

3. THE THEOREM OF MARCINKIEWICZ. — Return to the inequality (2.1) and suppose first that $1 \leq b < \infty$. Given any $y > 0$ denote by $E_y = E_y[h]$ the set of points of the space S where

$$|h| > y,$$

and write $\nu(E_y)$ for the ν measure of the set E_y. An immediate consequence of (2.1) is that

$$\nu(E_y[h]) \leq \left(\frac{M}{y}\|f\|_a\right)^b \tag{3.1}$$

(if no confusion arises we omit the symbols μ and ν in the notation for norms). An operation T which satisfies (3.1) will be said to be of *weak type* (a, b); the least value of M in (3.1) may be called the weak (a, b) norm of T. For the sake of emphasis, operations of type (a, b) will be occasionally called of *strong type* (a, b). Every operation of strong type (a, b) is also of weak type (a, b). The converse is not true: the operation $\tilde{f} = Tf$ mentioned above is not of strong type (1.1), though it is of weak type (1.1) since by a well known theorem of Kolmogoroff the Lebesgue measure $|E_y|$ of the set $E \subset (0, 2\pi)$ in which $|\tilde{f}(x)| > y$ satisfies an inequality

$$|E_y| \leq \frac{A}{y}\int_0^{2\pi} |f\, dx,$$

where A is an absolute constant.

Other examples of operations of weak type (a, b) will be considered below.

We have defined weak type (a, b) for $b < \infty$. We define weak type (a, ∞) as identical with strong type (a, ∞). Hence T is of (weak, strong) type (a, ∞) if

$$\text{ess sup}\, |h| \leq M\|f\|_a.$$

We no longer require that the operations T we consider be linear. An operation $h = Tf$ will be called *quasi-linear* if $T(f_1 + f_2)$ is uniquely defined whenever Tf_1 and Tf_2 are defined, and if

$$|T(f_1+f_2)| \leq \varkappa(|Tf_1|+|Tf_2|) \tag{3.2}$$

where $\varkappa$ is a constant independent of f_1 and f_2. The special case $\varkappa=1$ is not infrequent in applications and may be called *sublinear*.

We can now state the theorem of Marcinkiewicz as follows.

THEOREM 1. — *Let (α_1, β_1) and (α_2, β_2) be any two points of the triangle*

$$0 \leq \beta \leq \alpha \leq 1 \tag{Δ}$$

such that $\beta_1 \neq \beta_2$. Suppose that a quasi-linear operation $h=Tf$ is simultaneously of weak types $\left(\frac{1}{\alpha_1}, \frac{1}{\beta_1}\right)$ and $\left(\frac{1}{\alpha_2}, \frac{1}{\beta_2}\right)$, with norms M_1 and M_2 respectively. Then for any point (α, β) of the open segment

$$\alpha=(1-t)\alpha_1+t\alpha_2, \quad \beta=(1-t)\beta_1+t\beta_2 \quad (0<t<1) \tag{L}$$

the operation is of strong type $\left(\frac{1}{\alpha}, \frac{1}{\beta}\right)$ and we have

$$\|h\|_{\frac{1}{\beta}} \leq KM_1^{1-t}M_2^t \|f\|_{\frac{1}{\alpha}}, \tag{3.3}$$

where $K=K_{t,\varkappa,\alpha_1,\beta_1,\alpha_2,\beta_2}$ is independent of f and is bounded if α_1, β_1, α_2, β_2, $\varkappa$ are fixed and t stays away from 0 and 1 (²).

We postpone comments about the theorem, and in particular its comparison with the Riesz-Thorin theorem, until after the proof.

In the proof we shall repeatedly use the fact that for any non-negative and μ-measurable function f defined on R, and for any $p \geq 1$, we

(²) Marcinkiewicz [4] formulates his theorems for Lebesgue integrals only, and his spaces have finite measures. The extension to Stieltjes integrals, while routine and not requiring new ideas, is important for applications.

A comment of a different nature is indispensable here. In a letter written to me in June 1939, Marcinkiewicz [† 1940 (?)] indicated the idea of the proof of his main theorem of [4] in the case $(\alpha_1, \beta_1)=\left(\frac{1}{2}, \frac{1}{2}\right)$, $(\alpha_2, \beta_2)=(1, 1)$. The argument of the text is an elaboration of this idea. In [4] we also find theorems 2 and 3 of the present paper. The results of the sections 5 and 6 seem, however, to be essentially new.

have

$$\int_R f^p\, d\mu = -\int_0^\infty y^p\, d\, m(y) = p\int_0^\infty y^{p-1} m(y)\, dy,$$

where $m(y)$ is the distribution function of f, that is the μ-measure of the set of points where $f > y$.

We may suppose that

$$\alpha_1 \leqq \alpha_2.$$

Clearly $\alpha > 0$. Let $f \in L^{\frac{1}{\alpha}}$. Write

$$f = f' + f'',$$

where $f' = f$ whenever $|f| \leqq 1$, and $f' = 0$ otherwise; thus $|f''| > 1$ or else $f'' = 0$. Since f is in $L^{\frac{1}{\alpha}}$ the same holds for f' and f''; it follows that f' is in $L^{\frac{1}{\alpha_1}}$ and f'' in $L^{\frac{1}{\alpha_2}}$. Hence Tf' and Tf'' exist, by hypothesis, and so does $Tf = T(f' + f'')$.

We have to show that $h = Tf$ satisfies (3.3). We first consider the case when both β_1 and β_2 are different from zero. This implies that

$$\alpha_1 \neq 0, \qquad \alpha_2 \neq 0.$$

Denote the distribution functions of $|f|$ and $|h|$ by $m(y)$ and $n(y)$. Then

$$(3.4) \qquad \| h \|_b^b = b\int_0^\infty y^{b-1}\, n(y)\, dy = (2\varkappa)^b b\int_0^\infty y^{b-1} n(2\varkappa y)\, dy,$$

$\varkappa$ being the same as in (3.2).

For a fixed $z > 0$ we consider the decomposition

$$f = f_1 + f_2,$$

where $f_1 = f$ wherever $|f| \leqq z$; elsewhere we set $f_1 = z e^{i \arg f}$. It follows that

$$(3.5) \qquad |f_1| = \min(|f|, z), \qquad |f| = |f_1| + |f_2|.$$

Write $h_1 = Tf_1$, $h_2 = Tf_2$. The inequality (3.2) indicates that

$$|h| > 2\varkappa y$$

at those points at most at which either $|h_1| > y$ or $|h_2| > y$ (or both). Denote by $m_1(y)$, $m_2(y)$, $n_1(y)$, $n_2(y)$ the distribution functions of $|f_1|$, $|f_2|$, $|h_1|$, $|h_2|$ respectively. Then

$$(3.6) \qquad n(2\varkappa y) \leq n_1(y) + n_2(y) \leq \mathrm{M}_1^{b_1} y^{-b_1} \| f_1 \|_{a_1}^{b_1} + \mathrm{M}_2^{b_2} y^{-b_2} \| f_2 \|_{a_2}^{b_2},$$

by an application of (3.3) to f_1 and f_2. The right side here depends on z, and the main idea of the proof consists in defining z as a suitable monotone function of y, $z = z(y)$, to be determined later.

By (3.5),

$$\begin{aligned} m_1(y) &= m(y) && \text{for } 0 < y \leq z, \\ m_1(y) &= 0 && \text{for } y > z, \\ m_2(y) &= m(y+z) && \text{for } y > 0, \end{aligned}$$

the last equation being a consequence of the fact that wherever $f_2 \neq 0$ we must have $|f_1| = z$, and so the second equation (3.5) takes the form $|f| = z + |f_2|$.

It follows from (3.6) that the last integral in (3.4) does not exceed

$$\begin{aligned} (3.7) \quad & \mathrm{M}_1^{b_1} \int_0^\infty y^{b-b_1-1} \left\{ \int_R |f_1|^{a_1} d\mu \right\}^{\frac{b_1}{a_1}} dy + \mathrm{M}_2^{b_2} \int_0^\infty y^{b-b_2-1} \left\{ \int_R |f_2|^{a_2} d\mu \right\}^{\frac{b_2}{a_2}} dy \\ & = \mathrm{M}_1^{b_1} a_1^{k_1} \int_0^\infty y^{b-b_1-1} \left\{ \int_0^z t^{a_1-1} m(t)\, dt \right\}^{k_1} dy \\ & + \mathrm{M}_2^{b_2} a_2^{k_2} \int_0^\infty y^{b-b_2-1} \left\{ \int_z^\infty (t-z)^{a_2-1} m(t)\, dt \right\}^{k_2} dy, \end{aligned}$$

where

$$k_1 = \frac{b_1}{a_1}, \qquad k_2 = \frac{b_2}{a_2}$$

are not less than 1, by hypothesis.

Let us make initially instead of $\alpha_1 \leq \alpha_2$ the stronger assumption $\alpha_1 < \alpha_2$ (that is $a_2 < a_1$), and consider separately the two cases: 1° $\beta_1 < \beta_2$; 2° $\beta_2 < \beta_1$.

1° We have $b_2 < b < b_1$, and we set

$$z = \left(\frac{y}{\mathrm{A}} \right)^{\xi},$$

where A and ξ are positive numbers to be determined later.

Denote by P and Q the two double integrals last written. Then

$$
(3.8)\quad
\begin{cases}
P^{\frac{1}{k_1}} = \sup\limits_{\chi} \displaystyle\int_0^\infty y^{b-b_1-1}\left\{\int_0^z t^{a_1-1} m(t)\, dt\right\}\chi(y)\, dy \\
\qquad\qquad\qquad\text{for } \displaystyle\int_0^\infty y^{b-b_1-1}\chi^{k_1'}(y)\, dy \leq 1; \\
Q^{\frac{1}{k_2}} = \sup\limits_{\omega} \displaystyle\int_0^\infty y^{b-b_2-1}\left\{\int_z^\infty (t-z)^{a_2-1} m(t)\, dt\right\}\omega(y)\, dy \\
\qquad\qquad\qquad\text{for } \displaystyle\int_0^\infty y^{b-b_2-1}\omega^{k_2'}(y)\, dy \leq 1,
\end{cases}
$$

where k' denotes the exponent conjugate to k $\left(\frac{1}{k} + \frac{1}{k'} = 1\right)$.

The integral under the first sup sign is

$$
(3.9)\quad \int_0^\infty t^{a_1-1} m(t)\left\{\int_{At^{\frac{1}{\xi}}}^\infty y^{b-b_1-1}\chi(y)\, dy\right\} dt
$$

$$
\leq \int_0^\infty t^{a_1-1} m(t)\left\{\int_{At^{\frac{1}{\xi}}}^\infty y^{b-b_1-1}\, dy\right\}\left\{\int_{At^{\frac{1}{\xi}}}^\infty y^{b-b_1-1}\chi^{k_1'}(y)\, dy\right\} dt
$$

$$
\leq A^{\frac{b-b_1}{k_1}}(b_1-b)^{-\frac{1}{k_1}}\int_0^\infty t^{a_1-1+\frac{b-b_1}{k_1\xi}} m(t)\, dt,
$$

by Hölder's inequality and the condition for χ (observe that $b_1 > b$).

Similarly, substituting t^{a_2-1} for $(t-z)^{a_2-1}$, using Hölder's inequality and the condition for ω we see that the integral under the second sup sign in (3.8) does not exceed

$$
(3.10)\quad \int_0^\infty t^{a_2-1} m(t)\left\{\int_0^{At^{\frac{1}{\xi}}} y^{b-b_2-1}\omega(y)\, dy\right\} dt
$$

$$
\leq \int_0^\infty t^{a_2-1} m(t)\left\{\int_0^{At^{\frac{1}{\xi}}} y^{b-b_2-1}\, dy\right\}^{\frac{1}{k_2}}\left\{\int_0^{At^{\frac{1}{\xi}}} y^{b-b_2-1}\omega^{k_2'}(y)\, dy\right\}^{\frac{1}{k_2'}}
$$

$$
\leq A^{\frac{b-b_2}{k_2}}(b-b_2)^{-\frac{1}{k_2}}\int_0^\infty t^{a_2-1+\frac{b-b_2}{k_2\xi}} m(t)\, dt.
$$

Collecting results we find that

$$(3.11)\quad \| h \|_b^b \leq (2\varkappa)^b b \left\{ M_1^{b_1} a_1^{k_1} \frac{A^{b-b_1}}{b_1 - b} \left[\int_0^\infty t^{a_1 - 1 + \frac{b-b_1}{k_1 \xi}} m(t)\, dt \right] \right\}^{k_1} + \left\{ M_2^{b_2} a_2^{k_2} \frac{A^{b-b_2}}{b-b_2} \left[\int_0^\infty t^{a_2 - 1 + \frac{b-b_2}{k_2 \xi}} m(t)\, dt \right] \right\}^{k_2}$$

We now select ξ so that the exponents of t in both integrals are $a - 1$. This is possible, and we find

$$\xi = \frac{\alpha(\beta - \beta_1)}{\beta(\alpha - \alpha_1)}.$$

Next we set

$$A = M_1^\rho M_2^\sigma \| f \|_a^\tau,$$

and select ρ, σ, τ so that both terms in (3.11) contain the same powers of M_1, M_2 and $\|f\|_a$ respectively. A simple computation shows that

$$A = M_1^{\frac{b_1}{b_1 - b_2}} M_2^{\frac{b_2}{b_2 - b_1}} \| f \|_a^{\frac{a(k_2 - k_1)}{b_2 - b_1}},$$

and that both terms in (3.11) contain $(M_1^{1-t} M_2^t \| f \|_a)^b$, and we obtain (3.3) with

$$(3.12)\quad K^b = (2\varkappa)^b b \left\{ \frac{\left(\frac{a_1}{a}\right)^{k_1}}{b_1 - b} + \frac{\left(\frac{a_2}{a}\right)^{k_2}}{b - b_2} \right\}.$$

2° We now have $b_1 < b < b_2$. We set, as before, $z = \left(\frac{y}{A}\right)^\xi$, where A is positive and ξ *is negative*. In the first integrals in (3.9) and (3.10) the intervals of integration $\left(A t^{\frac{1}{\xi}}, \infty\right)$ and $\left(0, A t^{\frac{1}{\xi}}\right)$ are then interchanged, but otherwise the proof remains the same, and we again arrive at (3.3) with K^b given by (3.12), but with $b - b_1$ and $b_2 - b$ in the denominators for $b_1 - b$ and $b - b_2$ respectively.

We now consider the case $\alpha_1 = \alpha_2$ (it has no interesting applications, and the same holds for the case $\beta_1 = \beta_2$ not covered by the theorem). Suppose that $\alpha_1 = \alpha = \alpha_2$ and that, for example, $\beta_1 < \beta < \beta_2$. By hypothesis,

$$(3.13)\quad n(y) \leq (M_1 y^{-1} \| f \|_a)^{b_1}, \qquad n(y) \leq (M_2 y^{-1} \| f \|_a)^{b_2},$$

where $n(y)=\nu(E_y[h])$. We split the *first* integral (3.4) into two, extended over (o, A) and (A, ∞), where A is positive and to be determined in a moment. If we apply to the partial integrals the inequalities (3.13) respectively, and observe that $b_2 < b < b_1$, we see that $\|h\|_b^b$ is finite. Setting $A = M_1^\rho M_2^\sigma \|f\|_a^\tau$, and selecting ρ, σ, τ so that the exponents of M_1, M_2, $\|f\|_a$ in both integrals are, respectively, the same, we arrive at (3.3).

It remains to consider the case when one of the numbers β_1, β_2 is o. Suppose that $\beta_1 = o$. The proof of the theorem requires that after the decomposition $f=f_1+f_2$ we estimate $n_1(y)$ in terms of $\|h_1\|_{b_1} = \text{ess sup}\,|h_1|$. In general this cannot be done, unless we know that $|h_1| \leqq y$, in which case $n_1(y)=o$. Thus we must choose z so that $|h_1| \leqq y$. To be more specific, let us confine our attention to te case $o \leqq \alpha_1 < \alpha_2$ ($\beta_2 > \beta_1 = o$) and consider separately the two subcases : 1° $\alpha_1 = o$; 2° $\alpha_1 > o$.

1° Consider again the decomposition $f=f_1+f_2$ and the relations (3.5) and (3.6). Take $z=\frac{y}{M_1}$. Then

$$\text{ess sup}\,|h_1| \leqq M_1\,\text{ess sup}\,|f_1| = M_1\frac{y}{M_1} = y.$$

It follows that $n_1(y)=o$ in (3.6), and so of the two terms on the right of (3.11) only the second remains. Setting there $A=M_1$ we arrive at (3.3), where again K is given by (3.12) but with the first term in curly brackets omitted.

We easily verify that the choice or z and A conforms to the same pattern as in the general argument above.

2° We again select z so that $n_1(y)=o$, after which the proof proceeds as before. Suppose that

$$z=\left(\frac{y}{A}\right)^\xi$$

where

$$\xi=\frac{a_1}{a_1-a}, \qquad A=\lambda M_1\|f\|_a^{\frac{a}{a_1}}$$

and λ is a numerical factor to be determined presently. Except for the presence of this factor, the quantities z, ξ, A are given by the formulas above simplified by the hypothesis that $\beta_1 = 0$. It is therefore enough to show that for a suitable λ we have ess sup $|h_1| \leqq y$.

By hypothesis,

$$\text{ess sup } |h_1| \leqq M_1 \|f_1\|_a = M_1 \left\{ a_1 \int_0^z t^{a_1 - 1} m(t)\, dt \right\}^{\frac{1}{a_1}}.$$

It follows that we certainly have ess sup $|a_1| \leqq y$, if

$$M_1^{a_1} a_1 \int_0^z t^{a_1-1} m(t)\, dt \leqq \left(\lambda z^{\frac{1}{a}}\right)^{a_1} = \lambda^{a_1} z^{a_1 - a},$$

and *a fortiori* (since $a_1 > a$) if

$$M_1^{a_1} a_1 z^{a_1 - a} \int_0^\infty t^{a-1} m(t)\, dt \leqq \lambda^{a_1} M_1^{a_1} \|f\|_a^a z^{a_1 - a}.$$

Since the integral on the left is $a^{-1} \|f\|_a^a$, the inequality is satisfied if $\lambda^{a_1} \geqq \frac{a_1}{a}$. This completes the proof in the case 2°. The value of K is easily found.

The theorem is therefore completely established.

4. Remarks about the preceding theorem. — (I) The constant K in (3.3) tends, in general, to ∞ as t tends to 0 (or 1), for otherwise the operation T would be of strong type $\left(\frac{1}{\alpha_1}, \frac{1}{\beta_1}\right)$, which need not be the case. The drawback of the proof given above is that even if T is linear and of strong types $\left(\frac{1}{\alpha_1}, \frac{1}{\beta_1}\right)$ and $\left(\frac{1}{\alpha_2}, \frac{1}{\beta_2}\right)$ we cannot show that K is bounded as t tends to 0 or 1 [with one exception : if $\beta_1 = 0$, the first term in curly brackets in (3.12) is absent and K remains bounded as $\beta \to \beta_1$]. Hence the theorem of Marcinkiewicz is not a substitute for the Riesz-Thorin theorem completely. In compensation, the former applies to a number of cases when the latter does not; and then additional devices (if a direct appeal to the Riesz-Thorin theorem fails) may show that K is bounded.

We illustrate this by the example $Tf = \tilde{f}$ considered above. T is, as we have already observed, of weak type (1, 1). T is also of strong type (2, 2). Hence, by the theorem of the previous section, T is of type (p_0, p_0) for each $1 < p_0 < 2$. It is very easy to prove (*see* e. g. [10], p. 148) that this implies that T is also of type (p'_0, p'_0) where p'_0 is the exponent conjugate to p_0. Hence T is also of type (p, p), if $p_0 < p < p'_0$. The norm A_p in the inequality

$$\|\tilde{f}\|_p \leq A_p \|f\|_p \tag{4.1}$$

is bounded in every interval

$$p_0 + \varepsilon \leq p \leq p'_0 - \varepsilon \qquad (\varepsilon > 0,$$

and so also in every interval $1 + \varepsilon \leq p \leq \frac{1}{\varepsilon}$.

Since the fact that $Tf = \tilde{f}$ is of weak type (1, 1) can be proved by purely real methods, the proof just given of (4.1) is also purely real ([3]).

(II) In some cases it is important to know the order of magnitude of the constant K in (3.3) as t tends to 0 or 1. From (3.12) we see that

$$K = O\left(\frac{1}{t}\right), \qquad K = O\left(\frac{1}{1-t}\right) \tag{4.2}$$

as t tends to 0 and 1 respectively. Simple examples (in particular $Tf = \tilde{f}$) show that these are the right estimates.

(III) While in the theorem of Riesz-Thorin we interpolate in the square (Q) $0 \leq \alpha \leq 1$, $0 \leq \beta \leq 1$, in the theorem of Marcinkiewicz we interpolate only in the triangle $(\Delta)\, 0 \leq \beta \leq \alpha \leq 1$. Whether this restriction to Δ is essential we do not know but the problem is anyway of minor importance since there are no interesting operations of type $\left(\frac{1}{\alpha}, \frac{1}{\beta}\right)$ for (α, β) outside Δ. What is of some interest is that

([3]) This point is of considerable importance if we consider generalizations of the notion of the conjugate function to spaces of higher dimensions where complex methods cannot be applied. Such generalizations are considered in [1], where an analogue of (4.1) is proved by an argument which, implicitly, uses the same idea as the proof of theorem 1 in the case $(\alpha_1, \beta_1) = \left(\frac{1}{2}, \frac{1}{2}\right)$, $(\alpha_2, \beta_2) = (1, 1)$. An explicit application of theorem 1 would have shortened the proof.

the theorem of the preceding section is not affected if we consider purely real operations (i. e. if f and $h = Tf$ are real-valued functions), in which case the Riesz-Thorin theorem, or at least the part of it concerning the convexity of the logarithm of the norm of the operation, is known to be valid only in Δ and not in Q.

(IV) In a number of problems we are led to consider integrals of type

$$\int_R \varphi(|f|)\, d\mu,$$

where φ is not necessarily a power. The argument of the previous section makes it possible to "interpolate" the functions φ. Without aiming too much at generality we consider some special cases which are both illustrative and useful. The theorem which follows is, roughly speaking, interpolation along the hypothenuse $\beta = \alpha$ of the triangle Δ ([4]).

THEOREM 2. — *Suppose that $\mu(R)$ and $\nu(S)$ are both finite, and that a quasi-linear operation $h = Tf$ is of weak types (a, a) and (b, b), where $1 \leqq a < b < \infty$. Suppose also that $\varphi(u)$, $u \geqq 0$, is a continuous increasing function satisfying the conditions $\varphi(0) = 0$ and*

$$(4.3)\qquad \varphi(2u) = O\{\varphi(u)\},$$

$$(4.4)\qquad \int_u^\infty \frac{\varphi(t)}{t^{b+1}}\, dt = O\left\{\frac{\varphi(u)}{u^b}\right\},$$

$$(4.5)\qquad \int_1^u \frac{\varphi(t)}{t^{a+1}}\, dt = O\left\{\frac{\varphi(u)}{u^a}\right\}$$

for $u \to \infty$. Then $h = Tf$ is defined for every f with $\varphi(|f|)$ μ-integrable and

$$(4.6)\qquad \int_S \varphi(|h|)\, d\nu \leqq K \int_R \varphi(|f|)\, d\mu + K,$$

where K is independent of f ([5]).

([4]) The general case is considered in a forthcoming paper of W. Riordan.

([5]) Theorem 2 (and theorem 4 below) has an analogue in the case when $\mu(R)$ and $\nu(S)$ are not necessarily finite. In that case the behaviour of $\varphi(u)$ near $u = 0$ becomes important and the conditions (4.3)-(4.4) have to be supple-

The function

$$\varphi(u) = u^c \psi(u), \tag{4.7}$$

where $a < c < b$ and $\psi(u)$ is a *slowly varying* function is easily seen to satisfy the required conditions [a function $\psi \geq 0$ is said to be slowly varying if for each $\delta > 0$ the function $\psi(u)u^\delta$ is ultimately increasing and $\psi(u)u^{-\delta}$ ultimately decreasing as $u \to \infty$].

Return to the theorem. The proof that Tf exists is the same as in the case of theorem 1, and we only have to establish (4.6). The proof is similar to that of (3.3), but is in some respects simpler since we are dealing with the points of the hypothenuse of the triangle Δ. However, to avoid having to justify inverting the order of repeated Stieltjes integrals we confine ourselves to sums rather than integrals.

If $n(y)$ is the distribution function of $|h|$, then

$$\int_S \varphi(|h|)\, d\nu = -\int_0^\infty \varphi(y)\, dn(y) = \int_0^\infty n(y)\, d\varphi(y).$$

We write $\lambda = 2\varkappa$ and denote by K any positive constant independent of f. If η_j is the measure of the set in which $|h| \geq \lambda 2^j$ $(j = 1, 2, \ldots)$ then from (4.7) and $\nu(S) < \infty$ we deduce

$$\int_S \varphi(|h|)\, d\nu \leq K + \sum_{j=0}^{\infty} \eta_j \{\varphi(\lambda 2^{j+1}) - \varphi(\lambda 2^j)\} = K + \sum_0^\infty \eta_j \delta_j, \tag{4.8}$$

say. For each fixed j we write $f = f_1 + f_2$, where f_1 equals f or 0, according as $|f| \leq 2^j$ or $|f| > 2^j$. At the points where $|h| > \lambda 2^j$ we have either $|h_1| > 2^j$ or $|h_2| > 2^j$ $(h_i = Tf_i)$. Since f_1 and f_2 are in L^b and L^a respectively, we have

$$\eta_j \leq K \left\{ 2^{-jb} \int_R |f_1|^b\, d\mu + 2^{-ja} \int_R |f_2|^a\, d\mu \right\} \leq K \left\{ 2^{-jb} \sum_0^j 2^{ib} \varepsilon_i + 2^{-ja} \sum_{j+1}^\infty 2^{ia} \varepsilon_i \right\},$$

where ε_i $(i = 1, 2, \ldots)$ denotes the μ-measure of the set where $2^{i-1} < |f| \leq 2^i$, and ε_0 that of the set where $|f| \leq 1$. If we substitute

---

mented by similar conditions near $u = 0$. We omit the details which, can be supplied without difficulty. Perhaps the most interesting application here is an extension of theorem 3 to the case of the n-dimensional Hilbert transform discussed in [1].

this estimate of η_j in (4.8) and interchange the order of summation we are led to (4.6), provided we can prove that each of the sums

$$(4.9)\qquad \sum_{i=0}^{\infty} 2^{ib}\varepsilon_i \sum_{j=i}^{\infty} \delta_j 2^{-jb}, \qquad \sum_{i=1}^{\infty} \varepsilon_i 2^{ia} \sum_{j=0}^{i-1} \delta_j 2^{-ja}$$

is majorized by $K+K\int_R \varphi(|f|)\,d\mu$.

We may suppose that $\lambda \geqq 1$. Since $\delta_j \leqq \varphi(\lambda 2^{j+1})$, we have

$$\sum_{j=i}^{\infty} \delta_j 2^{-jb} \leqq K \sum_{j=i}^{\infty} 2^{-j(b+1)} \int_{\lambda 2^{j+1}}^{\lambda 2^{j+2}} \varphi(u)\,du$$

$$\leqq K \sum_{j=i}^{\infty} \int_{\lambda 2^{j+1}}^{\lambda 2^{j+2}} u^{-b-1}\varphi(u)\,du \leqq K \int_{2^i}^{\infty} u^{-b-1}\varphi(u)\,du \leqq K\varphi(2^i)\,2^{-ib},$$

by (4.3). Hence the first sum (4.9) does not exceed

$$K\sum_{i=0}^{\infty} \varepsilon_i \varphi(2^i) \leqq K\varepsilon_0\varphi(1) + K\sum_{i=1}^{\infty} \varepsilon_i \varphi(2^{i-1}) \leqq K + K\int_R \varphi(|f|)\,d\mu,$$

by an application of (4.2). Using (4.2), and (4.4) we obtain a similar estimate for the second sum (4.9), and theorem 2 is established.

In view of what we have said about the operation $Tf=\tilde{f}$, the theorem which follows is an immediate corollary of theorem 2.

THEOREM 3. — *If $\varphi(u)=u^r\psi(u)$, where $1<r<\infty$ and $\psi(u)$ is a positive slowly varying function, then*

$$\int_0^{2\pi} \varphi(|\tilde{f}|)\,dx \leqq K\int_0^{2\pi} \varphi(|f|)\,dx + K.$$

The theorem which follows is a modification of theorem 2 in the case when, say, (4.4) does not necessarily hold, that is when the growth of $\varphi(u)$ is "close" to that of u.

THEOREM 4. — *Suppose that $\mu(R)$ and $\nu(S)$ are finite, that $1\leqq a<b<\infty$, and that $h=Tf$ is a quasi-linear operation simultaneously of weak types (a, a) and (b, b). Let $\chi(u)$, $u\geqq 0$, be equal* *to* o *in a right-hand neighbourhood of* $u=0$, *say for* $u\leqq 1$, *positive and increasing elsewhere, satisfying* $\chi(2u)=O\{\chi(u)\}$ *for large* u. *Write*

$$\varphi(u)=u^a\int_1^u t^{-a-1}\chi(t)\,dt,$$

and suppose that φ *satisfies* (4.3). *Then* $h=Tf$ *is defined for all* f *such that* $\chi(|f|)$ *is integrable, and we have*

$$\int_S \varphi(|h|)\,d\nu\leqq K\int_R \chi(|f|)\,d\mu+K, \tag{4.10}$$

where K *is independent of* f.

The proof is a repetition of that of theorem 3 and need not be gone into.

The most interesting special case is when $\chi(u)=u^a$ for $u\geqq 1$ and $\chi=0$ otherwise. Then $\varphi(u)=u^a\log^+ u$ and (4.10) can be written

$$\int_S |h|^a\,d\nu\leqq K\int_R |f|^a\log^+|f|\,d\mu+K.$$

An extension of theorem 3 to this case ($a=1$) is immediate.

5. Application to Fourier coefficients. — Return to the inequality (2.4), which, using the notations (2.5) and (2.6), we write

$$\|h\|_{p,\nu}\leqq A_p M^{\frac{2-p}{p}}\|f\|_p. \tag{5.1}$$

Since the operation $h=Tf$ is of strong type (2, 2), (5.1) will follow if we show that T is of weak type (1, 1) [we know that it is not of strong type (1, 1)]. We show that

$$\nu\{E_y[h]\}\leqq\frac{2M}{y}\|f\|_1. \tag{5.2}$$

The left side here is Σn^{-2} extended over the n satisfying

$$|nc_n|>y.$$

For such n,

$$y<|nc_n|\leqq n\int_a^b |f\bar{\varphi}_n|\,dx\leqq nM\|f\|_1,$$

that is

$$n > \frac{y}{M} \|f\|_1.$$

If we set $\omega = \frac{y}{M} \|f\|_1$ and suppose that $\omega > 1$, then

$$\sum_{n>\omega} n^{-2} \leq 2\omega^{-1} = 2My^{-1} \|f\|_1,$$

which is (5.2). The latter is obvious $\left(\text{since } \sum n^{-2} = \frac{1}{6}\pi^2\right)$ if $\omega < 1$.

It is interesting to observe that the inequality dual to (5.1), namely

$$(5.3) \qquad \left(\int_0^1 |f|^q \, dx\right)^{\frac{1}{q}} = A_q M^{\frac{q-2}{q}} \left(\sum n^{q-2} |c_n|^q\right)^{\frac{1}{q}},$$

where now $q \geq 2$ and

$$f(x) = \lim \sum_1^n c_\nu \varphi_\nu(x),$$

cannot be obtained by a similar argument. In this case, T transforms a sequence $c_1, c_2, \ldots$ into a function f; the q-th norm of $\{c_n\}$ being defined as $(\Sigma |nc_n|^q n^{-2})^{\frac{1}{q}}$. This operation is again of type (2, 2), but is not of type (∞, ∞) since the boundedness of $\{nc_n\}$ does not imply the boundedness of the function f. It is well known, however, that (5.3) can be deduced directly and without difficulty from (5.1).

The above remarks apply easily to the inequalities we obtain from (5.1) and (5.3) by interchanging the roles of f and $\{c_n\}$

$$\left(\int_0^1 x^{p-2} |f|^p \, dx\right)^{\frac{1}{p}} \leq A_p M^{\frac{2-p}{p}} \left(\sum |c_n|^p\right)^{\frac{1}{p}},$$

$$\left(\sum |c_n|^q\right)^{\frac{1}{q}} \leq A_q M^{\frac{q-2}{q}} \left(\int_0^1 x^{q-2} |f|^q \, dx\right)^{\frac{1}{q}},$$

and in the same way we can prove more general results about unbounded orthonormal systems (*see* [5]).

Return to (5.1). Since the ν-measure of the set of all positive integers is finite, we can apply theorem 2, and considering for simplicity rather special functions φ, ψ we obtain, for example, the following result.

THEOREM 5. — *Suppose that* $\{\varphi_\nu\}$ *is orthonormal over* (a, b), *that* $b-a<\infty$, *and that* $|\varphi_\nu|\leq 1$ *for all* ν. *If* $1<p<2$, $\varphi(u)=u^p\varphi^\star(u)$, *where* $\varphi^\star(u)$ *is positive and slowly varying, then the Fourier coefficients of* f *satisfy an inequality*

$$\sum\varphi(|nc_n|)\,n^{-2}\leq K\int_a^b\varphi(|f|)\,dx+K. \tag{5.4}$$

If $2<q<\infty$, $\psi(u)=u^q\psi^\star(u)$, *where* $\psi^\star(u)$ *is positive and slowly varying, and if* $\Sigma\psi(|nc_n|)\,n^{-2}<\infty$, *then the* c_n *are the coefficients of an* f *such that*

$$\int_a^b\psi(|f|)\,dx\leq K\sum\psi(|nc_n|)\,n^{-2}+K. \tag{5.5}$$

Using theorem 4 we easily deduce the limiting case of (5.4) for $p=1$. The exclusion of the value 2 for p and q (for reasons explained in section 3) makes, however, the result somewhat incomplete.

It is natural to expect that (5.4) holds for $p=2$, provided $\varphi^\star$ is slowly varying and bounded above, and that (5.5) holds for $q=2$, provided $\psi^\star$ is slowly varying and bounded below (away from 0). A breakdown of (5.4) for $p=1$ is however only to be expected, and a substitute result is an immediate corollary of theorem 4. The special case $\chi(u)=u$ of the latter theorem gives then the inequality

$$\sum\frac{|c_n|}{n}\leq K\int_a^b|f|\log^+|f|\,dx+K, \tag{5.6}$$

which is well known (*see* [12], p. 235, ex. 6).

6. A THEOREM ON FRACTIONAL INTEGRATION. — In this section we "do not interpolate"; at least this is not our primary concern. Our main purpose is to show that operations of weak type (a, b) can occur in fractional integration.

We begin by a brief restatement of familiar facts about fractional integration of order γ, where $0<\gamma<1$.

Given a function $f(x)$, $-\infty < x < +\infty$, we define the fractional integral of order γ by the formula

$$(6.1) \qquad f_\gamma(x) = \frac{1}{\Gamma(\gamma)} \int_{-\infty}^{+\infty} \frac{f(t)}{|x-t|^{1-\gamma}} dt.$$

This is one of several existing definitions but what we say below applies equally well to all of them [the most prevalent definition is obtained from (6.1) by replacing the upper limit of integration on the right by x; it is also applicable when f is periodic, provided the integral of f over a period is 0]. A very well know result of Hardy and Littlewood (*see* e. g. [3], p. 290) asserts that if $f \in L^r(-\infty, +\infty)$, $1 < r < +\infty$, and if

$$(6.2) \qquad 1 < r < s < +\infty,$$

$$(6.3) \qquad \frac{1}{r} - \frac{1}{s} = \gamma,$$

then

$$(6.4) \qquad \|f_\gamma\|_s \leq A_{r,s} \|f\|_r,$$

where

$$\|f\|_r = \left(\int_{-\infty}^{+\infty} |f|^r dx \right)^{\frac{1}{r}}.$$

The result fails if either $r = 1$ or $s = \infty$ in (6.2), since simple examples show that if f is in L, f_γ need not be in $L^{\frac{1}{1-\gamma}}$, and if f is in $L^{\frac{1}{\gamma}}$, f_γ need not be bounded. There are several substitute results in these extreme cases, and we add one more by proving the following theorem.

THEOREM 6. — *The operation* $Tf = f_\gamma$ *is of weak type* $\left(1, \frac{1}{1-\gamma}\right)$. *In other words, if* $E_y(f)$ *is the set of points where* $|f_\gamma| > y > 0$, *and* $|E_y|$ *is the Lebesgue measure of* E_y, *then*

$$(6.5) \qquad |E_y(f_\gamma)| \leq A_\gamma \left(\frac{\|f\|_1}{y} \right)^{\frac{1}{1-\gamma}}.$$

The proof given below has some points in common with the real-variable proof of the fact that the operation $Tf = \tilde{f}$ is of weak type (1, 1). Without loss of generality we may suppose that

$$(6.6) \qquad f \geq 0, \qquad \|f\|_1 = 1.$$

We may also suppose that $f = 0$ in the neighbourhood of $\pm\infty$, since this special case leads to (6.5) in the general case by a simple passage to the limit.

Let

$$F(x) = \int_{-\infty}^{x} f(t)\,dt.$$

The function F is continuous, non-decreasing and bounded. We fix the y in (6.5), take a number $z > 0$ which will depend on y in a way to be specified later, and consider the set Q of points x such that

$$(6.7) \qquad \frac{F(x+h) - F(x)}{h} > z$$

for some $h = h_x > 0$. The set Q is open and, if not empty, consists (by a well known theorem of F. Riesz, *see* e. g. [3], p. 293) of a finite or denumerable number of disjoint open intervals (a_j, b_j) such that

$$(6.8) \qquad \frac{F(b_j) - F(a_j)}{b_j - a_j} = z$$

for each j. We also note that, since F is constant in the neighbourhood of $\pm\infty$, Q is bounded.

Denote by P the closed set complementary to Q; P is not empty. Since (6.7) is false for each x in P and each $h > 0$, we see that

$$(6.9) \qquad F' = f \leq z \quad \text{almost everywhere in P.}$$

Denote by $G(x)$ the continuous function which coincides with $F(x)$ in P and is linear in each of the intervals (a_j, b_j), and define $H(x)$ by the equation

$$(6.10) \qquad F(x) = G(x) + H(x).$$

Hence $H(x)$ vanishes in P. We investigate properties of $G(x)$ and $H(x)$.

The function G is, like F, non-decreasing. It also satisfies a

Lipschitz condition of order 1, i. e.

$$G(x+h) - G(x) \leq Ah \tag{6.11}$$

for all x and $h > 0$. For if x and $x+h$ are both in P, the left side of (6.11) equals $F(x+h) - F(x) \leq zh$. If x and $x+h$ are in the closure of the same interval (a_j, b_j), the left side of (6.11) equals zh. Finally, in the general case, if $x \leq x' \leq x'' \leq x+h$, where x' and x'' are the first and last points of P in $(x, x+h)$, splitting $(x, x+h)$ into the sum of (x, x'), (x', x''), $(x'', x+h)$, we prove (6.11) with $A = 3z$.

The derivative $G' = g$ exists almost everywhere, and so also does $H' = h$, and

$$f = g + h. \tag{6.12}$$

Since H vanishes in P, h vanishes almost everywhere in P. Hence $g = f$ almost everywhere in P. Clearly $g = z$ at each point of Q. These facts together with (6.9) imply that $0 \leq g \leq z$ almost everywhere. Since $g = 0$ in the neighbourhood of $\pm\infty$ [*see* (6.9)], g is integrable (in every power) over $(-\infty, +\infty)$. Hence $h = f - g$ is integrable over $(-\infty, \infty)$.

We have

$$f_\gamma = g_\gamma + h_\gamma.$$

Since

$$E_{2y}(f_\gamma) \subset E_y(g_\gamma) + E_y(h_\gamma),$$

the inequality (6.5) will follow if we show that both $|E_y(g_\gamma)|$ and $|E_y(h_\gamma)|$ are majorized by $A(y^{-1}\|f\|_1)^{\frac{1}{1-\gamma}}$, and the rest of the argument consists in verifying these two assertions. For the sake of simplicity we drop the factor $\frac{1}{\Gamma(\gamma)}$ in the definition (6.1).

The equations (6.8) and $\|f\|_1 = 1$ imply that the measure of Q is $\frac{1}{z}$. Denote by Q^* the set obtained by expanding each of the intervals $I_j = (a_j, b_j)$ concentrically twice. Hence

$$|Q| = \frac{1}{z}, \qquad |Q^*| \leq \frac{2}{z}. \tag{6.13}$$

Let P^* be the complement of Q^*. We first show that

$$\int_{P^*} h^{\frac{1}{1-\gamma}}\,dx \leq A_\gamma. \tag{6.14}$$

The definition of G implies that the integrals of f and g over each I_j are the same. Hence

$$\int_{I_j} |h|\,dt = \int_{I_j} |f-g|\,dt \leq \int_{I_j} f\,dt + \int_{I_j} g\,dt = 2\int_{I_j} f\,dt = 2|I_j|\xi.$$

Since H vanishes at the points a_j, b_j, integration by parts shows that

$$h_\gamma(x) = \sum_j \int_{I_j} h(t)\,|x-t|^{\gamma-1}\,dt = \sum_j \pm(\gamma-1)\int_{I_j} H(t)\,|x-t|^{\gamma-2}\,dt$$

for x in P. For x in P^* we therefore have

$$|h_\gamma(x)| \leq \sum_j \left(\tfrac{1}{2}|x-c_j|\right)^{\gamma-2} \int_{I_j} |H(t)|\,dt$$

$$\leq A_\gamma \sum_j |x-c_j|^{\gamma-2}\,|I_j|\,\underset{t\in I_j}{\mathrm{Max}}\,|H(t)| \leq A_\gamma \sum_j |x-c_j|^{\gamma-2}\,|I_j|^2.$$

By Hölder's inequality,

$$|h_\gamma(x)|^{\frac{1}{1-\gamma}} \leq A_\gamma \xi^{\frac{1}{1-\gamma}} \left\{ \sum_j \left(\frac{|I_j|}{|x-c_j|}\right)^{2-\gamma} |I_j|^\gamma \right\}^{\frac{1}{1-\gamma}} \tag{6.15}$$

$$\leq A_\gamma \xi^{\frac{1}{1-\gamma}} \left\{ \sum_j \left(\frac{|I_j|}{|x-c_j|}\right)^{\frac{2-\gamma}{1-\gamma}} \right\} \left(\sum_j |I_j|\right)^{\frac{\gamma}{1-\gamma}}.$$

Integrating this over P^* and observing that

$$\int_{P^*} |x-c_j|^{-\frac{2-\gamma}{1-\gamma}}\,dx \leq 2\int_{|I_j|} t^{-\frac{2-\gamma}{1-\gamma}}\,dt = A_\gamma |I_j|^{-\frac{1}{1-\gamma}}.$$

we obtain

$$\int_{P^*} |h_\gamma|^{\frac{1}{1-\gamma}}\,dx \leq A_\gamma \xi^{\frac{1}{1-\gamma}} \left(\sum_j |I_j|\right)^{1+\frac{\gamma}{1-\gamma}} = A_\gamma \xi^{\frac{1}{1-\gamma}} |Q|^{\frac{1}{1-\gamma}} = A_\gamma,$$

by (6.13), and (6.14) follows.

This implies that the subset of P^* where $|h_\gamma| > y$ has measure less than $A_\gamma y^{-\frac{1}{1-\gamma}}$. Since the complement Q^* of P^* has, by (6.13), measure not greater than $\frac{2}{z}$, we see that

$$|E_y(h_\gamma)| \leqq A_\gamma y^{-\frac{1}{1-\gamma}} + 2z^{-1}.$$

If we take $z = y^{\frac{1}{1-\gamma}}$, both terms on the right become, except for the coefficients, identical and we have

$$(6.16) \qquad |E_y(h_\gamma)| \leqq A_\gamma y^{-\frac{1}{1-\gamma}}.$$

It remains to estimate $|E_y(g_\gamma)|$. Take any r, s such that

$$1 < r < s < \infty, \qquad \frac{1}{r} - \frac{1}{s} = \gamma.$$

Since g is in L^r, we have, after (6.4),

$$(6.17) \qquad \| g_\gamma \|_s \leqq A_{rs} \| g \|_r.$$

We may, for example, suppose that

$$(6.18) \qquad \frac{1}{r} = \frac{1}{2} + \frac{1}{2}\gamma, \qquad \frac{1}{s} = \frac{1}{2} - \frac{1}{2}\gamma.$$

Then the A_{rs} in (6.17) becomes A_γ.

Since $0 \leqq g \leqq z$, we have

$$|E_y(g_\gamma)| \leqq y^{-s} \int_{-\infty}^{+\infty} g_\gamma^s \, dx$$

$$\leqq A_\gamma y^{-s} \left(\int_{-\infty}^{+\infty} g^r \, dx \right)^{\frac{s}{r}} \leqq A_\gamma y^{-s} z^{(r-1)\frac{s}{r}} \left(\int_{-\infty}^{+\infty} g \, dx \right)^{\frac{s}{r}}$$

$$= A_\gamma y^{-s} z^{(r-1)\frac{s}{r}} \left(\int_{-\infty}^{+\infty} f dx \right)^{\frac{s}{r}} = A_\gamma y^{-s} z^{(r-1)\frac{s}{r}},$$

and since (6.18) implies that $(r-1)\frac{s}{r} = 1$, we see that

$$(6.19) \qquad |E_y(g_\gamma)| \leqq A_\gamma y^{-\frac{1}{1-\gamma}}.$$

This, together with (6.16), completes the proof of theorem 6.

In the proof of this theorem we used (6.4), but only in the case $r \leqq 2 \leqq s$ [*see* (6.18)]. It is, however, well known that in the latter case the inequality (6.4) has little to do with fractional integration and, with suitable definitions, is valid for general uniformly bounded orthonormal systems (*see* [12], p. 232-233). It is then a corollary of (2.4) and the dual inequality (5.3). Using, therefore, (6.4) for $r = 2 < s$, and the fact that the operation $Tf = f_\gamma$ is of weak type $\left(1, \frac{1}{1-\gamma}\right)$, we can prove (6.4) for $2 \leqq r < s < \infty$. From this the case $1 < r < s \leqq 2$ follows by a very simple conjugacy argument, and we have thus obtained a new proof of (6.4) in the general case. While not simpler than other existing proofs, it may be of some interest.

2° Theorem 6 has an n-dimensional analogue. Let

$$x = (\xi_1, \xi_2, \ldots, \xi_n), \qquad y = (\eta_1, \eta_2, \ldots, \eta_n), \qquad \ldots$$

be points of the n-dimensional space R^n. For a function $f(x)$ in R^n we define the fractional integral (potential)

$$f_\gamma(x) = \int_{R^n} \frac{f(y)\,dy}{|x-y|^{n-\gamma}}$$

where $dy = d\eta_1 \ldots d\eta_n$ and $0 < \gamma < n$. *The operation* $Tf = f_\gamma$ *is then of weak type* $\left(1, \frac{1}{n-\gamma}\right)$. The proof is very similar to that of theorem 6 if instead of F. Riesz's lemma one uses its n-dimensional analogue (*see* [1], lemma 2). As in the case $n = 1$ we can then obtain, by interpolation, the (well known) inequality

$$\|f_\gamma\|_s \leqq A_{r,s} \|f\|_r,$$

where

$$1 < r < s < \infty, \qquad \frac{1}{r} - \frac{1}{s} = \frac{\gamma}{n}.$$

BIBLIOGRAPHY.

[1] A. P. Calderon and A. Zygmund, *On the existence of certain singular integrals* (*Acta Math.*, t. **88**, 1952, p. 85-139).

[2] A. P. Calderon and A. Zygmund, *On the theorem of Hausdorff-Young and its extensions* (*Ann. Math. Studies*, Princeton, 1950, No 25, p. 166-188).

[3] G. H. Hardy, J. E. Littlewood and G. Pólya, *Inequalities*, Cambridge University Press, 1952.

[4] J. Marcinkiewicz, *Sur l'interpolation d'opérations* (*C. R. Acad. Sc.*, t. **208**, 1939, p. 1272-1273).

[5] J. Marcinkiewicz and A. Zygmund, *Some theorems on orthogonal systems* (*Fund. Math.*, t. **28**, 1937, p. 309-335).

[6] M. Riesz, *Sur les maxima des formes billinéaires et sur les fonctionnelles linéaires* (*Acta Math.*, t. **49**, 1926, p. 465-497).

[7] R. Salem, *Convexity theorems* (*Bull. Amer. Math. Soc.*, t. **55**, 1949, p. 851-760).

[8] R. Salem, *Sur une extension du théorème de convexité de M. Marcel Riesz* (*Colloqium Mathematicum*, t. **1**, 1947, p. 6-8).

[9] J. D. Tamarkin and A. Zygmund, *Proof of a theorem of Thorin* (*Bull. Amer. Math. Soc.*, t. **50**, 1944, p. 279-282).

[10] G. O. Thorin, *An extension of a convexity theorem due to M. Riesz* (*Kungl. Fys. Saellskapets i Lund Forhaendliger*, t. **8**, 1939, No **14**).

[11] G. O. Thorin, *Convexity theorems*, Uppsala, 1948, p. 1-57.

[12] A. Zygmund, *Trigonometrical Series*, Monografje matematyczne, vol. V, Warsaw, 1935.

HILBERT TRANSFORMS IN E^n

A. ZYGMUND

In this lecture I present certain results obtained by Professor A. P. Calderón and myself within the last three years. [1])

1. *Hilbert transforms*. Let $f(x)$ be a function defined for all real x and belonging to one of the classes $L^p(-\infty, +\infty)$, $1 \leqq p < \infty$. By the *Hilbert transform* of f we mean the function

(1.1) $$\tilde{f}(x) = \frac{1}{\pi}\int_{-\infty}^{+\infty} \frac{f(t)}{x-t}\,dt,$$

where the integral is taken in the principal — value sense:

$$\frac{1}{\pi}\int_{-\infty}^{+\infty} \frac{f(t)}{x-t}\,dt = \lim_{\varepsilon\to 0} \int_{|x-t|>\varepsilon} \frac{f(t)}{x-t}\,dt.$$

It is very well known that the integral (1.1) exists almost everywhere. For $1 < p < \infty$ we also have the M. Riesz inequality

(1.2) $$\|\tilde{f}\|_p \leqq A_p \|f\|_p,$$

where A_p depends on p only, and $\|f\|_p = \left(\int_{-\infty}^{+\infty} |f|^p dx\right)^{1/p}$. For $p = 1$, (1.2) does not hold.

The inequality (1.2) can be strengthened in a way which will be useful later. Let

$$\tilde{f}_*(x) = \sup_{\varepsilon} \left| \frac{1}{\pi} \int_{|x-t|>\varepsilon} \frac{f(t)}{x-t}\,dt \right|.$$

Then again

(1.3) $$\|\tilde{f}_*\|_p \leqq A_p \|f\|_p$$

for $1 < p < \infty$. [2])

We shall now discuss an extension of the notion of Hilbert's transform to functions of several variables.

---

[1]) Some of the results have already been published; see papers **1**, **2**, **3**, **4** of the bibliography. For older literature see **5**.

[2]) The corresponding result for periodic functions is proved, by complex methods, in **6**, pp. 249—250, and the proof for f nonperiodic is similar. Another proof is given in **1**, p. 116.

Reprinted from *Proceedings of the International Congress of Mathematicians, 1954, Amsterdam*, Vol. 3, pp. 140–151 (1956).

We consider the n-dimensional Euclidean space E^n. The points of this space will be denoted by $x = (\xi_1, \xi_2, \ldots, \xi_n)$, $y = (\eta_1, \eta_2, \ldots, \eta_n), \ldots$ etc. By definition, $0 = (0, 0, \ldots, 0)$. We shall also treat E^n as a vector space by identifying the point x with the vector $0x$. Addition of vectors and multiplication of vectors by scalars are defined in the usual way, and we write $|x| = (\xi_1^2 + \xi_2^2 + \ldots + \xi_n^2)^{\frac{1}{2}}$. By L^p we mean the set of measurable functions $f(x)$ such that $\|f\|_p = \left(\int\limits_{E^n} |f|^p dx\right)^{1/p} < \infty$, $dx = d\xi_1 \ldots d\xi_n$ denoting the element of volume in E^n.

By Σ we mean the unit sphere $|x| = 1$. By x' we systematically denote the projection of x onto Σ, that is the intersection of Σ with the ray $0x$.

The Hilbert transform (1.1) on the straight line is the convolution of f wish the kernel $K(x) = 1/\pi x$. The Hilbert transform of $f(x)$ in E^n is defined as the convolution of f with a certain kernel K:

$$\tilde{f} = f * K = \int\limits_{E^n} f(y)\, K(x-y) dy. \tag{1.4}$$

The kernels we consider are suggested by various problems of the theory of the potential and, in the most important cases, are of the form

$$K(x) = \frac{\Omega(x')}{|x|^n},$$

where Ω is a function defined on the unit sphere Σ. The kernel K is in general non-integrable near $x = 0$, and the Hilbert transform (1.4) is taken in the principal-value sense

$$\tilde{f}(x) = \lim_{\varepsilon \to +0} \int\limits_{|x-y|>\varepsilon} f(y)\, K(x-y) dy.$$

Of course the integral under the limit sign on the right is improper, but under very general conditions (for example, if Ω is bounded) converegs absolutely for each $f \in L^p$, $1 \leqq p < \infty$ and each x.

We impose two conditions on the kernel K:

(i) The integral of Ω extended over Σ is zero;

(ii) Ω satisfies a Lipschitz condition of positive order, *i. e.*

$$|\Omega(x') - \Omega(y')| \leqq A\, |x' - y'|^\alpha.$$

Condition (ii) can in some cases be considerably relaxed (see § 2 below), but this matter, though of considerable intrinsic interest, is not of primary importance in applications, where Ω is usually a spherical harmonic. On the other hand, condition (i) is essential. If it is not satisfied, then we easily see that for any function f constant but not zero in a sphere D the integral (1.4) diverges to ∞ at every point of D. Condition (i) is satisfied in the case $n = 1$,

$K(x) = 1/x$, since the sphere Σ then consists of the points $x' = \pm 1$, and $\Omega(x') = \operatorname{sign} x'$.

While for $n = 1$ we have essentially only one Hilbert transform, in higher dimensions we have infinitely many, each characterized by the corresponding Ω. If $n = 2$, x' is an angle between 0 and 2π, and condition (i) means that the constant term of the Fourier series of $\Omega(x')$ is zero.

A few special cases deserve mention. If $n = 2$ and $\Omega(\varphi) \sim \Sigma' c_n e^{in\varphi}$ (the prime indicating that $c_0 = 0$), then

$$K(z) = \Sigma' c_k \frac{e^{ik\varphi}}{r^2},$$

where $z = r e^{i\varphi}$ is a complex number. This shows the particular importance of the special kernels

$$\frac{e^{ik\varphi}}{r^2} = \frac{z^k}{|z|^{k+2}}, \tag{1.5}$$

and among them of the kernels

$$z/|z|^3 \tag{1.6}$$

and

$$1/z^2. \tag{1.7}$$

The latter kernel is the only one among (1.6) which is analytic. It was considered in unpublished work of Beurling and we shall call it the *Beurling kernel.*

The kernel (1.7) is a special case of a kernel considered for general n by M. Riesz and Horwath, and defined by the formula

$$K(x) = \frac{x}{|x|^{n+1}}$$

This K is a vector kernel.

Theorem 1. *Suppose that K satisfies conditions (i) and (ii) above and write*

$$\tilde{f}_\varepsilon(x) = \int_{|x-y|>\varepsilon} f(y)\, K(x-y)dy.$$

Suppose that $f \in L^p$, $1 \leqq p < \infty$. Then

$$\tilde{f}(x) = \lim_{\varepsilon\to 0} \tilde{f}_\varepsilon(x)$$

exists almost everywhere. If $1 < p < \infty$, then also

$$\|\tilde{f}_\varepsilon\|_p \leqq A_p \|f\|_p, \quad \|\tilde{f}\|_p \leqq A_p \|f\|_p, \tag{1.8}$$

$$\lim_{\varepsilon\to +0} \|f - \tilde{f}_\varepsilon\|_p = 0, \tag{1.9}$$

(1.10) $$\|\bar{f}_*\|_p \leq A_p \|f\|_p,$$

where $$\bar{f}_*(x) = \sup_{\varepsilon>0} \left| \int_{|x-y|>\varepsilon} f(y) K(x-y) dy \right|.$$

The constants A_p depend on p and the kernel K only.

The proof of this and of other results of a similar nature can befound in **1**. Here we only mention (and the result is of importance in the theory of the potential) that if $F(e)$ is any countably additive mass distribution in E^n, then the Stieltjes integral

$$\int_{E^n} K(x-y)\, F(dy)$$

converges almost everywhere.

2. *A partial generalization of Theorem* 1. We call the function $\Omega(x')$ *even* if $\Omega(-x') = \Omega(x')$, and *odd* if $\Omega(-x') = -\Omega(x')$. The formula

$$\Omega(x') = \tfrac{1}{2}\{\Omega(x') + \Omega(-x')\} + \tfrac{1}{2}\{\Omega x') - \Omega(-x')\}$$

decomposes the kernel K into a sum of two analogous kernels with numerators respectively even and odd.

Theorem 2. *Suppose that Ω is odd that $\Omega(x')$ is integrable over Σ (in particular, the integral of Ω over Σ is zero). Suppose that $f \in L^p$, where p is strictly greater than* 1. *Then the integral*

$$\bar{f}(x) = f*K$$

converges almost everywhere. Moreover the function $\bar{f}_*$ of Theorem 1 *satisfies the inequality* (1.10).

Hence for Ω odd and $f \in L^p$, $p > 1$, condition (ii) is superfluous. The proof is a comparatively simple deduction from results in the one dimensional case. We give it here.

Let t be apoint on Σ; thus t is also a unit vector. Let L be any straight line parallel to t. Let $\epsilon = \epsilon(x)$ be a positive (and measurable) function of x. For any point x on L consider the expression

$$g_\varepsilon(x, t) = \int_{|\xi|>\varepsilon} \frac{f(x-\xi t)}{\xi} d\xi = \int_\varepsilon^\infty \frac{f(x-\xi t) - f(x-\xi t)}{\xi} d\xi,$$

which, for $\epsilon \to 0$, tends to the one-dimensional Hilbert transform of the function f confined to the line L. By (1.3),

$$\int_L |g_\varepsilon(x, t)|^p dx \leq A_p^p \int_L |f(x)|^p dx.$$

Imagine that the direction t, and so also line L, is parallel to one of the coordinate axes. Integrating with respect to the remaining coordinates we get

$$\int_{E^n} |g_\epsilon(x,t)|^p dx \leqq A_p^p \int_{E^n} |f(x)|^p dx$$

or, briefly,

$$\|g_\epsilon(x,t)\|_p \leqq A_p^p \|f(x)\|_p.$$

where t is any fixed unit vector.

It is easy to relate the functions $\tilde{f}_\epsilon(x)$ and $g_\epsilon(x,t)$. For, first of all,

$$\tilde{f}_\epsilon(x) = \int_{|y|>\epsilon} f(x-y)\,K(y)dy = \int_{|y|>\epsilon} f(x-y)\,\frac{\Omega(y')}{|y|^n}\,dy.$$

We now set $y = t\xi$, where t is an arbitrary unit vector and $\xi > 0$. In the last integral we integrate first on concentric spheres with radii ξ, and then integrate with respect to ξ over the interval (ϵ, ∞). The integral becomes

$$(2.1)\qquad \int_\epsilon^\infty \int_\Sigma f(x-t\xi)\,\frac{\Omega(t)}{\xi^n}\,\xi^{n-1}d\xi dt = \int_\Sigma \Omega(t)\left\{\int_\epsilon^\infty \frac{f(x-t\xi)}{\xi}\,d\xi\right\}dt.$$

If we replace here t by $-t$ and use the fact that Ω is odd we obtain

$$(2.2)\qquad -\int_\Sigma \Omega(t)\left\{\int_\epsilon^\infty \frac{f(x+t\xi)}{\xi}\,d\xi\right\}dt$$

Taking the semi-sum of the expressions we obtain

$$\tilde{f}_\epsilon(x) = \tfrac{1}{2}\int_\Sigma \Omega(t) g_\epsilon(x,t)dt,$$

a result which may be summarized as follows: if Ω is odd, then for fixed x the integral $\tilde{f}_\epsilon(x)$ is the average of $g_\epsilon(x,t)$ over the unit sphere Σ, with weight function Ω. From Minkowski's inequality

$$\left\| \int \lambda \right\| \leqq \int \|\lambda\|$$

we deduce

$$\|\tilde{f}_\epsilon(x)\|_p \leqq \tfrac{1}{2}\left\|\int_\Sigma \Omega(t) g_\epsilon(x,t)dt\right\| \leqq \tfrac{1}{2}\int_\Sigma |\Omega(t)|\,\|g_\epsilon(x,t)\|_p dt \leqq \tfrac{1}{2}A_p\|f\|_p \int_\Sigma |\Omega|\,dt,$$

or briefly

$$\|\tilde{f}_\epsilon(x)\|_p \leqq B_p\,\|f\|_p\,\|\Omega\|_1,$$

where B_p depends on p only.

By taking suitable $\epsilon = \epsilon(x)$ we are immediately led to (1.10). The existence of $\tilde{f}$ for almost all x is a consequence of (1.10), (and so are the inequalities (1.8) and the formula (1.9)).

The conclusion of Theorem 2 no longer holds if Ω is even. The sitution is then somewhat different and the proof much less simple. The following theorem is stated without proof.

Theorem 3. *Suppose that $\Omega(x') \log^+|\Omega(x')|$ is integrable over Σ and that the integral of Ω over Σ is zero. The conclusions of Theorem 1 then hold for any $f \in L^p$, $p > 1$.*

Examples show that the integrability of $|\Omega| \log^+|\Omega|$ is the best possible assumption and cannot be relaxed. In particular, the mere integrability of Ω is not enough.

The proofs of Theorems 2 and 3 break down when $p = 1$, so that these theorems are not a complete generalization of Theorem 1.

In certain problems we have to consider convolutions

$$\int_{E_n} f(x-y) \frac{\Omega(y')}{|y|^n} dy$$

where Ω depends not only on y' but also on x, *i.e.* $\Omega = \Omega_x(y')$. Going through the proof of Theorem 2 we see that it holds in this case if $\Omega_x(y')$ is odd in y', and if

$$\sup_x |\Omega_x(y')|$$

is integrable over Σ (in particular, if Ω is bounded in x and y').

3. *The periodic case.* The notion of a Hilbert transform can be extended, properly modified, to the case of periodic functions. In the one-dimensional case we have the *conjugate function*

$$f^*(x) = \frac{1}{\pi} \int_{-\pi}^{\pi} f(y) \tfrac{1}{2} \cot \tfrac{1}{2} (x-y) dy,$$

defined for f of period 2π and integrable.

Consider in E^n an orthogonal system of axes, and let $e_1, e_2, \ldots e_n$ be vectors of length 2π situated on the axes. Let

$$x_0 = 0,\ x_1, \ldots, x_k, \ldots$$

be the sequence of all lattice points generated by these vectors:

$$x_k = \mu_1 e_1 + \mu_2 e_2 + \ldots + \mu_n e_n,$$

where $k = 0, 1, \ldots$, and the μ's run through all integers, positive, negative and zero. Given any kernel $K(x)$ satisfying conditions (i) and (ii) above, we set

$$K^*(x) = K(x) + \sum_{k=1}^{\infty} \{K(x + x_k) - K(x_k)\}. \tag{3,1}$$

The series on the right is absolutely and uniformly convergent over any finite sphere. If $n = 1$, $K(x) = 1/x$, we obtain for $K^*(x)$ the function $\frac{1}{2}$ cot $\frac{1}{2}x$; if $n = 2$ and $K(z)$ is the Beurling kernel $1/z^2$, $K^*(z)$ is the Weierstrass function $\wp(z)$, etc.

The function $K^*(x)$ is periodic and has all the lattice vectors x_k as periods. Let $f(x) = f(\xi_1, \xi_2, \ldots, \xi_n)$ be any locally integrable function, of period 2π in each ξ_j. The convolution

$$(2\pi)^{-n} \int_R f(y)\, K^*(x - y) dy \tag{3.2}$$

of f and K^* will be called the *conjugate function of f*, with respect to the kernel K^*, and denoted by $f^*(x)$. R denotes the fundamental cube $|\xi_j| \leq \pi$, $j = 1, 2, \ldots, n$.

We see from (3.1) that $K^* - K$ is bounded in R. It follows from the corresponding results for $\tilde{f}$ that f^* exists almost everywhere. The following result is an easy consequence of Theorem 1.

Theorem 4. *If f is in $L^p(R)$ and $p > 1$, then f^* is also in $L^p(R)$ and*

$$\int_R |f(x)|^p dx \leq A_p^p \int_R |f(x)|^p dx.$$

Consider the Fourier coefficients

$$c_m = (2\pi)^{-n} \int_R f(x) e^{-i(m,x)}\, dx,$$

and the Fourier series

$$\Sigma\, c_m e^{i(m,x)}$$

of $f(x)$. Here $m = (\mu_1, \mu_2, \ldots, \mu_n)$ is an integral lattice point and $(m, x) = \mu_1\xi_1 + \ldots + \mu_n\xi_n$. Should the conjugate function f^* be integrable we denote its Fourier coefficients by c_m^*, so that

$$f^*(x) \sim \Sigma\, c_m^*\, e^{i(m,x)}$$

Denote by γ_m the Fourier coefficients of the (periodic) fuction K^*:

$$\gamma_m = (2\pi)^{-n} \int_R K^*(x) e^{-i(m,x)} dx.$$

These integrals exist in the principal — value sense, and a simple and classical argument shows that for $m \neq (0, 0, \ldots, 0)$ the Fourier coefficients of K^* coincide with the Fourier transform

$$\hat{K}(m) = (2\pi)^{-n} \int_R K(x) e^{-i(m,x)}\, dx$$

of K.

Since f^* is a convolution of K^* and f, we can anticipate that $c_m^* = \gamma_m c_m$, and the result is actually not difficult to prove. In particular,

$$c_m^* = \hat{K}(m)c_m \qquad (m \neq 0$$

From this and Theorem 4 we deduce the following result:

Theorem 5. *Let $\Sigma\, c_m e^{i(m,x)}$ be the Fourier series of an f in L^p, $p > 1$. Then $\Sigma'\, c_m \hat{K}(m) e^{i(m,x)}$, where the prime denotes that the term $m = 0$ is omitted in summation, is the Fourier series of a function g such that*

$$\int_R |g|^p dx \leqq A_p^p \int_R |f|^p dx.$$

4. *Discrete Hilbert transforms.* A discrete analogue of the transform (1.1) can be formulated as follows. Let

$$X = (\ldots, \xi_{-1}, \xi_0, \xi_1, \ldots, \xi_\nu, \ldots)$$

be a two-way infinite sequence of complex numbers, and let us define the sequence

$$\tilde{X} = (\ldots, \tilde{\xi}_{-1}, \tilde{\xi}_0, \tilde{\xi}_1, \ldots, \tilde{\xi}_\mu, \ldots)$$

by the equations

$$\tilde{\xi}_\mu = \sum_{\nu \neq \mu} \frac{\xi_\nu}{\mu - \nu}.$$

Then

$$(\Sigma\, |\tilde{\xi}_\mu|^2)^{\frac{1}{2}} \leqq A\, (\Sigma |\xi_\nu|^2)^{\frac{1}{2}}, \tag{4.1}$$

or

$$\|\, \tilde{X} \,\|_2 \leqq A \,\|\, X \,\|_2,$$

if we adopt the abbreviation

$$\|\, X \,\|_p = (\Sigma |\xi_\nu|^p)^{1/p}.$$

This result of Hilbert and Toeplitz was generalized by M. Riesz who showed that

$$\|\tilde{X}\|_p \leqq A_p \|X\|_p \tag{4.2}$$

for every $p > 1$, where A_p depends on p only. Theorem 1 leads to a generalization of M. Riesz' result. We state this generalization without proof.

Suppose that the kernel K satisfies conditions (i) and (ii) above. Let $e_1, e_2, \ldots, e_n$ be a system of n linearly independent vectors in E^n, and let $x_0 = 0, x_1, x_2, \ldots$ be the sequence of all lattice points in E^n generated by this system. Let $X = (\xi_0, \xi_1, \ldots)$ be any sequence of complex numbers. We define the transform $\tilde{X} = (\tilde{\xi}_0, \tilde{\xi}_1, \ldots)$ of X by the equations

$$\tilde{\xi}_\mu = \sum_{\nu \neq \mu} \xi_\nu K(x_\mu - x_\nu). \tag{4.3}$$

Theorem 6. *The transform* (4.3) *of* X *satisfies* (4.2), *where* A_p *depends on on* p *and* K.

5. *Radial kernels.* Certain problems of Analysis lead to kernels which are of a different type from Hilbert's though have close relation with the latter. Such kernels occur already in the one-dimensional case.

Theorem 7. *Let* $h(t)$ *be a Fourier-Stieltjes transform on the straight line,*

$$(h(t) = \int_{-\infty}^{+\infty} e^{itu} d\gamma(u), \tag{5.1}$$

where

$$\int_{-\infty}^{+\infty} |d\gamma(u)| = V < \infty.$$

Then for every f *of the class* $L^p(-\infty, +\infty)$, $p > 1$, *the integral*

$$F(x) = \int_{-\infty}^{+\infty} f(x+t) \frac{h(t)}{t} dt$$

converges almost everywhere and

$$||F||_p \leqq A_p V ||f||_p. \tag{5.2}$$

By (1.3) we have

$$\left\| \int_{|t|>\varepsilon} f(x+t) \frac{dt}{t} \right\|_p \leqq A_p ||f||_p,$$

where ϵ depends on x. Apply this not to $f(x)$ but to $f(x)\, e^{ixu}$, where u is fixed. We obtain

$$\left\| \int_{|t|>\varepsilon} f(x+t) \frac{e^{itu}}{t} dt \right\|_p \leqq A_p ||f||_p,$$

and, by Minkowski's inequality,

$$\left\| \int_{-\infty}^{+\infty} d\gamma(u) \int_{|t|>\varepsilon} f(x+t) \frac{e^{itu}}{t} dt \right\|_p \leqq A_p V ||f||_p.$$

If ε stays above a positive number ε_0, we can interchange the order of summation and integration and get

$$\left\| \int_{|t|>\varepsilon} f(x+t) \frac{h(t)}{t} dt \right\|_p \leqq A_p V ||f||_p.$$

Making now ε_0 tend to 0 we get the inequality when ε is any positive measurable function of x. From this the existence of $F(x)$, as well as the inequality (5.2), follows in a straightforward fashion.

To the integral $\xi(x)$ so obtained we may apply the argument of Section 2. It works if we assume that $h(t)$ is an *odd* function of t, and we get the following result.

Theorem 8. *Suppose that $f \in L^p(E^n)$, $p > 1$. Let $h(t)$, given by (5.1), be an odd function of t, and let $\Omega(x')$ be even and intebrable over Σ. Then the integral*

$$\int_{E^n} f(x+y)\frac{h(|y|)}{|y|^n}\Omega(y')dy$$

converges almost everywhere and its value $\tilde{f}(x)$ satisfies

$$\|\tilde{f}\|_p \leqq A_p V \|\Omega\|_1 \|f\|_p.$$

The same conclusion holds if Ω is odd and h even. The case $\Omega \equiv 1$ has some interesting applications. To arrive at them we first consider a problem in the one-dimensional case.

Suppose that $f(x) \in L^p(-\infty, +\infty)$, $p > 1$. The problem of the representation of f by its Fourier integral reduces to showing that the Dirichlet integral

$$f_w(x) = \frac{1}{\pi}\int_{-\infty}^{+\infty} f(x+t)\frac{\sin wt}{w}dt$$

converges to $f(x)$ as $w \to \infty$. We may consider here either pointwise convergence or convergence in L^p. As regards the former, no general results can be obtained unless we consider the limit of $f_w(x)$ by the method of arithemetic means of positive order. It is therefore quite remarkable (the result is due to M. Riesz) that, as regards convergence in L^p, the last integral does actually tend to f:

$$\|f(x) - f_w(x)\|_p \to 0,$$

as $w \to \infty$. That the norm $\|f_w(x)\|_p$ remains bounded as $w \to \infty$, follows immediately from Theorem 8. For if we set $h(t) = \sin wt$, then $h(t)$ is a Fourier-Stieltjes transform:

$$2i \sin wt = \int_{-\infty}^{+\infty} e^{itu}d\gamma(u),$$

where $\gamma(u)$ is a step-function having the jumps ± 1 at the points $\pm w$ as the only discontinuities. Thus the total variation of γ is 2 and, by Theorem 8, $\|f_w(x)\|_p \leqq A_p \|f\|_p$. The refinement to $\|f - f_w\|_p \to 0$ is straightforward.

The corresponding problem in n dimensions is more difficult. As regards the so called spherical means of the Fourier integral, it was shown by Bochner that for pointwise representations we must apply not the ordinary convergence but summability (C, α) where

$$\alpha > \lambda = \tfrac{1}{2}(n-1),$$

and, as in the case $n = 1$, this result cannot be improved. The question arises *whether the (C, λ) means converge to f in L^p.* It turns out that for n odd, $n = 1, 3, 5, \ldots$, we are led to integrals covered by the last theorem, and the problem does admit of an affirmative answer. Unfortunately the kernels which occur for n even are of somewhat different nature and the problem in that case is still open.

6. *Applications to potentials.* Suppose for simplicity that $n \geqq 3$ and consider the potential

$$u(x) = u(\xi_1, \ldots, \xi_n) = \int_{E^n} |x-y|^{2-n} f(y)dy$$

generated by a function f which, say, vanishes in the neighbourhood of infinity. The problem of the existence of the second derivatives, ordinary or generalized, of $u(x)$ is classical and has close connection with the existence of Hilbert transforms in E^n. It turns out that for the existence of *ordinary* second derivatives of u it is enough to assume that $|f| \log^+ |f|$ is integrable. The following theorem gives a somewhat stronger result.

Theorem 9. *If $|f| \log^+ |f|$ is integrable, then in almost every subspace $\xi_3 = \xi_3^0, \ldots, \xi_n = \xi_n^0$, the function $u(\xi_1, \xi_2, \xi_3^0, \ldots, \xi_n^0)$ is a continuous, indeed absolutely continuous, function of the variables ξ_1, ξ_2, and has a second differential almost everywhere in ξ_1, ξ_2.*

We say that a function $v(\xi_1, \xi_2)$ has a second differential at a point ξ_1^0, ξ_2^0 if $v(\xi_1^0 + h, \xi_2^0 + k) - v(\xi_1^0, \xi_2^0)$ is equal to

$$(\alpha_1 h + \alpha_2 k) + \tfrac{1}{2}(\alpha_{11}h^2 + 2\alpha_{12}hk + \alpha_{22}k^2) + o\,\{(h^2 + k^2)^{\frac{1}{2}}\}$$

for h and k tending to 0, $\alpha_1, \ldots, \alpha_{22}$ denoting constants.

Theorem 9 implies that under its hypotheses all second derivatives $u_{\xi_i\xi_k}$ exist almost everywhere, and it can easily be shown that they are given by the classical formulas. Results analogous to Theorem 9 but pertaining to first derivatives can be obtained for potentials of single or double layer of masses distributed on hyperplanes in E^n.

Bibliography

[1] A. P. Calderón and A. Zygmund, On the existence of certain integrals, Acta Math. 88 (1952), 85—139.

[2] A. P. Calderón and A. Zygmund, On a problem of Mihlin, Transactions of the American Math. Soc. 78 (1955), 209—224.

[3] A. P. Calderón and A. Zygmund, Singular integrals and periodic functions, Studia Math., 142 (1955).

[4] A. Zygmund. On the existence and properties of certain singular integrals, Lectures at the Kingston meeting of the American Math. soc., August 1953, mimeographed notes.

[5] S. Mihlin, Singular integral equations, Uspekhi Mat. Nauk, 3 (1948) 29–112 (in Russian).

[6] A. Zygmund, Trigonometrical series, Warszawa, 1935.

Reprinted from the Proceedings of the NATIONAL ACADEMY OF SCIENCES,
Vol. 42, No. 4, pp. 208–212. April, 1956.

ON THE LITTLEWOOD-PALEY FUNCTION $g^*(\theta)$

BY A. ZYGMUND*

Communicated by A. A. Albert, February 10, 1956

1. Let H^λ, $\lambda > 0$, denote the class of functions

$$\Phi(z) = \sum_0^\infty c_n z^n$$

regular in $|z| < 1$ and such that

$$\mathfrak{M}_\lambda[\Phi(re^{i\theta})] = \left\{(2\pi)^{-1} \int_0^{2\pi} |\Phi(re^{i\theta})|^\lambda d\theta\right\}^{1/\lambda} < \mathrm{M} < \infty$$

for some M independent of r. The partial sums of the series $\sum c_n e^{in\theta}$ we denote by $s_n(\theta)$, and the $(C, 1)$ means by $\sigma_n(\theta)$, and write $\Phi(e^{i\theta}) = \lim_{r\to 1} \Phi(re^{i\theta})$.

In their work[1] on Fourier series Littlewood and Paley introduced the functions

$$g(\theta) = g(\theta, \Phi) = \left\{\int_0^1 (1 - \rho)|\Phi'(\rho e^{i\theta})|^2 d\rho\right\}^{1/2}, \tag{1.1}$$

$$g^*(\theta) = g^*(\theta, \Phi) = \left\{\int_0^1 (1 - \rho)\chi^2(\rho, \theta)d\rho\right\}^{1/2}, \tag{1.2}$$

where

$$\chi(\rho, \theta) = \left\{\pi^{-1} \int_0^1 |\Phi'(\rho e^{i(\theta+t)})|^2 P(\rho, t)dt\right\}^{1/2}$$

and $P(\rho, t) = {}^1/_2 (1 - \rho^2)/(1 - 2\rho \cos t + \rho^2)$. They showed that

$$\mathfrak{M}_\lambda[g] \leq A_\lambda \mathfrak{M}_\lambda[\Phi(e^{i\theta})] \tag{1.3}$$

for all $\lambda > 0$ and that

$$\mathfrak{M}_\lambda[g^*] \leq A_\lambda \mathfrak{M}_\lambda[\Phi(e^{i\theta})] \tag{1.4}$$

for $\lambda = 2, 4, \ldots$, and raised the problem of the validity of relation (1.4) for other values of λ.

That relation (1.4) holds for all $\lambda > 1$ was shown in an earlier paper[2] by a rather laborious argument, which, moreover, left the case $\lambda = 1$ open (the example $\Phi(z) = 1/(1 - z)$ shows that the inequality fails for $\lambda < 1$; we omit the simple computation). It turns out, however, that the problem is easily reducible to one solved a long time ago, and this approach gives a more complete answer, namely, the following:

THEOREM 1. *We have*

$$\mathfrak{M}_\lambda[g^*] \le A_\lambda \, \mathfrak{M}_\lambda[\Phi(e^{i\theta})], \; \lambda > 1, \tag{1.5}$$

$$\mathfrak{M}[g^*] \le A \int_0^{2\pi} |\Phi(e^{i\theta})| \log^+ |\Phi(e^{i\theta})| \, d\theta + A, \tag{1.6}$$

$$\mathfrak{M}_\mu[g^*] \le A_\mu \, \mathfrak{M}[\Phi(e^{i\theta})], \; 0 < \mu < 1, \tag{1.7}$$

where $\mathfrak{M}$ *stands for* $\mathfrak{M}_1$.

Inequalities (1.6) and (1.7) seem to be new. The proof of Theorem 1 is based on the following simple observation that $g^*(\theta)$ is essentially identical with the function

$$\gamma(\theta) = \gamma(\theta, \Phi) = \left(\sum_0^\infty \frac{|s_n - \sigma_n|^2}{n+1} \right)^{1/2}, \tag{1.8}$$

which was known to satisfy inequalities analogous to relations (1.5), (1.6), and (1.7) (see an earlier paper[3] [Part I]). For first we observe that

$$g^*(\theta) = \left\{ \pi^{-1} \frac{1}{2} \int_0^1 (1+\rho)(1-\rho)^2 \, d\rho \int_0^{2\pi} \frac{|\Phi'(\rho e^{i(\theta+\psi)})|^2}{|1-\rho e^{i\psi}|^2} \, d\psi \right\}^{1/2}. \tag{1.9}$$

Next, with $z = \rho e^{i\psi}$,

$$\begin{aligned} \sum_0^\infty s_n'(\theta) z^n &= \frac{d}{d\theta} \sum s_n(\theta) z^n = \frac{d}{d\theta} \{(1-z)^{-1} \sum c_n e^{in\theta} z^n\} \\ &= \rho i (1-\rho e^{i\psi})^{-1} \, \Phi'(\rho e^{i(\theta+\psi)}) e^{i(\theta+\psi)} \end{aligned} \tag{1.10}$$

so that

$$\sum_1^\infty |s_n'(\theta)|^2 \rho^{2n} = (2\pi)^{-1} \rho^2 \int_0^{2\pi} |1-\rho e^{i\psi}|^{-2} |\Phi'(\rho e^{i(\theta+\psi)})|^2 d\psi.$$

If we multiply this by $\rho^{-2}(1-\rho)^2$ and integrate over $0 < \rho < 1$, we get

$$\sum_1^\infty \frac{|s_n'(\theta)|^2}{(2n-1)2n(2n+1)} = (2\pi)^{-1} \int_0^1 (1-\rho)^2 \, d\rho \int_{-\pi}^{\pi} \frac{|\Phi'(\rho e^{i(\theta+\psi)})|^2}{|1-\rho e^{i\psi}|^2} \, d\psi. \tag{1.11}$$

On the other hand,

$$\sum_1^\infty \frac{|s_n(\theta) - \sigma_n(\theta)|^2}{n+1} = \sum_1^\infty \frac{|s_n'(\theta)|^2}{(n+1)^3}. \tag{1.12}$$

If we compare this with the left side of equation (1.11) and compare equation (1.9) with the right side of equation (1.11), we notice that $g^*(\theta)/\gamma(\theta)$ is contained between two positive absolute constants, and if we use inequalities (1.5), (1.6), and (1.7) for γ,[4] Theorem 1 follows.

2. We now consider the function

$$g_\sigma^*(\theta) = \{ \int_0^1 \delta^\sigma \, d\rho \int_{-\pi}^{\pi} |1-\rho e^{i\psi}|^{-\sigma} |\Phi'(\rho e^{i(\theta+\psi)})|^2 d\psi \}^{1/2} \tag{2.1}$$

(introduced in an earlier paper[2]), where $\delta = 1 - \rho$. If $\sigma = 2$, this function reduces (except for the harmless factor $\frac{1}{2}\pi^{-1}(1+\rho)$) to $g^*(\theta)$, and from now on we use $g_2^*(\theta)$ for $g^*(\theta)$.

We shall denote by $H^\lambda \log^+ H$ the class of functions $\Phi(z) = \sum c_n z^n$, such that the integral

$$\int_0^{2\pi} |\Phi(re^{i\theta})|^\lambda \log^+ |\Phi(re^{i\theta})|\, d\theta$$

remains bounded for $r \to 1$. The following theorem generalizes Theorem 1.

THEOREM 2. *If $\Phi \in H^\lambda$, $0 < \lambda \le 1$, then $g_{2/\lambda}^*$ is finite almost everywhere, and*

$$\mathfrak{M}_{\eta\lambda}[g_{2/\lambda}^*] \le A_\eta\, \mathfrak{M}_\lambda[\Phi(e^{i\theta})], \qquad 0 < \eta < 1. \tag{2.2}$$

If $\Phi \in H^\lambda \log^+ H$, then

$$\mathfrak{M}_\lambda^\lambda[g_{2/\lambda}^*] \le A_\lambda \int_0^{2\pi} |\Phi(e^{i\theta})|^\lambda \log^+ |\Phi(e^{i\theta})|\, d\theta. \tag{2.3}$$

The theorem holds for $\lambda = 1$ (see Theorem 1), and we deduce from this the case $\lambda < 1$ by a familiar argument (the argument which follows is taken from an earlier paper,[2] where it was applied to a function G_σ which is the dominant part of g_σ^*).

Suppose first that Φ has no zeros, so that $\Psi = \Phi^\lambda$ is in H. Integral (2.1), with $\sigma = 2/\lambda$, becomes

$$\left(\frac{2}{\lambda}\right)^2 \int_0^1 \delta^{2/\lambda}\, d\rho \int_{-\pi}^{\pi} |1 - \rho e^{ix}|^{-2/\lambda} |\Psi(\rho e^{i(\theta+\psi)})|^{2(-1+1/\lambda)} |\Psi'|^2\, d\psi \tag{2.4}$$

Write $\xi(\theta) = \sup |h^{-1} \int_\theta^{\theta+h} |\Psi(e^{it})|\, dt|$ for all $h \ne 0$. It is well known[5] that

$$|\Psi(\rho e^{i(\theta+\psi)})| \le A\xi(\theta)\, \delta^{-1} |1 - \rho e^{i\psi}|, \tag{2.5}$$

so that integral (2.4) is majorized by

$$A_\lambda \{\xi(\theta)\}^{2(-1+1/\lambda)} \int_0^1 \delta^2\, d\rho \int_{-\pi}^{\pi} |1 - \rho e^{i\psi}|^{-2} |\Psi'(\rho e^{i(\theta+\psi)})|^2\, dx,$$

and we obtain the inequality

$$g_{2/\lambda}^*(\theta, \Phi) \le A_\lambda \{\xi(\theta)\}^{(1-\lambda)/\lambda} g_2^*(\theta, \Psi). \tag{2.6}$$

Since, by Theorem 1, $g_2^*(\theta, \Psi)$ is finite almost everywhere, the same holds for $g_{2/\lambda}^*(\theta, \Phi)$. Observe now that by the Hardy-Littlewood maximal theorem

$$\mathfrak{M}_\eta[\xi] \le A_\eta\, \mathfrak{M}[\Psi(e^{i\theta})], \qquad \mathfrak{M}[\xi] \le A \int_0^1 |\Psi(e^{i\theta})| \log^+ |\Psi|\, d\theta + A. \tag{2.7}$$

Raising relation (2.6) to the power $\eta\lambda$, applying Hölder's inequality with exponents $1/(1-\lambda)$ and $1/\lambda$, and also the first inequality (2.7) (for Ψ), we immediately obtain relation (2.2). To obtain relation (2.3), we use the second inequality (2.7) and inequality (1.6) (for Ψ). Then, writing Φ, Ψ for $\Phi(e^{i\theta})$, $\Psi(e^{i\theta})$, we have

$$\begin{aligned}
\int_0^{2\pi} g_{2/\lambda}^{*\lambda}[\Phi]\, d\theta &\le A_\lambda \int_0^{2\pi} \xi^{1-\lambda} g_2^{*\lambda}[\Psi]\, d\theta \le A_\lambda \left(\int_0^{2\pi} \xi\, d\theta\right)^{1-\lambda} \left(\int_0^{2\pi} g_2^*[\Psi]\, d\theta\right)^\lambda \\
&\le A_\lambda \left\{\int_0^{2\pi} |\Psi| \log^+ |\Psi|\, d\theta + A\right\}^{1-\lambda} \left\{A \int_0^{2\pi} |\Psi| \log^+ |\Psi|\, d\theta + A\right\}^\lambda \\
&= A_\lambda \int_0^{2\pi} |\Psi| \log^+ |\Psi|\, d\theta + A_\lambda \le A_\lambda \int_0^{2\pi} |\Phi|^\lambda \log^+ |\Phi|\, d\theta + A_\lambda
\end{aligned}$$

This completes the proof of Theorem 2 in the case when Φ has no zeros. In the general case we have, as is well known, $\Phi = \Phi_1 + \Phi_2$, where Φ_1 and Φ_2 have no zeros and satisfy the inequalities $|\Phi_1| \le 2|\Phi|$, $|\Phi_2| \le 2|\Phi|$ in $|z| < 1$. Since, by

Minkowski's inequality, $g_\sigma^*(\Phi_1 + \Phi_2) \leq g_\sigma^*(\Phi_1) + g_\sigma^*(\Phi_2)$, the general result is an immediate consequence of the one just proved.

I do not know whether Theorem 2 holds for $1 < \lambda < 2$. For $\lambda = 2$ (and a fortiori for $\lambda > 2$) it fails, since there are functions Φ regular in $|z| < 1$, continuous in $|z| \leq 1$ and such that $g_1^*(\theta, \Phi) = \infty$ for all θ. It can be shown that if $1 < \lambda < 2$ and $\sigma > 2/\lambda$, then the function $g_\sigma^*(\theta, \Phi)$ is finite almost everywhere and satisfies $\mathfrak{M}_\lambda[g_\sigma^*] \leq A_{\lambda\sigma}\, \mathfrak{M}_\lambda[\Phi]$. The proof is given in an earlier paper[2] (pp. 178–179) for the function G_σ^* mentioned above, but the result has no interesting application.

3. Let $s_n^\alpha = s_n^\alpha(\theta, \Phi)$ and $\sigma_n^\alpha = \sigma_n^\alpha(\theta, \Phi)$ be the Cesàro sums and Cesàro means for the series $\sum c_n z^n$, $z = e^{i\theta}$. Hence

$$\sigma_n^\alpha(\theta) = \frac{s_n^\alpha(\theta)}{A_n^\alpha} = \frac{1}{A_n^\alpha}\sum_{n=0}^{n} A_{n-\nu}^\alpha c_\nu e^{i\nu\theta},$$

where $A_n^\alpha = (\alpha + 1)(\alpha + 2) \ldots (\alpha + n)/n!$. A simple computation shows that

$$\sigma_n^\alpha - \sigma_n^{\alpha+1} = \frac{1}{A_n^\alpha}(\alpha + n + 1)^{-1}\sum_0^n A_{n-\nu}^\alpha\, \nu c_\nu e^{i\nu\theta} = -\frac{i}{A_n^\alpha}(\alpha + n + 1)^{-1}(s_n^\alpha(\theta))', \tag{3.1}$$

so that

$$\sum_0^\infty (n + 1)^{-1}|\sigma_n^\alpha - \sigma_n^{\alpha+1}|^2 \leq C_\alpha \sum_0^\infty (n + 1)^{-2\alpha-3}|s_n^{\alpha\prime}(\theta)|^2. \tag{3.2}$$

Observe that, if $z = \rho e^{i\psi}$, then

$$\sum_0^\infty s_n^\alpha(\theta) z^n = (1 - z)^{-\alpha-1}\sum_0^\infty c_n e^{in\theta} z^n = (1 - \rho e^{i\psi})^{-\alpha-1}\,\Phi(\rho e^{i(\theta+\psi)}).$$

Hence, differentiating and using Parseval's formula,

$$\sum_1^\infty |s_n^{\alpha\prime}(\theta)|^2\rho^{2n} = (2\pi)^{-1}\rho^{-2}\int_0^{2\pi} |1 - \rho e^{i\psi}|^{-2\alpha-2}|\Phi'(\rho e^{i(\theta+\psi)})|^2\, d\theta. \tag{3.3}$$

If we multiply this by $\rho^{-2}(1 - \rho)^{2\alpha+2}$ and integrate over $0 \leq \rho < 1$, we get

$$\sum (n + 1)^{-2\alpha-3}|s_n^{\alpha\prime}(\theta)|^2 \leq C_\alpha\, g_{2\alpha+2}^{*2}(\theta, \Phi),$$

which, after relation (3.2), gives

$$\sum (n + 1)^{-1}|\sigma_n^\alpha(\theta) - \sigma_n^{\alpha+1}(\theta)|^2 \leq C_\alpha g_{2\alpha+2}^{*2}(\theta, \Phi). \tag{3.4}$$

Denote the square root of the left side by $\gamma_\alpha(\theta) = \gamma_\alpha(\theta, \Phi)$; thus γ_0 is our former function γ. Put $\alpha = (1/\lambda) - 1$; we therefore have

$$\gamma_\alpha(\theta) \leq C_\alpha g_{2/\lambda}^*(\theta, \Phi),$$

and we obtain in view of Theorem 2 the inequalities

$$\mathfrak{M}_{\lambda_\eta}[\gamma_\alpha] \leq C_{\lambda,\eta}\, \mathfrak{M}_\lambda[\Phi],\ \alpha = \frac{1}{\lambda} - 1;\ 0 < \lambda \leq 1, \tag{3.5}$$

$$\mathfrak{M}_\lambda[\gamma_\alpha] \leq C_\lambda\left\{\int_0^{2\pi} |\Phi(e^{i\theta})|^\lambda \log^+|\Phi|\, d\theta + 1\right\}^{1/\lambda}. \tag{3.6}$$

In an earlier paper[3] (Part II) I proved that (*a*) if $\Phi = \sum c_n z^n$ is in H^λ, $0 < \lambda < 1$, then $\sum c_n e^{in\theta}$ is summable $(C, \alpha) = (C, (1/\lambda) - 1)$ almost everywhere. I also conjectured that (*b*) if $\Phi \in H^\lambda \log^+ H$, then $\sigma_*^\alpha(\theta) = \sup_r |\sigma_r^\alpha(\theta)|$ is in L^λ (I could prove the conjecture for $\lambda = 1/2, 1/3, 1/4, \ldots$ only, in which cases it is a relatively simple consequence of the result for $\lambda = 1$, established in Part I of an earlier paper[4]). Inequalities (3.5) and (3.6) contain the proofs of (*a*) and (*b*) for $0 < \lambda \leq 1/2$, but since we do not know whether Theorem 2 holds for $1 < \lambda < 2$, proposition (*b*) for $1/2 < \lambda < 1$ is still a conjecture.[6] I briefly indicate how inequality (3.5) implies (*b*) for $0 < \lambda \leq 1/2$.

We may suppose that Φ has no zeros, so that $\Phi = \Psi^2$, where $\Psi \in H^{2\lambda} \log^+ H$. Then

$$\sigma_n^\alpha[\Phi] = A_n^{-\alpha} s_n^\alpha[\Phi] = A_n^{-\alpha} \sum_{\nu=0}^{n} s_\nu^{1/2(\alpha-1)}[\Psi]\, s_{n-\nu}^{1/2(\alpha-1)}[\Psi],$$

$$|\sigma_n^\alpha[\Phi]| \leq (A_n^\alpha)^{-1} \sum_0^n |s_\nu^{1/2(\alpha-1)}[\Psi]|^2 \leq (A_n^\alpha)^{-1} \sum_0^n |\sigma_\nu^{1/2(\alpha-1)}[\Psi]|^2 (\nu+1)^{\alpha-1}$$

$$\leq (A_n^\alpha)^{-1} \sum_{\nu=0}^n |\sigma_\nu^{1/2(\alpha-1)}[\Psi] - \sigma_\nu^{1/2(\alpha+1)}[\Psi]|^2 (\nu+1)^{\alpha-1} + (A_n^\alpha)^{-1} \sum_0^n |\sigma_\nu^{1/2(\alpha+1)}[\Psi]|^2 (\nu+1)^{\alpha-1}.$$

Denote the upper bounds for $n = 0, 1, 2, \ldots$ of the last two terms by P and Q, respectively. Clearly,

$$P \leq C_\alpha \sum_1^\infty |\sigma_\nu^{1/2(\alpha-1)}[\Psi] - \sigma_\nu^{1/2(\alpha+1)}[\Psi]|^2 (\nu+1)^{-1} = C_\alpha \gamma_{1/2(\alpha-1)}{}^2[\Psi] \leq C_\alpha \gamma_{\alpha+1}{}^{*2}[\Psi]$$

by relation (3.4), so that, by relation (2.3) applied to Ψ,

$$\begin{aligned} \mathfrak{M}_\lambda[P] &\leq C_\alpha \mathfrak{M}_{2\lambda}{}^2[g_{\alpha+1}{}^*[\Psi]] = C_\alpha \mathfrak{M}_{2\lambda}{}^2[g_{1/\lambda}{}^*[\Psi]] \\ &\leq C_\alpha [\textstyle\int_0^{2\pi} |\Psi|^{2\lambda} \log^+ |\Psi|\, d\theta + 1]^{1/\lambda} \leq C_\alpha [\int_0^{2\pi} |\Phi|^\lambda \log^+ |\Phi|\, d\theta + 1]^{1/\lambda}. \end{aligned} \tag{3.7}$$

On the other hand, clearly, $Q \leq \sigma_*^{1/2(\alpha+1)}[\theta, \Psi]$. Since the index $1/2(\alpha+1)$ of summability is higher than the critical index $1/2(\alpha-1)$, it is well known that $\mathfrak{M}_\lambda[Q] \leq C_\lambda \mathfrak{M}_{2\lambda}{}^2[\Psi] = C_\lambda \mathfrak{M}_\lambda[\Phi]$, which, combined with relation (3.7), shows that $\mathfrak{M}_\lambda[\sigma_*^\alpha[\Phi]]$ is majorized by the right side of relation (3.6).

* The research resulting in this paper was supported in part by the Office of Scientific Research of the Air Force under Contract AF 18 (600)-1111.

[1] J. E. Littlewood and R. E. A. C. Paley, "Theorems on Fourier Series and Power Series," *J. London Math. Soc.*, 6, 230–233, 1931; *Proc. London Math. Soc.*, 42, 52–89, 1937.

[2] A. Zygmund, "On Certain Integrals," *Trans. Am. Math. Soc.*, 55, 170–204, 1944.

[3] A. Zygmund, "On the Convergence and Summability of Power Series on the Circle of Convergence," Part I, *Fund. Math.*, 30, 170–196, 1938; Part II, *Proc. London Math. Soc.*, 47, 326–350, 1942.

[4] See Part I, note 3.

[5] G. H. Hardy and J. E. Littlewood, "The Strong Summability of Fourier Series," *Fund. Math.*, 25, 162–189, 1935.

[6] Inequalities (3.5) and (3.6) can also be found in a recent paper of G. Sunouchi. My work was independent of his and was motivated by the desire to prove conjecture (*b*). I am presenting my partial results only because of the appearance of his work. ("On the Summability of Power Series and Fourier Series," *Tohoku Math. J.*, 7, 96–109, 1955.)

SINGULAR INTEGRAL OPERATORS AND DIFFERENTIAL EQUATIONS.* [1]

By A. P. CALDERÓN and A. ZYGMUND.

1. Introduction. Let $P(u)$ be a linear partial differential operator with smooth coefficients and of homogeneous order m. Then $P=H\Lambda^m$ where Λ is a square root of the Laplacian (see definition [1] below) and H is a singular integral operator (see Theorem 7). This fact seems to call for a closer study of the properties of singular integral operators in their connection with the operator Λ and supplies the subject matter of the present paper.

Our results can be briefly summarized as follows. With each singular integral operator there is associated a function (its "symbol" in the terminology of Giraud and Mihlin) in a one-to-one fashion. This correspondence is linear and pseudo-multiplicative in the sense that, modulo a class of regular operators, singular integral operators can be multiplied (in the sense of operator composition) by simply multiplying their symbols. The regular operators in that class have the property of remaining bounded after being multiplied on the left or on the right by Λ. An algebraic formulation of these facts will be found in Theorem 6. The reader familiar with the work of Giraud, Mihlin and Tricomi [1a] will recognize the similarity of some of our results with theirs. The main distinctive feature is that we are concerned with the operator Λ which they do not consider, and that our operators act on L^p, $1<p<\infty$, instead of on L^2 only. For many applications, though, it suffices to consider the case of L^2, and, in this respect and as far as mean convergence of the singular integrals goes, the paper is self-contained. To conclude these preliminary remarks, we want to stress the fact that many of the assumptions on which our results are obtained can be considerably relaxed. Since these improvements do not seem to be of particular relevance at the present time, we prefer not to burden the reader and postpone their discussion to another opportunity.

* Received June 12, 1957.

[1] This research was partly supported by the United States Air Force under contract No. AF18(600)-685 monitored by the Office of Scientific Research.

[1a] A description of the work of these authors can be found in the paper "Singular integral equations" by S. G. Mihlin, Uspekhi Matematicheskikh Nauk, No. 25 (1948), 29-112.

Reprinted from *Amer. J. Math.* 79, 901–921 (1957).

2. Definitions and notation. We will be concerned with functions defined in the k-dimensional Euclidean space E_k. Points in E_k will be denoted by $x=(x_1,\cdots,x_k)$, $y=(y_1,\cdots,y_k)$ etc. and we will use the following abbreviations $|x|=[\sum_1^k x_i^2]^{\frac{1}{2}}$, $\lambda x=(\lambda x_1,\cdots,\lambda x_k)$, $x'=x|x|^{-1}$, $x+y=(x_1+y_1,\cdots,x_k+y_k)$, $x\cdot y=\sum_1^k x_i y_i$. The sphere $|x|=1$ in E_k will be denoted by Σ, the element of surface area on Σ by $d\sigma$, and dx will stand for the volume element in E_k. By C_α, $\alpha\geqq 0$, we shall denote the class of complex valued continuous bounded functions on E_k with bounded continuous derivatives up to order $[\alpha]$ (integral part of α) and with derivatives of order $[\alpha]$ satisfying a (uniform) Hölder condition of order $\alpha-[\alpha]$. When dealing with functions depending on more than one argument, we will denote by C_α^∞ the class of functions in C_α which are in C^∞ with respect to the last argument and whose derivatives of all orders with respect to variables in the last argument are in C_α. Given a subclass of C_α or C_α^∞, we shall say that the subclass is uniform if the bounds and Hölder conditions on the functions and their derivatives are uniform in the subclass.

We shall also consider the class L_r^p of functions in $L^p(E_k)$ with derivatives up to order r in $L^p(E_k)$. The notion of derivative used here is that of Schwartz; that is, $g=\partial f/\partial x_i$ means $(f,\partial\phi/\partial x_i)=-(g,\phi)$ for every $\phi\in C^\infty$ vanishing outside a bounded set, where here, as in the rest of the paper, (f,g) stands for the integral of $f\bar{g}$ over E_k. By A and C we will denote constants, though they will not be necessarily the same in different occurrences.

3. In this section we shall establish some properties of expansions of functions in spherical harmonics. Let $Y_n(x')$ be a normalized real spherical harmonic of degree n, that is, such that

$$\int_\Sigma Y_n(x')^2 . d\sigma = 1$$

and $Y_{nm}(x')$, $m=1,2,\cdots$, a complete orthogonal system of normalized harmonics of degree n. Our first objective is to obtain bounds for the $Y_n(x')$ and their successive derivatives. Consider first the case $k\geqq 3$. Then we have the formula (see [4])

$$(1)\qquad \delta_{nm}Y_m(x')=\tfrac{1}{2}\Gamma(\lambda)(n+\lambda)/\pi^{\lambda+1}\int_\Sigma P_n^\lambda(x'\cdot y')Y_n(y')\,d\sigma$$

where $\lambda=\frac{1}{2}(k-2)$, δ_{nm} is Kronecker's delta and $P_n^\lambda(t)$ is the ultraspherical polynomial defined by

$$(1 - 2wt + w^2)^{-\lambda} = \sum_0^\infty w^n P_n^\lambda(t). \tag{2}$$

For each z, the function $P_n^\lambda(x' \cdot z')$ is a spherical harmonic of degree n, whence, replacing in (1) and setting $x' = z'$, we obtain

$$P_n^\lambda(1) = \tfrac{1}{2}\Gamma(\lambda)(n + \lambda)/\pi^{\lambda+1} \int_\Sigma P_n(y' \cdot z')^2 \, d\sigma.$$

On the other hand, from (2) it follows that

$$(1 - w)^{-k+2} = \sum_0^\infty w^n P_n^\lambda(1),$$

which implies that $P_n^\lambda(1)$ is of the order n^{k-3} as $n \to \infty$. Hence

$$\int_\Sigma P_n^\lambda(y' \cdot z')^2 \, d\sigma$$

is of order n^{k-4}, and Schwarz's inequality applied to (1) gives

$$|Y_n(x')| \leqq C n^{\frac{1}{2}(k-2)}, \qquad n \geqq 1, \tag{3}$$

where C is a constant depending only on k. In order to estimate the derivatives of $Y_n(x')$, let $P_n(x)$ denote temporarily the solid harmonic coinciding with $Y_n(x')$ on Σ. Then, if S denotes the sphere $|x| \leqq 1$ and $\partial P_n/\partial\nu$ is the derivative of P_n in the direction of the outer normal to the boundary Σ of S, we have

$$\int_\Sigma P_n(\partial P_n/\partial\nu) \, d\sigma = \int_S |\operatorname{grad} P_n|^2 \, dx.$$

Now P_n and $|\operatorname{grad} P_n|^2$ are homogeneous functions of degrees n and $2n - 2$, respectively, and from this it follows readily that the two integrals are respectively equal to

$$n \int_\Sigma P_n^2 \, d\sigma = n \text{ and } (2n + k - 2)^{-1} \int_\Sigma |\operatorname{grad} P_n|^2 \, d\sigma,$$

which implies that

$$\int_\Sigma |\partial P_n/\partial x_i|^2 \, d\sigma \leqq Cn^2.$$

But $\partial P_n/\partial x_i$ is a homogeneous harmonic polynomial of degree $n - 1$, and therefore we can write $\partial P_n/\partial x_i = \lambda P_{n-1}$, where P_{n-1} is a solid harmonic coinciding with a normalized spherical harmonic of degree $n - 1$ on Σ, and $|\lambda| \leqq Cn$.

Now we write $Y_n(x') = |x|^{-n} P_n(x)$, and by differentiating and applying (3) and the formula above to the successive derivatives of P_n, we obtain

(4) $$| D_r Y_n(x') | \leqq C n^{\frac{1}{2}(k-2)+r}, \qquad | x | \geqq 1,$$

where $D_r Y_n$ denotes a derivative of Y_n of order r and C is a constant depending on r and k. If $k=2$, a normalized spherical harmonic is the real part of $\pi^{-\frac{1}{2}} e^{i\theta}(w \mid w \mid^{-1})^n$, where $w = x_1 + ix_2$ and θ is a real number. Clearly (3) still holds in this case and (4) is obtained by differentiation.

Our next step will be to establish the formula (8) for the coefficients of the expansion of a function in spherical harmonics.

Let $F(x) = F(x')$, $G(x) = G(x')$ be two homogeneous functions of degree zero, and let S_ϵ be the spherical shell between the spheres of radii 1 $1+\epsilon$. Since F and G are homogeneous of degree zero, their normal derivatives at points of the boundary of S_ϵ are zero. Consequently, if we apply Green's formula to the pair F, G, the surface integral vanishes and we obtain

$$\int_{S_\epsilon} F \Delta G \, dx = \int_{S_\epsilon} G \Delta F \, dx;$$

dividing by ϵ and letting ϵ tend to zero it follows that

$$\int_\Sigma F \Delta G \, d\sigma = \int_\Sigma G \Delta F \, d\sigma.$$

If we define now

(5) $$L(F) = | x |^2 \Delta F(x),$$

then, since $| x | = 1$ on Σ, it follows that

$$\int_\Sigma F L(G) \, d\sigma = \int_\Sigma G L(F) \, d\sigma.$$

But if F is homogeneous of degree zero, so is $L(F)$, and a repeated application of the last formula gives

(6) $$\int_\Sigma F L^r(G) \, d\sigma = \int_\Sigma G L^r(F) \, d\sigma$$

which holds, of course, if F and G have sufficiently many continuous derivatives in $| x | > 0$.

Let us consider now a spherical harmonic $Y_{nm}(x')$. Then $| x |^n Y_{nm}(x')$ is a solid harmonic and its Laplacian vanishes. Since the gradients of $Y_{nm}(x')$ and $| x |^n$ are mutually orthogonal, we have that

$$0 = \Delta[| x |^n Y_{nm}(x')] = | x |^n \Delta Y_{nm} + n(n+k-2) | x |^{n-2} Y_{nm};$$

whence, we obtain

(7) $$L[Y_{nm}(x')] = -n(n+k-2) Y_{nm}(x').$$

Let now $F(x')$ be a homogeneous function of degree zero and let

$$F(x') = a_0 + \sum_{n \geqq 1} a_{nm} Y_{nm}(x')$$

be its expansion in spherical harmonics. Then

$$a_{nm} = \int_\Sigma F(x') Y_{nm}(x')\, d\sigma, \qquad n \geqq 1,$$

and an application of (7) and (6) to the last integral gives

$$a_{nm} = (-1)^r n^{-r}(n+k-2)^{-r} \int_\Sigma L^r(F) Y_{nm}(x')\, d\sigma, \qquad n \geqq 1. \tag{8}$$

Now we shall compute compute the Fourier transforms of homogeneous functions coinciding with a normalized spherical harmonic on Σ. Let us write

$$Y_{nm}(\epsilon, \delta, x) = \begin{cases} Y_{nm}(x') \,|\, x \,|^{-k} \text{ if } \epsilon \leqq |\, x \,| \leqq \delta, \\ 0 \text{ otherwise,} \end{cases}$$

(9)

$$Y_{nm}(\epsilon, x) = \begin{cases} Y_{nm}(x') \,|\, x \,|^{-k} \text{ if } \epsilon \leqq |\, x \,|, \\ 0 \text{ otherwise.} \end{cases}$$

Then

$$\hat{Y}_{nm}(\epsilon, \delta, x) = \int_{\epsilon \leqq |y| \leqq \delta} e^{i(x\cdot y)} Y_{nm}(y') \,|\, y \,|^{-k}\, dy$$

Now we set $r = |\, x \,|$, $\rho = |\, y \,|$ and denote by γ the angle between x and y, and the integral above becomes

$$\int e^{ir\rho\cos\gamma} Y_{nm}(y') \,|\, y \,|^{-k}\, dy$$

$$= \int_\epsilon^\delta d\rho/\rho \int_\Sigma e^{ir\rho\cos\gamma} Y_{nm}(y')\, d\sigma = \int_{\epsilon r}^{\delta r} ds/s \int_\Sigma e^{is\cos\gamma} Y_{nm}(y')\, d\sigma, \tag{10}$$

and since the integral of Y_{nm} over Σ is zero, the last integral can be written as

$$\int_{\epsilon r}^{\delta r} ds/s \int_\Sigma (e^{is\cos\gamma} - e^{-s}) Y_{nm}(y')\, d\sigma$$

$$= \int_\Sigma Y_{nm}(y') \left[\int_{\epsilon r}^{\delta r} (e^{is\cos\gamma} - e^{-s})/s\, ds. \right] d\sigma.$$

Now the inner integral can be estimated readily by integrating between ϵr and 1, and 1 and δr, and one verifies that it is dominated in absolute value by $\log C/|\cos\gamma|$ and that it converges as $\epsilon \to 0$ and $\delta \to \infty$. Hence, applying Schwarz's inequality to the last integral, we obtain

$$|\, \hat{Y}_{nm}(\epsilon, \delta, x) \,| \leqq C, \qquad |\, \hat{Y}_{nm}(\epsilon, x) \,| \leqq C, \tag{11}$$

where C depends only on k. Further, as $\epsilon \to 0$ and $\delta \to \infty$, both functions converge to the same limit which we shall denote by $\hat{Y}_{nm}(x)$.

Now we wish to obtain an explicit expression for $\hat{Y}_{nm}(x)$ (see also [2]). For this purpose, let us revert to (10) and assume first that $k \geqq 3$. Then from the expansion

$$e^{is\cos\gamma} = 2^{\lambda}\Gamma(\lambda)\sum_{n=0}^{\infty}(n+\lambda)i^n J_{n+\lambda}(s)/s^{\lambda}P_n^{\lambda}(\cos\gamma), \qquad \lambda = \tfrac{1}{2}(k-2),$$

where J_k is Bessel's function of order k and which converges uniformly in γ and s for s in any finite interval (see [5], p. 368), and from (1), we obtain

$$\hat{Y}_{nm}(\epsilon,\delta,x) = i^n(2\pi)^{k/2}\left[\int_{\epsilon r}^{\delta r} J_{n+\lambda}(s)/s^{1+\lambda}\,ds\right]Y_{nm}(x').$$

Letting ϵ tend to zero and δ tend to infinity and on account of the formula (see [5], p. 391)

$$\int_0^{\infty} J_{n+\lambda}(s)/s^{1+\lambda}\,ds = 2^{-\frac{1}{2}k}\Gamma(\tfrac{1}{2}n)/\Gamma(\tfrac{1}{2}[n+k]), \tag{12}$$

we obtain

$$\hat{Y}_{nm}(x) = i^n\pi^{\frac{1}{2}k}\Gamma(\tfrac{1}{2}n)/\Gamma(\tfrac{1}{2}[k+n])\,Y_{nm}(x'). \tag{13}$$

In the case $k=2$, we write $y_1 = \rho\cos\phi$, $y_2 = \rho\sin\phi$, $x_1 = r\cos\theta$, $x_2 = r\sin\theta$, and the inner integral on the right-hand side of (10) becomes

$$\int_0^{2\pi} e^{is\cos(\phi-\theta)}\cos n(\phi-\phi_0)\,d\phi = 2\pi J_n(s)i^n\cos n(\theta-\phi_0),$$

and integrating and applying (12) we obtain (13).

4. In this section we shall consider the Riesz transforms and the operator Λ.

The Riesz transforms are defined as follows. Let

$$R_{m\epsilon}(f) = -i\pi^{-\frac{1}{2}(k+1)}\Gamma[\tfrac{1}{2}(k+1)]\int_{\epsilon<|x-y|}(x_m-y_m)/|x-y|^{k+1}f(y)\,dy. \tag{14}$$

Then

$$R_m(f) = \operatorname*{l.i.m.}_{\epsilon\to 0} R_{m\epsilon}(f). \tag{15}$$

THEOREM 1. *If $f \in L^p$, $1<p<\infty$, the limit on the right of (15) exists as a limit in the mean of order p, and*

$$\| R_{m\epsilon}f\|_p \leqq A_p\|f\|_p, \tag{16}$$

where A_p depends only on p and k, and $\|f\|_p$ is the L^p-norm of f. If $f \in L_r^p$, $r \geqq 1$, then $R_m f \in L_r^p$

$$(\partial/\partial x_n)R_m = R_m(\partial/\partial x_n), \tag{17}$$

$$R_m(\partial/\partial x_n) = R_n(\partial/\partial x_m), \tag{18}$$

where $\partial/\partial x_n$ is the operator differentiation with respect to x_n.

Finally, R_m is selfadjoint in the sense that if $f \in L^p$ and $g \in L^q$, $1 < p < \infty$, $p^{-1} + q^{-1} = 1$, then $(R_m f, g) = (f, R_m g)$, and

$$\sum_{m=1}^{k} R_m^2 = I, \qquad R_m R_n = R_n R_m, \tag{19}$$

where I is the identity operator.

In order to establish these results it will be convenient to prove first the following

LEMMA 1. *Let $f \in L_r^p$, then there exists a sequence of functions f_n in C^∞, each vanishing outside a bounded set, such that $\| f_n - f \|_p \to 0$ and $\| D_j f_n - D_j f \|_p \to 0$ for each derivative $D_j f$ of f of order $j \leqq r$.*

Let $\phi(x)$ and $\psi(x)$ be two functions in C^∞ vanishing outside a bounded set. Assume that $\phi(x) = 1$ in a neighborhood of $x = 0$ and that

$$\int_{E_k} \psi(x)\, dx = 1.$$

Let $\phi_n = \phi(x/n)$ and $\psi_n = n^k \psi(nx)$. Then, if $\psi_n * f$ denotes the convolution of ψ_n and f, we have that

$$\| \psi_n * f - f \|_p \to 0 \tag{20}$$

and

$$\| \psi_n * D_j f - D_j f \|_p \to 0$$

as $n \to \infty$. Now, since ψ_n is in C^∞ and vanishes outside a bounded set, from the definition of derivatives of functions in L_r^p (see Section 2) and by differentiation under the integral sign, we obtain that

$$D_j(\psi_n * f) = (D_j \psi_n) * f = \psi_n * D_j f.$$

Consequently,

$$\| D_j(\psi_n * f) - D_j f \|_p \to 0. \tag{21}$$

Now since $\phi_n(x) \to 1$ for each x, and since each derivative of $\phi_n(x)$ converges uniformly to zero as $n \to \infty$, we have that

$$\begin{aligned} &\| D_j[\phi_n(\psi_n * f)] - \phi_n D_j(\psi_n * f) \|_p \to 0 \\ &\| \phi_n D_j(\psi_n * f) - D_j(\psi_n * f) \|_p \to 0. \end{aligned} \tag{22}$$

If we set now $f_n = \phi_n(\psi_n * f)$, the desired result follows from (20), (21) and (22).

We revert now to the proof of Theorem 1. If $f \in L^2$ the fact that $R_{m\epsilon}(f)$ converges in the mean of order 2 follows by taking Fourier transforms in (14). In the previous section, we showed that the Fourier transform of the kernel of the integral operator (14)[2] converges boundedly to $x_m |x|^{-1}$ as $\epsilon \to 0$, and this clearly implies the convergence of $R_{m\epsilon}(f)$ and (16).

In the general case, the convergence in the mean of $R_{m\epsilon}(f)$ and (16) follows from Theorem 1 in [3] (see also the remark on page 306 of the same paper).

In order to show that $R_m(f)$ belongs to L_r^p if f does, it will be sufficient to consider the case $r = 1$; the general case will follow from (17).

Let $f \in L_1^p$ and let f_n be a sequence of functions as in the preceding lemma. Then by differentiating under the integral sign, we obtain

$$(\partial/\partial x_l) R_{m\epsilon} f_n = R_{m\epsilon} (\partial/\partial x_l) f_n,$$

and, if g is in C^∞ and vanishes outside a bounded set,

$$(g, R_{m\epsilon}(\partial/\partial x_l) f_n) = (g, (\partial/\partial x_l) R_{m\epsilon} f_n) = -((\partial/\partial x_l) g, R_{m\epsilon} f_n)$$

and letting first ϵ tend to zero and then n tend to infinity, on account of (16) we obtain

$$(g, R_m(\partial/\partial x_l) f) = -((\partial/\partial x_l) g, R_m f),$$

which shows that $R_m f \in L_1^p$ and that (17) holds.

In order to establish (18), we observe that, for every g in C^∞ vanishing outside a bounded set, we have $R_m \partial g/\partial x_l = R_l \partial g/\partial x_m$ (as one readily sees by taking Fourier transforms), and replacing g by the f_n of Lemma 1 and passing to the limit, we obtain $R_m \partial f/\partial x_n = R_n \partial f/\partial x_m$ for every f in L_r^p, $r \geqq 1$.

Finally, if f and g are bounded and vanish outside a bounded set, we have $(R_m f, g) = (f, R_m g)$ by interchanging the order of integration, and

$$\sum_{m=1}^{k} R_m^2(f) = f, \qquad R_m R_n(f) = R_n R_m(f)$$

by taking Fourier transforms, whence the general case follows from the continuity of R_m in L^p, $1 < p < \infty$.

Definition 1. Let $f \in L_r^p$, $r \geqq 1$, $1 < p < \infty$. Then

$$\Lambda f = i \sum_1^k R_m \partial f/\partial x_m = i \sum_1^k (\partial/\partial x_m) R_m f.$$

[2] Observe that this kernel coincides, except for a numerical factor, with one of the functions $Y_{1,m}(\epsilon, x)$ in (9).

COROLLARY. *If $f \in L_r^p$, $r \geqq 1$ then $\Lambda f \in L_{r-1}^p$ and*

$$i\partial f/\partial x_n = R_n \Lambda f = \Lambda R_n f. \tag{23}$$

If $f \in L_r^p$, $r \geqq 2$ then

$$\Delta f = \sum_1^k \partial^2 f/dx_m^2 = -\Lambda^2 f. \tag{24}$$

The first assertion follows from the definition of Λ and the fact that the operators R_m preserve the classes L_r^p. The formulas (23) and (24) are obtained from the definition of Λ by using (18) and (19).

5. We proceed to present the main results of the paper. We begin with

THEOREM 2. *Let $h(x,z)$, $x, z \in E_k$, be a function in C_β^∞, $\beta \geqq 0$, homogeneous of degree $-k$ in z, that is, such that $h(x, \lambda z) = \lambda^{-k} h(x,z)$ for every $\lambda > 0$, and assume that $\int_\Sigma h(x,z)\,d\sigma = 0$ for every x, where Σ is the sphere $|z| = 1$. Let $a(x)$ be a function in C_β, and consider the operator*

$$H_\epsilon f = a(x) f(x) + \int_{|x-y|>\epsilon} h(x, x-y) f(y)\,dy \tag{25}$$

and its adjoint

$$H_\epsilon^* f = \bar{a}(x) f(x) + \int_{|x-y|>\epsilon} \bar{h}(y, y-x) f(y)\,dy. \tag{26}$$

Then

i) *H_ϵ and H_ϵ^* are defined for $f \in L^p$, $1 < p < \infty$ and as $\epsilon \to 0$, $H_\epsilon(f)$ and $H_\epsilon^*(f)$ converge in the mean of order p. If $H(f)$ and $H^*(f)$ denote their respective limits, we have*

$$\begin{aligned} \| Hf \|_p &\leqq \| f \|_p A_p \sup_{|z|=1} (|a(x)| + |h(x,z)|), \\ \| H^* f \|_p &\leqq \| f \|_p A_q \sup_{|z|=1} (|a(x)| + |h(x,z)|), \end{aligned} \tag{27}$$

where $p^{-1} + q^{-1} = 1$ and A_p depends only on p and k.

ii) *if $f \in L_r^p$, $1 < p < \infty$, with $r \leqq \beta$, then Hf and H^*f belong to L_r^p,*

iii) *if $f \in L^p$, $1 < p < \infty$, and is Hölder-continuous of order α, $0 < \alpha < \beta$, Hf and H^*f are Hölder continuous of the same order.*

Part of i) concerning the operators H_ϵ and H was proved in [3], Theorem 2. The estimate of $\| Hf \|_p$, which also holds for $\| H_\epsilon f \|_p$, is not given explicitly there but is contained in the proof of the theorem. We shall therefore concentrate in the case $p = 2$. We begin with

LEMMA 2. *Let*

$$(28) \qquad T_{nm\epsilon}f = \int_{|x-y|>\epsilon} Y_{nm}(\epsilon, x-y) f(y)\, dy,$$

where $Y_{nm}(\epsilon, x)$ is the function defined in (9). Then if $f \in L^p$, $1 < p < \infty$, there is a constant A_p depending on p and k such that

$$(29) \qquad \| T_{nm\epsilon} f \|_p \leqq A_p \| f \|_p,$$

and as $\epsilon \to 0$, $T_{nm\epsilon}f$ converges in the mean of order p to a limit $T_{nm}f$. If $f \in L_r^p$, then $T_{nm}f \in L_r^p$. The operators T_{nm} commute with the R_m of Theorem 1 and when acting on L_r^p, they also commute with the $\partial/\partial x_n$.

The proof of this lemma proceeds as that of Theorem 1 and we need not repeat it here.

Now we revert to the operator H_ϵ. Let us expand the function $h(x, z)$ in spherical harmonics

$$(31) \qquad h(x, z) = \Sigma a_{nm}(x) Y_{nm}(z') \,|\, z \,|^{-k},$$

where the $a_{nm}(x)$ can be calculated by means of formula (8) after replacing F by h. Since, for each n, the number of distinct spherical harmonics Y_{nm} is of the order n^{k-2} and since, according to (3), Y_{nm} has a bound of the order $n^{\frac{1}{2}(k-2)}$, it follows from the formula (8), by choosing r sufficiently large, that the series of absolute values of the terms of the series above is dominated by a multiple of $|\, z \,|^{-k}$. Consequently, given $f \in L^p$, we can replace h by the series in (25) and (26) and integrate term by term obtaining

$$(32) \qquad H_\epsilon f = a(x) f + \Sigma a_{nm}(x) T_{nm\epsilon} f, \qquad H_\epsilon^* f = \bar{a}(x) f + \Sigma (-1)^n T_{nm\epsilon}(\bar{a}_{nm} f).$$

Now the $a_{nm}(x)$ are dominated in absolute value by the terms of a convergent numerical series, and according to (29), the $T_{nm\epsilon}f$ are bounded in norm and converge in the mean as $\epsilon \to 0$. Hence $H_\epsilon f$ converges in the mean. Similarly, the functions $T_{nm\epsilon}(\bar{a}_{nm}f)$ converge in the mean and their norms are dominated by the terms of a convergent numerical series, which implies that $H_\epsilon^* f$ also converges.

Passing to the limit we obtain

$$(33) \qquad Hf = a(x) f + \Sigma a_{nm}(x) T_{nm} f, \qquad H^* f = \bar{a}(x) f + \Sigma (-1)^n T_{nm}(\bar{a}_{nm} f),$$

the series converging in the mean of order p. If f and g vanish outside a bounded set and are bounded, from absolute integrability, it follows that $(H_\epsilon f, g) = (f, H_\epsilon^* g)$ and from the continuity of the operators H_ϵ and H_ϵ^* in every L^p, $1 < p < \infty$, we obtain $(H_\epsilon f, g) = (f, H_\epsilon^* g)$ for $f \in L^p$ and $g \in L^q$, $p^{-1} + q^{-1} = 1$, which justifies the assertion that H_ϵ^* is the adjoint of H_ϵ.

By a passage to the limit we obtain that also H^* is the adjoint of H.

Our next step will be establishing (27) in the case $p=2$. In the first place, we have that

$$|a(x)|^2+\Sigma\,|a_{nm}(x)|^2 \leqq \sup_x\,[\,|a(x)|^2+\int_{|z|=1}|h(x,z)|^2\,d\sigma]=C^2,$$

and from (33) and Schwarz's inequality, we obtain

$$|Hf|^2 \leqq C^2[\,|f|^2+\Sigma\,|T_{nm}f|^2].$$

Integrating and applying Plancherel's theorem to the right hand side, we obtain (see formula (13))

$$\|Hf\|_2{}^2 \leqq C^2\int_{E_k}[1+\Sigma\pi^k\{\Gamma(\tfrac{1}{2}n)/\Gamma(\tfrac{1}{2}[k+n])\}^2Y_{nm}(x')^2]\,|\hat{f}(x)|^2\,dx.$$

Now the sum of the squares of the spherical harmonics of a given degree n is a constant equal to the number of such harmonics divided by the area of Σ (see [1], p. 242). Since this number is of the order n^{k-2}, the series in the last integral is absolutely convergent and represents a constant, and we obtain

$$\|H_f\|_2{}^2 \leqq C^2A_2{}^2\,\|\hat{f}\|_2{}^2=C^2A_2{}^2\,\|f\|_2{}^2,$$

where $A_2{}^2$ is the value of the series, which is the first inequality in (27). The second is obtained from the fact that H^* is the adjoint of H.

We turn now to the proof of ii). Since the $a_{nm}(x)$ can be calculated by means of formula (8), replacing the function F there by $h(x,z)$, it follows readily by choosing r sufficiently large, that the $a_{nm}(x)$ and their derivatives of order less than or equal to β are dominated in absolute value by the terms of a convergent numerical series. Now if $f\in L_r{}^p$, then $T_{nm}f$ and $T_{nm}(a_{nm}f)$ belong to $L_r{}^p$, and from (29) it follows that the series in (33) can be differentiated term by term, which establishes ii).

In order to establish iii), we shall first investigate the Hölder continuity of $T_{nm}f$. Write $h(x)=Y_{nm}(x')\,|x|^{-k}$ and let S be any set contained in a sphere of radius ρ with center at x. Then if $f(x)$ is Hölder continuous of order α, $0<\alpha<1$, and $|f(x_1)-f(x_2)|\leqq A\,|x_1-x_2|^\alpha$, inequality (3) gives

$$(34)\qquad \left|\int_S h(x-y)[f(y)-f(x)]dy\right| \leqq \int_{|x-y|\leqq\rho} CAn^{\frac{1}{2}(k-2)}\,|x-y|^{-k+\alpha}\,dy = C/\alpha\, An^{\frac{1}{2}(k-2)}\rho^\alpha.$$

In particular, we have

$$(35)\qquad \left|\int_{|x-y|\leqq\rho} h(x-y)f(y)\,dy\right| \leqq \left|\int_{|x-y|\leqq\rho} h(x-y)[f(y)-f(x)]dy\right| \leqq C/\alpha\, An^{\frac{1}{2}(k-2)}\rho^\alpha.$$

Let now x_1 and x_2 be two points. Set $\rho = 2\,|\,x_1 - x_2\,|$ and write

$$(T_{nm}f)(x_1) - (T_{nm}f)(x_2)$$
$$= \int_{|x_1-y|\leqq\rho} h(x_1-y)f(y)\,dy + \int_{|x_1-y|\geqq\rho} h(x_1-y)[f(y)-f(x_2)]\,dy$$
$$- \int_{|x_1-y|\leqq\rho} h(x_2-y)[f(y)-f(x_2)]\,dy + \int_{|x_1-y|\geqq\rho} h(x_2-y)[f(y)-f(x_2)]\,dy,$$

where the integrals over $|\,y-x_1\,|\geqq\rho$ are understood as the limit as $R\to\infty$ of the integrals extended over $\rho\leqq|\,y-x_1\,|\leqq R$. Then from (34) and (25), we obtain

$$|(T_{nm}f)(x_1)-(T_{nm}f)(x_2)|\leqq C/\alpha\, A n^{\frac{1}{2}(k-2)}\rho^{\alpha}$$
$$+\int_{|x_1-y|\geqq\rho} |\,h(x_1-h)-h(x_2-y)\,|\;|\,f(y)-f(x_2)\,|\,dy.$$

Now according to (4), $|\,h(x_1-y)-h(x_2-y)|\leqq Cn^{k/2}\,|\,x_1-x_2\,|\,|\,x_1-y\,|^{-k-1}$ in $|\,x_1-y\,|\geqq\rho$, and since $|\,f(y)-f(x_2)|\leqq A\,|\,y-x_2\,|^{\alpha}$ and $|\,y-x_2\,|\leqq 2\,|\,y-x_1\,|$ in $|\,x_1-y\,|\geqq\rho$, it follows that

$$(36)\qquad \begin{aligned}&|(T_{nm}f)(x_1)-(T_{nm}f)(x_2)|\leqq C/\alpha\, A n^{\frac{1}{2}(k-2)}\rho^{\alpha}\\ &+CAn^{k/2}\,|\,x_1-x_2\,|\int_{\rho}^{\infty} s^{\alpha-2}\,ds\leqq[1/\alpha+1/(1-\alpha)]CAn^{k/2}\,|x_1-x_2\,|^{\alpha},\end{aligned}$$

where the constant C in the last expression depends only on k. Further, since $\|\,T_{nm}f\,\|_p$ is bounded, as is readily verified, (36) implies that $(T_{nm}f)(x)$ has a bound of the order $n^{k/2}$. Now we can estimate $(Hf)(x_1)-(Hf)(x_2)$ from the series in (33). To obtain the desired result, it suffices to observe that, for every r, the functions $a_{nm}(x)n^r$ are uniformly bounded and uniformly Hölder continuous. This follows from formula (8) and allows us to estimate the preceding difference by estimating the corresponding differences of the terms of the series. The Hölder continuity of H^*f is established by a similar argument.

Remark. Infinite differentiability of $h(x,z)$ with respect to z is not indispensable for the validity of the preceding theorem, and, indeed, the argument used in its proof remains valid under weaker assumptions.

Another fact worth mentioning is this. Not only do the functions $H_\epsilon f$ and H_ϵ^*f converge in the mean of order p if f is in L^p, $1<p<\infty$, but they converge pointwise almost everywhere and are dominated in absolute value by functions in L^p. For $H_\epsilon f$, this result is contained in Theorem 2 of [3]. The result for H_ϵ^*f can be obtained by applying Theorem 1 in [3] [3] to each

[3] See also the remark on page 306 of the same paper.

of the terms of the expansion in (33) after observing that the $a_{nm}(x)$ are dominated in absolute value by the terms of a convergent numerical series.

Definition 2. A singular integral operator of type C_β^∞ is an operator such as the H of Theorem 2. The symbol of the operator H is the function

$$\sigma(H) = a(x) + \Sigma a_{nm}(x)\gamma_n Y_{nm}(z'), \tag{37}$$

where $a(x)$ *is the function in* (25), *the* $a_{nm}(x)$ *are the functions in* (31), *and*

$$\gamma_n = i^n \pi^{k/2}\Gamma(\tfrac{1}{2}n)/\Gamma(\tfrac{1}{2}[n+k]). \tag{38}$$

The reader will notice that, on account of (13) the summation sign in (37) represents in a sense the Fourier transform of $h(x,z)$ with respect to z. This fact could be used in defining the symbol $\sigma(H)$ of H, but in the present set-up we find the preceding definition more convenient.

THEOREM 3. *If H is a singular integral operator of type C_β^∞ its symbol is a homogeneous function of degree zero with respect to z and in C_β^∞ in $|z| \geqq 1$. Conversely, every function of x and z which is homogeneous of degree zero with respect to z and belongs to C_β^∞ in $|z| \geqq 1$ is the symbol of a unique operator of type C_β^∞. If M is a bound for the absolute value of $\sigma(H)$ and its derivatives with respect to the coordinates of z in $|z| \geqq 1$ of order $2k$, then*

$$\| Hf \|_p \leqq MA_p \| f \|_p \tag{39}$$

where A_p depends only on p and k.

According to formula (8), we have

$$\begin{aligned} a_{nm}(x) &= (-1)^r n^{-r}(n+k-2)^{-r} \int_\Sigma L^r[h(x,z')]Y_{nm}(z')d\sigma, \\ a_{nm}(x) &= (-1)^r \gamma_n^{-1} n^{-r}(n+k-2)^{-r} \int_\Sigma L^r[\sigma(H)(x,z')]Y_{nm}(z')d\sigma, \end{aligned} \tag{40}$$

where L is the operator defined in (5) and x is regarded as a parameter. From this representation it follows readily that, if $h(x,z)$ or $\sigma(H)(x,z)$ belongs to C_β^∞ in $|z| \geqq 1$, then, for each r, the functions $a_{nm}(x)n^r$ are uniformly in C_β. Conversely, if the latter holds, by taking (4) into account and differentiating the series (31) and (37) term by term, we obtain that both $h(x,z)$ and $\sigma(H)(x,z)$ belong to C_β^∞ in $|z| \geqq 1$.

In order to establish (39) we just set $r = k$ in the last formula and obtain $|a_{nm}(x)| \leqq CMn^{-\frac{1}{2}k}$ where C is a constant depending only on k.

This combined with (3), (31) and the fact that the number of distinct spherical harmonics of degree n is of the order n^{k-2} yields $|h(x,z)| \leqq CM$, where again C depends only on k. But on account of (27), this implies (39).

Definition 3. Let H, H_1 and H_2 be singular integral operators of type C_β^∞. We define $H^\#$ and $H_1 \circ H_2$ by the formulas

$$\sigma(H^\#) = \bar{\sigma}(H), \qquad \sigma(H_1 \circ H_2) = \sigma(H_1)\sigma(H_2),$$

where $\bar{\sigma}(H)$ is the complex conjugate of $\sigma(H)$.

THEOREM 4. *If the symbols of H, H_1 and H_2 are independent of x, then $H^\# = H^*$, $H_1 \circ H_2 = H_1 H_2 = H_2 H_1$ where $H_1 H_2$ is the composition product of H_1 and H_2.*

Let $f(x)$ be bounded and vanish outside a bounded set and let $\hat{f}$ be its Fourier transform. Then according to (32), we have $H_\epsilon f = af + \Sigma a_{nm} T_{nm\epsilon} f$, where a and a_{nm} are the same as those in (37) and are therefore assumed to be independent of x. Taking Fourier transforms and letting ϵ tend to zero, we obtain on account of (13)

$$(Hf)^\wedge = a\hat{f}(z) + \Sigma a_{nm}\gamma_n Y_{nm}(z')\hat{f}(z);$$

that is, $(Hf)^\wedge = \sigma(H)\hat{f}$, whence

$$(H^*f)^\wedge = \bar{\sigma}(H)\hat{f} = (H^\# f)^\wedge, \qquad [(H_1 \circ H_2)f]^\wedge = \sigma(H_1)\sigma(H_2)\hat{f} = (H_1 H_2 f)^\wedge.$$

which implies that $H^* f = H^\# f$ and $(H_1 \circ H_2)f = H_1 H_2 f$. From this and the continuity of H in L^p, the desired result follows.

COROLLARY. *If the symbol of the singular integral operator H is independent of x and does not vanish, then H has an inverse and the inverse is also a singular integral operator.*

Clearly the operator H_1 defined by $\sigma(H_1) = \sigma(H)^{-1}$ is an inverse of H.

The following theorem deals with the relationship between H^* and $H^\#$, and $H_1 H_2$ and $H_1 \circ H_2$ in the general case.

THEOREM 5. *Let H be an operator of type C_β^∞ with $\beta > 1$. Let M be a bound for $\sigma(H)(x,z)$ and its derivatives with respect to coordinates of z of order $2k$, the first derivatives of these with respect to the coordinates of x, and the Hölder constants of the latter. Then for every $f \in L_1^p$, $1 < p < \infty$, we have*

$$\begin{aligned} &\|(H\Lambda - \Lambda H)f\|_p \leqq A_p M \|f\|_p, \quad \|(H^*\Lambda - \Lambda H^*)f\|_p \leqq A_p M \|f\|_p, \\ (41)\qquad &\|(H^* - H^\#)\Lambda f\|_p \leqq A_p M \|f\|_p, \quad \|\Lambda(H^* - H^\#)f\|_p \leqq A_p M \|f\|_p, \end{aligned}$$

where A_p depends only on p, k and β. Further, if H_1 and H_2 are two operators in C_β^∞ and $f \in L_1^p$ then

$$(42)\qquad \begin{aligned} &\| (H_1 \circ H_2 - H_1 H_2)\Lambda f \|_p \leqq A_p M_1 M_2 \| f \|_p, \\ &\| \Lambda (H_1 \circ H_2 - H_1 H_2) f \|_p \leqq A_p M_1 M_2 \| f \|_p, \end{aligned}$$

where, again, A_p depends only on p, k and β and M_1 and M_2 are defined as above.

Let $c(x)$ be a function in C_β, T a singular integral operator with symbol independent of x, and f a function in C^∞ and vanishing outside a a bounded set. Denote by $Y(z)$ the kernel of the operator T, and consider the expression

$$(43)\qquad (cT - Tc) f_{x_i} = \lim_{\epsilon \to 0} \int_{|x-y|>\epsilon} [c(x) - c(y)] Y(x-y) f_{y_i}\, dy,$$

where f_{x_i} stands for $\partial f / \partial x_i$. Since $c(x) \in C_\beta$, $\beta > 1$, if $0 < \alpha \leqq \beta - [\beta]$, and $\alpha \leqq 1$, we have

$$| c_{x_j}(x) - c_{x_j}(y) | \leqq A \, | x - y |^\alpha,$$

$$c(x) - c(y) = \sum_j (x_j - y_j) c_{x_j}(x) + b(x, y),$$

where $| b(x,y) | \leqq kA \, | x - y |^{1+\alpha}$. Now we replace this expression in the integral in (43) and integrate by parts, obtaining

$$(44)\qquad \begin{aligned} &\int_{|x-y|>\epsilon} Y(x-y) c_{x_i}(y) f(y)\, dy \\ &+ \int_{\epsilon < |x-y| \leqq 1} \sum_j (x_j - y_j) c_{x_j}(x) Y_{x_i}(x-y) f(y)\, dy \\ &+ \int_{|x-y|>1} [c(x) - c(y)] Y_{x_i}(x-y) f(y)\, dy \\ &+ \int_{\epsilon < |x-y| \leqq 1} b(x,y) Y_{x_i}(x-y) f(y)\, dy \\ &+ \int_{|x-y|=\epsilon} [c(x) - c(y)] Y(x-y) f(y) \gamma_i\, d\sigma, \end{aligned}$$

where the last integral is extended over the surface of the sphere $| x - y | = \epsilon$ and γ_i is the i-th direction cosine of the normal to the spherical surface. Denote now by N_1 a bound for $| c(x) |$, $| c_{x_j} |$ and the Hölder constants for the derivatives c_{x_j}, and by N_2 a bound for $| Y(z) |$ and $| Y_{z_i}(z) |$ on $| z | = 1$, and observe that the functions $z_j Y_{z_i}(z)$, are homogeneous of degree $-k$ and their mean value over the sphere $| z | = 1$ is zero (otherwise, if $f(x) \neq 0$ and $c_{x_n} = \delta_{nj}$, the second integral in (44) would diverge as ϵ tends to zero while the remaining terms and the whole expression converge). Thus if we let ϵ

tend to zero in (44) and apply Theorem 1 in [3] (see also the remark on page 306 of the same paper) to the first two terms, we see that they represent a function of x whose L^p-norm does not exceed $AN_1N_2 \| f \|_p$ where A depends only on p and k. In the case $f \in L^2$, we apply Theorem 2 to the first term. The second term is not suitable for a direct application of Theorem 2, but the convolution integrals

$$\int_{\epsilon<|x-y|\leqq 1} (x_j - y_j) Y_{x_i}(x-y) f(y)\, dy$$

can be easily estimated by taking Fourier transforms and estimating the transform of $z_j Y_{x_i}(z)$ multiplied by the characteristic function of $\epsilon < |z| \leqq 1$ using the method used in Section 3.

Since $Y_{z_i}(z)$ is homogeneous of degree $-k-1$, it is absolutely integrable over $|z| \geqq 1$, and, on account of the estimate for $b(x,y)$, one sees readily that the third and fourth terms in (44) are dominated by convolutions of $|f(y)|$ with absolutely integrable functions. Hence, it follows from a theorem of Young that they represent functions whose L^p-norms are less than or equal to $\| f \|_p$ times the L^1-norms of those integrable functions, and an easy computation yields $AN_1N_2 \| f \|_p$ as a bound for the L^p-norms of those functions, where the constant A depends on k and α. Finally, one sees readily that, as ϵ tends to zero, the last term in (44) tends to a limit whose absolute value is dominated by $AN_1N_2 |f(x)|$, where A depends only on k. Collecting results and applying Lemma 1, we obtain

$$\| (cT - Tc) f_{x_i} \|_p \leqq A_p N_1 N_2 \| f \|_p, \tag{45}$$

for every f in $L_1{}^p$, $1 < p < \infty$, where A_p depends only on p, k and α (or β). Having established (45), we can proceed to prove the inequalities (41) and (42).

Consider the representation of Hf and H^*f given in (33). Write $a_0(x)$ for $a(x)$ and T_0 for the identity operator. Then since the $a_{nm}(x)$ are dominated in absolute value by a convergent numerical series and since the T_{nm} as operators on L^p, $1 < p < \infty$, have bounded norm for each fixed p, we may write

$$H = \sum_{n\geqq 0} a_{nm} T_{nm}, \qquad H^* = \sum_{n\geqq 0} (-1)^n T_{nm} \bar{a}_{nm}, \qquad H^{\#} = \sum_{n\geqq 0} (-1)^n \bar{a}_{nm} T_{nm},$$

where the series on the right converge in the operator norm. If f is in $L_1{}^p$, then

$$(\Lambda H - H\Lambda) f = \sum_{l=1}^{k} R_l \left(\sum a_{nm} T_{nm} f\right)_{x_l} - \sum a_{nm} T_{nm} \left(\sum R_l f_{x_l}\right)$$
$$= \sum_{l,n} R_l (a_{nm})_{x_l} T_{nm} f + \sum_{l,n} (R_l a_{nm} - a_{nm} R_l)(T_{nm} f)_{x_l}$$

since the $(a_{nm})_{x_l}$ are dominated by a convergent numerical series which justifies term by term differentiation of the first series above. Now we apply (45) to each of the terms of the last series, estimating N_1 each time from the second formula in (40), and use (29) obtaining readily the first inequality in (41). The second inequality in (41) follows by an argument almost identical to the preceding one. In order to prove the third inequality in (41), we write

$$(H^*-H^\#)\Lambda f=\sum_{l=1}^{k}(H^*-H^\#)(R_lf)_{x_l}$$
$$=\sum_{l,n}(-1)^n(T_{nm}\bar{a}_{nm}-\bar{a}_{nm}T_{nm})(R_lf)_{x_l},$$

and again (45) combined with (40) and (4) yields the desired result.

The last inequality in (41) is readily seen to be an immediate consequence of the preceding ones.

Let now $H_1=\Sigma\, b_{nm}T_{nm}$, $H_2=\Sigma\, c_{nm}T_{nm}$, and consider the series of operators

(46) $$\Sigma\, b_{nm}c_{\nu\mu}T_{nm}T_{\nu\mu},$$

the sum being extended over all indices, and the series of their symbols

(47) $$\Sigma\, b_{nm}(x)c_{\nu\mu}(x)Y_{nm}(z')Y_{\nu\mu}(z')\gamma_n\gamma_\nu$$
$$=(\Sigma\, b_{nm}(x)Y_{nm}(z')\gamma_n)(\Sigma\, c_{\nu\mu}(x)Y_{\nu\mu}(z')\gamma_\nu)=\sigma(H_1)\sigma(H_2)=\sigma(H_1\circ H_2).$$

From the estimate (4) of the successive derivatives of $Y_n(z')$ and the formula (40) applied to the coefficients b_{nm} and $c_{\nu\mu}$, it follows that the first series in (47) converges uniformly as well as the series obtained from it by differentiating its terms with respect to coordinates of z any number of times. But then Theorem 3 implies that (46) converges in the operator norm and that (47) is precisely the symbol of (46), or, equivalently, that (46) is precisely $H_1\circ H_2$. On the other hand, since the functions $b_{nm}(x)$ and $c_{nm}(x)$ are dominated by convergent numerical series we have

$$H_1H_2=\Sigma\, b_{nm}T_{nm}c_{\nu\mu}T_{\nu\mu},\qquad H_1\circ H_2-H_1H_2=\Sigma\, b_{nm}(c_{\nu\mu}T_{nm}-T_{nm}c_{\nu\mu})T_{\nu\mu}.$$

Thus

$$(H_1\circ H_2-H_1H_2)\Lambda f=\Sigma\, b_{nm}(c_{\nu\mu}T_{nm}-T_{nm}c_{\nu\mu})T_{\nu\mu}(R_lf)_{x_l}$$
$$=\Sigma\, b_{nm}(c_{\nu\mu}T_{nm}-T_{nm}c_{\nu\mu})(T_{\nu\mu}R_lf)_{x_l}.$$

If we now compute the $c_{\nu\mu}$ and b_{nm} by means of (40) and apply (45) using (3) and (4) in order to estimate the kernel of T_{nm} and its first order derivatives, and bear in mind that the $T_{\nu\mu}$ are uniformly bounded in L^p and that, for each n, the number of distinct T_{nm} is of the order n^{k-2}, we obtain the first inequality in (42). The second follows immediately from the first and the first inequality in (41). Theorem 5 is thus established.

THEOREM 6. *Let $\mathcal{A}_p$ be the algebra of bounded operators on L^p, $1 < p < \infty$ generated by all singular integral operators H of type C_β^∞, $\beta > 1$, and their adjoints H^* (see definition 2 and Theorem 2). Then there exists a homomorphism h_p of $\mathcal{A}_p$ onto the algebra of all functions $F(x,z)$ in C_β^∞ which are homogeneous of degree zero with respect to z, such that, for every singular integral operator H, the identities $h_p(H) = \sigma(H)$, $h_p(H^*) = \bar{\sigma}(H)$ hold. The kernel of h_p can be characterized as follows: K belongs to the kernel of h_p, or, equivalently, $h_p(K) = 0$, if and only if there exists a positive constant A depending on K such that $\| K\Lambda f \|_p \leqq A \| f \|_p$ for every $f \in L_1^p$. If $h_p(K)$ is bounded away from zero, then there exists $K' \in \mathcal{A}_p$ with a two sided inverse, such that $h_p(K) = h_p(K')$.*

Every bounded operator on L^p which commutes with every operator in $\mathcal{A}_p$ is a multiple of the identity operator.

The algebras $\mathcal{A}_p$, $\mathcal{A}_q$ corresponding to any two spaces L^p and L^q, $1 < p < q < \infty$, are isomorphic and there is a natural isomorphism ϕ between $\mathcal{A}_p$ and $\mathcal{A}_q$ such that $h_p = h_q\phi$.

We start with the following

LEMMA. *Let H be a singular integral operator of type C_β^∞, $\beta > 1$. Assume that for some positive constant A and every $f \in L_1^p$ the inequality $\| H\Lambda f \|_p \leqq A \| f \|_p$ holds. Then $H = 0$.*

Let $\mathcal{B}$ be the class of functions which are symbols of singular integral operators with the property that $H\Lambda$ is bounded in the sense just described. Then $\mathcal{B}$ is linear, and, on account of (42) and the definition 3, it is closed under multiplication by functions $F(x,z)$ in C_β^∞ which are homogeneous of degree zero with respect to z. Let u be a rotation of E_k about the point x_0. Then, if we denote $f[u(x)]$ by $f_u(x)$ and $\sigma(H)$ by $F(x,z)$, and define H_u and F_u by

$$F_u(x,z) = \sigma(H_u) = F[u(x), u(z+x_0) - x_0],$$

we have the following identities

$$\| f \|_p = \| f_u \|_p, \qquad (\Lambda f)_u = \Lambda f_u, \qquad (Hf)_u = H_u f_u.$$

Consequently, if $F(x,z) \in \mathcal{B}$ then

$$\| H_u \Lambda f_u \|_p = \| H_u (\Lambda f)_u \|_p = \| (H\Lambda f)_u \|_p = \| H\Lambda f \|_p \leqq A \| f \|_p = A \| f_u \|_p$$

which shows that $F_u \in \mathcal{B}$. Assume now that $F(x_0, z)$ is not identically zero in z. Then there exist finitely many rotations u_i of E_k about x_0 such that

$G(x,z) = \Sigma \mid F_{u_i}(x,z) \mid^2$ has the property that $G(x_0, z)$ does not vanish. Let now $a(x)$ be a function in C_β such that $a(x_0) \neq 0$ and vanishing outside a neighborhood of x_0 where $G(x,z)$ is bounded away from zero, and define $G_1(x,z) = a(x) G(x,z)^{-1}$, where $a(x) \neq 0$, and $G_1(x,z) = 0$ otherwise. Then $G_1 G z_1 \mid z \mid^{-1} = a(x) z_1 \mid z \mid^{-1}$, and since $F_{u_i} \in \mathcal{B}$, it follows that $a(x) z_1 \mid z \mid^{-1} \in \mathcal{B}$. Now $a(x) z_1 \mid z \mid^{-1}$ is the symbol of the operator $a(x) R_1$ and therefore, on account of (23), we have that

$$\| a(x) R_1 \Lambda f \|_p = \| i a(x) f_{x_1} \|_p \leqq A \| f \|_p$$

for every function f in $L_1{}^p$. But this is clearly impossible unless $a(x) = 0$ identically which contradicts our assumptions. Hence we must have $F(x_0, z) = 0$ for every z. Since the same argument applies to an arbitrary point x_0 we must have $F(x,z) = 0$ identically, and the only function in $\mathcal{B}$ is zero. This establishes the lemma.

Now let us revert to the proof of Theorem 6. Consider the class $\mathcal{B}$ of opeartors K in $\mathfrak{A}_p$ with the property that there exists a singular integral operator H and a constant A such that $\| (K - H) \Lambda f \|_p \leqq A \| f \|_p$ for every $f \in L_1{}^p$. Then $\mathcal{B}$ is clearly linear. Further, it follows from (41) and (42) that $\mathcal{B}$ is closed under multiplication. Now every singular integral operator belongs to $\mathcal{B}$, and the third inequality in (41) implies that their adjoints also belong to $\mathcal{B}$, that is, $\mathcal{B}$ is an algebra containing all singular integral operators and their adjoints. Hence $\mathcal{B}$ coincides with $\mathfrak{A}_p$.

The singular integral operator H associated with K in the manner described above is unique, as follows immediately from the preceding lemma. Now we define $h_p(K)$ to be $\sigma(H)$. Then h_p is clearly linear and, on account of (41) and (42), multiplicative. Since the mapping $H \to \sigma(H)$ is onto the class of all functions $F(x,z)$ in $C_\beta{}^\infty$ which are homogeneous of degree zero with respect to z, the same applies to h_p. The identity $h_p(H) = \sigma(H)$ for every singular integral operator H is clear, and $h(H^*) = \bar{\sigma}(H)$ follows from the third inequality in (41). Further, the fact that $h_p(K) = 0$ if and only if $K\Lambda$ is bounded in the sense described above is an immediate consequence of the definition of h_p.

Suppose now that $h_p(K) = F(x,z)$ is bounded away from zero. Let $z_0 \neq 0$ be a fixed point in E_k and set

$$a(x) = F(x, z_0) F(0, z_0)^{-1}, \qquad G(x,z) = F(x,z) a(x)^{-1} F(0,z)^{-1}.$$

Then $G(0,z) = G(x, z_0) = 1$ and this, as is readily verified, implies that $G(x,z)$ has an n-th root $G(x,z)^{1/n}$ in $C_\beta{}^\infty$. If we choose $G(x,z)^{1/n}$ so that $G(0,z)^{1/n} = 1$, the boundedness of the first order derivatives of $G(x,z)$ and the fact that $G(x, z_0) = 1$ imply that $G(x,z)^{1/n}$ converges uniformly to 1 and its derivatives with respect to coordinates of z of order $2k$ converge uniformly to zero in $|z| \geqq 1$. Thus theorem (3) implies that, for n sufficiently large, the operator H defined by $\sigma(H) = G(x,z)^{1/n}$ is close to the identity operator, or, more precisely, $\| I - H \| < \frac{1}{2}$, where I is the identity and the norm is taken in the sense of operator norm. But this implies that H has a two sided inverse. Further, define H_1 and H_2 by $\sigma(H_2) = F(0,z)$ and $\sigma(H_1) = a(x)$. Then since $a(x)$ is bounded away from zero, H_1 has an inverse, and since $F(0,z)$ does not vanish, according to the Corollary to Theorem 4, H_2 also has an inverse. Define now $K' = H_1 H_2 H^n$. Then K' has an inverse and

$$h_p(K') = h_p(H_1) h_p(H_2) h_p(H)^n$$
$$= \sigma(H_1)\sigma(H_2)\sigma(H)^n = a(x)F(0,z)G(x,z) = F(x,z) = h_p(K)$$

which establishes the corresponding statement in the Lemma.

Let now K be a bounded operator on L^p which commutes with all singular integral operators. In particular, K commutes with multiplication by functions in C_β, that is, if $a(x) \in C_\beta$ then $K(af) = aK(f)$ for every $f \in L^p$. Assume now that f is positive, continuous and in L^p and consider the function $\psi(x) = f^{-1}K(f)$. Then for every $g(x)$ of the form $g(x) = f(x)a(x)$ with $a(x) \in C_\beta$ we have $K(g) = K(af) = aK(f) = af[f^{-1}K(f)] = g(x)\psi(x)$. Since K is continuous on L^p and the functions of the form $a(x)f(x)$ are dense in L^p, it follows both that $\psi(x)$ is essentially bounded and that $K(g) = \psi g$ for every $g \in L^p$. Let now $\bar{x}$ be a point at which $\psi(x)$ is equal to the derivative of its indefinite integral and $f_n(x)$ a function which is constant on the sphere with center at $\bar{x}$ and radius $1/n$, vanishes outside this sphere and has integral equal to 1. If H is a singular integral operator with kernel $h(x-y)$, then

$$HKf_n = \lim_{\epsilon \to 0} \int_{|x-y|>\epsilon} h(x-y)\psi(y)f_n(y)\,dy$$

converges towards $h(x-\bar{x})\psi(\bar{x})$ for all x, $x \neq \bar{x}$. On the other hand,

$$KHf_n = \psi(x) \lim_{\epsilon \to 0} \int_{|x-y|>\epsilon} h(x-y)f_n(y)\,dy$$

converges towards $h(x-\bar{x})\psi(x)$. Since H and K commute we have $KHf_n = HKf_n$, and therefore $h(x-\bar{x})\psi(x) = h(x-\bar{x})\psi(\bar{x})$ almost everywhere. If we assume, as we may, that $h(x)$ does not vanish identically on a set of positive measure, it follows that $\psi(x) = \psi(\bar{x})$ almost everywhere, and this implies that K is a multiple of the identity.

In order to prove the last part of the theorem, we observe that the singular integral operators and their adjoints are defined in all spaces L^p, $1 < p < \infty$, and continuous in the corresponding topologies. Therefore an operator K in $\mathfrak{A}_p$ will map $L^p \cap L^q$ into itself, and since $L^p \cap L^q$ is dense in L^q, the restriction of K to $L^p \cap L^q$ will have a unique continuous extension to L^q, which we take as the definition of $\phi(K)$. Then one verifies that ϕ is the desired isomorphism between $\mathfrak{A}_p$ and $\mathfrak{A}_q$.

Theorem 6 is thus established.

THEOREM 7. *Let $\alpha = (\alpha_1, \cdots, \alpha_k)$ be a k-tuple of non-negative integers and write $D_\alpha u = \partial^{\alpha_1 + \cdots + \alpha_k} u / \partial x_1^{\alpha_1} \partial x_2^{\alpha_2} \cdots \partial x_k^{\alpha_k}$, $z^\alpha = z^{\alpha_1} z_2^{\alpha_2} \cdots z_n^{\alpha_n}$. Let $P(u) = \sum a_\alpha(x) D_\alpha u$ be a linear partial differential operator of homogeneous order m with coefficients $a_\alpha(x)$ in C_β, $\beta \geqq 0$. Then if $u \in L_m{}^p$, $P(u) = H\Lambda^m u$ where H is a singular integral operator of type $C_\beta{}^\infty$ and*

$$\sigma(H) = (-i)^m \sum_\alpha a_\alpha(x) z^\alpha \, |z|^{-m}, \text{ where } |z| = (z_1{}^2 + \cdots + z_k{}^2)^{\frac{1}{2}}.$$

According to (23), we have $\partial u / \partial x_n = -iR_n \Lambda u$ and therefore, since the R_n and Λ commute, it follows that

$$D_\alpha u = (-i)^m R_1{}^{\alpha_1} R_2{}^{\alpha_2} \cdots R_k{}^{\alpha_k} \Lambda^m u = (-i)^m R^\alpha \Lambda^m u$$

and

$$P(u) = \sum_\alpha a_\alpha(x) D_\alpha u = (-i)^m \sum a_\alpha(x) R^\alpha \Lambda^m u.$$

Now according to (14) and (37), $\sigma(R_n) = z_n \, |z|^{-1}$ and from this the desired expression for

$$\sigma(H) = \sigma[(-i)^m \sum_\alpha a_\alpha(x) R^\alpha]$$

follows. This proves the theorem.

MASSACHUSETTS INSTITUTE OF TECHNOLOGY,
CORNELL UNIVERSITY,
UNIVERSITY OF CHICAGO.

REFERENCES.

[1] H. Bateman, *Higher transcendental functions*, vol. 2, N. Y., 1953.

[2] S. Bochner, "Theta relations with spherical harmonics," *Proceedings of the National Academy of Sciences*, vol. 37 (1951), pp. 804-808.

[3] A. P. Calderón and A. Zygmund, "On singular integrals," *American Journal of Mathematics*, vol. 78 (1956), pp. 289-309.

[4] E. Heine, *Handbuch der Kugelfunktionen*, vols. I and II, 2d. ed., Berlin, 1878-1888.

[5] G. N. Watson, *A treatise on the theory of Bessel functions*, Cambridge University Press, 1922.

KONINKL. NEDERL. AKADEMIE VAN WETENSCHAPPEN – AMSTERDAM
Reprinted from Proceedings, Series A, 62, No. 1 and Indag. Math., 21, No. 1, 1959

MATHEMATICS

A NOTE ON SMOOTH FUNCTIONS[1])

BY

MARY WEISS AND ANTONI ZYGMUND

(Communicated by Prof. J. F. KOKSMA at the meeting of November 29, 1958)

§ 1. The present note arose out of an attempt to understand better the meaning and significance of the following theorem of SALEM [1] (see also [2]) in which $S[f]$ denotes the Fourier series of f, and $S_n(x)$, or $S_n[f]$ its partial sums; by periodic functions we mean functions of period 2π.

Theorem A. *Suppose that $f(x)$ is periodic, integrable, and satisfies the condition*

$$\frac{1}{2h}\int_0^h [f(x+t)-f(x-t)]\,dt = o\left\{\frac{1}{\log h}\right\} \quad (h\to+0) \tag{1}$$

uniformly in x. Then

(i) *$S[f]$ converges almost everywhere;*

(ii) *the convergence is uniform over every closed interval of points of continuity of f;*

(iii) *if f is in L^p, $p>1$, the function*

$$s^*(x) = \sup_n |S_n(x)| \tag{2}$$

belongs to L^p.

The main result of this section is that condition (1) alone implies that f is in L^p for *every* p, and the result is primarily a theorem about smooth functions (see below). This is a special case of the following theorem.

Theorem 1. *If $F(x)$ is periodic and for some $\beta>\frac{1}{2}$ satisfies*

$$\Delta^2 F(x,h) = F(x+h)+F(x-h)-2F(x) = O\left\{\frac{h}{|\log h|^\beta}\right\} \tag{3}$$

uniformly in x, then F is the indefinite integral of an f belonging to every L^p.

Functions F satisfying the condition $\Delta^2F(x,h)=o(h)$ for each x and $h\to 0$ are called *smooth*; (3) is a strengthening of the condition of smoothness. For the theory of smooth functions and some of their properties see e.g. [4], or [3_{I}], pp. 42 and 114.

It is of interest that Theorem 1 is false for $\beta=\frac{1}{2}$. For example, the Weierstrass type function

$$F(x) = \sum_{n=1}^{\infty} \frac{\cos 2^n x}{2^n n^{\frac{1}{2}}} \tag{4}$$

[1]) The reserch resulting in this paper was supported in part by the Office of Scientific Research of the Air Force under contract AF 18 (600) – 1111.

53

satisfies, as can easily be seen (see e.g. [3_I], p. 47) condition (3) with $\beta = \frac{1}{2}$ and is at the same time differentiable almost nowhere, since the last series when differentiated termwise is lacunary but not in L^2 (see [3_I], p. 206).

The following results makes possible an application of Theorem 1 to the proof of Theorem A.

Theorem 2. *Suppose that d periodic and continuous F satisfies the condition*

$$\Delta^2 F(x, h) = o\left\{\frac{h}{\log h}\right\} \tag{5}$$

and let s_n and σ_n be respectively the partial sums and $(C, 1)$ *means of* $S'[F]$ *(i.e.* $S[F]$ *differentiated termwise). Then*

$$s_n - \sigma_n \to 0 \tag{6}$$

uniformly in x.

In view of Theorem 1, $\sigma_n(x)$ converges almost everywhere to $f = F'$, and the convergence is uniform over every closed interval of points of continuity of f. This implies parts (i) and (ii) of Theorem A. It is also well known that if $f \in L^p$, then the function

$$\sigma^*(x) = \sup_n |\sigma_n(x)|$$

is also in L^p, so that part (iii) of Theorem A is immediate.

We now pass to the proof of Theorem 1.

Let $E_n[F]$ be the best approximation of F by trigonometric polynomials of order n. The hypothesis (3) implies that

$$E_n[F] = O\,\{n^{-1} (\log n)^{-\beta}\}.$$

This follows immediately if we e.g. consider Jackson's polynomials

$$J_n(x, F) = \frac{1}{k_n} \int_{-\pi}^{\pi} F(x+t) \left(\frac{\sin nt}{\sin t}\right)^4 dt,$$

where

$$k_n = \int_{-\pi}^{\pi} \left(\frac{\sin nt}{\sin t}\right)^4 dt \simeq A\, n^3,$$

and observe that

$$J_n(x, F) - F(x) = \frac{1}{k_n} \int_0^{\pi} \Delta^2 F(x, t) \left(\frac{\sin nt}{\sin t}\right)^4 dt =$$

$$= \frac{1}{k_n} \int_0^{1/n} O\left\{\frac{1}{n (\log n)^\beta}\right\} O(n^4)\, dt + \frac{1}{k_n} \int_{1/n}^{\pi} O\left\{\frac{t}{(\log t)^\beta}\right\} O\left(\frac{1}{t^4}\right) dt = O\left(\frac{1}{n \log^\beta n}\right).$$

On the other hand, it is also known that if S_n are the partial sums of $S[F]$ then the "delayed means"

$$\tau_n(x) = \frac{S_n + S_{n+1} + \ldots + S_{2n-1}}{n}$$

54

differ from $F(x)$ by not more than 4 $E_n[F]$ (see [3_{I}], p. 115). Hence, with the hypothesis (3),

$$F(x) - \tau_n(x) = O\{n^{-1} (\log n)^{-\beta}\}.$$

Write

$$F(x) = \tau_1 + (\tau_2 - \tau_1) + (\tau_4 - \tau_2) + \ldots + (\tau_{2^n} - \tau_{2^{n-1}}) + \ldots = \sum_{n=0}^{\infty} U_n,$$

say, so that $U_n = \tau_{2^n} - \tau_{2^{n-1}}$ for $n = 1, 2, \ldots$. Observe now that $\tau_m(x)$ is obtained by multiplying the k-th term of $S[F]$ by λ_k, where $\lambda_k = 1$ for $k \leqslant m$ and decreases linearly to 0 as k increases from m to $2m$. Hence the non-zero terms of $U_n = \tau_{2^n} - \tau_{2^{n-1}}$ are of ranks k satisfying the condition $2^{n-1} < k < 2^{n+1}$. It follows that the two series

$$U_1 + U_3 + U_5 + \ \ldots, \quad U_2 + U_4 + U_6 + \ \ldots$$

are non-overlapping. We show that if we differentiate these two series termwise we obtain Fourier series of functions belonging to every L^p. In view of the theorem of Littlewood and Paley[1] (see e.g. [3_{II}], p. 233) it is enough to prove that the two functions

$$(U_1'^2 + U_3'^2 + U_5'^2 + \ \ldots)^{\frac{1}{2}} \quad \text{and} \quad (U_2'^2 + U_4'^2 + U_6'^2 + \ \ldots)^{\frac{1}{2}}$$

are in every L^p. We shall show that under our hypotheses they are bounded. For by Bernstein's theorem on the derivatives of trigonometric polynomials,

$$\max_x |U_n'(x)| \leqslant 2^{n+1} \max_x |U_n(x)| = o(n^{-\beta}),$$

and since a series with terms $O(n^{-2\beta})$ converges, Theorem 1 is established.

We now pass to the proof of Theorem 2 and first show that under its hypotheses,

$$F - S_n(x) = o\left(\frac{1}{n}\right) \tag{7}$$

uniformly in x. To see this we write

$$F = \tau_m + \varrho_m, \text{ where } m = [\tfrac{1}{2}n].$$

Then

$$\begin{aligned} F - S_n[F] &= F - S_n[\tau_m] - S_n[\varrho_m] \\ &= (F - \tau_m) - S_n[\varrho_m] = \varrho_m - S_n[\varrho_m]. \end{aligned}$$

Since $\varrho_m = o(1/m \log m) = o(1/n \log n)$ and, using Lebesgue constants,

$$|S_n[\varrho_m]| = O\,(\log n) \max |\varrho_m(x)| = o(1/n),$$

(7) follows.

Write

$$F' = f \sim \sum_1^{\infty} (a_k \cos kx + b_k \sin kx) = \sum_1^{\infty} A_k(x)$$

---

[1]) The theorem asserts that if $f \in L^p$, $p > 1$, and if we decompose $S[f]$ into a series of blocks, $S[f] = \Sigma \Delta_k$, where Δ_k consists of the terms of rank ν satisfying $n_k \leqslant n < n_{k+1}$, with $1 < \alpha < n_{k+1}/n_k < \beta < \infty$, then the ratio of $\|f\|_p$ and $\|(\Sigma \Delta_k^2)^{\frac{1}{2}}\|_p$ is contained between two constants depending on α and β only.

55

and let $B_k = a_k \sin kx - b_k \cos kx$. Then (7) can be written

$$\sum_{k=n}^{\infty} \frac{B_k(x)}{k} = o\left(\frac{1}{n}\right).$$

In particular

$$\sum_{\frac{1}{2}N<k\leqslant N} k^{-1} B_k(x) = o(N^{-1})$$

for any positive number N, not necessarily an integer and, by Berstein's theorem,

$$\sum_{\frac{1}{2}N<k\leqslant N} A_k(x) = o(1).$$

Using Bernstein's inequality for conjugate trigonometric polynomials of order n ($\max|\tilde{T}'(x)| \leqslant n \max|T(x)|$) we have

$$\sum_{\frac{1}{2}N<k\leqslant N} kA_k(x) = o(N),$$

whence, replacing N by $\frac{1}{2}N, \frac{1}{4}N, \frac{1}{8}N, \ldots$ and adding,

$$\sum_{k=1}^{N} kA_k(x) = o(N),$$

which is (6), with N for n.

§ 2. In this section we consider a generalization of Theorem 2. A periodic $F(x) \in L^p$ will be said to satisfy condition Λ_p^* if

$$(8) \qquad \|\Delta^2 F(x,h)\|_p = \left(\int_0^{2\pi} |F(x+h) + F(x-h) - 2F(x)|^p dx\right)^{1/p} = o(h).$$

Replacing here 'O' by 'o' we obtain condition λ_*^p. It is well known that Λ_p^* and λ_p^* are the classes of functions which in the metric L^p can be approximated to by polynomials of order n with an error $O(1/n)$ and $o(1/n)$ respectively. While functions in Λ_*^1 can be discontinuous and even unbounded (for example the function equal to $\log|x|$ for $|x| \leqslant \pi$ and continued periodically is in Λ_*^1), the functions from Λ_*^p, $p>1$, are essentially continuous and even have absolutely convergent Fourier series. In addition to Λ_*^p we shall also consider the classes $\Lambda_{*.\beta}^p$ of functions satisfying the condition

$$\|\Delta^2 F(x,h)\|_p = O\left\{\frac{h}{|\log h|^\beta}\right\}.$$

Theorem 3. (i) *If $F \in \Lambda_{*.\beta}^p$, $1<p\leqslant 2$, $\beta>1/p$, then F is absolutely continuous and $F' \in L^p$. The result is false if $\beta = 1/p$.*

(ii) *If $F \in \Lambda_{*.\beta}^p$, $2\leqslant p<\infty$, $\beta>\frac{1}{2}$, then F is absolutely continuous and $F' \in L^p$. The result is false if $\beta=\frac{1}{2}$.*

(i) Let S_n and s_n be the partial sums of $S[F]$ and $S'[F]$ respectively and let (*assuming that the constant term of $S[F]$ is* 0)

$$S[F] = \sum_{k=1}^{\infty} k^{-1} B_k(x).$$

56

The hypothesis implies that the best approximation of F in the metric L^p by trigonometric polynomials of order n is $O(1/n \log^{\beta} n)$ (we may use Jackson's polynomials for the proof) and so, if $p>1$, leaving the case $p=1$ temporarily aside,

(9) $$\{\int_0^{2\pi} |F-S_n|^p\, dx\}^{1/p} = O\{n^{-1} (\log n)^{-\beta}\}.$$

Let Δ_n and δ_n denote the blocks of terms with indices $2^{n-1} \leqslant k < 2^n$ $(n=1, 2, \ldots)$ for $S[F]$ and $S'[F]$ respectively. From (9) we deduce that

(10) $$\|\Delta_n\|_p = O(2^{-n}\, n^{-\beta}),$$

and so, using Bernstein's inequality for the metric L^p,

(11) $$\|\delta_n\|_p = O(n^{-\beta}).$$

Applying the theorem of Littlewood and Paley, we see that the positive assertion of (i) will follow if we show that

$$\int_0^{2\pi} (\sum \delta_n^2)^{\frac{1}{2}p}\, dx$$

is finite. Since $p \leqslant 2$, the last integral is majorized by

$$\int_0^{2\pi} (\sum |\delta_n|^p)\, dx = \sum \|\delta_n\|_p^p = \sum O(n^{-\beta p}) < \infty,$$

if $\beta p > 1$.

The case $p=1$ must be treated slightly differently since it is no longer true that the approximation to F by the $S_n[F]$ is of the same order as the best approximation (in the metric L). The required modification has already been used in § 1. Let τ_n be the delayed means of $S[F]$. Then

$$F = \tau_1 + (\tau_2 - \tau_1) + (\tau_4 - \tau_2) + \ldots = U_0 + U_1 + U_2 + \ldots.$$

The two series $U_0+U_2+U_4+\ldots$ and $U_1+U_3+U_5+\ldots$ are non-overlapping and are respectively $S[F_1]$ and $S[F_2]$. It is enough to show that $S'[F_1]$ and $S'[F_2]$ are both Fourier series. Now

$$\|U_n\|_1 = O(2^{-n}\, n^{-\beta}), \quad \|U_n'\|_1 = O(n^{-\beta}),$$

and it is enough to observe that, say,

$$\sum \int_0^{2\pi} |U_{2k}'|\, dx = \sum O(k^{-\beta}) < \infty$$

if $\beta > 1$.

As regards counterexamples, suppose first that $1<p\leqslant 2$ and consider the periodic function

$$F(x) = |x|^{1/p'} \left(\log \frac{2\pi}{|x|}\right)^{-1/p} \quad (|x| \leqslant \pi),$$

where p' is the index conjugate to p. A simple computation which we

57

omit shows that F is in $\Lambda^p_{*,1/p}$ and that $F'(x)$, which is asymptotically equal to $x^{-1/p}(\log 1/x)^{-1/p}$ as $x \to +0$, is not in L^p.

Similarly the periodic function equal to

$$\log\log\frac{\pi}{|x|}$$

for $|x|<\pi$ is in $\Lambda^1_{*,1}$, but its derivative is not in L.

(ii) We pass to the case $2<p<\infty$, and suppose that $F\in\Lambda^p_{*,\beta}$ where $\beta>\frac{1}{2}$. Defining the blocks Δ_n and δ_n as before, we may write $S[F]=\sum\Delta_n$, $S'[F]=\sum\delta_n$.

We again have (9), (10) and (11). Since $p>2$, Minkowski's inequality gives

$$\{\int_0^{2\pi}(\sum\delta_n^2)^{\frac{1}{2}p}dx\}^{2/p}\leqslant\sum\{\int_0^{2\pi}(\delta_n^2)^{\frac{1}{2}p}dx\}^{2/p}=\sum\|\delta_n\|_p^2.$$

Since the terms of the last series are $O(n^{-2\beta})$, and $\beta>\frac{1}{2}$, the series converges, the integral on the left is finite and the Littlewood–Paley theorem shows that $S'[F]=\sum\delta_n$ is the Fourier series of a function in L^p. This completes the proof of the positive assertion in (ii). That the result is false for $\beta=\frac{1}{2}$ is seen by the example of the function (4), for which

$$\Delta^2F(x,h)=O\{h\log^{-\frac{1}{2}}(1/h)\},$$

so that $F\in\Lambda^p_{*,\frac{1}{2}}$, and which is differentiable only in a set of measure 0.

Remarks. a. SALEM localized his theorem to a subinterval (a,b) of a period. We can likewise generalize Theorem 2:

Theorem 2′. *If F is periodic, integrable, continuous in an interval (a,b), satisfies uniformly in that interval condition* (5), *and has Fourier coefficients $o(1/n)$, then $s_n-\sigma_n$ tends uniformly to 0 in every $(a+\varepsilon, b-\varepsilon)$, $\varepsilon>0$, where s_n and σ_n are the partial sums and $(C,1)$ means of $S'[F]$.*

The proof might, in principle, imitate that of Theorem 2, but since then a few non-trivial details would have to be attended to, we prefer to reduce Theorem 2′ to Theorem 2. If we could represent the F in Theorem 2′ as a sum F_1+F_2 of two periodic functions such that F_1 is everywhere continuous and satisfies a condition analogous to (5), and F_2 is integrable and zero in (a,b), the reduction would be immediate. For then, if $s_{1,n}$, $s_{2,n}$, $\sigma_{1,n}$, $\sigma_{2,n}$ are respectively the partial sums and $(C,1)$ means of $S'[F_1]$ and $S'[F_2]$, we would have

$$s_n-\sigma_n=(s_{1,n}-\sigma_{1,n})+(s_{2,n}-\sigma_{2,n}),$$

and since $s_{1,n}-\sigma_{1,n}$ tends uniformly to 0 it would be enough to show that $s_{2,n}-\sigma_{2,n}$ tends uniformly to 0 in $(a+\varepsilon, b-\varepsilon)$. But $s_{2,n}-\sigma_{2,n}$ is the n-th partial sum divided by $(n+1)$ of $\tilde{S}''[F_2]$, and since the coefficients of $\tilde{S}''[F_2]$ are $o(n)$, and $F_2=0$ in (a,b), the partial sums of $\tilde{S}''[F_2]$ would be $o(n)$ uniformly in $(a+\varepsilon, b-\varepsilon)$ ([3], p. 367) and the assertion would follow.

Whether a decomposition of the kind just described is possible we do

58

not know, and the problem of extending a smooth function outside the initial interval of definition is not obvious though possibly not difficult. For our purposes however it is enough to show that in (a, b) we can find points a_1, b_1 arbitrarily close to a and b respectively and such that F can be continued outside (a_1, b_1) with the preservation of (5). First we show that if F satisfies condition (5) in (a, b) then there is a dense set of points $\xi \in (a, b)$ such that $F'(\xi)$ exists and

$$\frac{F(\xi+h)-F(\xi)}{h} - F'(\xi) = o\left(\frac{1}{\log|h|}\right).$$

This is certainly true, with $F'(\xi)=0$, if ξ is an extremum of F, and subtracting from F linear functions we obtain a dense set – even one of the power of the continuum – of the points ξ, and if we take for a_1 and b_1 points ξ, and define F_1 as equal to F in (a_1, b_1) and equal to an arbitrary function of the class C' elsewhere, provided F_1 is continuous and differentiable at a_1 and b_1, then it is not difficult to see that F_1 is a required extension of F. (On a similar argument we might base an extension from (a_1, b_1) interior to (a, b) of a function which satisfies in (a, b) the condition $\Delta^2 F(x, h)=o(h)$ or $=O(h)$, not necessarily uniformly in x.)

b) There is an analogue of Theorem 2 for the metric L^p, $1 \leqslant p < \infty$. The cases $p=1$ and $1<p<\infty$ are slightly different. If $F \in \Lambda^1_{*,1}$, then

$$||s_n - \sigma_n||_1 \to 0,$$

and, slightly more generally, if $F \in \Lambda^1_{*,\beta}$, then $||s_n-\sigma_n||_p = o\{(\log n)^{1-\beta}\}$. If $1<p<\infty$, and if $F \in \Lambda^p_{*,\beta}$, then $||s_n-\sigma_n||=o\{(\log n)^{-\beta}\}$. The '$o$' can be replaced by '$O$' throughout.

De Paul University and the University of Chicago

REFERENCES

1. Salem, R., New theorems on the convergence of Fourier series, Proc. Kon. Ned. Akad. v. Wetensch., **A 57**, 550–557 (1954).
2. Tchereĭskaya, W. I., On the uniform convergence of trigonometric series (in Russian), Učenye zapiski Moscow University, (No. 181), 159–163 (1956).
3. Zygmund, A., Trigonometric series, Second Ed., vol. I, 1–383, vol. II, 1–354, Cambridge Univ. Press, (1959).
4. ———, Smooth functions, Duke Math. Journal, 12, 47–76 (1945).

STUDIA MATHEMATICA, T. XX. (1961)

Local properties of solutions of elliptic partial differential equations

by

A. P. CALDERÓN and A. ZYGMUND (Chicago)

Introduction. The purpose of this paper* is to establish *pointwise* estimates for solutions of elliptic partial differential equations and systems. The results presented here differ from the familiar ones in that they give information about the behavior of solutions at individual points. More specifically, we obtain two kinds of results. On the one hand, we establish inequalities for solutions and their derivatives at isolated individual points. On the other, we also obtain results of the „almost everywhere" type. Theorems 1 and 2 below summarize the main results.

1. We begin with notations and definitions.

By $x, y, \ldots$ we denote respectively points $(x_1, x_2, \ldots, x_n)$, $(y_1, y_2, \ldots, y_n), \ldots$ of the n-dimensional Euclidean space E_n; dimension n is fixed throughout the paper. The class of measurable functions $f(x)$ such that $\|f(x)\|_p = \left(\int |f(x)|^p dx\right)^{1/p} < \infty$ will be denoted by $L^p(E_n)$ or, simply, L^p; here and elsewhere dx stands for the element of volume in E_n, and $\int$ means $\int_{E_n}$. All functions we consider are complex-valued, unless otherwise stated. The symbols $x+y$ and λx, where λ is a scalar, have the usual meaning; we also use the notations $|x| = (x_1^2+x_2^2+\ldots+x_n^2)^{1/2}$, $x\cdot y = x_1y_1+x_2y_2+\ldots+x_ny_n$, and if $a = (a_1, a_2, \ldots, a_n)$, where the a_j are non-negative integers, we will write

$$|a| = a_1+a_2+\ldots+a_n, \qquad x^a = x_1^{a_1}x_2^{a_2}\ldots x_n^{a_n},$$

$$a! = a_1!a_2!\ldots a_n!, \qquad \left(\frac{\partial}{\partial x}\right)^a f = f_a = \left(\frac{\partial}{\partial x_1}\right)^{a_1}\left(\frac{\partial}{\partial x_2}\right)^{a_2}\ldots\left(\frac{\partial}{\partial x_n}\right)^{a_n} f.$$

Finally $f*g$ will stand for the convolution of f and g.

The symbol C with various subscripts will stand for a constant, not necessarily the same at each occurrence, which depends only (unless otherwise stated) on the variables displayed. Dependence on the dimension, though, will not be indicated. Thus C without subscripts will indicate either an absolute constant or a quantity depending on the dimension only.

* Research resulting in this paper was partly supported by the NSF, contract NSF G-8205, and the Air Force, contract AF-49 (638)-451.

Reprinted from *SM* 20, 171–225 (1961).

We denote by C_0^∞ the class of infinitely differentiable functions with compact support.

It seems that the notion of differentiability which is most suited to the treatment of the problems that concern us, is not the classical one. It appears that it is more convenient to estimate the remainder of the Taylor series in the mean with various exponents. This type of differentiability is much more stable than the classical one in the sense that it is preserved at individual points under various operations such as fractional integration and singular integral transformations.

Definition 1. Let u be any number $\geqslant -n/p$. By $T_u^p(x_0)$ we shall denote the class of functions $f(x)\in L^p(E_n)$ such that there exists a polynomial $P(x-x_0)$ of degree strictly less than u (in particular $P\equiv 0$ if $u\leqslant 0$) with the property that

$$(1.1)\qquad \Big(\varrho^{-n}\int_{|x-x_0|\leqslant\varrho}|f(x)-P(x-x_0)|^p\,dx\Big)^{1/p}\leqslant A\varrho^u \qquad (0<\varrho<\infty)$$

with A independent of ϱ. Here $1\leqslant p\leqslant\infty$; when $p=\infty$ the left side of (1.1) means, as usual, $\operatorname*{ess\,sup}_{|x-x_0|\leqslant\varrho}|f(x)-P(x-x_0)|$. Instead of $T_u^\infty(x_0)$ we shall simply write $T_u(x_0)$.

Definition 2. Let f be a function in $T_u^p(x_0)$. We shall say that f *belongs to* $t_u^p(x_0)$, if there exists a polynomial $P(x-x_0)$ of degree less than or equal to u such that

$$(1.2)\qquad \Big(\varrho^{-n}\int_{|x-x_0|\leqslant\varrho}|f(x)-P(x-x_0)|^p\,dx\Big)^{1/p}=o(\varrho^u)\quad\text{as}\quad \varrho\to 0.$$

As in definition 1, we shall denote $t_u^\infty(x_0)$ by $t_u(x_0)$.

In both definitions the polynomial P is unique, as easily seen by considering the difference P_1-P_2 of two such possible polynomials and observing that the inequalities (1.1) and (1.2) imply that it must vanish identically. Any function in L^p belongs to $T_u^p(x_0)$ with $u=-n/p$ for any x_0. The familiar result about the Lebesgue set of a function can be expressed by saying that if $f\in L^p$, $p<\infty$, then $f\in t_0^p(x_0)$ for almost all x_0.

The class $T_u^p(x_0)$ is a linear space in which we introduce a norm. It will be convenient to use the rather unorthodox notation $T_u^p(x_0,f)$ for the norm of an $f\in T_u^p(x_0)$. We define $T_u^p(x_0,f)$ as the sum of $\|f\|_p$, the moduli of the coefficients of the polynomial $P(x-x_0)$, and the least admissible value of A in (1.1).

In the definition of $T_u^p(x_0)$ or $t_u^p(x_0)$ the value of the function at x_0 is irrelevant. Nevertheless, if the function f in L^p belongs to $t_u^p(x_0)$ with $u\geqslant 0$ (and so also if f belongs to $T_u^p(x_0)$ with $u>0$) for each x_0 belonging to a set of positive measure, then at every x_0 in the Lebesgue set $f(x_0)$ coincides with the constant term of the corresponding $P(x-x_0)$; consequently be redefining f, if necessary, on a set of measure zero, we may assume that this condition is always satisfied.

Definition 3. Let Q be a closed set. We shall say that the *bounded function f belongs to the class $B_u(Q)$*, $u > 0$, if there exist bounded functions f_a, $|a| < u$, such that

$$f_a(x+h) = \sum_{|\beta| < u-|a|} \frac{h^\beta}{\beta!} f_{a+\beta}(x) + R_a(x, h)$$

for all x and $x+h$ in Q, with $|R_a(x, h)| \leqslant C|h|^{u-|a|}$. We say that *$f$ belongs to $b_u(Q)$*, $u \geqslant 0$, if there exist functions f_a with $|a| \leqslant u$ such that

$$f_a(x+h) = \sum_{|\beta| \leqslant u-|a|} \frac{h^\beta}{\beta!} f_{a+\beta}(x) + R_a(x, h)$$

for all x and $x+h$ in Q, with $|R_a(x, h)| \leqslant C|h|^{u-|a|}$ and, in addition, $R_a(x, h) = o(|h|)^{u-|a|}$ as $|h| \to 0$, uniformly in $x \epsilon Q$.

Definition 4. Given a non-negative integer k, L_k^p will denote the class of functions in L^p with distribution derivatives of orders less than or equal to k in L^p. The norm $\|f\|_{p,k}$ of a function f in L_k^p is, by definition, the sum of the norms in L^p of f and its derivatives of orders less than or equal to k.

Definition 5. Let $f = (f_1, f_2, \ldots, f_r)$ be a vector valued function and

$$\mathcal{L}f = \sum_{|a| \leqslant m} a_a(x) \left(\frac{\partial}{\partial x}\right)^a f = g,$$

where $g = (g_1, g_2, \ldots, g_s)$ is a vector valued function of s components, $s \geqslant r$, and $a_a(x)$ are $s \times r$ matrices. We shall say that the operator $\mathcal{L}$, or the equation $\mathcal{L}f = g$, is *elliptic* at the point x_0 if the characteristic matrix

$$\sum_{|a| = m} a_a(x_0)\, \xi^a \tag{1.3}$$

has the property that for $|\xi| = 1$

$$\det\left[\left(\sum_{|a|=m} a_a^*(x_0)\, \xi^a\right)\left(\sum_{|a|=m} a_a(x_0)\, \xi^a\right)\right] \geqslant \mu(x_0) > 0,$$

where a^* denotes the conjugate transpose of a. We shall call the largest admissible value of $\mu(x_0)$ the *ellipticity constant of $\mathcal{L}$ at x_0*.

THEOREM 1. *Let $\mathcal{L}f = g$ be an equation of order m with coefficients in $T_u(x_0)$, $u > 0$, which is elliptic at x_0 in the sense of definition 5. Let $1 < p < \infty$, $u \geqslant v \geqslant -n/p$ and v be non-integral. If $f \epsilon L_m^p$ and $g \epsilon T_v^p(x_0)$ in the sense that their components belong to these spaces, then*

(i) *for* $j = 1, 2, \ldots, r$ *and* $|a| \leqslant m$,

$$T^q_{v+m-|a|}\left(x_0, \left(\frac{\partial}{\partial x}\right)^a f_j\right) \leqslant C\left[\sum_{i=1}^{s} T^p_v(x_0, g_i) + \sum_{i=1}^{r} \|f_i\|_{p,m}\right], \tag{1.4}$$

where $1/p \geqslant 1/q \geqslant 1/p-(m-|a|)/n$ *if* $1/p-(m-|a|)/n > 0$, $p \leqslant q \leqslant \infty$ *if* $1/p < (m-|a|)/n$, *or* $p \leqslant q < \infty$ *if* $1/p = (m-|a|)/n$, *and* C *depends only on* v, p, r, s, $\mu(x_0)$ *and the least upper bound of the norms in* $T_u(x_0)$ *of the coefficients of* $\mathfrak{L}$.

(ii) *If in addition the leading coefficients of the equation are uniformly continuous and the equation is uniformly elliptic in the sense that the constant of ellipticity* $\mu(x)$ *is bounded away from zero, then the quantity* $\sum_{i=1}^{r}\|f_i\|_{p,m}$ *on the right of* (1.4) *can be replaced by* $C\left[\sum_{i=1}^{r}\|f_i\|_p + \sum_{i=1}^{s}\|g_i\|_p\right]$, *where* C *depends on* $\mathfrak{L}$.

(iii) *If* $g \in t^p_v(x_0)$, *then* $(\partial/\partial x)^a f$ *belongs to* $t^q_{v+m-|a|}$ *with the same* q *as in part* (i).

THEOREM 2. *Let* $\mathfrak{L}f = g$ *be an equation of order* m *which is elliptic in the sense of definition* 5 *at all points* x_0 *belonging to a set* Q *of positive measure, and whose coefficients belong to* $T_u(x_0)$, $u \geqslant 1$, *for all* x_0 *in* Q. *Let* v *be a positive integer not larger than* u. *Then, if* $g \in T^p_v(x_0)$, $1 < p < \infty$, *for all* x_0 *in* Q, *and* $f \in L^p_m$, *the functions* $(\partial/\partial x)^a f$, $|a| \leqslant m$, *belong to* $t^q_{v+m-|a|}(x_0)$ *for almost all* x_0 *in* Q, *where* q *is the same as in part* (i) *of Theorem* 1.

THEOREM 3. *Let* $\mathfrak{L}f = g$ *be an equation of order* m *with coefficients in* $T_u(x_0)$, $u > 0$, *for all* x_0 *in a closed set* Q. *Let the norms of the coefficients in* $T_u(x_0)$ *be bounded in* Q *and* $\mathfrak{L}$ *be uniformly elliptic in* Q, *that is, let the constant of ellipticity* $\mu(x_0)$ *of* $\mathfrak{L}$ *be bounded away from zero in* Q. *Then if* $g \in T^p_v(x_0)$ *for all* x_0 *in* Q *and* $\sum_{i=1}^{s} T^p_v(x_0, g_i)$ *is bounded in* Q, $1 < p < \infty$, v *is positive and non integral,* $-m < v \leqslant u$, *and* $f \in L^p_m$, *the function* f *belongs to* $B_{v+m}(Q)$.

If in addition $g \in t^p_v(x_0)$ *for all* $x_0 \in Q$, *then* $f \in b_{v+m}(Q)$.

It has already been observed (see [7]), and the idea is basic for the present paper, that, roughly speaking, a differential operator is the composition of fractional differentiation and a singular integral transformation. Accordingly, our method will consist in deriving estimates for solutions of differential equations from estimates for fractional integrals and singular integral transforms.

Definition 6. Let f be a tempered distribution in E_n; the fractional integral of order u of f, denoted by $J^u f$, is defined by

$$\widehat{J^u f} = (1+4\pi^2|x|^2)^{-u/2}\hat{f},$$

where $\hat{f}$ stands for the Fourier transform of f, that is,

$$\hat{f}(x) = \int_{E_n} e^{-2\pi i(x\cdot y)} f(y)\, dy,$$

if f is an integrable function.

This definition of fractional integration is different from the familiar one due to M. Riesz. It has some advantages, namely it is defined for any u real or complex and thus is a one-parameter group of operations, and furthermore, for $u > 0$ it is a bounded operation on L^p, $1 \leqslant p < \infty$. This notion of fractional integration was introduced in [1] and [5; I, page 25].

THEOREM 4. *Let* $u \geqslant -n/p$, $v > 0$, $u + v \neq 0, 1, 2, \ldots$, $1 < p \leqslant \infty$. *Then* J^v *maps continuously*

(i) $T_u^p(x_0)$ *into* $T_{v+u}^q(x_0)$, *provided*

(a) $$\frac{1}{p} \geqslant \frac{1}{q} \geqslant \frac{1}{p} - \frac{v}{n}, \quad \text{if} \quad p < \frac{n}{v};$$

(b) $$p \leqslant q \leqslant \infty, \quad \text{if} \quad \frac{n}{v} < p \leqslant \infty;$$

(c) $$p \leqslant q < \infty, \quad \text{if} \quad \frac{n}{v} = p;$$

(ii) $t_u^p(x_0)$ *into* $t_{u+v}^q(x_0)$, *with* u *and* v *as in* (i).

THEOREM 5. *Let* u *and* v *be non-negative integers. Then if* $f \in T_u^p(x_0)$, $1 < p < \infty$, *for all* x_0 *in a set* Q *of positive measure, we have* $J^v f \in t_{u+v}^q(x_0)$ *for almost all* $x_0 \in Q$, p *and* q *being related as in part* (i) *of Theorem* 4. *If* $u + v \geqslant 1$, *the assertion is valid for* $1 < p \leqslant \infty$.

This theorem asserts in particular that if u is a positive integer and $1 < p \leqslant \infty$ the condition $f \in T_u^p(x_0)$ implies $f \in t_u^p(x_0)$ almost everywhere; the case $u = 1$ and $p = \infty$ is the familiar result of Rademacher-Stepanov; the case $u > 1$, $p = \infty$ was proved by Oliver [12]; related results are Theorem 11 and 12 below which extend some known results (see [8] and [4]).

We shall now consider singular integral operators of the following form:

$$\mathcal{K}f = a(x)f(x) + \int k(x, x-y) f(y)\, dy,$$

where $a(x)$ is a bounded measurable function and $k(x, z)$ is homogeneous of degree $-n$ with respect to z, that is, such that $k(x, \lambda z) = \lambda^{-n} k(x, z)$ for all $\lambda > 0$, and further $k(x, z)$ has for each x mean value zero on $|z| = 1$. In addition we shall assume that $k(x, z)$ is infinitely differentiable with respect to z and is uniformly bounded for $|z| = 1$. The preceding integral must of course be interpreted as a principal value integral. Associated with the operator $\mathcal{K}$ is its *symbol* $\sigma(\mathcal{K})$ which is defined as

$$\sigma(\mathcal{K}) = a(x) + k(x, \hat{z}),$$

where $k(x, \hat{z})$ is the Fourier transform of $k(x, z)$ with respect to z.

Definition 7. An operator $\mathcal{K}$ as above is said *to belong to the class* $T_u(x_0)$, $u \geqslant 0$, if $\sigma(\mathcal{K})$ and its derivatives with respect to coordinates of z of orders $\leqslant 2n+u+1$ belong to $T_u(x_0)$ for each $z \neq 0$, uniformly in $|z| = 1$. The norm $T_u(x_0, \mathcal{K})$ of an operator of class $T_u(x_0)$ is, by definition, the least upper bound of the norms in $T_u(x_0)$ of $\sigma(\mathcal{K})$ and its derivatives with respect to z of orders less than or equal to $2n+u+1$ evaluated in $|z| = 1$. Operators of class $t_u(x_0)$ are analogously defined.

THEOREM 6. *Let $\mathcal{K}$ be a singular integral operator of class $T_u(x_0)$. If $1 < p < \infty$, and v is not equal to zero or a positive integer and is larger than or equal to $-n/p$, then $\mathcal{K}$ maps $T_v^p(x_0)$ continuously into $T_v^p(x_0)$, with norm less than or equal to $C_{p,u} T_u(x_0, \mathcal{K})$, provided $u \geqslant v$. The corresponding result for operators of class $t_u(x_0)$ and the spaces $t_v^p(x_0)$ is also valid.*

THEOREM 7. *Let u be a non negative integer and f a function belonging to $T_u^p(x_0)$, $1 < p < \infty$, for all x_0 in a set Q of positive measure. Then there exists a subset $\bar{Q}$ of Q such that $Q - \bar{Q}$ has measure zero and such that for every singular integral operator $\mathcal{K}$ belonging to $T_u(x_0)$, $x_0 \in \bar{Q}$, $\mathcal{K}f$ belongs to $T_u^p(x_0)$.*

2. In this section we establish certain properties of the spaces $T_u^p(x_0)$, $t_u^p(x_0)$ which we shall need later.

LEMMA 2.1. *If $-n/p \leqslant u \leqslant v$, $1 \leqslant p \leqslant \infty$, then $T_u^p(x_0) \supset T_v^p(x_0)$, and $T_u^p(x_0, f) \leqslant C T_v^p(x_0, f)$.*

Proof. Assume $u \geqslant 0$. Let P_u be the sum of the terms of degree $< u$ in the Taylor expansion of f, R_u the corresponding remainder, and let P_v and R_v be similarly defined. Then P_u is the sum of the terms of P_v of degree less than u. For $\varrho \leqslant 1$, we have $|P_v(h) - P_u(h)| \leqslant T_v^p(x_0, f)|h|^u$ and

$$\Big[\int_{|h| \leqslant \varrho} |R_u(h)|^p dh\Big]^{1/p} \leqslant \Big[\int_{|h| \leqslant \varrho} |P_v(h) - P_u(h)|^p dh\Big]^{1/p} + \Big[\int_{|h| \leqslant \varrho} |R_v(h)|^p dh\Big]^{1/q}$$
$$\leqslant C T_v^p(x_0, f) \varrho^{n/p+u} + T_v^p(x_0, f) \varrho^{n/p+v} \leqslant C T_v^p(x_0, f) \varrho^{n/p+u}.$$

And for $\varrho > 1$, since $|P_u(h)| \leqslant T_v^p(x_0, f)\varrho^u$ for $|h| \leqslant \varrho$, we have

$$\Big[\int_{|h| \leqslant \varrho} |R_u(h)|^p dh\Big]^{1/p} \leqslant \|f\|_p + \Big[\int_{|h| \leqslant \varrho} |P_u(h)|^p dh\Big]^{1/p}$$
$$\leqslant T_v^p(x_0, f) + C T_v^p(x_0, f) \varrho^{n/p+u} \leqslant C T_v^p(x_0, f) \varrho^{n/p+u},$$

and thus

$$\sup_{\varrho} \frac{1}{\varrho^{u+n/p}} \Big[\int_{|h|\leqslant\varrho} |R_u(h)|^p dh \Big]^{1/p} \leqslant CT_v^p(x_0, f).$$

This, as easily seen, implies the desired result.

For $u < 0$ we have, if $\varrho \leqslant 1$,

$$\frac{1}{\varrho^u} \Big[\frac{1}{\varrho^n} \int_{|h|\leqslant\varrho} |f(x_0+h)|^p dh \Big]^{1/p} \leqslant \frac{1}{\varrho^v} \Big[\frac{1}{\varrho^n} \int_{|h|\leqslant\varrho} |f(x_0+h)|^p dh \Big]^{1/p},$$

and, if $\varrho > 1$,

$$\frac{1}{\varrho^u} \Big[\frac{1}{\varrho^n} \int_{|h|\leqslant\varrho} |f(x_0+h)|^p dh \Big]^{1/p} \leqslant \|f\|_p,$$

which again implies the desired result.

LEMMA 2.2. *The spaces $T_u^p(x_0)$, $1 \leqslant p \leqslant \infty$, $u \geqslant -n/p$, are complete.*

Proof. Suppose that the sequence f_ν is such that $T_u^p(x_0, f_\nu - f_\mu) \to 0$ as ν and μ tend to infinity. Then, in the first place, f_ν converges in L^p to a limit f. Let $P = \lim_{\nu\to\infty} P_\nu$, where P_ν is the Taylor expansion of f_ν; P exists since the coefficients of P_ν converge. Then, for each ϱ,

$$\frac{1}{\varrho^u} \Big[\frac{1}{\varrho^n} \int_{|h|\leqslant\varrho} |[f(x_0+h) - f_\nu(x_0+h)] - [P(h) - P_\nu(h)]|^p dh \Big]^{1/p}$$

$$= \lim_{\mu\to\infty} \frac{1}{\varrho^u} \Big[\frac{1}{\varrho^n} \int_{|h|\leqslant\varrho} |(f_\mu - f_\nu) - (P_\mu - P_\nu)|^p dh \Big]^{1/p} \leqslant \varliminf_{\mu\to\infty} T_u^p(x_0, f_\mu - f_\nu) < \infty.$$

This shows that $f \in T_u^p(x_0)$, and as ν tends to infinity we find

$$\sup_{\varrho} \frac{1}{\varrho^u} \Big[\frac{1}{\varrho^n} \int_{|h|\leqslant\varrho} |(f - f_\nu) - (P - P_\nu)|^p dh \Big]^{1/p} \to 0.$$

From this it follows that $T_u^p(x_0, f - f_\nu) \to 0$ as $\nu \to \infty$.

LEMMA 2.3. *Let $-n/p \leqslant u$, $1 \leqslant p \leqslant \infty$.*

(i) *The space $t_u^p(x_0)$ is a closed subspace of $T_u^p(x_0)$.*

(ii) *If f is a function in $t_u^p(x_0)$ and $\varphi(x)$ is a function in C_0^∞ such that $\int \varphi(x) dx = 1$, then $f^\lambda = \lambda^n \varphi(\lambda x) * f(x)$ converges to f in $T_u^p(x_0)$ as λ tends to infinity.*

(iii) *The space C_0^∞ is dense in $t_u^p(x_0)$.*

Proof Suppose $f_\nu \epsilon t_u^p(x_0)$ and $T_u^p(x_0, f_\nu - f) \to 0$ as $\nu \to \infty$. Then, if R_ν and R are the respective remainders,

$$\sup \frac{1}{\varrho^u}\left[\frac{1}{\varrho^n}\int\limits_{|x-x_0|\leqslant\varrho} |R_\nu - R|^p dx\right]^{1/p} \leqslant T_u^p(x_0, f_\nu - f).$$

Consequently, the left-hand side is less than any preassigned ε if ν is sufficiently large. Since

$$\lim_{\varrho\to 0} \frac{1}{\varrho^u}\left[\frac{1}{\varrho^n}\int\limits_{|x-x_0|\leqslant\varrho} |R_\nu|^p dx\right]^{1/p} = 0,$$

it follows that

$$\overline{\lim_{\varrho\to 0}} \frac{1}{\varrho^u}\left[\frac{1}{\varrho^n}\int\limits_{|x-x_0|\leqslant\varrho} |R(x)|^p dx\right]^{1/p} \leqslant \varepsilon.$$

Since ε is arbitrary, this implies that $f \epsilon t_u^p(x_0)$.

To prove (ii) we will assume, without loss of generality, that $\varphi(x)$ vanishes for $|x| \geqslant 1$, and that $x_0 = 0$.

Denote by P^λ the sum of terms of the Taylor expansion of $f^\lambda(x)$ at $x_0 = 0$ of degree $\leqslant u$. Set $R^\lambda = f^\lambda - P^\lambda$. Then, by Hölder's inequality, we have

$$\frac{1}{\varrho^n}\int\limits_{|x|\leqslant\varrho} |R(x)|\, dx \leqslant C\left[\frac{1}{\varrho^n}\int\limits_{|x|\leqslant\varrho} |R(x)|^p dx\right]^{1/p} \leqslant C\varepsilon(\varrho)\,\varrho^u,$$

where R is the remainder in the expansion of f at $x_0 = 0$ and $\varepsilon(\varrho)$ is bounded and tends to zero as ϱ tends to zero. Without loss of generality we may assume that $\varepsilon(\varrho)$ decreases to 0.

We first show that the coefficients of P^λ converge to the coefficients of P. By differentiation we obtain

$$\left(\frac{\partial}{\partial x}\right)^a f^\lambda(0) = \int \lambda^n \varphi(-\lambda y) P_a(y)\, dy + \int \lambda^{n+|a|} \varphi_a(-\lambda y) R(y)\, dy,$$

where $f = P + R$, $P_a = (\partial/\partial x)^a P$ and $\varphi_a = (\partial/\partial x)^a \varphi$. The first integral converges to $P_a(0)$ as $\lambda \to \infty$, and for the second we have

$$\left|\lambda^{n+|a|}\int \varphi_a(-\lambda y) R(y)\, dy\right| \leqslant N\lambda^{n+|a|}\int\limits_{|y|\leqslant 1/\lambda} |R(y)|dy \leqslant CN\varepsilon\left(\frac{1}{\lambda}\right)\left(\frac{1}{\lambda}\right)^{u-|a|},$$

where N is a bound for $|\varphi_a|$, $|a| \leqslant u$, and so it tends to zero as $\lambda \to \infty$.

Since $\|f^\lambda - f\|_p$ tends to zero as $\lambda \to \infty$, it remains to show that

$$\sup_\varrho \frac{1}{\varrho^u}\left[\frac{1}{\varrho^n}\int\limits_{|x|\leqslant\varrho} |R^\lambda(x) - R(x)|^p\, dx\right]^{1/p}$$

tends to zero as $\lambda \to \infty$. Let $\eta > 0$ and let m_λ be the sum of the absolute values of the coefficients of $P_\lambda - P$. Then for $\varrho > \eta$ we have

$$\frac{1}{\varrho^u}\left[\frac{1}{\varrho^n}\int_{|x|\leqslant\varrho}|R^\lambda - R|^p dx\right]^{1/p} \leqslant \frac{1}{\eta^{n/p+u}}\|f^\lambda - f\|_p + \frac{1}{\varrho^u}\left[\frac{1}{\varrho^n}\int_{|x|\leqslant\varrho}|P^\lambda - P|^p dx\right]^{1/p}$$

$$\leqslant \frac{1}{\eta^{n/p+u}}\|f^\lambda - f\|_p + \frac{m_\lambda}{\varrho^{n/p+u}}\left[\int_{|x|\leqslant\varrho}(1+|x|^{up})\,dx\right]^{1/p},$$

and the right hand side tends to zero as $\lambda \to \infty$, uniformly in $\varrho > \eta$.

On the other hand, we have

$$R^\lambda(x) = \int\left[\lambda^n\varphi[\lambda(x-y)] - \sum_{|a|\leqslant u}\lambda^{n+|a|}\frac{x^a}{a!}\varphi_a(-\lambda y)\right][P(y)+R(y)]dy.$$

Since $\lambda^n\varphi(\lambda x) * P$ is a polynomial Q, the contribution of P to this integral is Q minus its Taylor expansion at $x = 0$, and so is zero. Thus P may be dropped in the preceding expression. If $\varrho \geqslant 1/\lambda$ we have, using Young's inequality,

$$\left[\int_{|x|\leqslant\varrho}|R^\lambda(x)|^p dx\right]^{1/p} \leqslant \left[\int_{|x|\leqslant 2\varrho}|R(x)|^p dx\right]^{1/p} +$$

$$+\sum_a \lambda^{n+|a|}\left|\int\varphi_a(-\lambda y)R(y)\,dy\right|\left[\int\!\!\int_{|x|\leqslant\varrho}|x|^{|a|p}dx\right]^{1/p}$$

$$\leqslant C_u\varepsilon(2\varrho)\,\varrho^{n/p+u} + C\sum N\varepsilon\left(\frac{1}{\lambda}\right)\left(\frac{1}{\lambda}\right)^{u-|a|}\varrho^{|a|+n/p}$$

$$\leqslant C_u\varepsilon(2\varrho)\,\varrho^{n/p+u}.$$

If $\varrho \leqslant 1/\lambda$ and if N now denotes a bound for $|\varphi_a(x)|$, $|a| = [u+1]$, we have

$$\left|\lambda^n\varphi[\lambda(x-y)] - \sum_{|a|\leqslant u}\lambda^{n+|a|}\frac{x^a}{a!}\varphi_a(-\lambda y)\right| \leqslant C_u N(\lambda|x|)^{[u+1]}\lambda^n,$$

whence

$$|R^\lambda(x)| \leqslant C_u N(\lambda|x|)^{[u+1]}\lambda^n\int_{|y|\leqslant 2/\lambda}|R(y)|\,dy \leqslant C_u\varepsilon\left(\frac{2}{\lambda}\right)\lambda^{[u+1]-u}|x|^{[u+1]},$$

and

$$\left[\int_{|x|\leqslant\varrho}|R^\lambda(x)|^p dx\right]^{1/p} \leqslant C_u\varepsilon\left(\frac{2}{\lambda}\right)\lambda^{[u+1]-u}\varrho^{n/p+[u+1]} \leqslant C_u\varepsilon\left(\frac{2}{\lambda}\right)\varrho^{n/p+u}.$$

This combined with the inequality obtained above gives

$$\frac{1}{\varrho^u}\left[\frac{1}{\varrho^n}\int_{|x|\leqslant\varrho}|R^\lambda(x)|^p dx\right]^{1/p} \leqslant C_u\left[\varepsilon(2\varrho)+\varepsilon\left(\frac{2}{\lambda}\right)\right]$$

for any ϱ and λ. Consequently,

$$\frac{1}{\varrho^u}\left[\frac{1}{\varrho^n}\int_{|x|\leqslant\varrho}|R^\lambda-R|^p dx\right]^{1/p}\leqslant C_u\left[\varepsilon(2\varrho)+\varepsilon(\varrho)+\varepsilon\left(\frac{2}{\lambda}\right)\right]\leqslant C_u\left[\varepsilon(2\varrho)+\varepsilon\left(\frac{2}{\lambda}\right)\right],$$

and thus the left hand side converges to zero uniformly in ϱ as $\lambda\to\infty$. This proves (ii). Part (iii) is a consequence of the fact that functions f in $t_u^p(x_0)$ with compact supports are dense in $t_u^p(x_0)$, and for each such f, f^λ belongs to C_0^∞.

LEMMA 2.4. *Let* $f\in T_u^p(x_0)$, $1\leqslant p\leqslant\infty$, $u\geqslant -n/p$, *and* $g\in T_v(x_0)$, $v\geqslant u$, $v\geqslant 0$. *Then* $fg\in T_u^p(x_0)$ *and* $T_u^p(x_0,fg)\leqslant CT_u^p(x_0,f)\,T_v(x_0,g)$. *If* $f\in t_u^p(x_0)$ *and* $g\in t_v(x_0)$, *then* $fg\in t_u^p(x_0)$.

Proof. If $u\leqslant 0$, our assertion is obvious. Suppose that $u>0$, and let r be the largest integer less than u. In view of Lemma 2.1, we may assume that $v=u$. Then $g(x_0+h)=P_1(h)+R_1(h)$, where $P_1(h)$ is polynomial of degree r and $|R_1(h)|\leqslant T_u(x_0,g)|h|^u$. On the other hand, $f(x_0+h)=P_2(h)+R_2(h)$, where

$$\left[\frac{1}{\varrho^n}\int_{|h|\leqslant\varrho}|R_2(h)|^p dh\right]^{1/p}\leqslant T_u^p(x_0,f)\,\varrho^u.$$

Let now $\bar P$ denote the sum of terms of degree $\leqslant r$ in P_1P_2. Then

$$f(x_0+h)g(x_0+h)=\bar P(h)+(R_1P_2+R_2g+P_1P_2-\bar P)=\bar P+\bar R,$$

say. Since $|g|\leqslant T_u(x_0,g)$ and $P_2(h)\leqslant T_u^p(x_0,f)$ for $|h|\leqslant 1$, and since the sum of the absolute values of the coefficients of P_1P_2 does not exceed $T_u(x_0,g)T_u^p(x_0,f)$, we have $|P_1P_2-\bar P|\leqslant |h|^{r+1}T_u(x_0,g)\,T_u^p(x_0,f)$ for $|h|\leqslant 1$. From this and the estimates for R_1 and R_2 we obtain, for $\varrho\leqslant 1$,

$$\varrho^{-u}\left[\frac{1}{\varrho^n}\int_{|h|\leqslant\varrho}|\bar R(h)|^p dh\right]^{1/p}\leqslant CT_u(x_0,g)\,T_u^p(x_0,f),\qquad \varrho\leqslant 1.$$

For $\varrho\geqslant 1$, $|h|\leqslant\varrho$, we use the inequalities $|\bar R|\leqslant|f|\,|g|+|\bar P|$ and $|\bar P(h)|\leqslant\varrho^r T_u(x_0,g)\,T_u^p(x_0,f)$, and obtain

$$\varrho^{-u}\left[\frac{1}{\varrho^n}\int_{|h|\leqslant\varrho}|\bar R(h)|^p dh\right]^{1/p}\leqslant\varrho^{-u}\|f\|_p\|g\|_\infty+C\varrho^{-u+r}T_u(x_0,g)\,T_u^p(x_0,f)$$
$$\leqslant CT_u(x_0,g)\,T_u^p(x_0,f).$$

The other terms that enter in the norm of fg are also majorized by $CT_u(x_0,g)\,T_u^p(x_0,f)$.

When dealing with the spaces $t_u^p(x_0)$ and $t_u(x_0)$ we take r to be the largest integer less than or equal to u. We then have $|R_1(h)|=o(|h|^u)$,

$|P_1P_2-\bar{P}| = o(|h|^u)$ and

$$\left[\frac{1}{\varrho^n}\int\limits_{|h|\leqslant\varrho}|R_2(h)|^p dh\right]^{1/p} = o(\varrho^u),$$

for $|h|$ and $\varrho \to 0$. From this the desired result follows at once.

LEMMA 2.5. *Let $f \in T_u^p(x_0)$, $1 \leqslant p \leqslant \infty$, $u \geqslant -n/p$, and let $g \in T_v(x_0)$, $v > 0$, $v \geqslant u$, and $g(x_0) = 0$. Then $fg \in T_w^p(x_0)$ and $T_w^p(x_0, fg) \leqslant CT_u^p(x_0, f)\, T_v(x_0, g)$, where*

(i) $w = \min(u+v, v)$ *if* $v \leqslant 1$,

(ii) $w = \min(u+1, v)$ *if* $v \geqslant 1$,

and

(iii) $w = u+1$ *if* $u > 0$, $v \geqslant 1$ *and* $f(x_0) = 0$.

Proof. The argument is parallel to that of the preceding proof. The case $u \leqslant 0$ is immediate. Assume therefore that $u > 0$, and let R_1, P_1, R_2, P_2 be as in the preceding proof, and r the largest integer less than w. Then

$$f(x_0+h)g(x_0+h) = \bar{P}(h)+(R_1P_2+R_2g+P_1P_2-\bar{P}) = \bar{P}+\bar{R},$$

where $\bar{P}$ is the sum of terms of P_1P_2 of degrees less than or equal to r.

We then have the inequalities

$$|g(x_0+h)| \leqslant T_v(x_0, g)|h|^v, \quad \text{if} \quad v \leqslant 1;$$

$$|g(x_0+h)| \leqslant T_v(x_0, g)\max(1, |h|), \quad \text{if} \quad v \geqslant 1;$$

$$|\bar{P}(h)| \leqslant T_v(x_0, g)T_u^p(x_0, f)\max(1, |h|^w).$$

Consequently, if $\varrho \geqslant 1$,

$$\left[\frac{1}{\varrho^n}\int\limits_{|h|\leqslant\varrho}|\bar{R}(h)|^p dh\right]^{1/p} \leqslant \left[\frac{1}{\varrho^n}\int\limits_{|h|\leqslant\varrho}|g(x_0+h)f(x_0+h)|^p dh\right]^{1/p} + $$

$$+\left[\frac{1}{\varrho^n}\int\limits_{|h|\leqslant\varrho}|\bar{P}(h)|^p dh\right]^{1/p} \leqslant CT_v(x_0, g)T_u^p(x_0, f)\varrho^w.$$

If $\varrho \leqslant 1$, we have

$$\left[\frac{1}{\varrho^n}\int\limits_{|h|\leqslant\varrho}|R_2(h)g(x_0+h)|^p dh\right]^{1/p} \leqslant CT_v(x_0, g)T_u^p(x_0, f)\varrho^w.$$

On the other hand, if $|h| \leqslant 1$ we have $|P_2(h)| \leqslant T_u^p(x_0, f)$ in the cases (i) and (ii), and $|P_2(h)| \leqslant T_u^p(x_0, f)|h|$ in the case (iii). Thus, since $|R_1(h)| \leqslant |h|^v T_v(x_0, g)$, we have

$$\left[\frac{1}{\varrho^n}\int\limits_{|h|\leqslant\varrho}|R_1(h)P_2(h)|^p dh\right]^{1/p} \leqslant CT_v(x_0, g)T_u^p(x_0, f)\varrho^w,$$

if $\varrho \leqslant 1$. Finally, since $|P_1P_2-\bar{P}| \leqslant T_v(x_0, g)T_u^p(x_0, f)|h|^{r+1}$ for $|h| \leqslant 1$, we have

$$\left[\frac{1}{\varrho^n}\int\limits_{|h|\leqslant\varrho} |P_1P_2-\bar{P}|^p dh\right]^{1/p} \leqslant CT_v(x_0, g)T_u^p(x_0, f)\,\varrho^w$$

for $\varrho \leqslant 1$. Collecting results we find that

$$\left[\frac{1}{\varrho^n}\int\limits_{|h|\leqslant\varrho} |\bar{R}(h)|^p dh\right]^{1/p} \leqslant CT_v(x_0, f)T_u^p(x_0, f)\,\varrho^w$$

for all ϱ. Since the sum of the moduli of the coefficients of $\bar{P}$ does not exceed $T_u^p(x_0, f)\ T_v(x_0, g)$, and since $\|gf\|_p \leqslant T_u^p(x_0, f)\,T_v(x_0, g)$, the lemma follows.

Lemma 2.6. *Given an integer m, $m \geqslant 0$, there exists a function $\varphi(x)$ infinitely differentiable with support in $|x| \leqslant 1$, such that for every $\lambda > 0$ and every polynomial P of degree $\leqslant m$,*

$$\int \lambda^n \varphi[\lambda(x-y)]P(y)\,dy = P(x)$$

holds.

Proof. Consider the class of all infinitely differentiable functions $\varphi(x)$ supported by $|x| \leqslant 1$, and the mapping of this linear space into the vector space V of points $\{\xi_a\}$, $0 \leqslant |a| \leqslant m$, given by

$$\xi_a = \int \varphi(x)x^a dx.$$

If the range of this mapping is not the entire space V, then there exist numbers η_a, $0 \leqslant |a| \leqslant m$ not all zero, such that

$$\sum \eta_a \xi_a = \int \varphi(x) \sum \eta_a x^a dx = 0$$

for all φ. In particular, if $\psi(x)$ is an infinitely differentiable function supported by $|x| \leqslant 1$ which is positive in $|x| < 1$, and $\varphi(x)$ is $\psi(x)\sum\eta_a x^a$, we obtain

$$\int \psi(x)\left|\sum \eta_a x^a\right|^2 dx = 0.$$

This implies that $\sum\eta_a x^a = 0$ in $|x| < 1$ and consequently $\eta_a = 0$, $0 \leqslant |a| \leqslant m$, which contradicts our assumption. That is, the range of the mapping is all of V, and therefore there exists a function $\varphi(x)$ such that

$$\int \varphi(x)\,dx = 1, \qquad \int \varphi(x)x^a dx = 0, \qquad 0 < |a| \leqslant m.$$

If Q is any polynomial of degree $\leqslant m$, then evidently $\int \varphi(z)Q(z)\,dz = Q(0)$. Given x and λ we change variables in the preceding integral by setting $z = \lambda x - \lambda y$ and replace $Q(z)$ by $Q(z) = P(x - z/\lambda)$, and the desired result follows.

The next proposition seems to be of independent interest.

THEOREM 8. *Let Q be a closed set in E_n and f a function in E_n, such that $f \in T_u^p(x_0)$, $1 \leqslant p \leqslant \infty$, $u > 0$, for all $x_0 \in Q$, with $T_u^p(x_0, f) \leqslant M < \infty$, for $x_0 \in Q$. Then $f \in B_u(Q)$. If in addition $f \in t_u^p(x_0)$ for all $x_0 \in Q$, then $f \in b_u(Q)$.*

Proof. Denote by $f_\alpha(x_0)$ the coefficient of $(x-x_0)^\alpha/\alpha!$ in the expansion of f at x_0, and let $x = x_0 + h$ be another point of Q. Let φ be the function of the preceding lemma with $m > u$, and $\varphi_\alpha(x) = (\partial/\partial x)^\alpha \varphi$. Then if $P(x)$ is any polynomial of degree $\leqslant m$ we have

$$\int \lambda^n \varphi[\lambda(x-y)]P(y)\,dy = P(x),$$

and by differentiating under the integral sign we obtain

$$\int \lambda^{n+|\alpha|}\varphi_\alpha[\lambda(x-y)]P(y)\,dy = \left(\frac{\partial}{\partial x}\right)^\alpha P(x).$$

Consider the case when $f \in T_u^p(x_0)$ for all $x_0 \in Q$. Then we have

$$f(y) = \sum_{|\alpha|<u} \frac{1}{\alpha!} f_\alpha(x_0)(y-x_0)^\alpha + R(x_0, y),$$

$$f(y) = \sum_{|\alpha|<u} \frac{1}{\alpha!} f_\alpha(x)(y-x)^\alpha + R(x, y),$$

where

$$\Big[\int_{|y-x_0|\leqslant\varrho} |R(x_0, y)|^p dy\Big]^{1/p} \leqslant M\varrho^{n/p+u},$$

and similarly for $R(x, y)$. Set now $\lambda^{-1} = |h| = |x-x_0|$ and given β, $0 \leqslant |\beta| < u$, consider the expression

$$I = \int \lambda^{n+|\beta|}\varphi_\beta[\lambda(x-y)]f(y)\,dy.$$

If we replace here $f(y)$ by the first of its expressions above we get

$$I = \sum_{|\alpha|<u} \frac{1}{\alpha!} f_\alpha(x_0)\left(\frac{\partial}{\partial x}\right)^\beta (x-x_0)^\alpha + \int \lambda^{n+|\beta|}\varphi_\beta[\lambda(x-y)]R(x_0, y)\,dy$$

$$= \sum_{|\gamma|<u-|\beta|} \frac{1}{\gamma!} f_{\beta+\gamma}(x_0)(x-x_0)^\gamma + \int \lambda^{n+|\beta|}\varphi_\beta[\lambda(x-y)]R(x_0; y)\,dy.$$

On the other hand, if we use the second expression for $f(y)$, we find

$$I = f_\beta(x) + \int \lambda^{n+|\beta|}\varphi_\beta[\lambda(x-y)]R(x, y)\,dy.$$

Consequently,

$$f_\beta(x) = \sum_{|\gamma|<u-|\beta|} \frac{1}{\gamma!} f_{\beta+\gamma}(x_0)(x-x_0)^\gamma + \int \lambda^{n+|\beta|}\varphi_\beta[\lambda(x-y)][R(x,y)- -R(x_0,y)]\,dy.$$

Since $\varphi(x)$ vanishes for $|x|\geqslant 1$, and $\lambda = |x-x_0|^{-1} = |h|^{-1}$, if N is a bound for $|\varphi_\beta|$ we find that the last integral is dominated by

$$\Big[\int_{|y-x_0|\leqslant 2|h|} |R(x_0,y)|\,dy + \int_{|y-x|\leqslant |h|} |R(x,y)|dy\Big]\,|h|^{-n-|\beta|}N,$$

and an application of Hölder's inequality shows that the last expression is less than or equal to $C_{\beta u}M|h|^{u-|\beta|}$.

A parallel argument gives the desired result in the case $f\in t_u^p(x_0)$. It is not difficult to see that in this case the theorem holds also if $u=0$.

3. The results of this section will only be needed in the proof of theorems 2, 3, 5 and 7. Essentially they are reformulations and slight extensions of theorems of Whitney [14] and Marcinkiewicz [10]. The proof of theorem 9 is taken from [5; II, p. 57].

LEMMA 3.1. *Let Q be a closed set in E_n and U the neighborhood of Q consisting of all points of E_n whose distance from Q is less than* 1. *Then there is a covering of $U-Q$ by means of non-overlapping closed cubes K with the property that $\frac{1}{2}\leqslant d_j/e_j\leqslant 1+\sqrt{n}$, where e_j is the length of the edge of K_j and d_j is the distance between K_j and Q.*

Proof. Consider the subdivisions π_l of E_n into the cubes $m_i/2^l\leqslant x_i \leqslant (m_i+1)/2^l$, $l=0,1,\ldots$ Select all cubes in π_0 which intersect $U-Q$ and are at distance not less than $\frac{1}{2}$ from Q. Having already selected cubes from π_{j-1}, select all cubes in π_j which are not contained in the previously selected cubes which intersect $U-Q$, and which are at distance no less than 2^{-j-1} from Q. The collection K_j of cubes thus obtained has the required properties. First of all, every $x_0\in U-Q$ is contained in one of the K_j; for if d is the distance between x_0 and Q, and $\bar{K}$ is the cube in π_ν, with $d>2^{-\nu+1}\sqrt{n}$, which contains x_0, then the distance between $\bar{K}$ and Q is not less than $2^{-\nu+1}\sqrt{n}-2^{-\nu}\sqrt{n}=2^{-\nu}\sqrt{n}$, whence it follows that if $\bar{K}$ was not selected at the νth step then $\bar{K}$ was contained in one of the previously selected cubes. It is clear that the distance between each K_j and Q is not less than one half the length of the edge of K_j. Finally consider K_j and let $e_j=2^{-\nu}$ be the length of its edge. Let $\bar{K}$ be the cube in $\pi_{\nu-1}$ which contains K_j. Since $\bar{K}$ was not selected the distance between $\bar{K}$ and Q must be less than $2^{-\nu}$, consequently the distance d_j between K_j and Q is less than $2^{-\nu}(1+\sqrt{n})$, whence $d_j/e_j<1+\sqrt{n}$.

LEMMA 3.2. *Let Q and U be as in the preceding lemma, and let $d(x)$ denote the distance between x and Q. Then there exists an infinitely differentiable function $\delta(x)$ defined in $U-Q$ and a positive number ε, $\varepsilon < 1$, such that*

$$d(x)\varepsilon \leqslant \delta(x) \leqslant \frac{1}{\varepsilon} d(x), \quad x \in U-Q,$$

$$\left|\left(\frac{\partial}{\partial x}\right)^{a} \delta(x)\right| \leqslant C_a d(x)^{1-|a|}, \quad x \in U-Q.$$

Proof. Let K_j be the covering of $U-Q$ of the preceding lemma, and $x^{(j)}$ the center of K_j. Let $\eta(x) \geqslant 0$ be an infinitely differentiable function which vanishes outside the cube

$$-\frac{1}{2}-\frac{1}{4\sqrt{n}} \leqslant x_j \leqslant \frac{1}{2}+\frac{1}{4\sqrt{n}}, \quad j = 1, 2, \ldots, n,$$

and which exceeds 1 in $-\frac{1}{2} < x_j \leqslant \frac{1}{2}$. Set

$$\delta(x) = \sum_{j=1}^{\infty} e_j \eta\left(\frac{x-x^{(j)}}{e_j}\right).$$

We shall show that $\delta(x)$ has the required properties. It is readily verified that $\eta\left(\frac{x-x^{(j)}}{e_j}\right)$ vanishes in Q. On the other hand, as we will presently prove, there is an integer m such that no more than m terms of the series are distinct from zero at each point.

For let $x_0 \in U-Q$ and let $\eta\left(\frac{x_0-x^{(j)}}{e_j}\right) \neq 0$. Then the distance between x_0 and K_j does not exceed $e_j/4$ and since $\frac{1}{2}e_j \leqslant d_j \leqslant (1+\sqrt{n})e_j$ we find that $d(x_0)$, that is the distance between x_0 and Q, satisfies the inequalities

$$(1+\sqrt{n})e_j + \frac{e_j}{4} + e_j\sqrt{n} \geqslant d(x_0) \geqslant \frac{e_j}{4},$$

where $e_j\sqrt{n}$ is the diameter of K_j. This means that a sphere with center at x_0 and radius $d(x_0)(1+4\sqrt{n})$ contains all the cubes K_j corresponding to terms of the series not vanishing at x_0. Since these cubes are disjoint and their edges are not less than $d(x_0)[5/4+2\sqrt{n}]^{-1}$, it follows that their number does not exceed a fixed integer m.

Let N_a be a bound for $(\partial/\partial x)^a \eta(x)$. Then differentiating the series term by term we find that

$$\left|\left(\frac{\partial}{\partial x}\right)^{a} \delta(x)\right| \leqslant \sum e_j^{1-|a|} N_a,$$

the sum being extended over all j for which $\eta\left(\frac{x-x^{(j)}}{e_j}\right)$ does not vanish. But for these terms we have $e_j \leqslant 4d(x)$, from which it follows that

$$\left|\left(\frac{\partial}{\partial x}\right)^{a} \delta(x)\right| \leqslant 4mN_{a}\, d(x)^{1-|a|},$$

if $|a| \leqslant 1$; and similarly for $|a| > 1$, since $e_j \geqslant Cd(x)$ (see below). Finally, if $x \epsilon K_j$, then clearly

$$\delta(x) \geqslant e_j \eta\left(\frac{x-x^{(j)}}{e_j}\right) = e_j,$$

and since $x \epsilon K_j$ implies that $d(x) \leqslant d_j + e_j\sqrt{n} \leqslant (1+2\sqrt{n})e_j$, it follows that $\delta(x) \geqslant d(x)(1+2\sqrt{n})^{-1}$. The lemma is thus established.

THEOREM 9. *Let f be a function in L^p, $1 \leqslant p \leqslant \infty$, such that $f \epsilon T_u^p(x_0)$ and $T_u^p(x_0, f) \leqslant M < \infty$ for all x_0 in a closed set Q. Then there exists a function $\bar{f}$ in $B_u(E_n)$ such that, for $|\beta| < u$, $(\partial/\partial x)^{\beta}\bar{f}(x_0) = f_{\beta}(x_0)$ for $x_0 \epsilon Q$. If in addition $f \epsilon t_u^p(x_0)$ for all $x_0 \epsilon Q$, then $\bar{f}$ can be chosen to be in $b_u(E_n)$ in such a way that $(\partial/\partial x)^{\beta}\bar{f}(x_0) = f_{\beta}(x_0)$ for $|\beta| \leqslant u$, and all $x_0 \epsilon Q$.*

Proof. Let U be a neighborhood of Q as in Lemma 3.2 and $\delta(x)$ the corresponding function. For $x \epsilon Q$ define $\bar{f}(x) = f(x)$, and for $x \epsilon U-Q$, set

$$\bar{f}(x) = \delta(x)^{-n} \int \varphi[(x-y)\,\delta^{-1}(x)] f(y)\,dy,$$

where $\varphi(x)$ is the function in Lemma 2.6.

Let us consider the case $f \epsilon T_u^p(x_0)$ for all $x_0 \epsilon Q$. We shall study first the behaviour of $\bar{f}$ near points x in $U-Q$. By differentiation under the integral sign it is readily seen that $\bar{f}(x)$ is infinitely differentiable in $U-Q$. Let x be given and $\bar{x}$ be a point in Q such that $|x-\bar{x}| = d(x)$. Then we can write

$$f(x) = \sum_{|a|<u} \frac{f_a(\bar{x})}{a!}(x-\bar{x})^a + R(\bar{x}, x).$$

Let $\Phi_{\beta}(x, y) = (\partial/\partial x)^{\beta}\{\delta(x)^{-n}\varphi[(x-y)\,\delta(x)^{-1}]\}$. By differentiation under the integral sign we obtain

$$\left(\frac{\partial}{\partial x}\right)^{\beta}\bar{f}(x) = \left(\frac{\partial}{\partial x}\right)^{\beta}\left[\delta(x)^{-n}\int \varphi[(x-y)\,\delta^{-1}(x)]\left[\sum_{|a|<u}\frac{f_a(\bar{x})}{a!}(y-\bar{x})^a\right]dy\right] + \\ + \int \Phi_{\beta}(x, y) R(\bar{x}, y)\,dy.$$

According to Lemma 2.6, the first term on the right is equal to

$$\left(\frac{\partial}{\partial x}\right)^{\beta}\left[\sum_{|a|<u}\frac{f_a(\bar{x})}{a!}(x-\bar{x})^a\right].$$

Consequently we have

$$(3.1)\qquad \left(\frac{\partial}{\partial x}\right)^{\beta} \bar{f}(x) = \sum_{|\gamma|<u-|\beta|} \frac{f_{\beta+\gamma}(\bar{x})}{\gamma!}(x-\bar{x})^{\gamma} + \int \Phi_{\beta}(x, y) R(\bar{x}, y) dy.$$

To estimate the remainder here, we need estimates for $\Phi_\beta(x, y)$. It is readily verified by induction that $\Phi_\beta(x, y)$ is a sum of terms of the form

$$\text{const} \cdot \delta^{-n-r}(x) \varphi_\gamma [(x-y)\delta^{-1}] \cdot p \cdot (x-y)^{\alpha},$$

where p is a product of derivatives of δ, such that if t is the number of factors in p and w the sum of its orders, then $r+w-t-|\alpha| = |\beta|$, and φ_γ is a derivative of φ with $|\gamma| \leqslant |\beta|$. Since $\varphi[(x-y)\delta^{-1}(x)]$ vanishes for $|x-y| > \delta(x)$, we have $|(x-y)^{\alpha}| \leqslant \delta(x)^{|\alpha|}$ on the support of $\Phi_\beta(x, y)$. From this and the estimates for $\delta(x)$ and its derivatives it follows that

$$|\Phi_\beta(x, y)| \leqslant C_\beta d(x)^{-n-|\beta|}.$$

Consequently we have

$$\left|\int \Phi_\beta(x, y) R(\bar{x}, y) dy\right| \leqslant C_\beta d(x)^{-n-|\beta|} \int |R(\bar{x}, y)| dy,$$

where the integral on the right is extended over $|y-x| \leqslant \delta(x)$, which is contained in the sphere $|y-\bar{x}| \leqslant cd(x)$, since $|x-\bar{x}| = d(x)$ and $\delta(x) \leqslant \leqslant \frac{1}{\varepsilon} d(x)$. Now

$$\int_{|y-\bar{x}|\leqslant cd(x)} |R(\bar{x}, y)| dy \leqslant C T_u^p(\bar{x}, f)[cd(x)]^{n+u} \leqslant CM[cd(x)]^{n+u},$$

and from this it follows that

$$(3.2)\qquad \left|\left(\frac{\partial}{\partial x}\right)^{\beta} \bar{f}(x) - \sum_{|\gamma|<u-|\beta|} \frac{f_{\beta+\gamma}(\bar{x})}{\gamma!}(x-\bar{x})^{\gamma}\right| \leqslant C_{\beta u} M |x-\bar{x}|^{u-|\beta|}.$$

Let us consider now any point $\bar{x}_1$, in Q. Given the assumptions on f, it follows from Theorem 8 that

$$\left|f_\alpha(\bar{x}) - \sum_{|\gamma|<u-|\alpha|} \frac{f_{\alpha+\gamma}(\bar{x}_1)}{\gamma!}(\bar{x}-\bar{x}_1)^{\gamma}\right| \leqslant C_{\alpha u} M |\bar{x}-\bar{x}_1|^{u-|\alpha|}.$$

If we replace these values of $f_\alpha(\bar{x})$ in (3.2) we obtain

$$\left|\left(\frac{\partial}{\partial x}\right)^{\beta} \bar{f}(x) - \sum_{|\gamma|<u-|\beta|} \frac{(x-\bar{x})^{\gamma}}{\gamma!} \sum_{|\eta|<u-|\beta+\gamma|} \frac{f_{\beta+\gamma+\eta}(\bar{x}_1)}{\eta!}(\bar{x}-\bar{x}_1)^{\eta}\right|$$

$$\leqslant C_{\beta u} M |x-\bar{x}|^{u-|\beta|} + \sum_{|\gamma|<u-|\beta|} C_{\beta+\gamma, u} M |x-\bar{x}|^{|\gamma|} |\bar{x}-\bar{x}_1|^{u-|\beta|-|\gamma|}.$$

Since $|x-\bar{x}| \leqslant |x-\bar{x}_1|$ and $|\bar{x}-\bar{x}_1| \leqslant |\bar{x}-x|+|x-\bar{x}_1| \leqslant 2|x-\bar{x}_1|$, using Taylor's expansion for polynomials this inequality gives

$$(3.3)\qquad \left|\left(\frac{\partial}{\partial x}\right)^{\beta} \bar{f}(x) - \sum_{|\gamma|<u-|\beta|} \frac{f_{\beta+\gamma}(\bar{x}_1)}{\gamma!}(x-\bar{x}_1)^{\gamma}\right| \leqslant C_{\beta u} M |x-\bar{x}_1|^{u-|\beta|},$$

which holds for any $x \in U-Q$ and any $\bar{x}_1 \in Q$. By Theorem 8 the inequality also holds if x and $\bar{x}_1$ belong both to Q. This shows that the functions $(\partial/\partial x)^{\beta}\bar{f}(x)$, $x \in U-Q$, are continuous bounded extensions of the functions $f_{\beta}(x)$, $x \in Q$, and that these extensions have a Taylor expansion at each point of Q. Since $\bar{f}$ is infinitely differentiable in $U-Q$, it follows that $\bar{f}$ has continuous bounded derivatives of orders less than u in U. It remains to show that the highest order derivatives of $\bar{f}$ satisfy a Hölder condition of appropriate order.

Let r be the largest integer less than u. Then, if $|\beta| = r$ and $x_1 \in Q$, the inequality (3.3) gives

$$\left|\left(\frac{\partial}{\partial x}\right)^{\beta} \bar{f}(x_1) - \left(\frac{\partial}{\partial x}\right)^{\beta} \bar{f}(x_2)\right| \leqslant C_u M |x_1-x_2|^{u-r}.$$

If both x_1 and x_2 belong to $U-Q$, we distinguish two cases. First, if $|x_1-x_2| \geqslant d(x_1)/2$, then if $\bar{x} \in Q$ is such that $|x_1-\bar{x}| \leqslant 2|x_1-x_2|$, we have $|x_2-\bar{x}| \leqslant |x_2-x_1|+|x_1-\bar{x}| \leqslant 3|x_1-x_2|$; and from

$$\left|\left(\frac{\partial}{\partial x}\right)^{\beta} \bar{f}(x_1) - \left(\frac{\partial}{\partial x}\right)^{\beta} \bar{f}(\bar{x})\right| \leqslant C_u M |x_1-\bar{x}|^{u-r},$$

$$\left|\left(\frac{\partial}{\partial x}\right)^{\beta} \bar{f}(x_2) - \left(\frac{\partial}{\partial x}\right)^{\beta} \bar{f}(\bar{x})\right| \leqslant C_u M |x_2-\bar{x}|^{u-r}$$

we obtain

$$\left|\left(\frac{\partial}{\partial x}\right)^{\beta} \bar{f}(x_1) - \left(\frac{\partial}{\partial x}\right)^{\beta} \bar{f}(x_2)\right| \leqslant C_u M |x_2-x_1|^{u-r}.$$

If, on the other hand, $|x_1-x_2| < d(x_1)/2$, we let $\bar{x}$ be a point in Q such that $|x_1-\bar{x}| = d(x_1)$, and represent $(\partial/\partial x)^{\beta}\bar{f}$ by means of (3.1). Using the mean value theorem we obtain

$$\left|\left(\frac{\partial}{\partial x}\right)^{\beta} \bar{f}(x_1) - \left(\frac{\partial}{\partial x}\right)^{\beta} \bar{f}(x_2)\right| \leqslant |x_1-x_2| \sum_{j=1}^{n} \int \left|\frac{\partial}{\partial x_j} \Phi(x_0, y)\right| |R(\bar{x}, y)|\, dy,$$

where x_0 is a point of the segment x_1, x_2.

Now $\partial\Phi(x_0, y)/\partial x_j$ vanishes outside the sphere $|y-x_0| \leqslant c d(x_0)$ and is dominated in absolute value by $C_{\beta} d(x_0)^{-n-|\beta|-1} = C_{\beta} d(x_0)^{-n-r-1}$, whence

$$(3.4)\qquad \left|\left(\frac{\partial}{\partial x}\right)^{\beta} f(x_1) - \left(\frac{\partial}{\partial x}\right)^{\beta} \bar{f}(x_2)\right|$$

$$\leqslant C_{\beta} |x_1-x_2| d(x_0)^{-n-r-1} \int_{|y-x_0| \leqslant c d(x_0)} |R(\bar{x}, y)|\, dy.$$

Since $d(x_0) \geqslant d(x_1) - |x_1 - x_2| > d(x_1)/2 > |x_1 - x_2|$, the sphere $|y - x_0| \leqslant cd(x_0)$, is contained in the sphere $|y - \bar{x}| \leqslant cd(x_0) + |x_0 - \bar{x}| \leqslant cd(x_0) + |x_1 - x_2| + d(x_1) \leqslant (c+3)d(x_0)$, and therefore

$$\int_{|y-x_0|\leqslant cd(x_0)} R(\bar{x}, y)\,dy \leqslant \int_{|y-\bar{x}|\leqslant (c+3)d(x_0)} |R(\bar{x}, y)|\,dy \leqslant M[(c+3)d(x_0)]^{n+u}.$$

This, combined with $d(x_0) > |x_1 - x_2|$ shows that the left-hand side of (3.4) is dominated by $C_u M |x_1 - x_2|^{u-r}$.

The function $\bar{f}$ as constructed above is defined only in U; finding an $\bar{f}$ which is defined everywhere now offers little difficulty: we merely multiply the $\bar{f}$ already obtained by an infinitely differentiable function $\eta(x)$ with bounded derivatives of all orders and which is equal to 1 for $d(x) \leqslant \frac{1}{2}$ and vanishes for $d(x) \geqslant \frac{3}{4}$.

The case when $f \in t_u^p(x_0)$ is treated in exactly the same way, and further explanations do not seem to be necessary.

COROLLARY. *Let $f \in T_u^p(x_0)$, $T_u^p(x_0, f) \leqslant M < \infty$, $1 \leqslant p < \infty$, for all x_0 in a closed set Q. Then $f = f_1 + f_2$, where $f_1 \in B_u(E_n)$ and $f_2 \in T_u^p(x_0)$ for all $x_0 \in Q$, and*

$$\left[\frac{1}{\varrho^n} \int_{|x-x_0|\leqslant\varrho} |f_2(x)|^p\,dx\right]^{1/p} \leqslant C_u M \varrho^u$$

for all $x_0 \in Q$ and all $\varrho > 0$. If, in addition, $f \in t_u^p(x_0)$ for all $x_0 \in Q$, then the left-hand side of the inequality above is $o(\varrho^u)$ as $\varrho \to 0$.

This is merely a reformulation of the preceding theorem with $f_1 = \bar{f}$.

THEOREM 10. *Let $f \in L^p$, $1 \leqslant p \leqslant \infty$, be such that*

$$\left[\frac{1}{\varrho^n} \int_{|x-x_0|\leqslant\varrho} |f(x)|^p\,dx\right]^{1/p} = O(\varrho^u), \quad u > 0,$$

for all x_0 in a measurable set S. Then

$$\text{(i)} \int \frac{|f(x)|}{|x-x_0|^{n+u}}\,dx < \infty, \quad \text{(ii)} \left[\frac{1}{\varrho^n} \int_{|x-x_0|\leqslant\varrho} |f(x)|^p\,dx\right]^{1/p} = o(\varrho^u), \quad \varrho \to 0,$$

for almost all x_0 in S.

Proof. We may assume without loss of generality that the set S is bounded. Given $\varepsilon > 0$ we can find a closed subset Q of S such that $S - Q$ has measure less than ε and that

$$\left[\frac{1}{\varrho^n} \int_{|x-x_0|\leqslant\varrho} |f(x)|^p\,dx\right]^{1/p} \leqslant M\varrho^u, \quad M < \infty,$$

for all x_0 in Q and all $\varrho > 0$. Let U be a neighborhood of Q and K_j a covering of $U - Q$ as in Lemma 3.1. Let $d(x)$ denote the distance between

x and Q. Since the distance between the complement $\bar{U}$ of U and Q is 1, it follows that

$$\int_{\bar{U}} \frac{|f(x)|}{|x-x_0|^{n+u}}\, dx < \infty$$

for all $x_0 \in Q$. On the other hand,

$$(3.5) \qquad \int_Q dx_0 \int_{\bar{U}} \frac{|f(x)|}{|x-x_0|^{n+u}}\, dx = \int_Q dx_0 \sum_j \int_{K_j} \frac{|f(x)|}{|x-x_0|^{n+u}}\, dx$$
$$= \sum_j \int_{K_j} |f(x)| \left[\int_Q \frac{dx_0}{|x-x_0|^{n+u}}\right] dx.$$

If e_j denotes the edge of K_j then, according to Lemma 3.1, $|x-x_0| \geqslant e_j/2$. Further, since the distance between K_j and Q does not exceed $(1+\sqrt{n})e_j$, if $\bar{x}$ in Q is within that distance from K_j we have, setting $\varrho = (1+2\sqrt{n})e_j$,

$$\int_{K_j} |f(x)|\, dx \leqslant e_j^{n/q} \Big[\int_{K_j} |f(x)|^p\, dx\Big]^{1/p} \leqslant e_j^{n/q} \Big[\int_{|x-\bar{x}|\leqslant \varrho} |f(x)|^p\, dx\Big]^{1/p}$$
$$\leqslant e_j^{n/q} [(1+2\sqrt{n})e_j]^{n/p+u} M, \quad \text{where} \quad q = \frac{p}{p-1}.$$

Consequently,

$$\int_{K_j} |f(x)|\, dx \leqslant C M e_j^{n+u},$$

and since $|x-x_0| \geqslant e_j/2$ for $x \in K_j$,

$$\int_Q \frac{dx_0}{|x-x_0|^{n+u}} \leqslant C_u e_j^{-u},$$

which combined with the previous inequality gives

$$\int_{K_j} |f(x)| \left[\int_Q \frac{dx_0}{|x-x_0|^{n+u}}\right] dx \leqslant C_u M e_j^n.$$

Summing over j we find that the left-hand side of (3.5) is less than $C_u M \sum_j e_j^n = C_u M |U-Q|$, which is finite. Hence the inner integral there is finite for almost all $x_0 \in Q$, and (i) is established.

To prove (ii) in the case $p < \infty$, we merely apply (i) to the function $g = |f|^p$ and conclude that for almost all x_0 in Q we have

$$\int \frac{g(x)}{|x-x_0|^{n+pu}}\, dx = \int \frac{|f(x)|^p}{|x-x_0|^{n+pu}}\, dx < \infty.$$

At every point where this holds (ii) is evidently valid. If $p = \infty$, then $|f(x)| \leqslant M d(x)^u$, and (ii) is satisfied at every point of density of Q.

4. We pass to the study of the properties of the fractional integration introduced in definition 5.

LEMMA 4.1. *Let* $0 < u < n+1$, *and let* f *be a tempered distribution. Then* $J^u f = G_u * f$, *where*

$$G_u(x) = \gamma(u) e^{-|x|} \int_0^{-\infty} e^{-|x|t}\left(t+\frac{t^2}{2}\right)^{(n-u-1)/2} dt,$$

$$\gamma(u)^{-1} = (2\pi)^{(n-1)/2} 2^{u/2} \Gamma\left(\frac{u}{2}\right) \Gamma\left(\frac{n-u+1}{2}\right).$$

Proof. The inverse Fourier transform of $(1+4\pi^2|x|^2)^{-u/2}$, $u > n \geqslant 2$, can be calculated in polar coordinates by

$$\int_{E_n} (1+4\pi^2|y|^2)^{-u/2} e^{2\pi i(x\cdot y)} dy = \int_\Sigma \int_0^\infty (1+4\pi^2\varrho^2)^{-u/2} e^{2\pi i r\varrho\cos\theta} \varrho^{n-1} d\varrho\, d\sigma,$$

where $|y| = \varrho$, $|x| = r$, $x\cdot y = r\varrho\cos\theta$, Σ is the sphere $|y| = 1$ and $d\sigma$ is the element of area of Σ. If we set

$$\varphi(s) = \int_0^\pi e^{is\cos\theta}(\sin\theta)^{n-2} d\theta,$$

and denote by ω_n the area of the unit sphere in E_n, the last integral becomes

$$\omega_{n-1}\int_0^\infty (1+4\pi^2\varrho^2)^{-u/2}\varrho^{n-1}\varphi(2\pi r\varrho)\, d\varrho.$$

Using successively the formulas (6) page 48, (2) page 434 and (4) page 172 of [13], and setting $t = s+1$ in the integral in the last formula we find that the Fourier transform of $(1+4\pi^2|x|^2)^{-u/2}$ is the function $G_u(x)$ of the lemma, provided $u < n+1$.

Consider now the function $G_u(x)$ for $0 < u < n+1$. Then, since

$$e^{-|x|}\int_0^\infty e^{-|x|t}\left(t+\frac{t^2}{2}\right)^{(n-u-1)/2} dt$$

$$\leqslant e^{-|x|}\int_0^1 \left(t+\frac{t^2}{2}\right)^{(n-u-1)/2} dt + C_u e^{-|x|}\int_1^\infty e^{-|x|t} t^{n-u-1} dt$$

$$\leqslant C_u e^{-|x|}(1+|x|^{-n+u}+|\log|x||),$$

it follows that $G_u(x)$ is integrable for $0 < u < n+1$. In addition $G_u(x)$ is, for each fixed x, an analytic function of u in $0 < R(u) < n+1$, which in a neighborhood of each u of the strip is majorized by an integrable function of x independent of u. From this it follows that the Fourier transform of $G_u(x)$ is also an analytic function of u in $0 < R(u) < n+1$, and consequently it coincides with $(1+4\pi^2|x|^2)^{-u/2}$ there. The assertion of the lemma follows now from convolution theorem for distributions.

The formula for G_u is still valid in the case $n = 1$, for $0 < u < 2$. This follows as before if we use the formulas (28), page 14, of $[2_{\text{I}}]$ and formula (19), page 82, of $[3_{\text{II}}]$.

LEMMA 4.2. *The function $G_u(x)$ of Lemma 1 is non-negative, has integral over E_n equal to 1 and satisfies the following inequalities:*

$$G_u(x) \leqslant C_u e^{-|x|}(1+|x|^{-n+u}), \qquad \textit{for} \qquad 0 < u < n;$$

$$G_u(x) \leqslant Ce^{-|x|}\left(1+\overset{+}{\log}\frac{1}{|x|}\right), \qquad \textit{for} \qquad u = n;$$

$$\left|\left(\frac{\partial}{\partial x}\right)^a G_u(x)\right| \leqslant C_{u,a} e^{-|x|}(1+|x|^{-n+u-|a|}), \qquad |a| > 0, \qquad 0 < u < n+1.$$

Proof. We have

$$G_u(x) = \gamma(u)e^{-|x|}\int_0^\infty e^{-|x|t}\left(t+\frac{t^2}{2}\right)^{(n-u-1)/2} dt$$

$$\leqslant C_u e^{-|x|}\left[\int_0^1 \left(t+\frac{t^2}{2}\right)^{(n-u-1)/2} dt + \int_1^\infty e^{-|x|t} t^{n-u-1} dt\right],$$

from which the first two inequalities follow. Differentiating the expression for $G_u(x)$ we find by induction that $(\partial/\partial x)^a G_u(x)$ is a sum of terms of the form

$$e^{-|x|} g_r(x) \int_0^\infty e^{-|x|t} t^s \left(t+\frac{t^2}{2}\right)^{(n-u-1)/2} dt$$

where $g_r(x)$ is a homogeneous function of degree $-r$ and $r+s \leqslant |a|$. The desired estimate follows now by decomposing the integral as before.

Definition 8. Let $\mathcal{R}_j$ be the operator on L^p defined by

$$\mathcal{R}_j f = -i\pi^{-(n+1)/2}\Gamma\left(\frac{n+1}{2}\right)\lim_{\varepsilon\to 0}\int_{|x-y|>\varepsilon}\frac{x_j-y_j}{|x-y|^{n+1}} f(y)dy.$$

Then Λ is the operator on L_1^p given by

$$\Lambda f = i\sum_{j=1}^n \mathcal{R}_j \frac{\partial}{\partial x_j} f.$$

LEMMA 4.3. *If $f \in L^p$, $1 \leqslant p < \infty$, $\mathcal{R}_j f$ is defined almost everywhere as an ordinary limit. The operation $\mathcal{R}_j$ transforms L_k^p continuously into*

L^p_k, $k \geqslant 0, 1 < p < \infty$. *The operation* Λ *transforms* L^p_{k+1} *continuously into* L^p_k, $k \geqslant 0$, $1 < p < \infty$. *Furthermore we have*

$$\frac{\partial}{\partial x_j}\mathcal{R}_k = \mathcal{R}_k\frac{\partial}{\partial x_j}; \qquad \frac{\partial}{\partial x_j} = -i\mathcal{R}_j\Lambda.$$

Proof. According to [7], Theorem 1, $\mathcal{R}_j$ is defined almost everywhere and is continuous in L^p, and transforms L^p_k into L^p_k. The continuity of $\mathcal{R}_j$ in L^p_k is an immediate consequence of the fact that

$$\frac{\partial}{\partial x_i}\mathcal{R}_j f = \mathcal{R}_j\frac{\partial}{\partial x_i} f \quad \text{for} \quad f \in L^p_1$$

which was established loc. cit. The identity $\partial/\partial x_j = -i\mathcal{R}_j\Lambda$ was proved in [7], p. 309. Since evidently $\partial/\partial x_j$ maps L^p_{k+1} continuously in L^p_k, the same holds for Λ in view of its definition.

LEMMA 4.4. *Let* $f \in L^p$, $1 \leqslant p \leqslant \infty$, *and let* $g_j = \partial G_1/\partial x_j$, *then the ordinary limit*

$$\lim_{\varepsilon\to 0}\int_{|x-y|>\varepsilon} g_j(x-y)f(y)\,dy$$

exists for almost all x.

Proof. Differentiating with respect to x_j the expression for G_1 in Lemma 4.1 and setting $s = t+1$ in the integral one obtains

$$\frac{\partial}{\partial x_j}G_1(x) = g_j(x) = -\gamma(1)2^{(2-n)/2}\frac{x_j}{|x|}\int_1^\infty e^{-|x|s}s(s^2-1)^{(n-2)/2}\,ds.$$

Now, for $s \geqslant 1$ we have $s(s^2-1)^{(n-2)/2} = s^{n-1}+O(s^{n-2})$ and consequently,

$$g_j(x) = C\frac{x_j}{|x|^{n+1}}+r(x),$$

where $r(x) = O(|x|^{-n+1})$ as $|x|\to 0$ if $n > 1$, or $r(x) = O(\log|x|)$ if $n = 1$. In either case r is locally integrable.

We may assume that f has compact support. Then

$$\int_{|x-y|>\varepsilon} g_j(x-y)f(y)\,dy = C\int_{|x-y|>\varepsilon}\frac{x_j-y_j}{|x-y|^{n+1}}f(y)\,dy+\int_{|x-y|>\varepsilon} r(x-y)f(y)\,dy.$$

The second integral on the right is absolutely convergent for $\varepsilon = 0$ and almost all x. On the other hand, according to Lemma 4.3, the first integral has a limit as ε tends to zero for almost all x.

LEMMA 4.5. *If* $f \in L^2_1$, *then* $\widehat{\Lambda f} = 2\pi|x|\hat{f}$.

Proof. It is well known, see e. g., [9] or [7], p. 914, that if $g \in L^2$, then

$$\widehat{\mathcal{R}_j g} = -\frac{x_j}{|x|} g,$$

and this combined with the definition of Λ gives the desired identity.

In what follows we will write systematically J for J^1.

LEMMA 4.6. *The operator J transforms L^p continuously into L_1^p, $1 < p < \infty$. Furthermore for $m \geqslant 0$ we have*

$$\Lambda J = I + a_1 J^2 + a_2 J^4 + \ldots + a_m J^{2m} - Q_m,$$

where $a_j < 0$ for all j, $\sum_{j=1}^{\infty} a_j = -1$ and the operation Q_m is convolution with a positive integrable function with derivatives which are integrable up to order $2m+1$, and bounded and continuous up to order $2m+1-n$, if $2m+1 \geqslant n$. In particular, $I - \Lambda J$ is convolution with a positive integrable function of integral 1, and with integrable first order derivatives.

Proof. Let f be a function in C_0^∞, that is, f is infinitely differentiable and has compact support. According to Lemmas 4.1 and 4.2, $Jf = G_1 * f$, where G_1 is a positive integrable function of integral equal to 1. By differentiating $G_1 * f$ under the integral sign it follows that Jf is in L_1^2. Consequently, by Lemma 4.5, we have

$$\widehat{\Lambda J f} = 2\pi |x| (1 + 4\pi^2 |x|^2)^{-1/2} \hat{f}.$$

Set $u = (1 + 4\pi^2 |x|^2)^{-1/2}$. Then

$$2\pi |x| (1 + 4\pi^2 |x|^2)^{-1/2} = \sqrt{1 - u^2} = 1 + a_1 u^2 + a_2 u^4 + \ldots + a_m u^{2m} + \ldots$$
$$= 1 + a_1 u^2 + \ldots + a_m u^{2m} - R_m(u).$$

The coefficients a_j are all negative and $\sum_{j=1}^{\infty} a_j = -1$; and $0 \leqslant R_m(u) \leqslant u^{2m+2}$ for $0 \leqslant u \leqslant 1$. Consequently

$$1 - 2\pi |x| (1 + 4\pi^2 |x|^2)^{-1/2} = -\sum_{j=1}^{\infty} a_j [(1 + 4\pi^2 |x|^2)^{-1/2}]^{2j},$$

and this shows that $I - \Lambda J$ is convolution with a positive integrable function of integral equal to 1. Furthermore,

$$R_m[(1 + 4\pi^2 |x|^2)^{-1/2}] = -\sum_{j=m+1}^{\infty} a_j [(1 + 4\pi^2 |x|^2)^{-1/2}]^{2j},$$

and this in turn shows that Q_m is also a convolution with a positive integrable function h_m. We shall now show that h_m has the properties stated in the lemma.

We have in fact $\hat{h}_m(x) = R_m(u) \leqslant u^{2m+2}$, where $u = (1+4\pi^2|x|^2)^{-1/2}$. Consequently $x^a \hat{h}_m(x)$ is integrable for $|a| \leqslant 2m+1-n$ and therefore $h_m(x)$ has continuous bounded derivatives up to order $2m+1-n$. Further, from the last inequality of Lemma 4.2, it follows that $\partial G_u(x)/\partial x_j$ is integrable for $1 < u < n+1$, that is $x_j(1+4\pi^2|x|^2)^{-u/2}$ has an integrable inverse Fourier transform. Since $(1+4\pi^2|x|^2)^{-1/2}$ also has an integrable Fourier transform we find that the inverse Fourier transform of $x^a(1+4\pi^2|x|^2)^{-m-1}$ is integrable for $|a| \leqslant 2m+1$. From this it follows that $(\partial/\partial x)^a h_m$, whose Fourier transform is

$$(2\pi i x)^a (1+4\pi^2|x|^2)^{-m-1} \sum_{m+1}^{\infty} a_j (1+4\pi^2|x|^2)^{-j+m+1},$$

is integrable for $|a| \leqslant 2m+1$.

It remains to show that J transforms L^p into L_1^p. If $f \in C_0^\infty$, then $Jf = G_1 * f \in L_1^p$ as we already saw, and according to Lemma 4.3, $\partial Jf/\partial x_j = -i\mathcal{R}_j \Lambda Jf$. Since $\Lambda J - I$ is convolution with an integrable function and $\mathcal{R}_j$ is continuous in L^p, $1 < p < \infty$, it follows that $\|\partial Jf/\partial x_j\|_p \leqslant C_p \|f\|_p$. If f is now a function in L^p and f_n is a sequence of functions in C_0^∞ converging to f in L^p, Jf_n converges in L_1^p. This shows that $Jf \in L_1^p$.

LEMMA 4.7. *If m is an integer, $m \geqslant -k$, and $1 < p < \infty$ then J^m. transforms L_k^p continuously onto L_{k+m}^p.*

Proof. Since J transforms L^p continuously into L_1^p (see Lemma 4.6) and since $(\partial/\partial x)^a J = J(\partial/\partial x)^a$, as seen by taking Fourier transforms, it follows that J transforms L_k^p continuously into L_{k+1}^p. On the other hand, since $J^{-1} = (I-\Delta)J$ and since $(1-\Delta)$ maps L_{k+2}^p continuously into L_k^p we find that J^{-1} maps L_{k+1}^p continuously into L_k^p. From this the lemma follows.

LEMMA 4.8. *If $f \in L_k^p$ then $f \in L^q$ with $1/q = 1/p - k/n$ if $1 \leqslant p < n/k$, or q is any number $p \leqslant q < \infty$ if $p = n/k$, or $q = \infty$ if $p > n/k$.*

Proof. The case $1 \leqslant p < n/k$ is an immediate consequence of Soboleff's theorem which also holds for $p = 1$ (see [11]). The case $p = n/k$ follows from the fact that $f = J^k g = G_k * g$, where $g \in L^p$, from the inequalities for G_k given in Lemma 4.2 and Young's theorem on convolutions. The case $p > n/k$ is obtained by applying Hölder's inequality to $G_k * g$.

Proof of Theorem 4. Let $f \in T_u^p(x_0)$ and assume for simplicity that $x_0 = 0$. Then $f = P + R$ where P is a polynomial of degree $< u$ if $u > 0$, or zero if $u \leqslant 0$ and $R(x)$ is such that

$$\left[\frac{1}{\varrho^n} \int_{|x| \leqslant \varrho} |R(x)|^p dx\right]^{1/p} \leqslant T_u^p(x_0, f)\, \varrho^u. \tag{4.1}$$

We shall first consider the case where $0 < v < n$. We have

$$J^v f = G_v * f = G_v * P + G_v * R.$$

Since, according to Lemma 4.2, $G_v(x)$ decreases exponentially at infinity, both convolutions on the right of the preceding equation are meaningful. Furthermore, by differentiating under the integral sign one sees readily that $G_v * P$ is a polynomial of degree $< u$ whose coefficients are dominated in absolute value by $C_{u,v} T_u^p(x_0, f)$.

Consider now the integrals

$$\int_{|x|\leqslant\varrho} |R(x)|\, |x|^{-r} dx, \qquad \int_{|x|\geqslant\varrho} |R(x)|\, |x|^{-r} dx.$$

If we set

$$\varphi(\varrho) = \int_{|x|\leqslant\varrho} |R(x)|\, dx$$

and use Hölder's inequality and (4.1) we obtain

$$\varphi(\varrho) \leqslant C\Big[\int_{|x|\leqslant\varrho} |R(x)|^p dx\Big]^{1/p} \varrho^{\frac{n(p-1)}{p}} \leqslant C T_u^p(x_0, f)\, \varrho^{n+u}.$$

Hence, if $n+u-r>0$

$$\int_{\varepsilon<|x|<\varrho} |R(x)|\, |x|^{-r} dx = \int_\varepsilon^\varrho s^{-r} d\varphi(s) \leqslant r\int_0^\varrho \varphi(s) s^{-r-1} ds + \varphi(\varrho)\varrho^{-r}$$

$$\leqslant C T_u^p(x_0, f)\Big[r\int_0^\varrho s^{n+u-r-1} ds + \varrho^{n+u-r}\Big] \leqslant C_{r,u} T_u^p(x_0, f)\, \varrho^{n+u-r},$$

that is, if $n+u-r>0$, then

$$\int_{|x|\leqslant\varrho} |R(x)|\, |x|^{-r} dx \leqslant C_{r,u} T_u^p(x_0, f)\, \varrho^{n+u-r}. \tag{4.2}$$

Similarly, if $n+u-r<0$, then

$$\int_{|x|\geqslant\varrho} |R(x)|\, |x|^{-r} dx = \int_\varrho^\infty s^{-r} d\varphi(s) \leqslant r\int_\varrho^\infty s^{-r-1}\varphi(s)\, ds,$$

so that

$$\int_{|x|\geqslant\varrho} |R(x)|\, |x|^{-r} dx \leqslant C_{r,u} T_u^p(x_0, f)\, \varrho^{n+u-r}. \tag{4.3}$$

Let us write g for G_v and g_a for $(\partial/\partial x)^a G_v$, and assume that $2|x|\leqslant\varrho$. Then, by the mean-value theorem,

$$(G_v * R)(x) = g*R = \int g(x-y)R(y)\,dy = \int_{|y|\leqslant\varrho} g(x-y)R(y)\,dy +$$

$$+\sum_{|a|\leqslant u+v}\frac{x^a}{a!}\int g_a(-y)R(y)\,dy - \sum_{|a|\leqslant u+v}\frac{x^a}{a!}\int_{|y|\leqslant\varrho} g_a(-y)R(y)\,dy + \tag{4.4}$$

$$+\sum_{|a|=[u+v]+1}\frac{x^a}{a!}\int_{|y|>\varrho} g_a(\Theta x-y)R(y)\,dy \ (^1),$$

(1) If $u+v<0$ we merely decompose the integral of $g(x-y)R(y)$ into two, extended over $|y|<\varrho$ and $|y|>\varrho$ respectively, and the argument simplifies.

where $0 < \Theta < 1$. Now, according to Lemma 4.1, $|g_a(-y)| \leqslant C_{a,v}[1 + |y|^{-n+v-|a|}]e^{-|y|} \leqslant C_{av}|y|^{-n+v-|a|}$ whence setting $r = n - v + |a|$ in (4.2) it follows that for $|a| < u + v$ the integral

$$\int g_a(-y)R(y)\,dy \tag{4.5}$$

is absolutely convergent and is dominated in absolute value by $C_{uv}T_u^p(x_0, f)$. Furthermore

$$\Big| \int_{|y|\leqslant \varrho} g_a(-y)R(y)\,dy \Big| \leqslant C_{uv}T_u^p(x_0, f)\,\varrho^{u+v-|a|} \tag{4.6}$$

for $|a| \leqslant u + v$.

For the functions $g_a(\Theta x - y)$, $|a| = [u+v]+1$, we have $|g_a(\Theta x - y)| \leqslant C_{av}|\Theta x - y|^{-n+v-|a|} \leqslant C_{av}|y|^{-n+v-|a|}$ if $|y| \geqslant \varrho \geqslant 2|x|$. Consequently it follows from (4.3) on setting $r = n - v + |a|$,

$$\Big| \int_{|y|>\varrho} g_a(\Theta x - y)R(y)\,dy \Big| \leqslant C_{u,v}\,T_u^p(x_0, f)\,\varrho^{u+v-|a|}. \tag{4.7}$$

It remains to estimate the first term of the right-hand side of (4.4). If $1/p - v/n < 0$ the inequality $|g(y)| \leqslant C_v|y|^{-n+v}$, (4.1) and Hölder's inequality give, with $q = p/(p-1)$,

$$\begin{aligned} \Big| \int_{|y|\leqslant\varrho} g(x-y)R(y)\,dy \Big| &\leqslant \Big[\int_{|y|\leqslant\varrho} |R(y)|^p dy \Big]^{1/p} \Big[\int_{|y|\leqslant 2\varrho} |g(y)|^q dy \Big]^{1/q} \\ &\leqslant C_{pv}T_u^p(x_0, f)\,\varrho^{u+v}. \end{aligned} \tag{4.8}$$

If, on the other hand, $1/p - v/n > 0$, replacing $g(x-y)$ by $C_v|y|^{-n+v}$ and applying Soboleff's theorem with $1/q = 1/p - v/n$ we obtain

$$\begin{aligned} \Big[\int dx \Big| \int_{|y|\leqslant\varrho} g(x-y)R(y)\,dy \Big|^q \Big]^{1/q} &\leqslant C_{pv}\Big[\int_{|y|\leqslant\varrho} |R(y)|^p dy \Big]^{1/p} \\ &\leqslant C_{pv}T_u^p(x_0, f)\,\varrho^{n/p+u} = C_{pv}T_u^p(x_0, f)\,\varrho^{n/q+u+v}. \end{aligned} \tag{4.9}$$

It now follows from the estimates (4.6) to (4.9) that the assertion of the theorem is valid if $f \in T_u^p(x_0)$ with $1/q = 1/p - v/n$ if $1/p - v/n > 0$, or $q = \infty$ if $1/p - v/n < 0$, provided $0 < v < n$. This result can now be extended to general v by repeated application of the case $0 < v < n$ using the group properties of J^v.

To cover the case $1/p = v/n$ and the other values of q in the other cases we argue as follows.

Suppose that $f \in T_u^p(x_0)$ and $J^v f \in T_{u+v}^r(x_0)$ with $r \geqslant p$. Then $J^v f \in T_{u+v}^s(x_0)$ for all s, $p \leqslant s \leqslant r$, and

$$T_{u+v}^s(x_0, J^v f) \leqslant C[T_{u+v}^r(x_0, J^v f) + \|f\|_p].$$

This is an immediate consequence of the inequalities

$$\left[\frac{1}{\varrho^n}\int_{|x-x_0|\leqslant\varrho}|R(x)|^s dx\right]^{1/s}\leqslant C\left[\frac{1}{\varrho^n}\int_{|x-x_0|\leqslant\varrho}|R(x)|^r dx\right]^{1/r},$$

$$\|J^v f\|_s\leqslant\|J^v f\|_r^{\Theta}\|J^v f\|_p^{1-\Theta}\leqslant\|J^v f\|_r^{\Theta}\|f\|_p^{1-\Theta}\leqslant\|J^v f\|_r+\|f\|_p,$$

which are obtained by applying Hölder's inequality, and where R is the remainder of the expansion if $J^v f$ at x_0, and $0\leqslant\theta\leqslant 1$.

This combined with the results already obtained gives: if $1<p\leqslant\infty$ and $u+v\neq 0,1,2,\ldots$, then J^v maps $T_u^p(x_0)$ continuously into $T_{u+v}^q(x_0)$ provided

(a) $\dfrac{1}{p}\geqslant\dfrac{1}{q}\geqslant\dfrac{1}{p}-\dfrac{v}{n}$, if $\dfrac{1}{p}>\dfrac{v}{n}$,

(b) $p\leqslant q\leqslant\infty$, if $\dfrac{1}{p}<\dfrac{v}{n}$,

(c) $p\leqslant q<\infty$, if $\dfrac{1}{p}=\dfrac{v}{n}$.

Only (c) requires additional explanation. If $1/p=v/n$ and $0<\varepsilon<v$, then, according to (a), J^ε maps $T_u^p(x_0)$ continuously into $T_{u+\varepsilon}^p(x_0)$ and $J^{v-\varepsilon}$ maps $T_{u+\varepsilon}^p(x_0)$ continuously into $T_{u+\varepsilon}^q(x_0)$ with $1/p\geqslant 1/q\geqslant 1/p-(v-\varepsilon)/n=\varepsilon/n$. Thus $J^v=J^{v-\varepsilon}J^\varepsilon$ maps $T_u^p(x_0)$ continuously into $T_{u+v}^q(x_0)$ for all q such that $1/p\geqslant 1/q\geqslant\varepsilon/n$.

To prove that $J^v f\in t_{u+v}^q(x_0)$ if $f\in t_u^p(x_0)$ it is enough to observe that if $f\in C_0^\infty$ then $J^v f$ is infinitely differentiable and thus belongs to $t_{u+v}^q(x_0)$. Since, according to Lemma 2.3, C_0^∞ is dense in $t_u^p(x_0)$ and $t_{u+v}^q(x_0)$ is a closed subspace of $T_{u+v}^q(x_0)$ the desired conclusion is obtained by a passage to the limit.

Remark. If $u+v$ is a non-negative integer then the preceding argument shows that $J^v f\in T_{u+v}^q(x_0)$ (or $t_{u+v}^q(x_0)$) if we assume in addition that

$$\int\frac{|R(y)|}{|y-x_0|^{n+u}}dy<\infty.$$

This validates (4.4) for $|\alpha|=u+v$, which is what fails otherwise if $u+v$ is an integer. An alternative assumption could be $f(x_0+h)=f(x_0-h)$ if $u+v$ is odd, or $f(x_0+h)=-f(x_0-h)$ if $u+v$ is even. This makes the left-hand side of (4.6) vanish if $|\alpha|=u+v$.

The proof of Theorem 5 will be based on Theorem 11 below. The latter is analogous to the special case of Theorem 4 when $v=1$ and makes stronger assumptions, but in return its conclusions are also stronger: the case $p=1$ is included, and the only exceptional u is $u=-1$.

THEOREM 11. *Let f have first order derivatives f_j in L^p, and let $f_j \in T^p_u(x_0)$, $j = 1, 2, \ldots, n$, $-n/p \leqslant u \neq -1$. If*

1° $1 \leqslant p < n$, *and* $f \in L^q$ (2) *with* $1/q = 1/p - 1/n$, *then* $f \in T^q_{u+1}(x_0)$ *and*

$$T^q_{u+1}(x_0, f) \leqslant C_{pu} \sum_{j=1}^{n} T^p_u(x_0, f_j);$$

2° *if* $n < p \leqslant \infty$ *and* $f \in B(E_n)$, *then* $f \in T_{u+1}(x_0)$ *and*

$$T_{u+1}(x_0, f) \leqslant C_{pu}\Big[B(f) + \sum_{j=1}^{n} T^p_u(x_0, f_j)\Big],$$

where $B(f)$ *is the essential least upper bound of* $|f|$;

3° *if* $f \in L^r$ *with* $1/p \geqslant 1/r > 1/p - 1/n$, *then* $f \in T^r_{u+1}(x_0)$ *and*

$$T^r_{u+1}(x_0, f) \leqslant C_{ru}\Big[\|f\|_r + \sum_{j=1}^{n} T^p_u(x_0, f_j)\Big];$$

4° *the preceding statements hold with the spaces T replaced by the spaces t.*

Proof. We will first prove the inequality in 3° with $p = r$ assuming that $f \in C_0^\infty$. Let

$$k_j(x) = \frac{1}{\omega_n}\,\frac{x_j}{|x|^n},$$

where ω_n is the surface area of the unit sphere $|x| = 1$. Then

$$\sum_{1}^{n}\left(\frac{\partial}{\partial x_j}\right) k_j(x) = 0,$$

and consequently if $f \in C_0^\infty$ using Green's formula we obtain

$$\sum_{j=1}^{n} \int_{|x-y| \geqslant \varrho} k_j(x-y) f_j(y)\, dy = \frac{1}{\omega_n} \int_{|y-x| = \varrho} \frac{f(y)}{|x-y|^{n-1}}\, d\sigma,$$

where $d\sigma$ is the area element of the sphere $|x - y| = \varrho$. As ϱ tends to zero the right-hand side tends to $f(x)$. Thus we have the following representation of $f(x)$:

$$f(x) = \sum_{j=1}^{n} \int k_j(x-y) f_j(y)\, dy.$$

---

(2) The assumption $f \in L^q$ is almost superfluous. For if $f_j \in T^p_u(x_0)$, then $f_j \in L^p$ and using Soboleff's Theorem one can show that f differs by an additive constant from a function in L^q. Our assumption makes it certain that this constant is zero.

Let us consider first the case $-n/p \leqslant u < -1$. Let $|x| \leqslant \varrho$ and set

$$f(x) = \sum \int_{|x_0-y|\leqslant 2\varrho} k_j(x-y)f_j(y)\,dy + \\ + \sum_j \int_{|x_0-y|\geqslant 2\varrho} k_j(x-y)f_j(y)\,dy = f_1(x)+f_2(x),$$

say. Then an application of Young's theorem on convolution gives

$$\left[\frac{1}{\varrho^n}\int_{|x_0-y|\leqslant\varrho} |f_1(y)|^p\,dy\right]^{1/p} \leqslant \sum\left[\int_{|y|\leqslant 4\varrho} |k_j(y)|\,dy\right]\times \tag{4.10}$$
$$\times\left[\frac{1}{\varrho^n}\int_{|x_0-y|\leqslant 2\varrho} |f_j(y)|^p\,dy\right]^{1/p} \leqslant C\varrho\sum_{j=1}^{n} T_n^p(x_0,f_j)\,\varrho^u.$$

To estimate $f_2(x)$ we proceed as follows. First we observe that

$$\frac{1}{\varrho^n}F_j(\varrho) = \frac{1}{\varrho^n}\int_{|x_0-y|\leqslant\varrho} |f_j(y)|\,dy \leqslant C\left[\frac{1}{\varrho^n}\int_{|x_0-y|\leqslant\varrho} |f_j(y)|^p\,dy\right]^{1/p} \leqslant CT_u^p(x_0,f_j)\,\varrho^u,$$

whence

$$\left|\int_{|x_0-y|\geqslant 2\varrho} k_j(x-y)f_j(y)\,dy\right| \leqslant C\int_{2\varrho}^{\infty}\frac{dF_j(s)}{s^{n-1}} \leqslant C_u T_u^p(x_0,f_j)\,\varrho^{u+1}.$$

Consequently,

$$\left[\frac{1}{\varrho^n}\int_{|x_0-y|\leqslant\varrho} |f_2(y)|^p\,dy\right]^{1/p} \leqslant C_u\sum T_u^p(x_0,f_j)\,\varrho^{u+1}. \tag{4.11}$$

From this and (4.10) we obtain

$$\left[\frac{1}{\varrho^n}\int_{|x_0-y|\leqslant\varrho} |f(y)|^p\,dy\right]^{1/p} \leqslant C_u\sum T_u^p(x_0,f_j)\,\varrho^{u+1}. \tag{4.12}$$

We next prove a similar inequality for $-1 < u < 0$. We have again

$$f(x)-f(x_0) = \sum\int_{|x_0-y|\leqslant 2\varrho} k_j(x-y)f_j(y)\,dy + \\ + \sum\int_{|x_0-y|\geqslant 2\varrho} [k_j(x-y)-k_j(x_0-y)]f_j(y)\,dy - \\ - \sum\int_{|x_0-y|\leqslant 2\varrho} k_j(x_0-y)f_j(y)\,dy \\ = f_1(x)+f_2(x)+\varepsilon(\varrho),$$

say. Then, as before, $f_1(x)$ satisfies (4.10). If $F_j(\varrho)$ is defined as in the previous case, then, since $|k_j(x-y)-k(x_0-y)| \leqslant C\varrho|x_0-y|^{-n}$ for $|x_0-y| \geqslant 2\varrho$, we see readily that $f_2(x)$ satisfies (4.11). Finally,

$$\Big|\int_{|x_0-y|\leqslant 2\varrho} k_j(x_0-y)f_j(y)\Big| = C\int_0^r \frac{\varrho\, dF_j(s)}{s^{n-1}} \leqslant C_u T_u^p(x_0, f_j)\varrho^{u+1}.$$

Combining these results we obtain

$$\left[\frac{1}{\varrho^n}\int_{|x_0-y|\leqslant\varrho} |f(y)-f(x_0)|^p dy\right]^{1/p} \leqslant C_u \sum T_u^p(x_0, f_j)\varrho^{u+1}. \tag{4.13}$$

We now consider the case $u \geqslant 0$. Let P denote the sum of the terms of degree less than $u+1$ of the Taylor expansion of f at x_0; set $\bar f = f-P$ and $\bar f_j = f_j - P_j$. Then, using polar coordinates,

$$\Big[\int_{|y-x_0|\leqslant\varrho} |\bar f(y)|^p dy\Big]^{1/p} \leqslant \Big[\int_{r\leqslant\varrho} r^{n-1}\Big|\int_0^r \sum \bar f_j(s,\omega)\, a_j\, ds\Big|^p dr\, d\omega\Big]^{1/p}$$

$$\leqslant C\sum_j \Big[\int_{r\leqslant\varrho} r^{n+p-2}\int_0^r |\bar f_j(s,\omega)|^p ds\, dr\, d\omega\Big]^{1/p}$$

$$\leqslant C\varrho^{(n+p-1)/p}\sum_j\Big[\int_{s\leqslant\varrho} |\bar f_j(s,\omega)|^p ds\, d\omega\Big]^{1/p}$$

$$= C\varrho^{(n+p-1)/p}\sum_j\Big[\int_{|y-x_0|\leqslant\varrho} \frac{|\bar f_j(y)|^p}{|y-x_0|^{n-1}}dy\Big]^{1/p}.$$

On the other hand, setting

$$\int_{|y-x_0|\leqslant\varrho} |\bar f_j(y)|^p dy = F_j(\varrho),$$

we find that $F_j(\varrho) \leqslant T_u^p(x_0, f_j)^p \varrho^{n+pu}$ and

$$\int_{|y-x_0|\leqslant\varrho} \frac{|\bar f_j(y)|^p}{|y-x_0|^{n-1}} dy = \int_0^\varrho \frac{dF_j(s)}{s^{n-1}} \leqslant CT_u^p(x_0, f_j)\varrho^{pu+1}.$$

This combined with the inequality obtained previously gives

$$\left[\frac{1}{\varrho^n}\int_{|y-x_0|\leqslant\varrho} |f(y)-P(y)|^p dy\right]^{1/p} \leqslant C\sum_j T_u^p(x_0, f_j)\varrho^{u+1}. \tag{4.14}$$

Since all the coefficients of P except the constant term enter in the $T_u^p(x_0, f_j)$ it remains to estimate $P(x_0) = f(x_0)$. Let $\varphi(x)$, $|\varphi|\leqslant 1$, be a function in C_0^∞ which is equal to 1 in $|x| \leqslant 1$ and vanishes in $|x| \geqslant 2$. Set $\psi(x) = \varphi(x - x_0)$. Then

$$f(x_0) = f(x_0)\psi(x_0) = \sum_j \int k_j(x_0 - y)\psi(y) f_j(y)\,dy + \sum_j \int k_j(x_0 - y) f(y)\psi_j(y)\,dy,$$

whence

$$|f(x_0)| \leqslant \sum_j \int_{|y - x_0| \leqslant 2} \frac{|f_j(y)|\,dy}{|y - x_0|^{n-1}} + C \int_{|y - x_0| \leqslant 2} |f(y)|\,dy.$$

Let $u > -1$ and $v = \min(u, 0)$; then, by Lemma 2.1,

$$F_j(\varrho) = \int_{|y - x_0| \leqslant \varrho} |f_j(y)|\,dy \leqslant C\varrho^n \left[\frac{1}{\varrho^n} \int_{|y - x_0| \leqslant \varrho} |f_j(y)|^p\,dy\right]^{1/p} \leqslant C T_v^p(x_0, f_j)\,\varrho^{n+v}$$
$$\leqslant C T_u^p(x_0, f_j)\,\varrho^{n+v}.$$

Consequently,

$$|f(x_0)| \leqslant \sum_j \int_0^2 \frac{dF_j(s)}{s^{n-1}} + C \int_{|y - x_0| \leqslant 2} |f(y|\,dy,$$

(4.15)
$$|f(x_0)| \leqslant C_u \sum T_u^p(x_0, f_j) + C\|f\|_p.$$

Now the inequalities (4.12) to (4.15) clearly imply that

(4.16)
$$T_{u+1}^p(x_0, f) \leqslant C_u \sum T_u^p(x_0, f_j) + C\|f\|_p.$$

The factor C_u on the right tends to infinity as u tends to -1, but is bounded away from $u = -1$. The argument given above covers also the case $p = \infty$.

To prove 1° let us denote $f(x) - P(x)$ by $\bar{f}(x)$, and let $\varphi(x)$ be again a function which is equal to 1 for $|x| < 1$ and vanishes for $|x| \geqslant 2$. Then

$$\frac{\partial}{\partial y_j}\left[\bar{f}(y)\varphi\left(\frac{y - x_0}{\varrho}\right)\right] = \bar{f}_j(y)\varphi\left(\frac{y - x_0}{\varrho}\right) + \frac{1}{\varrho}\bar{f}(y)\varphi_j\left(\frac{y - x_0}{\varrho}\right),$$

and if $1 \leqslant p < n$, Soboleff's theorem, which is also valid when $p = 1$ (see [11]), gives, with $1/q = 1/p - 1/n$,

$$\left[\frac{1}{\varrho^n} \int_{|y - x_0| \leqslant \varrho} |\bar{f}(y)|^q\,dy\right]^{1/q} \leqslant \left[\frac{1}{\varrho^n} \int \left|\bar{f}(y)\varphi\left(\frac{y - x_0}{\varrho}\right)\right|^q dy\right]^{1/q}$$
$$\leqslant C_p\varrho \sum_{j=1}^n \left[\frac{1}{\varrho^n} \int \left|\bar{f}_j(y)\varphi\left(\frac{y - x_0}{\varrho}\right)\right|^p dy\right]^{1/p} +$$
$$+ C_p \sum_{j=1}^n \left[\frac{1}{\varrho^n} \int \left|\bar{f}(y)\varphi_j\left(\frac{y - x_0}{\varrho}\right)\right|^p dy\right]^{1/p},$$

and applying (4.12), (4.13) or (4.14) to the second sum, as the case may be, we obtain

$$(4.17)\qquad \left[\frac{1}{\varrho^n}\int_{|y-x_0|\leqslant\varrho}|f(y)-P(y)|^q dy\right]^{1/q}\leqslant C_{pu}\sum_{j=1}^{n}T_u^p(x_0,f_j)\,\varrho^{u+1}.$$

On the other hand, since $f_j\in L^p$ it follows from Soboleff's theorem that $f\in L^q$ and $\|f\|_q\leqslant C_p\sum\|f_j\|_p$. Furthermore it is easy to verify that (4.15) holds with $\|f\|_p$ replaced by $\|f\|_q$. Combining this with (4.17) we obtain

$$(4.18)\qquad T_{u+1}^q(x_0,f)\leqslant C_{pu}\sum_{j=1}^{n}T_u^p(x_0,f_j).$$

If $p>n$, then instead of applying Soboleff's theorem we use the representation

$$(4.19)\ \bar f(x)\varphi\left(\frac{x-x_0}{\varrho}\right)=\sum_j\int k_j(x-y)\left[\bar f_j(y)\varphi\left(\frac{y-x_0}{\varrho}\right)+\frac{1}{\varrho}\bar f(y)\varphi_j\left(\frac{y-x_0}{\varrho}\right)\right]dy$$

and Hölder's inequality, and obtain instead of (4.17)

$$\operatorname*{ess\,sup}_{|y-x_0|\leqslant\varrho}|f(y)-P(y)|\leqslant C_{pu}\sum_{j=1}^{n}T_u^p(x_0,f_j)\,\varrho^{u+1},$$

and thus, if $p>n$,

$$T_{u+1}(x_0,f)\leqslant C_{pu}\sum_{j=1}^{n}T_u^p(x_0,f_j)+CB(f).$$

To prove the inequality in 3° for general r we use the representation (4.19) for $\bar f(x)\varphi\left(\frac{x-x_0}{\varrho}\right)$ and apply to it Young's theorem on convolution with exponent r for the left side and exponent s, $1/r=1/p+1/s-1$, for k_j on the right. Then using (4.12), (4.13) or (4.14), as the case may be, we obtain

$$\left[\frac{1}{\varrho^n}\int_{|y|\leqslant\varrho}|f(y)-P(y)|^r dy\right]^{1/r}\leqslant C_{ru}\sum_j T_u^p(x_0,f_j)\,\varrho^{u+1}.$$

This combined with (4.15) where we can replace $\|f\|_p$ on the right by $\|f\|_r$, gives the inequality in 3°.

So far we have been considering functions in C_0^∞. We now extend these results to functions in $t_u^p(x_0)$. Let f have compact support and have first order derivatives in $t_u^p(x_0)$, and let $f^\lambda=\lambda^n\varphi(\lambda x)*f$ where φ is in C_0^∞ and has integral equal to 1. Then $f^\lambda\in C_0^\infty$ and $f_j^\lambda=\lambda^n\varphi(\lambda x)*f_j$, and, by Lemma 2.3, f_j^λ converges in $T_u^p(x_0)$ to f_j. Applying to f^λ the inequalities we have obtained for functions in C_0^∞ we see that under the various hypothesis of the theorem, f^λ converges in the corresponding space $T_{u+1}^q(x_0)$ with appropriate q (according to Lemma 2.2, $T_{u+1}^q(x_0)$ is complete). On the other hand, f^λ converges to f in L^q, consequently $f \in t^q_{u+1}(x_0)$, since it is the limit in $T^q_{u+1}(x_0)$ of functions in C_0^∞ which belong to $t^q_{u+1}(x_0)$ (see Lemma 2.3). In passing to the limit we see that the inequalities also hold for f.

We now consider the case of general f. Let $\varphi(x)$ be a function of C_0^∞ which is equal to 1 in $|x| < 1$ and vanishes in $|x| \geqslant 2$. We consider the function $f^\varepsilon(x) = f(x)\varphi[\varepsilon(x-x_0)]$. If $f_j \in t^p_u(x_0)$ then also $f^\varepsilon_j \in t^p_u(x_0)$, provided f belongs to L^p on bounded sets. Now this is part of our hypothesis in all cases. Furthermore, f^ε_j converges to f_j in $T^p_u(x_0)$. To see this we must verify that on the one hand $\|f^\varepsilon_j - f_j\|_p \to 0$ as $\varepsilon \to 0$, and that on the other

$$\sup_\varrho \frac{1}{\varrho^u}\left[\frac{1}{\varrho^n}\int_{|x-x_0|\leqslant\varrho} |R^\varepsilon_j(x) - R_j(x)|^p\,dx\right]^{1/p} \tag{4.18}$$

tends to zero with ε, where R_j and R^ε_j are the remainders of f_j and f^ε_j respectively. Now

$$R^\varepsilon_j(x) - R_j(x) = f^\varepsilon_j(x) - f_j(x) = -f_j(x)\big[1-\varphi[\varepsilon(x-x_0)]\big] + \varepsilon f(x)\varphi_j[\varepsilon(x-x_0)],$$

and the first term on the right clearly converges to zero in L^p. For the second term we have

$$\left[\int |\varepsilon f\varphi_j[\varepsilon(x-x_0)]|^p\,dx\right]^{1/p} \leqslant C\varepsilon\left[\int_{1/\varepsilon<|x-x_0|<2/\varepsilon} |f(x)|^p\,dx\right]^{1/p}$$

$$\leqslant C\varepsilon\left[\int_{1/\varepsilon\leqslant|x-x_0|\leqslant 2/\varepsilon} |f(x)|^r\,dx\right]^{1/r}\left(\frac{1}{\varepsilon^n}\right)^{1/p-1/r} = o(\varepsilon^{1-n/p+n/r}) = o(1),$$

where r is the exponent of the class to which f belongs (we have, in all cases, $1-n/p+n/\lambda > 0$). This shows that $\|f^\varepsilon_j - f_j\|_p \to 0$ as $\varepsilon \to 0$. Since $R^\varepsilon_j(x) - R_j(x)$ vanishes for $|x| < 1/\varepsilon$ this also shows that (4.18) tends to zero with ε. Consequently f^ε converges to f in $T^q_{u+1}(x_0)$ and 4° is established.

There only remains the case of $f_j \in T^p_u(x_0)$. By Lemma 2.1 evidently $f_j \in t^p_{u-\varepsilon}(x_0)$ and $T^p_{u-\varepsilon}(x_0, f_j) \leqslant CT^p_u(x_0, f_j)$ for every sufficiently small positive ε(3). Consequently $f \in t^q_{u+1-\varepsilon}(x_0)$ with appropriate q and we have that the inequalities of theorem 11 hold for f with u replaced by $u-\varepsilon$ on the left and in the constants. These constants are bounded functions of u for u away from -1. Now it is easy to see that if $T^q_{u+1-\varepsilon}(x_0, f) \leqslant M$ for sufficiently small positive ε, then $f \in T^q_{u+1}(x_0)$ and $T^q_{u+1}(x_0, f) \leqslant M$. Thus we can pass to the limit in the inequalities by letting ε tend to zero. This completes the proof of the theorem (4).

THEOREM 12. *If $f \in L^p_k$, $1 \leqslant p < \infty$, $k = 0, 1, 2, \ldots$ then $f \in t^q_k(x_0)$ for almost all x_0 with $1/p \geqslant 1/q \geqslant 1/p - k/n$ if $p < n/k$; $p \leqslant q \leqslant \infty$ if $p > n/k$, and $p \leqslant q < \infty$ if $p = n/k$.*

(3) This presupposes that $u > -n/p$. Observe, however, that $T^p_{-n/p}(x_0) = t^p_{-n/p}(x_0)$.

(4) Examples showing that the theorem is false for $u = -1$ can be easily constructed by means of the function $\log|x|$.

Proof. The case $k = 0$ is the familiar theorem about the Lebesgue set of a function in L^p. The general case is obtained by induction on k using Theorem 11 and noting that in the first place $f_j \in L^p_{k-1}$, secondly that, according to Lemma 4.8, f is bounded if $1/p < k/n$ and $f \in L^r$ for all r, $p \leqslant r < \infty$, if $1/p = n/k$.

LEMMA 4.9. *Let $f \in t^p_k(x_0)$, $1 < p \leqslant \infty$, and let*

$$\int \frac{|f(x)|}{|x-x_0|^{n+k}}\,dx < \infty.$$

Then, for $v \geqslant 0$, we have $J^v f \in t^q_{u+v}(x_0)$, where $1/p \geqslant 1/q \geqslant 1/p - v/n$ if $p < n/v$, $p \leqslant q \leqslant \infty$ if $p > n/v$, or $p \leqslant q < \infty$ if $p = n/v$.

Proof. The case $v = 0$ is obvious. The finiteness of the integral above implies that the Taylor expansion of f reduces to the remainder $R(x)$, and therefore

$$\int \frac{|R(x)|}{|x-x_0|^{n+k}}\,dx < \infty.$$

Now the assertion of the lemma follows from the remark to the proof of Theorem 4.

Proof of Theorem 5. We shall distinguish two cases namely, $u \geqslant 1$ and $u = 0$.

Suppose that $f \in T^p_u(x_0)$, $u \leqslant 1$, $1 < p \leqslant \infty$, for all x_0 in a set S of positive measure. We may assume without loss of generality that $T^p_u(x_0, f)$ is bounded on S, and that S itself is closed and bounded. This presupposes the measurability of $T^p_u(x_0, f)$ as a function of x_0; we assume this for the moment. Then according to the corollary of Theorem 9, f can be written as $f_1 + f_2$, where $f_1 \in B_u(E_n)$ and f_2 satisfies the hypothesis of Theorem 10 on S. Further f_1 can be chosen to have compact support. But then, if $v \geqslant 0$, $J^v f_1$ belongs to L^p_{u+v} for all p, and Theorem 12 asserts that $J^v f_1 \in t_{u+v}(x_0)$ for almost all x_0. Since f_1 has compact support, $J^v f_1 \in L^q$ and thus $J^v f_1 \in t^q_{u+v}(x_0)$ for almost all x_0. On the other hand, since f_2 satisfies the hypothesis of Theorem 10 on S, $f \in t^p_u(x_0)$ for almost all x_0 in S and, by Lemma 4.9, $J^v f_2 \in t^q_{u+v}(x_0)$ for almost all $x_0 \in S$; therefore $J^v f \in t^q_{u+v}(x_0)$ at every point x_0 where $J^v f_1 \in t^q_{u+v}(x_0)$ and $J^v f_2 \in t^q_{u+v}(x_0)$.

There remains the case $u = 0$. If $v = 0$, then since $f \in L^p$, $1 < p < \infty$, it follows that

$$\left[\frac{1}{\varrho^n}\int\limits_{|h|\leqslant\varrho} |f(x_0+h) - f(x_0)|^p\,dh\right]^{1/p} = o(1), \qquad \varrho \to 0,$$

for almost all x_0, that is $f \in t^p_0(x_0)$ for almost all x_0. If $v \geqslant 1$, then $J^v f \in L^p_v$ and Theorem 12 gives the desired result.

We now sketch briefly the proof of the measurability of $T_u^p(x_0, f)$ as a function of x_0. Let φ be the function of Lemma 2.6 and let

$$f_\lambda(x) = \int \lambda^n \varphi[\lambda(x-y)]f(y)\,dy.$$

Then if $f \in T_u^p(x_0)$ and $f_a(x_0)$ is one of the coefficients of the Taylor expansion of f at x_0 we have

$$\left(\frac{\partial}{\partial x}\right)^a f_\lambda(x_0) = f_a(x_0) + \int \lambda^{n+|a|}\varphi_a[\lambda(x_0-y)]R(y)\,dy, \quad \text{where} \quad \varphi_a = \left(\frac{\partial}{\partial x}\right)^a \varphi,$$

and R is the remainder in the Taylor expansion of f. The integral above is majorized by

$$\lambda^{n+|a|} C_a \int_{|y-x_0|\leqslant 1/\lambda} |R(y)|\,dy = O(\lambda^{|a|-u}),$$

which tends to zero as $\lambda \to \infty$. Consequently the function $f_a(x_0)$ are limits of infinitely differentiable functions on the set where $T_u^p(x_0, f)$ is finite, which was assumed to be measurable, and therefore are measurable. From this the measurability of $T_u^p(x_0, f)$ follows without difficulty.

We conclude this section with a theorem which may be interpreted as an extension of the well known theorem of Lusin on the structure of measurable functions.

THEOREM 13. *Let $f \in L_k^p$, $1 \leqslant p < \infty$, then given $\varepsilon > 0$, there is a function $g(x)$ with continuous derivatives of orders $\leqslant k$, such that $f(x) = g(x)$ outside a set of measure $\leqslant \varepsilon$.*

Proof. According to Theorem 12, $f \in t_k^q(x_0)$ for almost all x_0, where q is some exponent larger than or equal to 1. Since $T_k^q(x_0, f)$ is a measurable function, given ε we can find an open set O such that $T_k^q(x_0, f)$ is bounded outside O, and whose measure is less than ε, and applying Theorem 9 to f and the complement of O the desired result follows.

5. In this section we study the effect of singular integral operators in the classes $T_u^p(x_0)$ and $t_u^p(x_0)$. We will use properties of singular integral operators which were established in [6].

LEMMA 5.1. *Let $\mathcal{K}$ be a convolution singular integral operator defined by*

$$\mathcal{K}f = \lim_{\varepsilon\to 0} \int_{|x-y|>\varepsilon} k(x-y)f(y)\,dy,$$

where $k(x)$ is homogeneous of degree $-n$, is infinitely differentiable in $|x| \neq 0$, and has mean value zero on $|x| = 1$.

Then if $1 < p < \infty$, $-n/p \leqslant u \neq 0, 1, 2, \ldots$, and $f \in T_u^p(x_0)$, we also have $\mathcal{K}f \in T_u^p(x_0)$ and

$$T_u^p(x_0, \mathcal{K}f) \leqslant C_{up} M\, T_u^p(x_0, f),$$

where M is a bound for the absolute values $|(\partial/\partial x)^a k(x)|$ on $|x| = 1$, $0 \leqslant |a| \leqslant u+1$, if $u > 0$ and $|a| = 0$ if $u \leqslant 0$.

If, in addition, $f \in t_u^p(x_0)$, then $\mathcal{K} f \in t_u^p(x_0)$.

Proof. We assume for simplicity that $x_0 = 0$. We choose once for all a function φ in C_0^∞ which is equal to 1 for $|x| \leqslant 1$, and we set $f = f_1 + f_2$ with $f_1 = \varphi P$ where P is the Taylor expansion of f at 0. Then $f_1 \in T_u^p(x_0)$ and it is not difficult to verify that $T_u^p(x_0, f_1) \leqslant C T_u^p(x_0, f)$; consequently f_2 also belongs to $T_u^p(x_0)$ and $T_u^p(x_0, f_2) \leqslant C T_u^p(x_0, f)$. We will apply $\mathcal{K}$ to f_1 and f_2 separately.

First let us observe that if ψ is a function in C_0^∞ which vanishes outside $|x| \leqslant 2$, then we have

$$\mathcal{K}\psi = \lim_{\varepsilon \to 0} \int_{|y| > \varepsilon} k(y)\psi(x-y)\,dy = \lim_{\varepsilon \to 0} \int_{|y|>\varepsilon} k(y)[\psi(x-y) - \psi(x)]\,dy,$$

which shows that the first integral converges uniformly as $\varepsilon \to 0$ and that

$$|\mathcal{K}\psi| \leqslant C_\psi M,$$

where C_ψ depends on ψ. From the uniform convergence of the integral it follows that

$$\frac{\partial}{\partial x_j}\mathcal{K}\psi = \mathcal{K}\left(\frac{\partial}{\partial x_j}\right)\psi,$$

and thus $\mathcal{K}\psi$ is an infinitely differentiable function and

$$\left|\left(\frac{\partial}{\partial x}\right)^a \mathcal{K}\psi\right| \leqslant C_{a\psi} M.$$

Furthermore $(\mathcal{K}\psi)(x) \leqslant MC_\psi |x|^{-n}$ for $|x| \geqslant 3$. This and the preceding inequalities show that $\|\mathcal{K}\psi\|_p \leqslant MC_{p,\psi}$. It is easy to see now that $\mathcal{K}\psi \in T_u^p(x_0)$ and that

$$T_u^p(x_0, \mathcal{K}\psi) \leqslant C_{up\psi} M.$$

Applying this result to $x^a\varphi(x)$ we find that

$$T_u^p(x_0, \mathcal{K}x^a\varphi) \leqslant C_{p,a} M$$

since the function φ is fixed. Consequently, if $P = \sum a_a x^a$, we have

$$T_u^p(x_0, \mathcal{K}f_1) \leqslant \sum_{|a|<u} |a_a| T_u^p(x_0, \mathcal{K}x^a\varphi) \leqslant C_{up} M T_u^p(x_0, f).$$

This is of course trivial if $u \leqslant 0$, since then $f_1 = 0$.

Consider next the function f_2. Its Taylor expansion vanishes and therefore the inequality $T_u^p(x_0, f_2) \leqslant C T_u^p(x_0, f)$ implies that

$$\left[\frac{1}{\varrho^n} \int_{|y| \leqslant \varrho} |f_2(y)|^p\,dy\right]^{1/p} \leqslant C T_u^p(x_0, f)\,\varrho^u. \tag{5.1}$$

From this we obtain the following inequalities, which are analogues of (4.2) and (4.3) in the proof of Theorem 4:

$$\int_{|y|\leqslant \varrho} |f_2(y)|\,|y|^{-r}dy \leqslant C_{ru}T_u^p(x_0,f)\,\varrho^{n+u-r} \tag{5.2}$$

if $n+u-r>0$, and

$$\int_{|y|\geqslant \varrho} |f_2(y)|\,|y|^{-r}dy \leqslant C_{ru}T_u^p(x_0,f)\,\varrho^{n+u-r} \tag{5.3}$$

if $n+u-r<0$. Expanding k by Taylor's formula we can write, if $u\geqslant 0$,

$$\begin{aligned}\int k(x-y)f_2(y)\,dy = &\int_{|y|\leqslant\varrho} f_2(y)k(x-y)\,dy+\\ &+\sum_{|a|\leqslant u}\frac{x^a}{a!}\int f_2(y)k_a(-y)\,dy+\\ &+\sum_{u<|a|\leqslant u+1}\frac{x^a}{a!}\int_{|y|\geqslant\varrho} f_2(y)k_a(\Theta x-y)\,dy-\\ &-\sum_{|a|\leqslant u}\frac{x^a}{a!}\int_{|y|\leqslant\varrho} f_2(y)k_a(-y)\,dy,\end{aligned} \tag{5.4}$$

where the first two integrals are taken in the principal value sense. Since $|k_a(-y)|\leqslant M|y|^{-n-|a|}$ and on account of (5.2), the integrals in the first sum are absolutely convergent near zero. Further $\|f_2\|_p\leqslant CT_u^p(x_0,f)$ and using Hölder's inequality we see that those integrals are also absolutely convergent at infinity. Combining this with (5.2) we see that the first sum on the right of (5.4) is a polynomial $P(x)$ whose coefficients are dominated by $C_{pu}MT_u^p(x_0,f)$.

If we assume that $2|x|\leqslant\varrho$, then $|k_a|(\Theta x-y)|\leqslant C_uM|y|^{-n-|a|}$ for $|y|\geqslant\varrho$. Consequently, it follows from (5.3) that the second sum in (5.4) is dominated by $C_uMT_u^p(x_0,f)\,\varrho^{u-|a|}|x|^{|a|}$.

From (5.2) and the inequality for k_a it follows that the last sum is dominated by $C_uMT_u^p(x_0,f)\,\varrho^{u-|a|}|x|^{|a|}$.

Finally, let us consider the first term on the right of (5.4). From the remark on page 306 of [6] it follows that the norm in L^p of this term is dominated by

$$C_pM\Big[\int_{|y|\leqslant\varrho}|f_2(y)|^p\,dy\Big]^{1/p}\leqslant C_pMT_u^p(x_0,f)\,\varrho^{u+n/p}.$$

Combining these results we finally obtain

$$\Big[\int_{|x|<\varrho/2}|\mathcal{K}f_2-P(x)|^p\,dx\Big]^{1/p}\leqslant C_{up}MT_u^p(x_0,f)\,\varrho^{u+n/p}.$$

Since $\|\mathcal{K}f_2\|_p \leqslant C_p M \|f_2\|_p$ (see [6], loc. cit.), combining this with estimates for the coefficients of $P(x)$ we find that $T_u^p(x_0, \mathcal{K}f_2) \leqslant C_{up} M T_u^p(x_0, f)$. Since the same inequality has been established for $\mathcal{K}f_1$, the proof of the part of the lemma concerning $T_u^p(x_0)$ is complete if $u \geqslant 0$.

If $u < 0$, instead of (5.4) we write

$$\int k(x-y)f_2(y)\,dy = \int_{|y|\leqslant\varrho} k(x-y)f_2(y)\,dy + \int_{|y|\geqslant\varrho} k(x-y)f_2(y)\,dy.$$

The first term on the right can be estimated as before, and for $|x| \leqslant \varrho/2$ and $|y| \geqslant \varrho$ we have $|k(x-y)| \leqslant CM|y|^{-n}$, and thus on account of (5.3) with $r = n$ we find that the second term is dominated by $C_u M T_u^p(x_0, f)\varrho^u$.

To prove that $\mathcal{K}f \epsilon t_u^p(x_0)$ it is enough to observe that if $f \epsilon C_0^\infty$ then $\mathcal{K}f$ is infinitely differentiable and thus belongs to $t_u^p(x_0)$. Since $\mathcal{K}$ is continuous in $T_u^p(x_0)$ and, according to Lemma 2.3, C_0^∞ is dense in $t_u^p(x_0)$ we conclude that $\mathcal{K}$ maps $t_u^p(x_0)$ into itself.

LEMMA 5.2. *If u is a non-negative integer and the other assumptions of Lemma 5.1 are satisfied, then $\mathcal{K}f \epsilon T_u^p(x_0)$ provided*

$$\int \frac{|f(x)|}{|x-x_0|^{n+u}}\,dx = N < \infty. \tag{5.5}$$

Furthermore,

$$T_u^p(x_0, \mathcal{K}f) \leqslant C_{up} M T_u^p(x_0, f) + MN.$$

If u is odd and $f(x_0+x)$ and $k(x)$ are of the same parity, that is, they are both even or both odd functions of x, then the conclusions of Lemma 5.1 hold without the additional assumption (5.5). If u is even we get the same conclusion provided $f(x_0+x)$ and $k(x)$ are of opposite parity.

Proof. We merely observe that the proof of Lemma 5.1 applies also if u is a non-negative integer, except at one point: the integrals of the functions

$$f_2(y)k_\alpha(-y), \qquad |\alpha| = u,$$

are no longer convergent. However, under the additional assumption (5.5) of the present lemma the Taylor expansion P of f must vanish and thus $f_2 = f$ and the integrals above converge. Under the other hypotheses f_2 has the same parity as f, if we assume, as we may, that φ is even, and the integrals above can be dropped.

LEMMA 5.3. *Let $u = 0$ and let the other assumptions of Lemma 5.1 be satisfied. Let*

$$(\mathcal{K}f)^*(x_0) = \sup_\varepsilon \Big| \int_{|y-x_0|\geqslant\varepsilon} k(x_0-y)f(y)\,dy \Big| < \infty.$$

Then $\mathcal{K}f \epsilon T_0^p(x_0)$ and

$$T_0^p(x_0, \mathcal{K}f) \leqslant C_p M T_0^p(x_0, f) + C(\mathcal{K}f)^*(x_0).$$

Proof. Let $\varrho > 0$ be given and let $f = f_1 + f_2$ where $f_1(x) = f(x)$ if $|x - x_0| \leqslant 2\varrho$, $f_1(x) = 0$ otherwise. Let $g_i = \mathcal{K} f_i$, $i = 1, 2$; $g = \mathcal{K} f$. Then

$$\Big[\int\limits_{|x-x_0|\leqslant\varrho} |g(x)|^p\,dx\Big]^{1/p} \leqslant \Big[\int\limits_{|x-x_0|\leqslant\varrho} |g_1(x)|^p\,dx\Big]^{1/p} + \Big[\int\limits_{|x-x_0|\leqslant\varrho} |g_2(x)|^p\,dx\Big]^{1/p},$$

and since $\mathcal{K}$ is continuous in L^p (see [6], loc. cit.) the right-hand side is less than or equal to

$$(5.6)\qquad C_p M \|f_1\|_p + \Big[\int\limits_{|x_0-x|\leqslant\varrho} dx \Big| \int\limits_{|y-x_0|\geqslant 2\varrho} [k(x-y) - k(x_0-y)] f(y)\,dy\Big|^p\Big]^{1/p} + \\ + C\varrho^{n/p} (\mathcal{K} f)^*(x_0).$$

We have here $|k(x-y) - k(x_0-y)| \leqslant CM|x_0-y|^{-n-1}|x-x_0|$. Let

$$F(\varrho) = \int\limits_{|x-x_0|\leqslant\varrho} |f(x)|\,dx.$$

Then Hölder's inequality shows that $F(\varrho) \leqslant C T_0^p(x_0, f)\varrho^n$. Consequently,

$$\Big|\int\limits_{|y-x_0|\geqslant 2\varrho} [k(x-y) - k(x_0-y)] f(y)\,dy\Big| \leqslant CM|x-x_0| \int\limits_{2\varrho}^{\infty} \frac{dF(s)}{s^{n+1}}$$

$$\leqslant CM T_0^p(x_0, f)|x-x_0|\,\varrho^{-1}.$$

Substituting the left-hand side in (5.6) and observing that $\|f_1\|_p \leqslant \varrho^{n/p} T_0^p(x_0, f)$, we obtain the inequality of the lemma.

Lemma 5.4. *Let $f \in L_k^p$, $1 \leqslant p \leqslant \infty$, and denote $\left(\frac{\partial}{\partial x}\right)^a f$ by f_a. Then*

$$T_k^p(x_0, f) \leqslant C_{kp}\Big[\sum_{|a|\leqslant k-1} \|f_a\|_p + \sum_{|a|=k} T_0^p(x_0, f_a)\Big]$$

for almost all x_0.

Proof. The statement is obtained at once by induction on k from 3° of Theorem 11 with $r = p$ if $p < \infty$, or from 2° if $p = \infty$.

Definition 9. We denote by $\mathcal{R}_{lm}$ a convolution singular integral operator whose kernel is of the form $Y_{lm}(x|x|^{-1})|x|^{-n}$, where Y_{lm} is a complete orthonormal system of spherical harmonics and m denotes the degree of the harmonic.

Lemma 5.5. *With the notation of Lemma 5.3 we have, for $1 < p < \infty$ and $-n/p \leqslant u \neq 0, 1, 2, \ldots$,*

$$\|\mathcal{R}_{lm} f\|_p \leqslant C_p \|f\|_p;\quad \|(\mathcal{R}_{lm} f)^*\|_p \leqslant C_p \|f\|_p.$$

$$T_u^p(x_0, \mathcal{R}_{lm} f) \leqslant C_{up} m^v T_u^p(x_0, f),$$

where $v = (n-2)/2 + [u+1]$ if $u \geqslant 0$ and $v = (n-2)/2$ if $u < 0$.

Proof. The first two inequalities follow immediately from [6], p. 306 and 307, Remark, taking into account that the Y_{lm} are normalized. The third inequality is a consequence of Lemma 5.1 and of the inequalities

$$\left|\left(\frac{\partial}{\partial x}\right)^{a} Y_{lm}(x|x|^{-1})|x|^{-n}\right| \leqslant C_a m^{(n-2)/2+|a|}, \qquad |x| \geqslant 1,$$

which follow immediately from the inequality (4) of [7].

The theorem which follows has Theorem 6 as an immediate corollary.

THEOREM 14. *Let $\mathcal{K}$ be a singular integral operator of class $T_u(x_0)$, $u \geqslant 0$, with kernel $k(x, x-y)$ and symbol $h(x, z)$. Let*

$$(5.7) \qquad a_{lm}(x) = (-1)^v \gamma_m^{-1}[m(m+n-2)]^{-v} \int_{|z|=1} Y_{lm}(z) L^v h(x, z) d\sigma,$$

where $Lg(z) = |z|^2 \Delta g(z)$, $\gamma_m = i^m \pi^{m/2}\Gamma(\frac{1}{2}m)/\Gamma(\frac{1}{2}m+\frac{1}{2}n)$, $v = [n+(u+1)/2]$. Then

(i) *$a_{lm}(x) \in T_u(x_0)$ and $T_u(x_0, a_{lm}) \leqslant Cm^{\frac{1}{2}n-2v} T_u(x_0, \mathcal{K})$;*

(ii) *$\mathcal{K} = a(x) + \sum a_{lm}(x)\mathcal{R}_{lm}$ in the following senses:*

(a) *for $f \in L^p$, $1 < p < \infty$, and for almost every x, the principal value integrals in $\mathcal{K}f$ and $\mathcal{R}_{lm}f$ exist and the series $a(x)f(x) + \sum a_{lm}(x)\mathcal{R}_{lm}f$ converges absolutely to $\mathcal{K}f$.*

(b) *$\mathcal{K}$ is a bounded operator in L^p and $T_u^p(x_0)$, provided $u \neq 0, 1, 2, \ldots$, and the series converges to $\mathcal{K}f$ absolutely in the operator norm. The norms of $\mathcal{K}$ in L^p and $T_u^p(x_0)$ do not exceed $C_p T_u(x_0, \mathcal{K})$ and $C_{up} T_u(x_0, \mathcal{K})$ respectively.*

(iii) *If $\mathcal{K} \in t_u(x_0)$, then $a_{lm}(x) \in t_u^p(x_0)$ and if $f \in t_u^p(x_0)$, $u \neq 0, 1, 2, \ldots$, the function $\mathcal{K}f$ also belongs to $t_u^p(x_0)$.*

Proof. The $a_{lm}(x)$ are nothing but the coefficients of the expansion of the kernel $k(x, z)$ in series of spherical harmonics $Y_{lm}(z)$ on $|z| = 1$. The expression (5.7) for the $a_{lm}(x)$ given above was obtained in [7], p. 913. Part (i) of the theorem follows from (5.7) taking into account that for each z, $|z| = 1$, the derivatives $\left(\frac{\partial}{\partial z}\right)^{a} h(x, z)$, $|a| \leqslant 2n+u$, belong to $T_u(x_0)$ and their norms in $T_u(x_0)$ are dominated by $T_u(x_0, \mathcal{K})$ (see definition 6).

To prove part (a) of (ii) let us consider $(\mathcal{R}_{lm}f)^*$. Then since, by Lemma 5.5, $\|(\mathcal{R}_{lm}f)^*\|_p \leqslant C_p\|f\|_p$ and since for each m the number of distinct spherical harmonics of degree m is of the order m^{n-2}, it follows from (i) that $\sum |a_{lm}(x)| \leqslant CT_u(x_0, \mathcal{K})$ and consequently the series

$$(5.8) \qquad \sum |a_{lm}(x)| (\mathcal{R}_{lm}f)^*$$

is finite almost everywhere. On the other hand, since $|Y_{lm}(z)| \leqslant Cm^{(n-2)/2}$

(see [7], p. 903], and since the number of distinct spherical harmonics of degree m is of the order m^{n-2}, it again follows from (i) that

$$\sum |a_{lm}(x)|\,|Y_{lm}(z)| \leqslant CT_u(x_0, \mathcal{K}).$$

But $k(x, z)|z|^n = \sum a_{lm}(x) Y_{lm}(z)$ and $f(y)|x-y|^{-n}$ is, by Hölder's inequality, integrable in $|x-y| > \varepsilon$, whence

$$(5.9)\qquad \int\limits_{|x-y|>\varepsilon} k(x, x-y) f(y)\,dy = \sum a_{lm}(x) \int\limits_{|x-y|>\varepsilon} Y_{lm}\left(\frac{x-y}{|x-y|}\right) |x-y|^{-n} f(y)\,dy.$$

If x is point where $\mathcal{R}_{lm}f$ exists as principal value integral for all l and m, and where (5.8) is finite, then, since the terms on the right of (5.9) are dominated by the corresponding terms of (5.8), we can pass to the limit termwise on the right of (5.9), as ε tends to zero. Thus (ii)(a) is established.

To prove (ii)(b) we merely have to observe that, by Lemma 5.5, the norm in L^p of $a_{lm}(x)\mathcal{R}_{lm}$ is $\leqslant C_p \sup_x |a_{lm}(x)| \leqslant C_p T_u(x_0, a_{lm})$ and thus, by (i), the series of the norms of these operators is finite.

Similarly, by Lemmas 5.5 and 2.4, the norm of $a_{lm}(x)\mathcal{R}_{lm}$ in $T_u^p(x_0)$ is less than or equal to $C_{up} m^v T_u(x_0, a_{lm})$, where v is as in Lemma 5.5. It now follows again from (i) that the series of the norms in $a_{lm}(x)\mathcal{R}_{lm}$ in $T_u(x_0)$ is finite.

The proof of (iii) is merely a repetition of the preceding one and rests on the completeness of $t_u^p(x_0)$.

Proof of Theorem 7. Since, as it was shown in the proof of Theorem 5, the norm $T_u^p(x_0, f)$ is a measurable function of x_0 on Q, we may assume without loss of generality that Q is compact and that $T_u^p(x_0, f)$ is a bounded function of x_0 on Q. We first assume that u is a positive integer. Then according to the corollary of Theorem 9 and Theorem 10 we can decompose f as $f = f_1 + f_2$, where $f_1 \in B_u(E_n)$ and has compact support and f_2 is such that

$$(5.10)\qquad \int \frac{|f_2(x)|}{|x-x_0|^{n+u}}\,dx = N_2(x_0) < \infty$$

for almost all x_0 in Q. Since the function f_1 belongs to $B_u(E_n)$ and has compact support it belongs to L_u^p. Let $f_{1a} = (\partial/\partial x)^a f_1$. Then since, according to Lemma 5.5, $\|(\mathcal{R}_{lm} f_{1a})^*\|_p \leqslant C_p \|f_{1a}\|_p$ and the number of distinct $\mathcal{R}_{lm}$ for each m is of the order m^{n-2}, the sum

$$\sum_{|a|=u} \sum_{l,m} m^{-n} (\mathcal{R}_{lm} f_{1a})^*(x_0) = N_1(x_0)$$

is finite for almost all x_0. Let now x_0 be a point where $N_1(x_0)$ and $N_2(x_0)$ are finite and in addition $f_{1\alpha} \in T_0^p(x_0)$, $|\alpha| = u$, and let $\mathcal{K}$ be an operator in $T_u(x_0)$. We will show that $\mathcal{K}f \in T_u^p(x_0)$ and that

$$(5.11) \qquad T_u^p(x_0, \mathcal{K}f) \leqslant C_{up} T_u(x_0, \mathcal{K}) \Big[T_u^p(x_0, f_2) + \sum_{|\alpha|<u} \|f_{1\alpha}\|_p + \\ + \sum_{|\alpha|=u} T_0^p(x_0, f_{1\alpha}) + N_1(x_0) + N_2(x_0) \Big].$$

For this purpose it will be enough to show that the sum of the norms in $T_u^p(x_0)$ of the terms of the series

$$\mathcal{K}f = \sum a_{lm}(x) \mathcal{R}_{lm} f_1 + \sum a_{lm}(x) \mathcal{R}_{lm} f_2$$

is finite.

We have

$$T_u^p(x_0, a_{lm} \mathcal{R}_{lm} f_2) \leqslant C T_u(x_0, a_{lm}) T_u^p(x_0, \mathcal{R}_{lm} f_2) \\ \leqslant C m^{n/2-2n-2[(u+1)/2]} T_u(x_0, \mathcal{K}) M_m [C_{up} T_u^p(x_0, f_2) + N_2(x_0)].$$

The first inequality follows from Lemma 2.4. The second inequality, in which M_m is the constant associated with $\mathcal{R}_{lm}$ as in Lemma 5.1, follows from part (i) of Theorem 14, and Lemma 5.2. We have

$$M_m \leqslant C m^{(n-2)/2+[u+1]}$$

(see, e. g., [7], p. 904, formula (4)), and thus

$$\sum T_u^p(x_0, a_{lm} \mathcal{R}_{lm} f_2) \leqslant C T_u(x_0, \mathcal{K}) [C_{up} T_u^p(x_0, f_2) + N_2(x_0)].$$

On the other hand,

$$T_u^p(x_0, a_{lm} \mathcal{R}_{lm} f_1) \leqslant C T_u(x_0, a_{lm}) T_u^p(x_0, \mathcal{R}_{lm} f_1) \\ \leqslant C m^{n/2-2n-2[(u+1)/2]} T_u(x_0, \mathcal{K}) T_u^p(x_0, \mathcal{R}_{lm} f_1).$$

Now, by Lemma 5.4 we have

$$T_u^p(x_0, \mathcal{R}_{lm} f_1) \leqslant C_{up} \left[\sum_{|\alpha|<u} \left\| \left(\frac{\partial}{\partial x}\right)^\alpha \mathcal{R}_{lm} f_1 \right\|_p + \sum_{|\alpha|=u} T_0^p\left(x_0, \left(\frac{\partial}{\partial x}\right)^\alpha \mathcal{R}_{lm} f_1\right)\right],$$

and since $\frac{\partial}{\partial x_j} \mathcal{R}_{lm} = \mathcal{R}_{lm} \frac{\partial}{\partial x_j}$ (the proof is the same as in the case $m = 1$; see Lemma 4.3) it follows from Lemmas 5.5 and 5.3 that

$$T_u^p(x_0, \mathcal{R}_{lm} f_1) \leqslant C_{up} \Big[\sum_{|\alpha|<u} \|f_{1\alpha}\|_p + C_p M_m \sum_{|\alpha|=u} T_0^p(x_0, f_{1\alpha}) + C \sum_{|\alpha|=u} (\mathcal{R}_{lm} f_{1\alpha})^*(x_0) \Big],$$

where M_m is the same as before. Consequently,

$$\sum T_u^p(x_0, a_{lm}\mathcal{R}_{lm}f_1) \leqslant \sum C_{up} m^{n/2-2n-2[(u+1)/2]} T_u(x_0, \mathcal{K})\Big[\sum_{|\alpha|<u} \|f_{1\alpha}\|_p + $$
$$+ C_p M_m \sum_{|\alpha|=u} T_0^p(x_0, f_{1\alpha}) + C \sum_{|\alpha|=u} (\mathcal{R}_{lm} f_{1\alpha})^*(x_0)\Big]$$
$$\leqslant C_{up} T_u(x_0, \mathcal{K})\Big[\sum_{|\alpha|<u} \|f_{1\alpha}\|_p + \sum_{|\alpha|=u} T_0^p(x_0, f_{1\alpha}) + N_1(x_0)\Big].$$

This completes the proof of the theorem if $u > 0$.

If $u = 0$ we set $f_2 = 0$ and argue with $f_1 = f$ as above, and (5.11) simplifies to

$$T_0^p(x_0, \mathcal{K}f) \leqslant C_p T_0(x_0, \mathcal{K})[T_0^p(x_0, f) + N_1(x_0)].$$

A special variant of Theorems 6 and 7 is the following

THEOREM 15. *The operators* $\dfrac{\partial}{\partial x_j} J$ *map* $T_u^p(x_0)$ *and* $t_u^p(x_0)$, $1 < p < \infty$, $-n/p \leqslant u \neq 0, 1, 2, \ldots$, *continuously into themselves. If, on the other hand, u is a non-negative integer and* $f \in T_u^p(x_0)$ $1 < p < \infty$, *for all x_0 in a set Q of positive measure, then* $\dfrac{\partial}{\partial x_j} Jf$ *belongs to* $T_u^p(x_0)$ *for almost all x_0 in Q.*

Proof. According to Lemma 4.3 we have

$$\frac{\partial}{\partial x_j} = -i\mathcal{R}_j \Lambda.$$

Consequently,

$$\frac{\partial}{\partial x_j} J = -i\mathcal{R}_j \Lambda J,$$

and an account of Theorems 6 and 7 it will be enough to show that ΛJ maps $T_u^p(x_0)$ and $t_u^p(x_0)$ continuously into themselves for $u \geqslant -n/p$.

We first observe that J maps $T_u^p(x_0)$ continuously into itself. If u is not an integer, this is an immediate consequence of Theorem 4 and of Lemma 2.1. If u is an integer then for the same reasons $J^{1/2}$ maps $T_u^p(x_0)$ continuously into itself. Consequently $J = J^{1/2}J^{1/2}$ has the same property.

According to Lemma 4.6,

$$\Lambda J = I + a_1 J^2 + a_2 J^4 + \ldots + a_m J^{2m} - Q_m.$$

By what we have just shown, it only remains to prove that Q_m maps $T_u^p(x_0)$ continuously into $t_u^p(x_0)$. If m is large enough, Q_m is convolution with a bounded integrable kernel with bounded integrable derivatives up to order $[u]+1$. Hence, in the first place, $\|Q_m f\|_p \leqslant C_u \|f\|_p$ and $Q_m f$ has continuous derivatives up to order $[u]+1$ majorized by $C_u\|f\|_p$. Thus $Q_m f \in t_u^p(x_0)$ and $T_u^p(x_0, Q_m f) \leqslant C_u \|f\|_p \leqslant C_u T_u^p(x_0, f)$. This completes the proof of the theorem.

6. This section will be devoted to proving our theorems on differential equations.

LEMMA 6.1. *Let $\mathcal{L}$ be a system of differential operators with constant coefficients of homogeneous order m which is elliptic in the sense of Definition 5. Then*

$$\mathcal{L} = \mathcal{K}\Lambda^m, \tag{6.1}$$

where $\mathcal{K}$ is an $s\times r$, $s \geqslant r$, matrix of convolution singular integral operators, with the property that there exists an $r\times s$ matrix $\mathcal{H}$ of operators of the same kind which is a left inverse of $\mathcal{K}$, that is $\mathcal{H}\mathcal{K}$ is the identity operator. The norm of $\mathcal{H}$ in the various spaces L^p, $T_u^p(x_0)$ depends only on the space, the least upper bound of the absolute values of the coefficients of $\mathcal{L}$ and the constant of ellipticity μ of $\mathcal{L}$.

Proof. According to [7], Theorem 7, the operator $\mathcal{L}$ has the representation (6.1) and the matrix of symbols of operators in $\mathcal{K}$ is precisely the matrix

$$\sigma(\mathcal{K}) = (-i)^m \sum a_a z^a |z|^{-m},$$

where $\sum a_a \xi^a$ is the characteristic matrix of $\mathcal{L}$. The assumption of ellipticity implies that $\sigma(\mathcal{K})$ is of rank r for all $z \neq 0$; consequently if $\sigma(\mathcal{K})^*$ denotes the conjugate transposed matrix, then $\sigma(\mathcal{K})^*\sigma(\mathcal{K})$ is a positive self-adjoint $r\times r$ matrix, which consequently is invertible for all $z \neq 0$. Consider now the matrix $\sigma(\mathcal{H}) = [\sigma(\mathcal{K})^*\sigma(\mathcal{K})]^{-1}\sigma(\mathcal{K})^*$. The entries of this matrix are functions of z which are homogeneous of degree zero, and so according to Theorem 3 in [7], there exists an $r\times s$ matrix $\mathcal{H}$ of convolution singular integral operators whose symbols coincide with $\sigma(\mathcal{H})$. Thus, according to [7], Theorem 4, $\sigma(\mathcal{H}\mathcal{K}) = \sigma(\mathcal{H})\sigma(\mathcal{K}) = I$, where I is the identity $r\times r$ matrix, and consequently $\mathcal{H}\mathcal{K}$ is the identity operator. The norm of the operator $\mathcal{H}$ in L^p, $T_u^p(x_0)$ can be estimated using Theorem 14. In what follows we will deal with matrices of operators as a single operator acting on vector valued functions, and we shall apply to this case our previous results without further explanations. Here we only add that by the norm of a vector-valued function we mean the sum of the norms of its components.

LEMMA 6.2. *Let $\mathcal{L}$ be a differential operator of order m which is elliptic at x_0 in the sense of Definition 5 and has coefficients in $T_u(x_0)$, $u>0$. Let $\mathcal{L}_0$ be the operator obtained from $\mathcal{L}$ by evaluating its leading coefficients at x_0. Let $f \in L_m^p$ and let $h = (1-\Delta)^{m/2}f$ if m is even and $h = (1-\Delta)^{(m-1)/2}(i+\Lambda)f$ if m is odd, where Δ is the Laplacian operator. Then, if $\mathcal{K}\Lambda^m = \mathcal{L}_0$ is the representation of $\mathcal{L}_0$ described in Lemma 6.1, $\mathcal{H}$ is the left inverse of $\mathcal{K}$ and $\mathcal{L}f = g$, we have*

$$h = \mathcal{H}g + \mathcal{H}(\mathcal{L}_0 - \mathcal{L})f + (\mathcal{M}_1 + \Lambda\mathcal{M}_2)f, \tag{6.2}$$

where $\mathcal{M}_1$ and $\mathcal{M}_2$ are differential operators with constant coefficients of orders $m-1$ and $m-3$ respectively which depend only on m. When m is even, the order of $\mathcal{M}_1$ is $m-2$ and $\mathcal{M}_2 = 0$.

Proof. We have

$$\mathcal{L}f = \mathcal{L}_0 f + (\mathcal{L}-\mathcal{L}_0)f = g,$$

and representing $\mathcal{L}_0$ as in Lemma 6.1 and multiplying the equation by the left inverse $\mathcal{H}$ or $\mathcal{K}$ we obtain

$$\Lambda^m f = \mathcal{H}g + \mathcal{H}(\mathcal{L}_0 - \mathcal{L})f.$$

Now it follows immediately from the definition of Λ that $\Lambda^2 = -\Delta$ (see also [7], p. 909, Corollary) and thus we have

$$h = (1-\Delta)^{m/2} f = \mathcal{H}g + \mathcal{H}(\mathcal{L}_0-\mathcal{L})f + [(1-\Delta)^{m/2} - (-\Delta)^{m/2}]f$$

if m is even; and if m is odd,

$$\begin{aligned} h &= (1-\Delta)^{(m-1)/2}(i+\Lambda)f \\ &= \mathcal{H}g + \mathcal{H}(\mathcal{L}_0-\mathcal{L})f + [(1-\Delta)^{(m-1)/2}(i+\Lambda) - (-\Delta)^{(m-1)/2}\Lambda]f. \end{aligned}$$

The lemma is thus established.

LEMMA 6.3. *Under the assumptions of Lemma 6.2, if $u \geqslant v \geqslant w \geqslant -n/p$, $v-w \leqslant \min(u, 1)$ and if in addition neither v nor w is an integer, then*

$$T_v^p(x_0, h) \leqslant T_v^p(x_0, \mathcal{H}g) + C_{pvwm}(1+MN)T_w^p(x_0, h),$$

where M is a bound for the norms in $T_u(x_0)$ of the coefficients of $\mathcal{L}$, and N is the norm of $\mathcal{H}$ as an operator on $T_v^p(x_0)$.

Proof. Referring to formula (6.2) we have

$$T_v^p(x_0, \mathcal{M}_1 f) \leqslant C_m \sum_{|\beta| \leqslant m-1} T_v^p(x_0, f_\beta), \quad \text{where} \quad f_\beta = \left(\frac{\partial}{\partial x}\right)^\beta f. \tag{6.3}$$

Since $\Lambda = i\sum \mathcal{R}_j \partial/\partial x_j$ it follows from Theorem 6 that

$$T_v^p(x_0, \Lambda\mathcal{M}_2 f) \leqslant C_{pvm} \sum_{|\beta| \leqslant m-2} T_v^p(x_0, f_\beta). \tag{6.4}$$

Similarly, if S_1 denotes the sum of terms of order $\leqslant m-1$ of $\mathcal{L}_0 - \mathcal{L}$, then

$$T_v^p(x_0, \mathcal{H}S_1 f) \leqslant N T_v^p(x_0, S_1 f),$$

and, according to Lemma 2.4,

$$T_v^p(x_0, S_1 f) \leqslant CM \sum_{|\beta| \leqslant m-1} T_v^p(x_0, f_\beta).$$

Thus

(6.5) $$T_v^p(x_0, \mathcal{H}S_1 f) \leqslant CMN \sum_{|\beta| \leqslant m-1} T_v^p(x_0, f_\beta).$$

Let now S_2 denote the sum of terms of order m of $\mathcal{L}_0 - \mathcal{L}$. Since the coefficients of S_2 vanish at x_0 and belong to $T_u(x_0)$, Lemma 2.5 gives

$$T_v^p(x_0, S_2 f) \leqslant CM \sum_{|\beta|=m} T_w^p(x_0, f_\beta).$$

Thus

(6.6) $$T_v^p(x_0, \mathcal{H}S_2 f) \leqslant NT_v^p(x_0, S_2 f) \leqslant CMN \sum_{|\beta|=m} T_w^p(x_0, f_\beta).$$

Combining (6.3) to (6.6) we obtain

(6.7) $$T_v^p(x_0, h)$$
$$\leqslant T_v^p(x_0, \mathcal{H}g) + C_{vpm}\Big[(1+MN) \sum_{|\beta| \leqslant m-1} T_v^p(x_0, f_\beta)\Big] + CMN \sum_{|\beta|=m} T_w^p(x_0, f_\beta).$$

Now, if m is even, then $h = (1-\Delta)^{m/2} f = J^{-m} f$ or $f = J^m h$. Thus

$$\left(\frac{\partial}{\partial x}\right)^a f = f_a = \left(\frac{\partial}{\partial x} J\right)^a J^{m-|a|} h,$$

whence by Theorem 15, Lemma 2.1 and Theorem 4 we have if $|a| < m$,

(6.8) $$T_v^p(x_0, f_a) \leqslant C_{pvm} T_v^p(x_0, J^{m-|a|}h) \leqslant C_{pvm} T_{w+m-|a|}^p(x_0, J^{m-|a|}h)$$
$$\leqslant C_{pvwm} T_w^p(x_0, h),$$

and, if $|a| = m$,

(6.9) $$T_w^p(x_0, f_a) \leqslant C_{pwm} T_w^p(x_0, h).$$

If on the other hand m is odd, then, since $\Lambda^2 = -\Delta$,

$$h = (1-\Delta)^{(m-1)/2}(i+\Lambda)f, \quad J^{m-1}h = (i+\Lambda)f,$$

$$J^{m+1}h = J^2(i+\Lambda)f = (i+\Lambda)J^2 f, \quad (-i+\Lambda)J^{m+1}h = (1-\Delta)J^2 f = f,$$

and on account of the definition of Λ and of Lemma 4.3 we have

$$\left(\frac{\partial}{\partial x}\right)^a f = \left[i \sum_{j=1}^{m} \mathcal{R}_j \left(\frac{\partial}{\partial x_j} J\right) - iJ\right] \left(\frac{\partial}{\partial x} J\right)^a J^{m-|a|} h.$$

From this and Lemma 2.1 we find that the inequalities (6.8) and (6.9) also hold in this case.

Combining (6.7) with (6.8) and (6.9) the inequality of the lemma follows.

The next lemma states a known result.

LEMMA 6.4. *Let $\mathcal{L}$ be a uniformly elliptic differential operator of order m with bounded coefficients whose leading coefficients are uniformly continuous. Then there exists a constant A depending on $\mathcal{L}$ and p such that*

$$\|f\|_{p,m} \leqslant A[\|\mathcal{L}f\|_p + \|f\|_p]$$

for every f in L_m^p, $1 < p < \infty$.

Proof. We assume first that the leading coefficients of $\mathcal{L}$ have bounded continuous derivatives of the first two orders. Let now $\mathcal{L}_0$ denote the principal part of the operator $\mathcal{L}$, that is, the sum of its terms of order m. Then according to Theorem 7 of [7] we have $\mathcal{L}_0 = \mathcal{K}\Lambda^m$ where $\mathcal{K}$ is a singular integral operator whose symbol is given by

$$\sigma(\mathcal{K}) = (-i)^m \sum a_a(x) z^a |z|^{-m},$$

where $\sum a_a(x)\xi^a$ is the characteristic matrix of $\mathcal{L}$. The assumption of uniform ellipticity implies that the matrix $\sigma(\mathcal{K})$ has a left inverse $\sigma(\mathcal{H})$ which is the symbol of an operator of class C_2^∞ (see Definition 2 in [7]). Now we have

$$\mathcal{L}f = \mathcal{L}_0 f + (\mathcal{L}-\mathcal{L}_0)f = \mathcal{K}\Lambda^m f + (\mathcal{L}-\mathcal{L}_0)f = g,$$

and multiplying the equation on the left by $\mathcal{H}$ we get

$$\mathcal{H}\mathcal{K}\Lambda^m f = \mathcal{H}g - \mathcal{H}(\mathcal{L}-\mathcal{L}_0)f,$$

$$\Lambda^m f = \mathcal{H}g - \mathcal{H}(\mathcal{L}-\mathcal{L}_0)f + [(I-\mathcal{H}\mathcal{K})\Lambda]\Lambda^{m-1}f.$$

Since $\sigma(\mathcal{H})\sigma(\mathcal{K}) = I$ it follows from Theorem 5 of [7] that the operator in square brackets on the right is bounded on L^p. On the other hand, $\mathcal{H}$ is also bounded on L^p, and so taking norms in the last equation we find that

$$\|\Lambda^m f\|_p \leqslant C_{\mathcal{L},p}\Big(\sum_{|\beta|\leqslant m-1} \|f_\beta\|_p + \|\Lambda^{m-1} f\|_p\Big) + \|\mathcal{H}g\|_p.$$

Now, according to Lemma 4.3,

$$\left(\frac{\partial}{\partial x}\right)^a = (-i)^a \mathcal{R}^a \Lambda^a, \qquad \mathcal{R}^a = \mathcal{R}_1^{a_1}\mathcal{R}_2^{a_2}\ldots\mathcal{R}_n^{a_n},$$

which implies that for $|a| = m$ we have

$$\|f_a\|_p \leqslant C_{ap}\|\Lambda^m f\|_p.$$

On the other hand,

$$\Lambda = i\sum \mathcal{R}_j \frac{\partial}{\partial x_j},$$

and this implies that

$$\|\Lambda^{m-1} f\|_p \leqslant C_{p,m} \sum_{|\beta|=m-1} \|f\|_\beta .$$

This combined with the previous inequalities gives

$$(6.10) \qquad \sum_{|\alpha|=m} \|f_\alpha\|_p \leqslant C_{\mathcal{L}p} \|f\|_{p,m-1} + \|\mathcal{H} g\|_p .$$

Now we use an inequality given in [11], p. 125, namely

$$\|f_\beta\|_p \leqslant C_{pm} \Big(\sum_{|\alpha|=m} \|f_\alpha\|_p \Big)^\theta \|f\|_p^{1-\theta} = C_{pm} \Big(\varepsilon \sum_{|\alpha|=m} \|f_\alpha\|_p \Big)^\theta (\varepsilon^{-\theta/(1-\theta)} \|f\|_p)^{1-\theta}$$

$$\leqslant C_{pm} \Big[\varepsilon \sum_{|\alpha|=m} \|f_\alpha\|_p + \varepsilon^{-\theta/(1-\theta)} \|f\|_p \Big],$$

where $\theta = (m-|\beta|)/m$, $|\beta| < m$ and ε is an arbitrary positive number. Consequently we have

$$(6.11) \qquad \|f\|_{p,m-1} \leqslant \varepsilon \sum_{|\alpha|=m} \|f_\alpha\|_p + C_{p,\varepsilon,m} \|f\|_p ,$$

and from this and (6.10) we obtain

$$(6.12) \qquad \sum_{|\alpha|=m} \|f_\alpha\|_p \leqslant C_{\mathcal{L}p} \|f\|_p + 2 \|\mathcal{H} g\|_p .$$

In the general case, given an elliptic operator $\mathcal{L}$ whose leading coefficients are merely uniformly continuous, we approximate $\mathcal{L}$ by an operator $\bar{\mathcal{L}}$ which has the same lower order coefficients but whose leading coefficients have continuous bounded derivatives of the first two orders. Then Theorem 3 in [7] and the uniform ellipticity of $\mathcal{L}$ permit us to assert that it is possible to approximate the coefficients of $\mathcal{L}$ by those of $\bar{\mathcal{L}}$ uniformly, by less than any preassigned number ε, keeping at the same time the norm in L^p of the operator $\mathcal{H}$ associated with $\bar{\mathcal{L}}$ bounded, say, less than N. Thus from (6.12) we will have

$$\bar{\mathcal{L}} f = g + (\bar{\mathcal{L}} - \mathcal{L}) f ,$$

$$\sum_{|\alpha|=m} \|f_\alpha\|_p \leqslant C_{\bar{\mathcal{L}}p} \|f\|_p + 2 \|\mathcal{H} g\|_p + 2 \|\mathcal{H} (\bar{\mathcal{L}} - \mathcal{L}) f\|_p$$

$$\leqslant C_{\bar{\mathcal{L}}p} \|f\|_p + 2N \|g\|_p + 2N \|(\bar{\mathcal{L}} - \mathcal{L}) f\|_p$$

$$\leqslant C_{\bar{\mathcal{L}}p} \|f\|_p + 2N \|g\|_p + 2N\varepsilon \sum_{|\alpha|=m} \|f_\alpha\|_p ,$$

and choosing ε so that $N\varepsilon < \frac{1}{4}$ we obtain finally

$$(6.12a) \qquad \sum_{|\alpha|=m} \|f_\alpha\|_p \leqslant C_{\mathcal{L}p} \|f\|_p + 4N \|g\|_p .$$

This combined with (6.11) gives the desired result.

Remark. It may be worth noting that the coefficient $4N$ of $\|g\|_p$ in (6.12a) merely depends on the bounds for the leading coefficients of $\mathcal{L}$ and the bounds for uniform ellipticity.

Proof of Theorem 1. We refer to Lemmas 6.2 and 6.3. We will show first that if $g \in T_v^p(x_0)$, $1 < p < \infty$, v not an integer and the coefficients of $\mathcal{L}$ are in $T_u(x_0)$, $u > 0$, $u \geqslant v$, then $h \in T_v^p(x_0)$ and satisfies an appropriate inequality.

Since f is assumed to belong to L_m^p, the function h belongs to L^p, and thus also to $T_{-n/p}^p(x_0)$, and

$$T_{-n/p}^p(x_0, h) \leqslant 2\|h\|_p \leqslant C_{mp}\|f\|_{p,m}.$$

The last inequality is obvious from the definition of h if m is an even integer. If m is odd we use the fact stated in Lemma 4.3 that Λ maps L_m^p continuously into L_{m-1}^p. Let now k be an integer such that for $\nu = 1, 2, \ldots, k$

$$-\frac{n}{p} + \frac{\nu}{k}\left(v + \frac{n}{p}\right) = -\frac{n}{p} + \nu\delta$$

is never integral, and

$$\delta = \frac{1}{k}\left(v + \frac{n}{p}\right) \leqslant \min(u, 1).$$

Then the inequality of Lemma 6.3 gives

$$T_{-n/p+(\nu+1)\delta}^p(x_0, h) \leqslant T_{-n/p+(\nu+1)\delta}^p(x_0, \mathcal{H}g) + C_{pvm}(1+NM)T_{-n/p+\nu\delta}^p(x_0, h),$$

$\nu = 0, 1, \ldots, k-1$. On the other hand, we have

$$T_{-n/p+(\nu+1)\delta}^p(x_0, \mathcal{H}g) \leqslant CT_v^p(x_0, \mathcal{H}g) \leqslant CMT_v^p(x_0, g),$$

where M is, as in Lemma 6.3, the norm of the operator $\mathcal{H}$ in $T_v^p(x_0)$. Thus

$$T_{-n/p+(\nu+1)\delta}^p(x_0, h) \leqslant CMT_v^p(x_0, g) + C_{pvm}(1+NM)T_{-n/p+\nu\delta}^p(x_0, h).$$

If we write $a_\nu = T_{-n/p+\nu\delta}^p(x_0, h)$, $\beta = CMT_v^p(x_0, g)\gamma$ and $\gamma = [C_{pvm}(1+$ $+NM)]^{-1}$ the last inequality can be written as

$$\gamma a_{\nu+1} \leqslant \beta + a_\nu, \qquad \gamma^{\nu+1}a_{\nu+1} \leqslant \gamma^\nu\beta + \gamma^\nu a_\nu,$$

and summing from $\nu = 0$ to $\nu = k-1$ we find

$$\gamma^k a_k \leqslant \beta\frac{1-\gamma^k}{1-\gamma} + a_0.$$

If we replace C_{pvm} by $2 + C_{pvm}$ then $\gamma \leqslant \frac{1}{2}$ and

$$a_k \leqslant [2\beta + a_0]\gamma^{-k};$$

consequently,

$$T_v^p(x_0, h) \leqslant C_{pvm}[MT_v^p(x_0, g)(1+NM)^{k-1}+T^p_{-n/p}(x_0, h)(1+NM)^k]$$
$$\leqslant C_{pvm}(1+NM)^k[T_v^p(x_0, g)+\|f\|_{p,m}].$$

Now we use the identities in the proof of Lemma 6.3,

$$f_a = \left(\frac{\partial}{\partial x}\right)^a f = J^{m-|a|}\left(\frac{\partial}{\partial x}J\right)^a h \tag{6.13}$$

if m is even, and

$$f_a = \left(\frac{\partial}{\partial x}\right)^a f = J^{m-|a|}\left[i\sum_{j=1}^{n}\mathcal{R}_j\left(\frac{\partial}{\partial x_j}\right) - iJ\right]\left(\frac{\partial}{\partial x}J\right)^a h \tag{6.14}$$

if m is odd, and using Theorems 4, 6, 15 we finally obtain, with q as in Theorem 1,

$$T^q_{v+m-|a|}(x_0, f_a) \leqslant C_{pvm}[1+NM]^k[N^{-1}T_v^p(x_0, g)+\|f\|_{p,m}].$$

Combining these results with Lemma 6.4 we obtain parts (i) and (ii) of Theorem 1.

To show (iii) we return to equation (6.2) and show that under the present assumptions $h \in t_v^p(x_0)$. First we observe that the argument given above shows not only that $f_a \in T^q_{v+m-|a|}(x_0)$ but also $f_a \in T^p_{v+m-|a|}(x_0) \subset t_v^p(x_0)$ if $|a| < m$. On the other hand, on account of Theorem 6, $\mathcal{H}g \in t_v^p(x_0)$. Finally, the leading terms of $(\mathcal{L}_0-\mathcal{L})f$ have coefficients which vanish at x_0; since $f_a \in T_v^p(x_0)$ it follows from Lemma 2.5 that these terms represent functions in $t_v^p(x_0)$. Consequently $(\mathcal{L}_0-\mathcal{L})f \in t_u^p(x_0)$ and by Theorem 6 again $\mathcal{H}(\mathcal{L}_0-\mathcal{L})f \in t_u^p(x_0)$. Thus all terms on the right of equation (6.2) are functions in $t_u^p(x_0)$, and $h \in t_u^p(x_0)$. Using the representation of f_a in terms of h given above and applying the same theorems we conclude that $f_a \in t^q_{v+m-|a|}(x_0)$. Theorem 1 is thus established.

Proof of Theorem 2. We refer again to equation (6.2). We may suppose without loss of generality that Q is bounded and we will show that $h \in T_v^p(x_0)$ for almost all x_0 in Q. If $g \in T_v^p(x_0)$ and v is an integer, then also $g \in T_w^p(x_0)$ with w slightly smaller than v and non-integral and, as was seen in the proof of Theorem 1, $f_a \in T^p_{w+m-|a|}(x_0) \subset T_v^p(x_0)$ for $|a| < m$, and $f_a \in T_w^p(x_0)$ for all $|a| \leqslant m$. Let $P(x_0, x) = \sum_{|\beta|<w+m} a_\beta(x_0)(x-x_0)^\beta$ be the Taylor expansion of f at x_0. Then $(\partial/\partial x)^a P(x_0, x)$ is the Taylor expansion of $f_a(x_0)$ at every point x_0 where $g \in T_w^p(x_0)$ and where, consequently, also $f \in T^p_{w+m}(x_0)$ and $f_a \in T^p_{w+m-|a|}(x_0)$. To see this we merely have to consider the functions $f^\lambda = \lambda^n\varphi(\lambda x) * f$ where φ is a function of C_0^∞ of integral equal to 1; then, according to Lemma 2.3, f^λ converges to f in $T^p_{w+m-\varepsilon}(x_0)$ as λ tends to infinity and $(\partial/\partial x)^a f^\lambda = \lambda^n\varphi(\lambda x) * f_a$ converges to f_a in $T^p_{w+m-|a|-\varepsilon}(x_0)$. This clearly implies that the coefficients of $(\partial/\partial x)^a P(x_0, x)$ are the coefficients of the

Taylor expansion of f_a. Let now ψ be a function in C_0^∞ which is equal to 1 on Q, and let us rewrite equation (6.2) as follows

$$h = \mathcal{H}g + \mathcal{H}(\mathcal{L}_0 - \mathcal{L})[f - P(x_0, x)\psi(x)] + \\ + \mathcal{H}(\mathcal{L}_0 - \mathcal{L})P(x_0, x)\psi(x) + (\mathcal{M}_1 + \Lambda\mathcal{M}_2)f,$$

or, setting now $P(x_0, x) = \sum_{|\beta| < m+w} b_\beta(x_0) x^\beta$,

$$(6.15)\qquad h = \mathcal{H}g + \mathcal{H}(\mathcal{L}_0 - \mathcal{L})[f - P(x_0, x)\psi(x)] + \\ + \sum_{|\beta| < m+w} b_\beta(x_0)\mathcal{H}(\mathcal{L}_0 - \mathcal{L})x^\beta\psi(x) + (\mathcal{M}_1 + \Lambda\mathcal{M}_2)f.$$

Evidently $(\partial/\partial x)^a x^\beta \psi(x) \in T_v^p(x_0)$ for all x, a, and β, and $\mathcal{L}\{x^\beta\psi(x)\} \in T_v^p(x_0)$ for all $x_0 \in Q$. Consequently, by Theorem 7, there exists a subset $\bar{Q}_1$ of Q of full measure such that $\mathcal{H}(\mathcal{L}_0 - \mathcal{L})x^\beta\psi(x) \in T_v^p(x_0)$ for all $x_0 \in \bar{Q}_1$, regardless of the choice of the operator $\mathcal{H}$, which in our case depends on x_0. Similarly, for x_0 in another subset $\bar{Q}_2$ of Q of full measure we have $\mathcal{H}g \in T_v^p(x_0)$ for all $x_0 \in \bar{Q}_2$. On the other hand, since as we saw $f_a \in T_v^p(x_0)$ for $|a| < m$ and all x_0 in Q, we have $\mathcal{M}_1 f \in T_v^p(x_0)$ for $x_0 \in Q$, and using the definition of Λ and Theorem 7 again we conclude that $\Lambda\mathcal{M}_2 f \in T_v^p(x_0)$ for almost all $x_0 \in Q$.

There remains the second term on the right of (6.15). We note first that

$$(\mathcal{L}_0 - \mathcal{L})[f - P(x_0, x)\psi(x)] = \sum_{|\beta| \leqslant m} c_\beta(x) \left(\frac{\partial}{\partial x}\right)^\beta [f - P(x_0, x)\psi(x)],$$

where the $c_\beta(x)$ are functions in $T_u(x_0)$ which *vanish* at x_0, and $(\partial/\partial x)^\beta[f - P(x_0, x)]$ are functions in $T_w^p(x_0)$ which also vanish at x_0, for every x_0 in Q. Consequently by Lemma 2.5, part (iii), these functions belong to $T_{w+1}^p(x_0)$ for all $x_0 \in Q$. Now, since $w+1$ is not an integer, Theorem 6 guarantees that

$$\mathcal{H}(\mathcal{L}_0 - \mathcal{L})[f - P(x_0, x)\psi(x)]$$

belongs to $T_{w+1}^p(x_0) \subset T_v^p(x_0)$ for all x_0 in Q. Summarizing we have found that $h \in T_v^p(x_0)$ for almost all x_0 in Q.

Using now the representation of f_a given in (6.13) and (6.14) and applying theorems 15, 7, 5 we obtain the desired result.

Proof of Theorem 3. This theorem is an immediate consequence of Theorem 1. For under the given assumptions, it follows that $f \in T_{v+m}^p(x_0)$ for all $x_0 \in Q$, and that $T_{v+m}^p(x_0, f)$ is bounded on Q; and an application of Theorem 9 gives the desired result. If $g \in t_v^p(x_0)$ for all $x_0 \in Q$, then $f \in t_{v+m}^p(x_0)$ for all $x_0 \in Q$ and Theorem 9 asserts that $f \in b_{v+m}(Q)$.

7. In this section we make additional observations about the solutions of the equation $\mathcal{L}f = g$.

It is well known that Theorem 1 is not true when v is a non-negative integer (Theorem 2 is a substitute result), and the simplest illustration is the equation $\Delta f = g$: if g is merely continuous, f is not necessarily twice differentiable. The only thing we can then say is that f has continuous first order derivatives $f_j = \partial f/\partial x_j$ $(j = 1, 2, \ldots, n)$ which in turn satisfy a Hölder condition of any order less than 1, or more precisely, have a modulus of continuity $\omega(h) = o\left(|h| \log \frac{1}{|h|}\right)$.

It is however not difficult to see, and this result is a special case of Theorem 16 below, that if g is continuous at a given point x_0, then any solution f of $\Delta f = g$ has continuous derivatives f_j near x_0 and the f_j satisfy at that point the condition of "smoothness":

$$f_j(x_0+h)+f_j(x_0-h)-2f_j(x_0) = o(|h|), \quad |h| \to 0. \tag{7.1}$$

(this notion was first introduced by Riemann).

This result can be interpreted as the differentiability at $h = 0$ of the "even part" $\frac{1}{2}[f_j(x_0+h)+f_j(x_0-h)]$ of $f_j(x_0+h)$, and from this it is easy to deduce that the "odd part" $\frac{1}{2}[f(x_0+h)-f(x_0-h)]$ of $f(x_0+h)$ has a second differential at $h = 0$. This fact gives a clue to the situation in the general case. We may add, and the result is familiar enough (see e. g., 15_{I}, p. 44), that if a continuous function $g(x)$ satisfies the condition $g(x+h)+g(x-h)-2g(x) = o(|h|)$ uniformly in a domain, then in every compact subdomain g has modulus of continuity $o\left(|h| \log \frac{1}{|h|}\right)$, so that the result about the modulus of continuity of the derivatives f_j in the case of the equation $\Delta f = g$ is a corollary of 7.1.

Definition 10. We will say that $f(x)$ *belongs to the class* $\Lambda_u^p(x_0)$, where u is a non-negative integer, if $\frac{1}{2}[f(x_0+h)-(-1)^u f(x_0-h)]$ belongs to $T_u^p(0)$ as a function of h. We will say that $f(x)$ *belongs to* $M_u^p(x_0)$, where u is again a non-negative integer, if $\frac{1}{2}[f(x_0+h)+(-1)^u f(x_0-h)]$ belongs to $T_u^p(0)$ as a function of h. By replacing $T_u^p(0)$ by $t_u^p(0)$ we obtain the definitions of $\lambda_u^p(x_0)$ and $\mu_u^p(x_0)$.

THEOREM 16. *Let $\mathfrak{L}f = g$ be an equation of order m with coefficients in $T_v(x_0)$, which is elliptic at x_0 in the sense of Definition 5. Let u, $u < v$, be a non-negative integer. Then, if m is even and $g \in T_w^p(x_0)$, $1 < p < \infty$, $w > u-1$, and $g \in \Lambda_u^p(x_0)$, the function $f_\alpha = (\partial/\partial x)^\alpha f$ belongs to $\Lambda_{u+m-|\alpha|}^q(x_0)$, where p and q are related as in Theorem 1. If m is odd the same conclusion holds with the classes $\Lambda_u^p(x_0)$ replaced by $M_u^p(x_0)$. Analogous conclusions hold with classes Λ replaced by λ, or M by μ, as the case may be.*

We will omit proving in detail the preceding statement, and will confine ourselves to merely sketching the main line of the argument. Referring to Lemma 6.2, one shows first that the function h on the left-hand side of formula (6.2) belongs to $\Lambda_u^p(x_0)$ or $M_u^p(x_0)$ as the case may be. According to Theorem 1, and the hypothesis made above, f and its derivatives of orders less than m belong to $T_{w+1}^p(x_0)$ and the derivatives of order m belong to $T_w^p(x_0)$. Since the leading coefficients of $\mathcal{L}-\mathcal{L}_0$ vanish at x_0 and belong to $T_v(x_0)$ where $v>u$, it follows that $(\mathcal{L}_0-\mathcal{L})f$ belongs to a class $T_r^p(x_0)$, with $r>u$ (see Lemma 2.5) and by Theorem 6 the same is true of $\mathcal{H}(\mathcal{L}_0-\mathcal{L})f$. Consequently all terms on the right of (6.2) except $\mathcal{H}g$ belong to $T_u^p(x_0)$ and a fortiori, also to $\Lambda_u^p(x_0)$, or $M_u^p(x_0)$.

Consider now the term $\mathcal{H}g$. If m is even then the symbol of the operator $\mathcal{H}$ is an even function and thus also the kernel of the operator is an even function. Let $O_u f$ be the function of x given by $\frac{1}{2}[f(x_0+x)-(-1)^u f(x_0-x)]$ and thus, since the kernel of $\mathcal{H}$ is even we find that $O_u(\mathcal{H}g)=\mathcal{H}(O_u g)$ and applying Lemma 5.2 we conclude that $O_u(\mathcal{H}g)$ belongs to $T_u^p(x_0)$, that is $\mathcal{H}g\in\Lambda_u^p(x_0)$. A similar argument shows that $\mathcal{H}g\in M_u^p(x_0)$ if m is odd. Consequently h belongs to $\Lambda_u^p(x_0)$ or $M_u^p(x_0)$ according to the parity of m. Once this has been established we obtain f and its derivatives f_a from h by means of the identities (6.13) and (6.14). In case m is even we obtain

$$O_{m+u}f = O_{m+u}J^m h = J^m O_{m+u}h,$$

and taking into account the remark to the proof of Theorem 4 (p. 198), we find that $O_{m+u}f\in T_{u+m}^q(x_0)$, that is $f\in\Lambda_{u+m}^q(x_0)$. A similar argument employing Theorem 15 (for which a remark analogous to that to Theorem 4 holds) gives the desired result for the derivatives of f. The case of odd m is treated similarly.

Inequalities for the norms can also be obtained by this argument.

Bibliography

[1] N. Aronszajn and K. T. Smith, *Theory of Bessel potentials*, Part I, Studies in Eigenvalue Problems, Technical Report 22, University of Kansas, Department of Mathematics (1959), p. 1-113.

[2] H. Bateman, *Tables of integral transforms*, New York 1954.

[3] — *Higher transcendental functions*, New York 1953.

[4] A. P. Calderón, *On the differentiability of absolutely continuous functions*, Rivista de Math. Univ. Parma 2 (1951), p. 203-212.

[5] — *Singular integrals*, Course notes of lectures given at the M. I. T., 1958-59.

[6] A. P. Calderón and A. Zygmund, *On singular integrals*, Amer. Jour. of Math. 78. 2 (1956), p. 290-309.

[7] — *Singular integral operators and differential equations*, ibidem 79 (1957), p. 901-921.

[8] L. Cesàri, *Sulle funzioni assolutamente continue in due variabili*, Annali di Pisa 10 (1041), p. 91-101.

[9] J. Horváth, *Sur les fonctions conjuguées à plusieurs variables*, Indagationes Math. (1953), p. 17-29.

[10] J. Marcinkiewicz, *Sur les séries de Fourier*, Fund. Math. 27 (1936), p. 38-69.

[11] L. Nirenberg, *On elliptic partial differential equations*, Annali della Scuola Norm. Sup. Pisa 13 (1959), p. 116-162.

[12] W. H. Oliver, a) *Differential properties of real functions*, Ph. D. Dissertation, Univ. of Chicago, 1951, p. 1-109; b) *An existence theorem for the n-th Peano differential*, Abstract, Bull. Amer. Math. Soc. 57 (1951), p. 472.

[13] G. H. Watson, *A treatise on the theory of Bessel functions*, Cambridge 1944.

[14] H. Whitney, *Analytic extensions of differentiable functions defined in closed sets*, Trans. Amer. Math. Soc. 36 (1934), p. 63-89.

[15] A. Zygmund, *Trigonometric Series*, vols I and II, Cambridge 1959.

UNIVERSITY OF CHICAGO

Reçu par la Rédaction le 1. 10. 1960

STUDIA MATHEMATICA, T. XXIII. (1964)

On the differentiability of functions

by

E. M. STEIN and A. ZYGMUND (Chicago)

Dedicated to E. Hille
on the occasion of his 70-th birthday

Chapter I

1. In this paper* we extend and generalize the main results of our paper [9]. The knowledge of [9], however, is not indispensable here.

In this chapter we formulate the main results of the paper. Their proofs are given in Chapter II. Chapter III contains some additional results.

Let $F(x)$ be a function defined in the neighborhood of the point x_0 (in what follows we consider only measurable sets and functions). The two functions

$$\varphi_{x_0}(t) = \varphi_{x_0}(t; F) = \tfrac{1}{2}[F(x_0+t)+F(x_0-t)],$$

$$\psi_{x_0}(t) = \psi_{x_0}(t; F) = \tfrac{1}{2}[F(x_0+t)-F(x_0-t)],$$

whose sum is equal to $F(x_0+t)$, will be called respectively the *even* and *odd* part of $F(x_0+t)$; we shall also use the expression the *even* and *odd part of F at x_0*.

These parts are of importance in certain problems of the Theory of Functions and, in particular, in Fourier series. Let $S[F]$ denote the Fourier series of a periodic function F (by "periodic" we shall always mean "of period 2π") and $\tilde{S}[F]$ the conjugate series. By $S^{(k)}[F]$ and $\tilde{S}^{(k)}[F]$ we shall mean the series $S[F]$ and $\tilde{S}[F]$ differentiated termwise k times. It is a familiar fact that for the summability (and, in particular, convergence) of $S[F]$ at a given point x_0 decisive is the behavior of the even part $\varphi_{x_0}(t; F)$ near $t = 0$. The same holds for the summability of $S^{(k)}[F]$ if k is even and the summability of $\tilde{S}^{(k)}[F]$ if k is odd. Similarly, the behavior of $\psi_{x_0}(t; F)$ near $t = 0$ is decisive for the summability of $S^{(k)}[F]$ if k is odd and of $\tilde{S}^{(k)}[F]$ if k is even.

* Research resulting in this paper was partly supported by the Air Force contract AF-AFOSR-62-118 and the NSF contract GP-574.

Reprinted from *SM* 23, 247–283 (1963/1964).

In the present paper we are primarily interested in the problems of the differentiability of the even and odd parts of a function. The problems belong essentially to Real Variable, but the methods we use lean heavily on the theory of Fourier series and integrals and Complex Variable; in view of the remarks just made about Fourier series this is rather natural.

2. To make the picture more clear we begin with the case of derivatives of order 1.

The differentiability of the odd part $\psi_{x_0}(t; F)$ at $t = 0$ is the same thing as the existence of the first symmetric derivative

$$\lim_{t\to 0} \frac{F(x_0+t)-F(x_0-t)}{2t}$$

of F at x_0. The differentiability of the even part $\varphi_{x_0}(t; F)$ at $t = 0$ is clearly equivalent to the relation

$$F(x_0+t)+F(x_0-t)-2F(x_0) = o(t). \tag{1}$$

The latter relation is usually called the *smoothness* of the function F at the point x_0 and was first considered by Riemann in his memoir on trigonometric series. Functions which are continuous and smooth at each point have a number of interesting properties (see [12] or [11_{I}], p. 42 sqq.). If (1) holds we shall also say that *F satisfies condition* λ *at* x_0. If we merely have

$$F(x_0+t)+F(x_0-t)-2F(x_0) = O(t),$$

we shall say that *F satisfies condition* Λ at x_0 (¹).

(¹) The following reflexion upon the significance of smooth functions may be not totally out of place. In view of the fact that smooth functions play important role in certain problems of the Theory of Functions one may ask about the origin of this importance. The answer is not immediately obvious and one may be easily led to irrelevant notions and generalizations. For example, it may appear that the expression

$$\omega_2(h) = \operatorname*{Max}_{x,\,|t|\leqslant h} |F(x+t)+F(x-t)-2F(x)|$$

is merely an analogue of the modulus of continuity

$$\omega_1(h) = \operatorname*{Max}_{x,\,|t|\leqslant h} |F(x+t)-F(x)|$$

of F, and one is naturally led to considering expressions $\omega_k(h)$ defined in a similar way but using the k-th differences. Such expressions are interesting and useful, but after $k = 1$ only the case $k = 2$ seems to be of real importance, the reason being that the behavior of $\omega_2(h)$ expresses a property of the *even part* of the function. Here, it seems, lies the source of significance of smooth functions. A good illustration are applications of smooth functions to elliptic differential equations discussed in [1].

It is obvious that the differentiability of F at x_0 implies the differentiability of both $\varphi_{x_0}(t)$ and $\psi_{x_0}(t)$ at $t=0$. It is equally clear that neither the differentiability of $\varphi_{x_0}(t)$ at $t=0$, nor that of $\psi_{x_0}(t)$ implies the existence of $F'(x_0)$. It is natural, however, to ask about theorems of the "almost everywhere" type. It turns out that the roles played here by the even and odd parts of the function are completely different. We list a few known results.

THEOREM A. *If F has a first symmetric derivative at each point of a set E, then F is differentiable almost everywhere in E.*

THEOREM B. *There exist continuous functions which satisfy condition* (1) *everywhere, even uniformly in x, and which are differentiable in sets of measure* 0 *only* (²).

If, however, we strengthen condition (1) somewhat, the function becomes differentiable. The precise result is as follows:

THEOREM C. *Let $\varepsilon(t)$ be a function defined in some interval $0 < t \leqslant \eta$, monotonically decreasing to* 0 *with t and such that the integral*

$$\int_0^\eta \frac{\varepsilon^2(t)}{t}\,dt \tag{2}$$

is finite. If for each x belonging to a set E we have

$$F(x+t)+F(x-t)-2F(x) = O\{t\varepsilon(t)\} \qquad (t\to 0), \tag{3}$$

not necessarily uniformly in x, then F is differentiable almost everywhere in E.

(²) Theorems A and B clearly show the difference between $\varphi_{x_0}(t)$ and $\psi_{x_0}(t)$ as regards differentiability. The picture is a little different for the continuity of the even and odd part of F. The problems here are easier and we state a few facts.

The continuity at $t=0$ of the even and odd parts of $F(x_0+t)$ means respectively

(a) $F(x_0+t)+F(x_0-t)-2F(x_0)\to 0$,

(b) $F(x_0+t)-F(x_0-t)\to 0$

for $t\to 0$. It is not difficult to show (see Lemma 9 in Chapter II) that if we have either (a) or (b) at each point of a set E, then F is continuous at almost all points of E. Thus there is no difference between the continuity of the even and odd part of F. (In particular, if F satisfies condition Λ at each point of E, then F is continuous almost everywhere in E).

The result just stated can be generalized as follows. Let $a_1, a_2, \dots, a_k$ be a sequence of real numbers all different, and let $\beta_1, \beta_2, \dots, \beta_k$ be another sequence such that $\sum\beta_i = 0$. Suppose that at some point x_0 we have

(c) $\sum\beta_i F(x_0+a_i t)\to 0$

or, what is the same thing, (c′) $\sum\beta_i\{F(x_0+a_i t)-F(x_0)\}\to 0$, as $t\to 0$. We may then say that F is *conditionally* continuous at x_0 (relative to the sequences $\{a_i\}$ and $\{\beta_i\}$). It can be shown that if (c) holds for $t\to +0$ (or $t\to -0$) at each $x_0\in E$, then F is continuous almost everywhere in E. This stems from the fact that F is anyway approximately continuous almost everywhere in E and this coupled with condition (c) gives the desired result. Similarly, if at each $x_0\in E$ the left side of (c) is ultimately bounded as $t\to +0$, then F is bounded in the neighborhood of almost all points of E.

That this result is, in a sense, best possible is shown by the following

THEOREM D. *Let* $\varepsilon(t)$, $0 < t \leqslant \eta$, *be a function monotonically decreasing with* t, *satisfying the condition*

$$\varepsilon(2t)/\varepsilon(t) \to 1$$

for $t \to 0$ *and such that the integral* (2) *diverges. Then there is a continuous function* $F(x)$ *satisfying for all* x *the condition*

$$|F(x+t)+F(x-t)-2F(x)| \leqslant t\varepsilon(t) \quad (0 < t \leqslant \eta) \tag{4}$$

and differentiable in a set of measure 0 *only.*

Theorem A is an old result of Khintchin [3]. The function

$$F(x) = \sum_{n=1}^{\infty} \frac{\sin 2^n x}{2^n n^{1/2}} \tag{5}$$

can be taken for the function of Theorem B (see e. g., [11_{I}], p. 47-48). It is known that continuous functions which are smooth at each point must necessarily have points of differentiability so that the exceptional set of measure 0 in Theorem B cannot be empty (see [11_{I}], p. 43). Theorem C is the main result of paper [1], and it is indicated there that if $\varepsilon(t)$ satisfies the hypotheses of Theorem D, then the continuous function

$$F(x) = \sum \varepsilon\left(\frac{1}{2^n}\right)\frac{\sin 2^n x}{2^n} \tag{6}$$

is differentiable in a set of measure 0 only and satisfies condition (4) if multiplied by a suitable positive constant.

That the function (6) is differentiable in a set of measure 0 only follows from the fact that the divergence of the integral (2) is equivalent to the divergence of $\sum \varepsilon^2(2^{-n})$ so that the lacunary series $\sum \varepsilon(2^{-n})\cos 2^n x$ obtained by the termwise differentiation of (6) is not in L^2 and therefore, as is well known (see [11_{I}], p. 203) cannot be summable by any given linear method of summation except, perhaps, in a set of measure 0 only; in particular, F' can exist in a set of measure 0 only.

The remaining part of the conclusion of Theorem D is easy to verify by a familiar argument. We have

$$F(x+t)+F(x-t)-2F(x) = -4\sum_{1}^{\infty} \varepsilon(2^{-n})2^{-n}\sin 2^n x \sin^2 \tfrac{1}{2}2^n t$$

$$= -4\sum_{1}^{N} -4\sum_{N+1}^{\infty} = P+Q,$$

say, where N is determined by the condition $2^{-N-1} \leqslant t < 2^{-N}$, $N = 1, 2, \ldots$ Clearly,

$$|Q| \leqslant 4 \sum_{N+1}^{\infty} \varepsilon(2^{-n})2^{-n} \leqslant 4 \cdot 2^{-N} \varepsilon(2^{-N-1}) \leqslant 8t\varepsilon(t),$$

$$|P| \leqslant t^2 \sum_{1}^{N} \varepsilon(2^{-n})2^{n} = t^2 \sum_{1}^{N} \varepsilon(2^{-n})(\tfrac{3}{2})^n(\tfrac{4}{3})^n,$$

say. Since $(\frac{3}{2})^n \varepsilon(2^{-n}) \leqslant (\frac{3}{2})^{n+1}\varepsilon(2^{-n-1})$ for n large enough, there is a constant A independent of N such that $(\frac{3}{2})^n\varepsilon(2^{-n}) \leqslant A(\frac{3}{2})^N\varepsilon(2^{-N})$ for $n = 1, 2, \ldots, N$. It follows that $|P| \leqslant O(t^2)\varepsilon(2^{-N})2^N = O\big(t\varepsilon(t)\big)$. Hence, collecting results we see that $P+Q = O\big(t\varepsilon(t)\big)$ and Theorem D follows. (We easily see from the proof that the condition $\varepsilon(2t)/\varepsilon(t) \to 1$ can be replaced by $\limsup_{t\to 0}\{\varepsilon(2t)/\varepsilon(t)\} = \gamma < 2$; for $\gamma = 2$ the result is false.)

It may be observed that (5) is essentially a special case of (6) with $\varepsilon(t) = \left(\log \frac{1}{t}\right)^{-(1/2)}$, and that in this case the integral (2) diverges.

3. In what follows we shall sometimes say that the function $\varepsilon(t)$ defined in a right-hand side neighborhood of $t = 0$ *satisfies condition* N if $\varepsilon^2(t)/t$ is integrable over some interval $(0, \eta)$.

If we have (3) and the integral (2) is finite, then the function

$$\frac{[F(x_0+t)+F(x_0-t)-2F(x_0)]^2}{t^3} \tag{7}$$

is integrable near $t = 0$. Conversely, the integrability of the function (7) near $t = 0$ implies that we have a relation (3) with $\varepsilon(t)$ satisfying condition N. But it is important to observe that in the latter case the function $\varepsilon(t)$ may, first, depend on x_0 (that is, $\varepsilon(t) = \varepsilon_{x_0}(t)$) and, second, that it does not necessarily tend to 0 with t; and even if it does, it need not tend to 0 monotonically.

Theorem E which follows clearly generalizes Theorem D; it is one of the main results of the paper.

THEOREM E. *If $F(x)$ satisfies condition Λ at each point x_0 of a set E, and if for each $x_0 \in E$ the function* (7) *is integrable near $t = 0$, then $F'(x)$ exists almost everywhere in E.*

The integrability of the function (7) near $t = 0$ was first considered by Marcinkiewicz [4] who proved the following theorem:

THEOREM F. *Suppose that F is differentiable at each point of a set E. Then at almost all $x_0 \in E$ the function* (7) *is integrable near $t = 0$.*

Since the differentiability of F at a point implies that F satisfies condition Λ (even, λ) at that point, Theorems E and F can be combined in the following single theorem:

THEOREM G. *Suppose that $F(x)$, defined in an interval, satisfies condition Λ at each point of a set E. Then the necessary and sufficient condition for F to be differentiable almost everywhere in E is that for almost all $x_0 \in E$ the function*

$$\varepsilon_{x_0}(t) = \frac{F(x_0+t)+F(x_0-t)-2F(x_0)}{2t} \tag{8}$$

satisfies condition N.

The following result is merely a variant of Theorem G:

THEOREM G′. *Suppose that $F \in L^2(-\infty, +\infty)$ and satisfies condition Λ at each point of a set E. Then F is differentiable almost everywhere in E if and only if for almost all $x \in E$ the expression*

$$\mu(x) = \mu(x, F) = \left\{\int_0^\infty \frac{[F(x+t)+F(x-t)-2F(x)]^2}{t^3}\, dt\right\}^{1/2} \tag{9}$$

if finite.

This is clear since, if $F \in L^2(-\infty, +\infty)$, the part of the integral in (9) which extends over any interval $\eta \leqslant t < \infty$ $(\eta > 0)$ is always finite.

The integral (9) is sometimes called *the integral of Marcinkiewicz*.

4. The question naturally arises what are the necessary and sufficient conditions for the existence of the integral (9) almost everywhere in E if we no longer assume that F satisfies condition Λ in E. To answer this question we must generalize the notion of derivative.

Let $1 \leqslant p \leqslant \infty$ and suppose that F belongs to L^p in the neighborhood of the point x_0. Suppose further that there is a polynomial

$$P(t) = \sum_{j=0}^{k} \frac{a_j}{j!} t^j \tag{10}$$

of degree k such that

$$\left\{\frac{1}{2h}\int_{-h}^{h} |F(x_0+t)-P(t)|^r\, dt\right\}^{1/r} = o(h^k) \tag{11}$$

as $h \to +0$; the polynomial P, if it exists, is unique. We shall then say that F is *differentiable of order k* at x_0 in L^p. The polynomial $P(t)$ may be called the *k-th differential* of F at x_0, and the number a_k the *k-th derivative* of F at x_0, both in the metric L^p (these notions were introduced in [1]). It is clear that the existence of the k-th differential implies that of the $(k-1)$-th differential. If $p = \infty$ the left side of (11) means, of course, ess sup $|F(x_0+t)-F(x_0)|$ for $|t| \leqslant h$, and modifying F in a set of measure 0 we then have

$$F(x_0+t) = P(t)+o(t^k) \quad (t \to 0),$$

so that F has at x_0 a k-th derivative, or k-th differential, in the classical sense (of Peano). In this case, the coefficient a_j of P we shall occasionally denote by $F_{(j)}(x_0)$.

We can now answer the question raised above.

THEOREM H. *Suppose that $F(x)$ is defined in an interval. The necessary and sufficient condition for*

$$\varepsilon_{x_0}(t) = \frac{F(x_0+t)+F(x_0-t)-2F(x_0)}{2t}$$

to satisfy condition N *almost everywhere in the set E is that F has a derivative in L^2 at almost all points of E.*

THEOREM H'. *Let $F \epsilon L^2(-\infty, +\infty)$. The necessary and sufficient condition for $\mu(x, F)$ to be finite for almost all $x \epsilon E$ is that F be differentiable in L^2 almost everywhere in E.*

These two results do not differ essentially. Observe that the integrability of $\varepsilon_{x_0}^2(t)/t$ near $t = 0$ implies the integrability of $[F(x+t)+ +F(x_0-t)-2F(x_0)]^2$ near $t = 0$, and it can be shown without much difficulty that the latter implies the integrability of F^2 near almost every point $x_0 \epsilon E$. The version H' is useful in some cases.

5. Theorems G, G', H, H' can be extended to higher derivatives. Suppose that k is *even* and the *even* part of F has a k-th Peano derivative at x_0, or that k is *odd* and the *odd* part of F has a k-th derivative at x_0. In the former case,

$$\varphi_{x_0}(t) = a_0 + \frac{1}{2!}a_2t^2 + \ldots + \frac{1}{k!}a_kt^k + o(t^k),$$

and in the latter,

$$\psi_{x_0}(t) = a_1t + \frac{1}{3!}a_3t^3 + \ldots + \frac{1}{k!}a_kt^k + o(t^k),$$

as $t \to 0$. In either case F has a symmetric k-th derivative at x_0 equal to a_k, that is

$$\lim_{t\to 0}\left\{t^{-k}\sum_{j=0}^{k}\binom{k}{j}(-1)^{k-j}F\big(x_0+(j-\tfrac{1}{2}k)t\big)\right\} = a_k. \tag{13}$$

For if, e. g., k is even, the sum $\sum$ here is equal to

$$\sum_{j=0}^{k}\binom{k}{j}(-1)^{k-j}\varphi_{x_0}\big((j-\tfrac{1}{2}k)t\big)$$

and (13) is a consequence of the formula for $\varphi_{x_0}(t)$ and the formulas for the k-th differences of the function t^j. On the other hand, it is known (see [7]) that if a function has a k-th symmetric derivative in a set E, then it has a k-th differential almost everywhere in E. Hence we have the following analogue of Theorem A: *If for each x_0 in a set E the even part of F has a k-th differential at $t=0$ of even order k, or the odd part has a k-th differential of odd order k, then almost everywhere in E the function F itself has a differential of order k.*

The question is what happens if we interchange the roles of the even and odd parts. We shall say that F is *smooth of order* k at x_0 if k is even and $\psi_{x_0}(t;F)$ has a k-th differential at $t=0$, or if k is odd and $\varphi_{x_0}(t;F)$ has a k-th differential at $t=0$. In the former case,

$$\psi_{x_0}(t) = a_1 t + \frac{1}{3!}a_3 t^3 + \ldots + \frac{1}{(k-1)!}a_{k-1}t^{k-1} + o(t^k), \tag{14}$$

and in the latter,

$$\varphi_{x_0}(t) = a_0 + \frac{1}{2!}a_2 t^2 + \ldots + \frac{1}{(k-1)!}a_{k-1}t^{k-1} + o(t^k). \tag{15}$$

Either of these conditions will also be called λ_k, and condition Λ_k will be defined by replacing here o by O. Clearly, conditions λ and Λ introduced previously correspond to the case $k=1$.

A function may satisfy condition λ_k, even uniformly in x, and have a k-th differential in a set of measure 0 only. A simple example is obtained by integrating the series (5) term-by-term $k-1$ times. Since the function (5) satisfies condition λ uniformly in x, the sum of the integrated series satisfies, as one can easily verify, condition λ_k uniformly in x. At each point where the sum of the integrated series has a k-th differential, the series obtained by termwise differentiation $k-1$ times, that is, the series $\sum n^{-(1/2)}\cos 2^n x$, is summable by a linear method of summation, and this can occur only in a set of measure 0.

If at each point of a set E the function F satisfies condition λ_k, or even only Λ_k, then the last term in (14) or (15) (as the case may be) is $o(t^{k-1})$, and since the parity of $k-1$ is opposite to that of k the function F has, by the result stated above, a $(k-1)$-th differential, that is,

$$F(x_0+t) = \sum_{j=0}^{k-1}\frac{1}{j!}a_j(x_0)t^j + o(t^{k-1}) \tag{16}$$

at almost all points of E, and the problem is to find when F has a k-th differential almost everywhere in E. The theorem which follows is an extension of Theorem E.

THEOREM 1. *Suppose that at each point $x_0 \in E$ the function F satisfies condition Λ_k, i. e., we have either*

(17) $\psi_{x_0}(t)$

$$= a_1(x_0)t + \frac{1}{3!}a_3(x_0)t^3 + \ldots + \frac{1}{(k-1)!}a_{k-1}(x_0)t^{k-1} + \frac{\varepsilon_{x_0}(t)}{k!}t^k \quad (k\text{—}even)$$

or

(18) $\varphi_{x_0}(t)$

$$= a_0(x_0)+\frac{1}{2!}a_2(x_0)t^2+\ldots+\frac{1}{(k-1)!}a_{k-1}(x_0)t^{k-1}+\frac{\varepsilon_{x_0}(t)}{k!}t^k \qquad (k\text{–}odd)$$

where $\varepsilon_{x_0}(t)$ is bounded near $t=0$. Suppose, moreover, that $\varepsilon_{x_0}(t)$ satisfies condition N *at each point of E. Then F has a k-th differential almost everywhere in E.*

That the result is best possible can be proved by means of the function obtained by integrating the series (6) termwise $k-1$ times, where $\varepsilon(t)$ satisfies the hypotheses of Theorem D. The resulting function will satisfy conditions (17) or (18) uniformly in x_0, with $|\varepsilon_{x_0}(t)|\leqslant A\varepsilon(t)$, and will have a k-th differential in a set of measure 0 only. We shall not dwell on this point.

6. The next theorem is an analogue of Theorem F.

THEOREM 2. *Suppose that $F(x)$ is defined in an interval and at each point of a set E has a k-th differential (and so, in particular, is smooth of order k). Then at almost all points $x_0\in E$ the function $\varepsilon_{x_0}(t)$ in* (17) *or* (18) *satisfies condition* N.

This theorem is proved in [13]. It is included in Theorem 4 below which asserts that it is enough to assume that at each point of E the function F has a k-th differential in L^2.

A corollary of Theorem 1 and 2 is the following Theorem 3:

THEOREM 3. *If $F(x)$ is defined in an interval and at each point of E satisfies condition Λ_k (and, in particular, has a $(k-1)$-th differential almost everywhere in E) then F has a k-th differential almost everywhere in E if and only if the function $\varepsilon_{x_0}(t)$ in* (17) *or* (18) *satisfies condition* N *almost everywhere in E*

or

THEOREM 3′. *If $F\in L^2(-\infty,+\infty)$ and satisfies condition Λ_k at each point $x\in E$, then F has a k-th differential almost everywhere in E if and only if the expression*

$$\mu_k(x_0)=\mu_k(x_0,F)=\left\{\int_0^\infty \frac{\varepsilon_{x_0}^2(t)}{t}\,dt\right\}^{1/2},$$

where $\varepsilon_{x_0}(t)$ is given by (17) *or* (18), *is finite almost everywhere in E.*

We now pass to the case when F is no longer supposed to satisfy condition Λ_k in E, so that $\varepsilon_{x_0}(t)$ in (17) or (18) need no longer be bounded as $t\to 0$, $x_0\in E$. We want, of course, the developments (17) and (18) to be unique, and conditions like, e. g.,

$$\varepsilon_{x_0}(t)=o(1/t) \qquad (t\to 0)$$

are certainly sufficient. Another condition guaranteeing the same result would be that F has at x_0 a $(k-1)$-differential in some sense, for example, in the metric L^p, $1 \leq p < \infty$. On the other hand, theorems can be formulated in such a way that we do not need the assumption of the existence of the $(k-1)$-th differential of F in any sense, and we will follow this approach (³).

THEOREM 4. *Suppose that at each point $x_0 \in E$ we have* (17) *or* (18). *Then* a) *if $\varepsilon_{x_0}(t)$ satisfies condition* N *everywhere in E, the function F has a k-th differential in L^2 almost everywhere in E. Conversely,* b) *if at each point $x_0 \in E$ the function F has a k-th differential $\sum_0^k a_j(x_0)t^j/j!$ in L^2, then the function $\varepsilon_{x_0}(t)$ defined by* (17) *or* (18), *as the case may be, satisfies condition* N *almost everywhere in E.*

THEOREM 4′. *Suppose that $F \in L^2(-\infty, +\infty)$ and satisfies* (17) *or* (18) *at each point $x_0 \in E$. Then,* a) *if $\mu_k(x, F) < \infty$ everywhere in E, the function F has a k-th differential in L^2 almost everywhere in L^2, and, conversely,* b) *if F has a k-th differential in L^2 in E, then $\mu_k(x, F) < \infty$ almost everywhere in E.*

Chapter II

1. In this chapter we prove the theorems stated in Chapter I. All the theorems are essentially of local character and in their proofs we may assume that the functions under consideration are periodic of period 2π and use the properties of their Fourier series and Poisson integrals.

Given a periodic and integrable function $F(\theta)$ we denote by $F(\varrho, \theta)$ its Poisson integral

$$F(\varrho, \theta) = \frac{1}{\pi} \int_{-\pi}^{\pi} F(t)P(\varrho, \theta - t)\,dt,$$

where

$$P(\varrho, \theta) = \frac{1}{2} + \sum_1^\infty \varrho^\nu \cos\nu\theta = \frac{1}{2}\,\frac{1-\varrho^2}{1-2\varrho\cos\theta+\varrho^2}$$

is the Poisson kernel.

(³) Consider, e. g., the case of Theorem 4′ below, and denote by P_{k-1} the polynomial on the right of (17) or (18). The hypothesis that $\varepsilon_{x_0}(t)$ satisfies condition N implies that

$$\Big\{h^{-1}\int_0^h [\psi_{x_0}(t)-P_{k-1}(t)]^2\,dt\Big\}^{1/2} \quad \text{or} \quad \Big\{h^{-1}\int_0^h [\varphi_{x_0}(t)-P_{k-1}(t)]^2\,dt\Big\}^{1/2}$$

is $o(h^{k-1})$, as the case may be. It is known that this implies the existence of the $(k-1)$-th differential of F, in L^2 and at almost all points of E (see also Chapter II, Section 11) but we prefer not to use this fact.

Given any number $0<\sigma<1$, we shall denote by $\Omega_\sigma(\theta)$ the convex domain limited by the two tangents from $\zeta=e^{i\theta}$ to the circumference $|\zeta|=\sigma$ and the arc of this circumference between the points of contact; if σ is fixed we shall write $\Omega(\theta)$ for $\Omega_\sigma(\theta)$.

LEMMA 1. *Let k be a positive integer and suppose that a periodic and quadratically integrable $F(\theta)$ satisfies at the point θ_0 a relation*

$$\varphi_{\theta_0}(t;F)=a_1t+\frac{a_3}{3!}t^3+\ldots+\frac{a_{k-1}}{(k-1)!}t^{k-1}+\frac{\varepsilon(t)}{k!}t^k \tag{1}$$

if k is even, or

$$\varphi_{\theta_0}(t;F)=a_0+\frac{a_2}{2!}t^2+\ldots+\frac{a_{k-1}}{(k-1)!}t^{k-1}+\frac{\varepsilon(t)}{k!}t^k \tag{2}$$

if k is odd, where, in either case, $\varepsilon(t)$ satisfies condition N. *Then for any $0<\sigma<1$ we have*

$$\iint\limits_{\Omega_\sigma(0)}\left[\frac{\partial^{k+1}}{\partial\theta^{k+1}}F(\varrho,\theta_0+\xi)+\frac{\partial^{k+1}}{\partial\theta^{k+1}}F(\varrho,\theta_0-\xi)\right]^2\varrho\,d\varrho\,d\xi<\infty. \tag{3}$$

We shall denote, for brevity, the l-th derivative of $P(\varrho,\theta)$ with respect to θ by $P_l(\varrho,\theta)$. We shall also systematically write δ for $1-\varrho$. We need the following two inequalities, valid for $l=0,1,\ldots$:

$$|P_l(\varrho,\theta)|\leqslant A_l\delta^{-l-1}, \tag{4}$$

$$|P_l(\varrho,\theta)|\leqslant A_l\delta|\theta|^{-l-2}. \tag{5}$$

Here and in what follows (except when otherwise stated), A with various subscripts will mean constants (not always the same) depending only on parameters displayed in subscripts; A without subscripts will mean an absolute constant.

The inequality (4) follows by differentiating the series $\frac{1}{2}+\sum\varrho^\nu\cos\nu\theta$ termwise l times and observing that $\sum\varrho^\nu\nu^l=O(\delta^{-l-1})$. To prove (5), we use the formula

$$f_\theta=i(\zeta f_\zeta-\bar\zeta f_{\bar\zeta})\quad(\zeta=\varrho e^{i\theta}),$$

where $f_\zeta=\frac{1}{2}(f_\xi-if_\eta)$, $f_{\bar\zeta}=\frac{1}{2}(f_\xi+if_\eta)$, if $\zeta=\xi+i\eta$. If f is, say, a rational function of ζ and $\bar\zeta$, then f_ζ is obtained by formal differentiation with respect to ζ, treating $\bar\zeta$ as constant, and similarly for $f_{\bar\zeta}$. Hence, by observing that

$$2P(\varrho,\theta)(1-\varrho^2)^{-1}=\frac{1}{\zeta-\bar\zeta}\left(\frac{1}{1-\zeta}-\frac{1}{1-\bar\zeta}\right),$$

and that in the neighborhood of $\zeta=1$ both $|\zeta-\bar\zeta|$ and $|1-\zeta|$ majorize $A|\theta|$, we obtain (5).

The proof of (3) is the same for k even and odd; for the sake of definiteness we assume that k is even. Then

$$(6)\quad \frac{\partial^{k+1}}{\partial\theta^{k+1}}F(\varrho,\theta_0+\xi)+\frac{\partial^{k+1}}{\partial\theta^{k+1}}F(\varrho,\theta_0-\xi)$$
$$=-\frac{1}{\pi}\int_{-\pi}^{\pi}F(\theta_0+t)[P_{k+1}(\varrho,t-\xi)+P_{k+1}(\varrho,t+\xi)]dt$$
$$=-\frac{1}{\pi}\int_{-\pi}^{\pi}F(\theta_0+t)[P_{k+1}(\varrho,\xi+t)-P_{k+1}(\varrho,\xi-t)]dt$$
$$=-\frac{2}{\pi}\int_{0}^{\pi}\psi_{\theta_0}(t)[P_{k+1}(\varrho,\xi+t)-P_{k+1}(\varrho,\xi-t)]dt.$$

By hypothesis, $\psi_{\theta_0}(t)$ consists of an odd polynomial of degree $\leqslant k-1$ in t and a remainder $\varepsilon(t)t^k/k!$. Integration by parts shows (see (5)) that if l is odd and $l\leqslant k$, then

$$\int_0^{\pi}t^l\{P_{k+1}(\varrho,\xi+t)-P_{k+1}(\varrho,\xi-t)\}dt=-\int_{-\pi}^{\pi}t^lP_{k+1}(\varrho,\xi+t)dt$$
$$=O(\delta)+(-1)^l l!\int_{-\pi}^{\pi}P_{k+1-l}(\varrho,\xi+t)dt=O(\delta).$$

Hence the last term of (6) is

$$(7)\quad -\frac{2}{\pi k!}\int_0^{\pi}\varepsilon(t)t^k\{P_{k+1}(\varrho,\xi+t)-P_{k+1}(\varrho,\xi-t)\}dt+O(\delta).$$

In showing that the integral in (3) is finite, we may restrict our considerations to that part of $\Omega_\sigma(0)$ which is in the neighborhood of the point $z=1$. If $\varrho e^{it}\in\Omega_\sigma(0)$ and $\delta=1-\varrho$ is small enough, we have $|\xi|\leqslant\varkappa\delta$, where $\varkappa=\varkappa(\sigma)$. Let Ω' be the part of $\Omega_\sigma(0)$ where $|\xi|\leqslant\varkappa\delta$, $\delta\leqslant\delta_0$ and δ_0 is so small that $2\varkappa\delta_0\leqslant1-\sigma$.

In estimating the integral of the square of the expression (7) over Ω' we may omit the term $O(\delta)$ whose contribution is finite. We split the integral in (7) into two parts extended over $0\leqslant t\leqslant 2\varkappa\delta$ and $2\varkappa\delta\leqslant t\leqslant\pi$, and denote the resulting expressions respectively by $S(\delta,\xi)$ and $T(\delta,\xi)$. We have, by (4),

$$|S|\leqslant A_k\delta^{-k-2}\int_0^{2\varkappa\delta}|\varepsilon(t)|t^kdt\leqslant A_{k\sigma}\delta^{-2}\int_0^{2\varkappa\delta}|\varepsilon(t)|dt,$$

and, by (5),

$$|T|\leqslant A_k\delta\int_{2\varkappa\delta}^{\pi}|\varepsilon(t)|\frac{t^k}{|t-\xi|^{k+3}}dt\leqslant A_k\delta\int_{2\varkappa\delta}^{\pi}|\varepsilon(t)|\frac{t^k}{(\frac{1}{2}t)^{k+3}}\leqslant A_k\delta\int_{2\varkappa\delta}^{\pi}|\varepsilon(t)|t^{-3}dt.$$

It is enough to show that both S^2 and T^2 are integrable over Ω'. Schwarz's inequality gives

$$S^2(\delta,\xi) \leqslant A_{k\sigma}\delta^{-3}\int_0^{2\varkappa\delta} \varepsilon^2(t)\,dt,$$

$$T^2(\delta,\xi) \leqslant A_k\delta^2 \int_{2\varkappa\delta}^{\pi} \frac{\varepsilon^2(t)\,dt}{t^3}\int_{2\varkappa\delta}^{\infty}\frac{dt}{t^3} \leqslant A_{k\sigma}\int_{2\varkappa\delta}^{\pi}\varepsilon^2(t)t^{-3}\,dt,$$

so that

$$\iint_{\Omega} S^2(\delta,\xi)\varrho d\theta\, d\xi \leqslant \int_0^{\delta_0} d\delta \int_{-\varkappa\delta}^{\varkappa\delta} S^2(\delta,\xi)\,d\xi \leqslant \int_0^{\delta_0} d\delta\cdot 2\varkappa\delta\cdot A_{k\sigma}\delta^{-3}\int_0^{2\varkappa\delta}\varepsilon^2(t)\,dt$$

$$\leqslant A_{\varkappa\sigma}\int_0^{2\varkappa\delta_0}\varepsilon^2(t)\,dt\int_{t/2\varkappa}^{\delta_0}\delta^{-2}d\delta \leqslant A_{\varkappa\sigma}\int_0^{2\varkappa\delta_0} t^{-1}\varepsilon^2(t)\,dt < \infty,$$

and

$$\iint_{\Omega'} T^2(\delta,\xi)\varrho d\varrho\, d\theta \leqslant A_{\varkappa\sigma}\int_0^{\delta_0}\delta d\delta\int_{2\varkappa\delta}^{\pi} t^{-3}\varepsilon^2(t)\,dt \leqslant A_{\varkappa\sigma}\int_0^{\delta_0}\delta d\delta \int_{2\varkappa\delta}^{2\varkappa\delta_0} t^{-3}\varepsilon^2(t)\,dt + O(1)$$

$$= A_{\varkappa\sigma}\int_0^{2\varkappa\delta_0} t^{-3}\varepsilon^2(t)\,dt\int_0^{t/2\varkappa}\delta d\delta + O(1) \leqslant A_{\varkappa\sigma}\int_0^{2\varkappa\delta_0} t^{-1}\varepsilon^2(t)\,dt + O(1) < \infty.$$

This completes the proof of Lemma 1.

2. LEMMA 2. *If $U(x,y)+iV(x,y)$ is holomorphic in the interior of a finite rectilinear triangle D, and U^2 is integrable over D, so is V^2.*

This is a special case of a general theorem of Friedrichs [2] valid for a much wider class of domains D. If V, which contains an arbitrary additive constant, is suitably normalized we even have an inequality

$$\iint_D V^2\,dx\,dy \leqslant A_D\iint_D U^2\,dx\,dy,$$

but the weaker statement of the lemma is sufficient for our purposes.

LEMMA 3. *Let $\Phi(\zeta)=\Phi(\varrho e^{i\theta})$ be regular in $|\zeta|<1$, and suppose that for each θ_0 belonging to a set E of positive measure there exists a $\sigma=\sigma(\theta_0)$, $0<\sigma<1$, such that the integral*

$$\iint_{\Omega_\sigma(\theta_0)} |\Phi'(\zeta)|^2\,\varrho\, d\varrho\, d\theta$$

is finite. Then $\Phi(\zeta)$ has a finite non-tangential limit almost everywhere in E.

For the proof see, e. g., [$11_{\rm II}$], p. 207.

LEMMA 4. *If the integral $\int_0^{\eta} t^{-1}\varepsilon^2(t)\,dt$ is finite for some $\eta>0$, then $\int_0^u |\varepsilon(t)|\,dt = o(u)$.*

The hypothesis implies that $\int_0^u \varepsilon^2(t)dt = o(u)$ and it is enough to apply Schwarz's inequality.

LEMMA 5. *Suppose that a periodic and integrable $F(\theta)$ satisfies at $\theta = \theta_0$ the relation* (1) *or* (2), *where*

$$\int_0^u |\varepsilon(t)|\,dt = o(u). \tag{8}$$

Then $\tilde{S}^{(k)}[F]$ is summable $(C, k+1)$ at θ_0 if and only if it is summable A.

Let $\tilde{\sigma}_n^\alpha(\theta)$ be the (C, α) means of $\tilde{S}[F]$. It is well known that if F satisfies condition λ_k at θ_0, that is, if $\varepsilon(t)$ in (1) or (2), as the case may be, tends to 0 with t, then

$$\frac{d^k}{d\theta^k}\tilde{\sigma}_n^{k+1}(\theta_0) - \left(-\frac{2}{\pi}\int_{1/n}^{\infty}\frac{\varepsilon(t)}{t}dt\right) \to 0 \quad (n \to \infty) \tag{9}$$

(see [11_{II}], p. 63), but a glance at the proof shows that the conclusion holds under the weaker condition (8). The proof of (9) is exclusively based on the following estimates for the conjugate $(C, k+1)$ kernel $\tilde{K}_n^{k+1}(\theta)$:

$$\left|\frac{d^k}{d\theta^k}\tilde{K}_n^{k+1}(\theta)\right| \leqslant A_k n^{k+1},$$

$$\left|\frac{d^k}{d\theta^k}\left\{\frac{1}{2}\cot\frac{1}{2}\theta - \tilde{K}_n^{k+1}(\theta)\right\}\right| \leqslant A_k n^{-1}\theta^{-k-2} \quad (n^{-1} \leqslant \theta \leqslant \pi).$$

But the conjugate Poisson kernel

$$Q(\varrho, \theta) = \sum \varrho^\nu \sin\nu\theta = \varrho\sin\theta/(1-2\varrho\cos\theta+\varrho^2)$$

satisfies analogous inequalities, with n replaced by $1/\delta$. For, clearly $|d^kQ/d\theta^k| \leqslant \sum\varrho^\nu\nu^k \leqslant A_k\delta^{-k-1}$, and from the formula

$$\frac{\sin\theta}{2(1-\cos\theta)} - Q(\varrho, \theta) = \frac{1-\varrho}{1+\varrho}P(\varrho, \theta)\cot\frac{1}{2}\theta,$$

using (5), we obtain

$$\left|\frac{d^k}{d\theta^k}\left\{\frac{1}{2}\cot\frac{1}{2}\theta - Q(\varrho, \theta)\right\}\right| \leqslant A_k\delta^2\theta^{-k-3} \leqslant A_k\delta\theta^{-k-2},$$

where $\delta \leqslant \theta \leqslant \pi$. Hence, if $\tilde{F}(\varrho, \theta)$ is the harmonic function conjugate $F(\varrho, \theta)$, we have, under the hypotheses of the lemma, the following analogue of (9):

$$\frac{d^k}{d\theta^k}\tilde{F}(\varrho, \theta_0) - \left(-\frac{2}{\pi}\right)\int_\delta^\infty \frac{\varepsilon(t)}{t}dt \to 0 \quad (\delta = 1-\varrho \to 0).$$

Observing that (8) implies the relation $\int_{1/(n+1)}^{1/n} t^{-1}|\varepsilon(t)|\,dt \to 0$, we obtain from (9) and (10)

$$\frac{d^k}{d\theta^k}\tilde{F}(\varrho,\theta_0)-\frac{d^k}{d\theta^k}\tilde{\sigma}_n^{k+1}(\theta_0)\to 0 \qquad \left(n=\left[\frac{1}{1-\varrho}\right]\right),$$

which completes the proof of the lemma.

LEMMA 6. *If a trigonometric series S with coefficients $o(n^l)$ ($l = 0,1,2,\ldots$) is summable (C, l) at a point θ_0 to sum s, and if $F(\theta)$ is the sum of the series obtained by integrating S termwise $l+2$ times, then $\frac{1}{2}\{F(\theta_0+t)+(-1)^l F(\theta_0-t)\}$ has at $t=0$ an $(l+2)$-nd Peano derivative equal to s.*

For a proof, see [3_{II}], p. 66.

LEMMA 7. *If a trigonometric series S is summable (C,α), $\alpha > -1$, in a set E, then the conjugate series $\tilde{S}$ is summable (C,α) almost everywhere in E.*

The proof may be found in [8]. Only the case $\alpha = 0,1,2,\ldots$ is needed here.

3. In the preceding sections of this chapter we gave a number of lemmas pertaining to trigonometric series and harmonic and analytic functions. We are now going to give lemmas about functions of real variables. While most of these lemmas are known, and we shall be satisfied with giving references, the lemma which follows is essentially new and is basic for the proofs of theorems of Chapter I.

LEMMA 8. *Let $f(x,y)$ be defined and real-valued in the open half-plane $y>0$, and be in L^2 near each point of this open half-plane. Suppose also, for simplicity, that $f=0$ for $y>\omega>0$. Then, if there is a set P on the axis $y=0$ such that, for each $x\in P$,*

$$\int_0^\infty y f^2(x,y)\,dy < \infty \tag{11}$$

and, with some $a=a_x>0$,

$$\iint_{y\geqslant a|t|} [f(x+t,y)+f(x-t,y)]^2\,dy\,dt < \infty, \tag{12}$$

then for almost all $x\in P$ we have

$$\iint_{y\geqslant \beta|t|} f^2(x+t,y)\,dt\,dy < \infty \tag{13}$$

no matter how small is $\beta>0$.

It is easy to see that for each fixed a the integral in (12) is a lower-continuous, possibly infinite, function of x, and so is certainly measurable.

Also measurable, as a function of x, is the integral in (11). The domain of integration in (12) decreases as a increases, and so, by considering the sequence of values $a = 1, 2, 3, \ldots$ and the corresponding subsets of P we may suppose that 1° a is fixed troughout P; 2° the integrals in (11) and (12) are bounded on P:

$$(14)\quad \int_0^\infty y f^2(x, y)\,dy \leqslant M; \quad \iint_{y \geqslant a|t|} \{f(x+t, y)+f(x-t, y)\}^2\,dy\,dt \leqslant M \quad (x \in P);$$

3° P is closed, bounded and of positive measure. It is enough to prove that (13) holds almost everywhere in P.

Integrating the second inequality (14) over P we have

$$\int_P dx \iint_{y \geqslant a|t|} \{f(x+t, y)+f(x-t, y)\}^2\,dy\,dt \leqslant M|P|,$$

or, making the change of variables $x+t = u$, $x-t = v$,

$$\tfrac{1}{2} \iint_{\frac{1}{2}(u+v) \in P} du\,dv \int_{y \geqslant \frac{1}{2}a|u-v|} \{f(u, y)+f(v, y)\}^2\,dy \leqslant M|P|.$$

If we reduce the domain of integration by restricting the variable v to P we have, a fortiori,

$$(15)\quad \iint_{\frac{1}{2}(u+v) \in P,\ v \in P} du\,dv \int_{y \geqslant \frac{1}{2}a|u-v|} \{f(u, y)+f(v, y)\}^2\,dy \leqslant 2M|P|.$$

The main idea of the proof of the lemma consists a) in showing that, with our hypotheses, the integral

$$(16)\quad I = \iint_{\frac{1}{2}(u+v) \in P,\ v \in P} du\,dv \int_{y \geqslant \frac{1}{2}a|u-v|} f^2(v, y)\,dy$$

is finite, so that, in view of (15), the integral

$$(17)\quad J = \iint_{\frac{1}{2}(u+v) \in P,\ v \in P} du\,dv \int_{y \geqslant \frac{1}{2}a|u-v|} f^2(u, y)\,dy$$

is finite and then, b) deducing from the latter fact the inequality (13) for almost all $x \in P$.

Since $f(x, y) = 0$ for $y > \omega$, we may, if need be, restrict our integration in (16) or (17) to subdomains of the strip $0 < y \leqslant \omega$. Hence the values of u in (16) or (17) are actually confined to a finite interval. Dropping the condition $\frac{1}{2}(u+v) \in P$ we obtain from (16) that

$$I \leqslant \int_{-\infty}^{+\infty} du \int_P dv \int_{y \geqslant \frac{1}{2}a|u-v|} f^2(v, y)\,dy = \int_P dv \int_0^\infty f^2(v, y) \Big\{ \int_{y \geqslant \frac{1}{2}a|u-v|} du \Big\}\,dy$$

$$\leqslant \int_P dv \int_0^\infty f^2(v, y) \frac{4y}{a}\,dy \leqslant \frac{4M}{a}|P|,$$

the last inequality being a cosequence of the first condition (14). From this and (15), (16), (17) we deduce that

$$(18)\qquad J = \iint\limits_{\frac{1}{2}(u+v)\in P,\ v\in P} du\,dv \int\limits_{y\geqslant\frac{1}{2}a|u-v|} f^2(u,y)\,dy \leqslant 4\left(1+\frac{2}{a}\right) M|P|.$$

Clearly,

$$(19)\qquad J = \int_{-\infty}^{+\infty} du \int_0^{\infty} f^2(u,y)\,dy \Big\{ \int\limits_{\substack{\frac{1}{2}(u+v)\in P,\ v\in P \\ y\geqslant\frac{1}{2}a|u-v|}} dv \Big\} = \int_{-\infty}^{+\infty}\int_0^{\infty} f^2(u,y)\mu(u,y)\,du\,dy,$$

where $\mu(u,y)$ denotes the integral in curly brackets. For fixed $y>0$ and u, $\mu(u,y)$ is the measure of the set of points v on the real axis which lie in the interval

$$(20)\qquad u-\frac{2y}{a} \leqslant v \leqslant u+\frac{2y}{a}$$

and which, in addition, satisfy the conditions

$$(21)\qquad v\in P, \quad \tfrac{1}{2}(u+v)\in P.$$

We claim that if u_0 is any point of density of P, then, as the point (u,y) approaches $(u_0,0)$ non-tangentially from the upper half-plane, $\mu(u,y)$ is asymptotically equal to $4y/a$, that is the length of the interval (20).

Suppose, e. g., that $u_0=0$ is a point of density of P. The non-tangential approach in this case means that

$$(22)\qquad y \geqslant \varepsilon|u|$$

for some $\varepsilon>0$. Let $\varphi(v)$ be the characteristic function of the set P and $\psi(v)=1-\varphi(v)$ that of the complementary set. Then

$$\mu(u,y) = \int_{u-2y/a}^{u+2y/a} \varphi(v)\varphi\left(\tfrac{1}{2}(u+v)\right)dv = \int_{u-2y/a}^{u+2y/a} \{1-\psi(v)\}\left\{(1-\psi\left(\tfrac{1}{2}(u+v)\right)\right\}dv$$

$$= \frac{4y}{a} - \int_{u-2y/a}^{u+2y/a}\psi(v)\,dv - \int_{u-2y/a}^{u+2y/a}\psi\left(\frac{u+v}{2}\right)dv + \int_{u-2y/a}^{u+2y/a}\psi(v)\psi\left(\frac{u+v}{2}\right)dv,$$

and it is enough to show that each of the last three integrals is $o(y)$.

The first integral is, in view of (22), dominated by

$$(23)\qquad \int_{-(2a^{-1}+\varepsilon^{-1})y}^{(2a^{-1}+\varepsilon^{-1})y} \psi(v)\,dv = o(y).$$

The second integral can be written

$$2\int_{u-y/a}^{u+y/a}\psi(w)\,dw$$

and so is, likewise, $o(y)$. Finally, the third integral, being dominated by the first, is $o(y)$. Hence, actually, $\mu(u, y) \simeq 4a^{-1}y$ as (u, y) tends non-tangentially to any point of density of P.

It follows that there is a closed subset P_0 of P, with $|P-P_0|$ arbitrarily small, and a $\delta > 0$ such that if $u_0 \epsilon P_0$, then

$$\mu(u, y) > \frac{2y}{a} \tag{24}$$

provided $0 < y \leqslant \delta$, $y \geqslant \beta|u-u_0|$, where β is any fixed positive number.

In particular, denoting by $\Delta_\beta(u_0)$ the set of points (u, y) satisfying these two conditions, and by U_β the union of the $\Delta_\beta(u_0)$ for $u_0 \epsilon P_0$, we obtain

$$\frac{2}{a}\iint_{U_\beta} f^2(u, y)y\,du\,dy \leqslant J. \tag{25}$$

Let now $g(u, y)$ be equal to $f^2(u, y)$ in U_β and to 0 elsewhere. Then

$$\int_{-\infty}^{+\infty} dx \int_{\Delta_\beta(x)} g(u, y)\,du\,dy$$

$$= \int_{-\infty}^{+\infty} dx \iint_{0<y\leqslant\delta,\ y\geqslant\beta|x-u|} g(u, y)\,du\,dy = \int_{-\infty}^{+\infty}\int_0^\delta g(u, y)\,du\,dy\Big\{\int_{\beta|x-u|\leqslant y} 1\cdot dx\Big\}$$

$$= 2\beta^{-1}\int_{-\infty}^{+\infty}\int_0^\delta g(u, y)y\,du\,dy = 2\beta^{-1}\iint_{U_\beta} f^2(u, y)y\,du\,dy \leqslant a\beta^{-1}J,$$

by (25). Since, by (18), J is finite, the integral $\iint_{\Delta_\beta(x)} g(u, y)\,du\,dy$ is finite for almost all x. It follows that the integral

$$\iint_{\Delta_\beta(x_0)} f^2(u, y)\,du\,dy$$

is finite for almost all $x_0 \epsilon P_0$. Since $f^2(u, y)$ is locally integrable in the interior of the upper half-plane, and since δ and $|P-P_0|$ can be arbitrarily small, the lemma follows.

4. We add a few remarks about the lemma.

(i) In certain cases important for applications, condition (11) is, essentially, a corollary of condition (12). Suppose, for example, that $f(x, y)$ is harmonic in the closed triangle

$$a|x| \leqslant y, \quad 0 \leqslant y \leqslant \omega, \tag{Δ}$$

except, possibly, at the vertex $(0, 0)$. Let $\beta = (1+a^2)^{1/2}$. The circle with center $(0, y)$ and radius $r = \beta y$ is tangent to two sides of Δ and is contained in Δ provided y is sufficiently small, $0 < y < \omega' < \omega$. Let $\chi_y(\xi, \eta)$

be the characteristic function of the disc limited by the circle. If $y \leqslant \omega'$, we have, by the familiar property of harmonic functions,

$$f(0, y) = \frac{1}{\pi r^2} \iint\limits_{\xi^2+(\eta-y)^2 \leqslant r^2} f(\xi, \eta)\, d\xi\, d\eta$$

$$= \frac{1}{\pi r^2} \iint\limits_{\Delta} \frac{1}{2}\big(f(\xi, \eta)+f(-\xi, \eta)\big)\chi_y(\xi, \eta)\, d\xi\, d\eta,$$

$$f^2(0, y) \leqslant \frac{1}{4}\, \frac{1}{\pi r^2} \iint\limits_{\Delta} \{f(\xi, \eta)+f(-\xi, \eta)\}^2 \chi_y(\xi, \eta)\, d\xi\, d\eta$$

$$= \frac{A_a}{y^2} \iint\limits_{\Delta} \{f(\xi, \eta)+f(-\xi, \eta)\}^2 \chi_y\, d\xi\, d\eta.$$

Hence, multiplying by y, integrating over $0 < y \leqslant \omega'$ and changing the order of integration,

$$\int\limits_0^{\omega'} y f^2(0, y)\, dy \leqslant A_a \iint\limits_{\Delta} \{f(\xi, \eta)+f(-\xi, \eta)\}^2 \Big\{\int\limits_0^{\omega'} y^{-1}\chi_y(\xi, \eta)\, dy\Big\} d\xi\, d\eta.$$

But, for a fixed (ξ, η), we can have $\chi_y(\xi, \eta) = 1$ at most if $y - r \leqslant \eta \leqslant$ $\leqslant y + r$, that is to say, if y is contained between two fixed multiples (depending only on a) of η. Hence the integral in brackets is majorized by A_a and

$$\int\limits_0^{\omega'} y f^2(0, y)\, dy \leqslant A_a \iint\limits_{\Delta} \{f(\xi, \eta)+f(-\xi, \eta)\}^2 d\xi\, d\eta.$$

Since the hypotheses of Lemma 8 imply that for almost all x the integral $\int\limits_{\varepsilon}^{\infty} y f^2(0, y)\, dy$ is finite, no matter how small $\varepsilon > 0$, we immediately see that if $f(x, y)$ is harmonic in some strip $0 < y < \eta$, condition (11) may be dropped without affecting the conclusions of the lemma.

(ii) It is clear that the quadratic integrabilities in Lemma 8 are not essential. If $1 \leqslant p < \infty$, f is locally in L^p in the interior of the upper half-plane, and if the integrands in (11) and (12) are replaced respectively by $y\,|f(x, y)|^p$ and $|f(x+t, y)+f(x-t, y)|^p$ we still have the conclusion (13) with f^2 replaced by $|f|^p$ (this holds even for $0 < p < \infty$). We may also assume that f is complex-valued. If f is harmonic in some strip $0 < y < \eta$ we may again omit the analogue of condition (11), etc.

Certain problems lead to integrals analogous to (12) with sum in the integrand replaced by the difference, but the conclusion of the lemma still holds; instead of the sum in (12) we could take a linear combination with constant coefficients, and even some more general expressions.

(iii) This case corresponding to $p = \infty$ in (ii) is of interest, though it will not be needed in this paper. *Suppose that* $f(x, y)$ *is defined for* $y > 0$ *and that for each* $x \in P$, $|P| > 0$, *there is an* $a = a_x$ *such that* $\frac{1}{2}\{f(x+t, y)+ +f(x-t, y)\}$ *tends to a finite limit* $g(x)$ *as* $y \to 0$ *and* $0 < a|t| \leqslant y$. *Then at almost all points* $x \in P$ *we have* $f(x+t, y) \to g(x)$ *as* $y \to 0$ *and* $\beta|t| \leqslant y$, *no matter what* $\beta > 0$. The proof resembles that of Lemma 8 but the details are simpler. The hypotheses imply, in particular, that $f(x, y) \to \to g(x)$ for each $x \in P$ as $y \to 0$. We give a sketch of the proof.

We may suppose that a is fixed, that the convergence of the semi-sum $\frac{1}{2}\{f(x+t, y)+f(x-t, y)\}$ is uniform on P, that P is closed, and that $g(x)$ (being measurable on P) is continuous on P. We set, as before, $x+t = u$, $x-t = v$ and we deduce that

$$\frac{1}{2}\{f(u, y)+f(v, y)\}-g\left(\frac{u+v}{2}\right) \tag{26}$$

tends uniformly to 0 if $\frac{1}{2}(u+v) \in P$, $y \geqslant \frac{1}{2}a|u-v|$, $y \to 0$. Let us augment these conditions by the requirement that $v \in P$. We then have conditions (21) and (22) satisfied and, as we have shown, if u_0 is any point of density of P and the point (u, y) tends non-tangentially to $(u_0, 0)$, the set of points v such that $v \in P$, $\frac{1}{2}(u+v) \in P$, $\frac{1}{2}a|u-v| \leqslant y$ has, asymptotically, measure $4y/a$ and so is (this is the only thing we need here) non-empty if y is small enough. To sum up, if (u, y) tends non-tangentially to $(u_0, 0)$ and y is small enough, then there is a $v \in P$ such that $\frac{1}{2}(u+v) \in P$, $\frac{1}{2}a|u-v| \leqslant y$. Applying this information to the expression (26) and using the facts that $g\left(\frac{u+v}{2}\right) \to g(u_0)$ and $f(v, y) = g(v)+\{f(v, y)-g(v)\} \to g(u_0)$, we see that $f(u, y) \to g(u_0)$ and the assertion is established.

5. LEMMA 9 (4). (i) *Suppose that a function* $F(x)$ *is defined in an interval and that at each point* x_0 *of a set* E *we have*

$$\lim_{t\to 0} \{F(x_0+t)-F(x_0-t)\} = 0, \tag{27}$$

then F *is continuous at almost all points of* E. (ii) *The conclusion holds if* (27) *is replaced by*

$$\lim_{t\to 0} \{F(x_0+t)+F(x_0-t)-2F(x_0)\} = 0. \tag{28}$$

(4) This lemma is certainly known but since we cannot give an adequate quotation we give the proof.

Added in proof. The conclusion for assumption (28) is an immediate consequence of Lemma 8, Remark (iii), this page: it is enough to take $f(x, y) = F(x)$, $g(x) = F(x)$. Similarly for assumption (27).

Results analogous to Lemma 9 are discussed in a paper of C. J. Neugebauer which is to appear in the Duke Journal.

re (i). Let E_n be the set of $x_0 \epsilon E$ such that $|F(x_0+t)-F(x_0-t)| \leqslant n^{-1}$ for $0 < t \leqslant 1/n$. We have $E_1 \subset E_2 \subset \ldots \subset E_n \ldots$, $E = \sum E_n$, and it is enough to show that at almost all points of each set E_n we have $\limsup_{t\to 0} |F(x_0+t)-F(x_0)| \leqslant n^{-1}$. In accordance with our general assumptions, F and E are supposed to be measurable but we do not assume the measurability of the E_n. We fix $n = n_0$. Since F is measurable, there is a closed set P, with $|E-P|$ arbitrarily small, such that F is continuous on, and with respect to, P. If we show that at each point x_0 which is a point of external density for E_{n_0} and at the same time a point of density of P we have $\limsup_{t\to 0}|F(x_0+t)-F(x_0)| \leqslant n_0^{-1}$, part (i) of the lemma will be established. Suppose for the sake of simplicity that $x_0 = 0$, $F(x_0)=0$, and that $x \to +0$.

The set of points $y \epsilon P$ such that $0 < y < x$ has, asymptotically, measure x as $x \to 0$. The set of points y, $0 < y < x$, such that $z = \frac{1}{2}(x+y) \epsilon E_{n_0}$ has, as one can easily see, external measure asymptotically equal to x as $x \to 0$. Hence, if x is small enough, there is a $y \epsilon P$ such that $\frac{1}{2}(x+y) \epsilon E_{n_0}$. It follows that, with $h = \frac{1}{2}(x-y)$,

$$F(x) = F(y)+\{F(x)-F(y)\} = F(y)+\{F(z+h)-F(z-h)\}.$$

Since $|F(z+h)-F(z-h)| \leqslant n_0^{-1}$ for x sufficiently small and $F(y) \to F(0) = 0$, we see that $\limsup_{x\to 0}|F(x)| \leqslant n_0^{-1}$ and (i) is established.

re (ii). Let E_n be the set of x such that $|F(x+t)+F(x-t)-2F(x)| \leqslant n^{-1}$ for $0 < t \leqslant n^{-1}$, and let P have the same meaning as before. It is enough to show that at every point x_0 which is a point of external density for E_n and a point of density for P we have $\limsup_{t\to 0}|F(x_0+t)-F(x_0)| \leqslant n^{-1}$. We fix $n = n_0$ and assume that $x_0 = 0$, $F(x_0) = 0$, $x \to +0$. For a fixed x, the external measure of the set of the y's in $(0, x)$ such that $\frac{1}{2}(x+y) \epsilon E_{n_0}$ is, asymptotically, x. The measure of the set of y's in $(0, x)$ such that $y \epsilon P$, and $z = \frac{1}{2}(x+y) \epsilon P$ is also, asymptotically, x. Hence, if x is small enough, there is in $(0, x)$ a point y such that y is in P and $z = \frac{1}{2}(x+y)$ is both in P and E_{n_0}. Since, with $h = \frac{1}{2}(x-y)$,

$$F(x) = F(z+h)+F(z-h)-2F(z)+2F(z)-F(y),$$

and since the last two terms tend to 0 as $x \to 0$, it follows that $\limsup_{x\to +0} |F(x)| \leqslant n_0^{-1}$, and the proof of (ii) is complete.

6. LEMMA 10. *If for each x_0 in a set E the even part $\varphi_{x_0}(t)$ of a function F has a k-th differential at $t = 0$ of even order k, or the odd part $\psi_{x_0}(t)$ has at $t = 0$ a k-th differential of odd order k, then almost everywhere in E the function F itself has a k-th differential.*

This result was already stated and used in Section 6 of Chapter I. As indicated above, a proof may be found in [7].

LEMMA 11. *Let $F(x)$ be a function defined in an interval and equal to 0 in a set E. Suppose, moreover, that at each point of E we have, with $k = 1, 2, \dots$, independent of x,*

$$F(x+t)-F(x-t) = O(t^k) \tag{29}$$

or

$$F(x+t)+F(x-t) = O(t^k). \tag{30}$$

Then $F_{(k)}(x)$ exists almost everywhere in E.

This result is essentially contained in [10], but since now the assumptions are somewhat different we give the proof here. We note the basic difference between Lemmas 10 and 11: a possible reversal of the roles of k even and odd and, correspondingly, additional assumptions in Lemma 11 about the behavior of F on E (these assumptions could be considerably relaxed).

Consider first (29). Let E_n be the subset of E consisting of points x such that $|F(x+t)-F(x-t)| \leqslant nt^k$ for $0 \leqslant t \leqslant 1/n$. It is enough to show that $F(x_0+t) = o(t^k)$ at each point $x_0 \in E_n$ which is a point of external density for E_n (and so also a point of density for E). Suppose, for simplicity, that $x_0 = 0$, $t > 0$, and let ε be an arbitrarily small but fixed positive number. If t is small enough, in particular $t \leqslant 1/n$, we can find in the interval $((1-\varepsilon)t, t)$ a point $\xi \in E$ such that $u = \frac{1}{2}(\xi+t) \in E_n$. Then, with $h = \frac{1}{2}(t-\xi)$, we have

$$|F(t)| = |F(u+h)-F(x-h)| \leqslant nh^k \leqslant n(\tfrac{1}{2}\varepsilon t)^k,$$

and since ε can be as small as we please, $F(t) = o(t^k)$. For assumption (30) the proof is the same.

7. LEMMA 12. *Let $f(x)$ be defined in a finite interval and suppose that f has a k-th differential at each point of a set E, $|E| > 0$. Then, for any $\varepsilon > 0$ we can find a closed subset P of E, with $|E-P| < \varepsilon$, and a decomposition*

$$f(x) = g(x)+h(x),$$

where $g \in C^k$, $g = f$ on P, and, except possibly for a finite number of intervals contiguous to P,

$$|h(x)| \leqslant C\{\delta(x)\}^k,$$

$\delta(x)$ denoting the distance of x from the set P, and C a constant independent of x.

For a proof see [5] or [11_{II}], p. 73-77.

LEMMA 13. *Suppose that $F(x)$ is defined in an interval and that for each x_0 in a set E there exists a number $h = h_{x_0} > 0$ such that the integral $\int_0^h [F(x_0+t)+F(x_0-t)]^2 dt$ is finite. Then F^2 is integrable near almost all points of E. The same conclusion holds if in the assumption we replace $F(x_0+t)+F(x_0-t)$ by $F(x_0+t)-F(x_0-t)$.*

The proof follows the usual pattern. We may assume that the interval (a, b) of definition of F is finite and denote by E_n the set of points $x_0 \in E$ such that $|F(x_0)| \leqslant n$ and $\int_0^{1/n} [F(x_0+t)+F(x_0-t)]^2 dt \leqslant n$ (hence the distance of x_0 from both a and b is $\geqslant 1/n$). Thus $E_1 \subset E_2 \subset E_3 \subset, \ldots$, $E = \sum E_n$. We fix n and integrate the last inequality over E_n which we denote by $\mathscr{E}$. Setting $x_0+t = u$, $x_0-t = v$, we have

$$\iint\limits_{\substack{v\in\mathscr{E},\ \frac{1}{2}(u+v)\in\mathscr{E} \\ 0\leqslant\frac{1}{2}(u-v)\leqslant 1/n}} [F(u)+F(v)]^2 du\,dv \leqslant 2n|\mathscr{E}|,$$

and in particular, since $|F(v)| \leqslant n$, the integral

$$\int_a^b F^2(u)\Big\{\int\limits_{\substack{v\in\mathscr{E},\ \frac{1}{2}(u+v)\in\mathscr{E} \\ 0\leqslant\frac{1}{2}(u-v)\leqslant 1/n}} dv\Big\} du = \int_a^b F^2(u)\,\xi(u)\,du, \tag{31}$$

say, is finite. If u is situated in $(a+2/n, b)$ if $0 < \eta \leqslant 2/n$, and if φ is the characteristic function of $\mathscr{E}$, then

$$\xi(u) = \int_{u-2/n}^{u} \varphi(v)\varphi\{\tfrac{1}{2}(u+v)\}dv \geqslant \int_{u-\eta}^{u} \varphi(v)\varphi\{\tfrac{1}{2}(u+v)\}dv.$$

But if u_0 is any point of density of $\mathscr{E}$ and if $|u-u_0| \leqslant \eta$, the last integral is asymptotically equal to η as $\eta \to 0$. Hence $\xi(u)$ is bounded below by a positive number in the neighborhood of any point of density of $\mathscr{E}$ that is situated in the interior of the interval $(a+2/n, b)$, and the finiteness of the integral (31) implies that F^2 is integrable in the neighbourhood of every such point. From this we deduce that F^2 is integrable in the neighbourhood of almost all points of the set E. This is the first part of the lemma and the proof of the second part is similar.

8. LEMMA 14 (5). *Let a be a positive number. Suppose that a function $F(x)$ is defined in an interval, vanishes on a set E, $|E| > 0$, and at each point $x \in E$*

(5) This is a special case of a more general result of Dr. Mary Weiss, [14]. For $a = 0$ it reduces to Lemma 13.

(32) $$\int_0^h [F(x+t)+F(x-t)]^2 dt = O(h^a) \quad (h \to 0).$$

Then at almost all points $x \in E$ we have

(33) $$\int_{-h}^h F^2(x+t)\,dt = O(h^a).$$

If $a \geqslant 1$ we can replace the O in the last equation by o. The conclusions hold if the integrand in (32) *is* $[F(x+t)-F(x-t)]^2$.

We write $E = \sum E_n$, where E_n is the set of points $x \in E$ such that

(34) $$\int_0^h [F(x+t)+F(x-t)]^2 dt \leqslant nh^a \quad \text{for} \quad 0 < h \leqslant 1/n,$$

and we will show that (33) holds at the points of density of each E_n. We fix n, write $E_n = \mathscr{E}$, and suppose, for example, that $x = 0$ is a point of density of $\mathscr{E}$. Let $\mathscr{E}(h)$ be the part of $\mathscr{E}$ situated in the interval $(-h, h)$. Integrating (34) over $\mathscr{E}(h)$ we have (for $h \leqslant 1/n$)

$$\int_{\mathscr{E}(h)} dx \int_{-h}^h [F(x+t)+F(x-t)]^2 dt = O(h^{a+1}).$$

Hence, with $x+t = u$, $x-t = v$,

$$\iint_{\frac{1}{2}|u-v| \leqslant h,\ \frac{1}{2}(u+v) \in \mathscr{E}(h)} [F(u)+F(v)]^2 du\,dv = O(h^{a+1}).$$

We adjoin on the left the condition $v \in \mathscr{E}(h)$ and restrict u (which is, anyway, contained in $(-2h, 2h)$) to the interval $(-h, h)$. Since then $F(v) = 0$, we obtain, a fortiori,

(35) $$\int_{-h}^h F^2(u) \Big\{ \int_{\frac{1}{2}|u-v| \leqslant h,\ v \in \mathscr{E}(h),\ \frac{1}{2}(u+v) \in \mathscr{E}(h)} dv \Big\} du = O(h^{a+1}).$$

But if $u \in (-h, h)$ and $v \in \mathscr{E}(h)$, then, necessarily, $\frac{1}{2}|u-v| \leqslant h$, so that the latter condition can be dropped in the last formula and the the cofactor of $F^2(u)$ in the integrand can be written

$$\int_{-h}^h \varphi(v)\varphi\big(\tfrac{1}{2}(u+v)\big)\,dv,$$

where φ designates the characteristic function of the set $\mathscr{E}$. But if $u \in (-h, h)$ the last integral is asymptotically equal to $2h$ so that (35) implies $\int_{-h}^h F^2(u)\,du = O(h^a)$. Hence (33) holds almost everywhere in E.

Suppose now that $a \geqslant 1$, and let P be any closed subset of E where (33) holds uniformly; in other words,

$$\int_{-h}^{h} F^2(x+t)\,dt \leqslant Mh^a \quad \text{for} \quad 0 < h \leqslant \eta,\ x \in P, \tag{36}$$

with M independent of x. At every point x_0 which is a point of density of P we necessarily have $\int_{-h}^{h} F^2(u)\,du = o(h^a)$. For if $0 < h \leqslant \eta$, the left side of (36), with $x = x_0$, is majorized by $M\sum(b_n - a_n)^a$, where (a_n, b_n) are the intervals contiguous to P which overlap with $(x_0 - h, x_0 + h)$. Since $a \geqslant 1$, we have $\sum(b_n - a_n)^a \leqslant \{\sum(b_n - a_n)\}^a = o(h^a)$, x_0 being a point of density of P. It is now enough to observe that $|E - P|$ can be arbitrarily small.

Remark. Lemma 14 holds, of course, if the exponent 2 on the left is replaced by any $p \geqslant 1$.

Lemma 15. *Let P be a closed set situated in a finite interval (a, b). Let $\chi(x)$ be the function equal to 0 in P and equal to $\beta - \alpha$ for x in (α, β), if (α, β) is any interval contiguous to P. Let $\delta(t)$ denote the distance of the point t from P. Then for any $\lambda > 0$ the integrals*

$$\int_a^b \frac{\chi^\lambda(t)}{|x-t|^{\lambda+1}}\,dt, \qquad \int_a^b \frac{\delta^\lambda(t)}{|x-t|^{\lambda+1}}\,dt$$

converge at almost all points $x \in P$.

This is well known; see e. g., [11_1], p. 130. If x is a point of density of P, the convergence of either integral is equivalent to that of $\sum[(b_n - a_n)/d_n(x)]^{\lambda+1}$ where $d_n(x)$ is the distance of (a_n, b_n) from x.

The lemma which follows is an analogue of Lemma 12 for derivatives in L^2. Its proof may be found in [1], p. 186-189, Theorem 9 and Corollary. The formulation there is very general, valid for derivatives in L^p and functions of n variables. The special case we need is as follows:

Lemma 16. *Let $F(x)$ be a function defined in a finite interval and suppose that at each point x of a set E, $|E| > 0$, F has a k-th derivative in L^2. Then for every $\varepsilon > 0$ we can find a closed subset P of E with $|E - P| < \varepsilon$, a positive number η, and a decomposition $F(x) = G(x) + H(x)$ such that $G \in C^k$, $G = F$ on P, and*

$$\int_{-h}^{h} H^2(x+t)\,dt \leqslant Mh^{2k+1} \quad \textit{for} \quad x \in P,\ 0 \leqslant h \leqslant \eta,$$

with M independent of x.

9. We can now pass to the proofs of the theorems enunciated in Chapter I, beginning with Theorem 1. For the sake of definiteness, we assume that k is odd; for k even the proof is similar.

Let T denote the series obtained by integrating $S^{(k)}$ termwise $k+3$ times, and let Φ be the sum of T (observe that T converges absolutely and uniformly). By Lemmas 6 and 10, Φ has at almost all points of E, a differential of order $k+3$. Clearly, Φ is a third integral of F.

By Lemma 12, for any $\varepsilon > 0$ there is a closed subset P of E, with $|E-P| < \varepsilon$, and a decomposition

$$\Phi = \Psi + X \tag{37}$$

where $\Psi = \Phi$ on P, $\Psi \in C^{k+3}$ and

$$|X(x)| \leqslant C\{\delta(x)\}^{k+3} \quad (\delta(x) = \delta(x, P)) \tag{37a}$$

except, perhaps, for a finite number of intervals contiguous to P.

Without loss of generality we may assume that $\Phi_{(k+3)}(x)$ exists everywhere in P; hence $X_{(k+3)}(x)$ exists everywhere in P. The inequality for X shows that $X_{(j)}(x) = 0$ for $j \leqslant k+3$ at each point of density of P. We may also assume that $F_{(k-1)}(x)$ exists everywhere in P.

Let us differentiate the equation $\Phi = \Psi + X$ three times. Since Φ is a third integral of F, and F is continuous in $E \supset P$, we have $\Phi'''(x) = F(x)$ in P (6). Hence, with $\Psi''' = G \in C^k$ and $H = F - G$, we have the identity

$$F(x) = G(x) + H(x) \tag{38}$$

where $H(x) = X'''(x)$ in P.

The function H satisfies, like F, condition Λ_k at each point of P. It also has, like F, a differential of order $k-1$ everywhere in P. Since, clearly, X is a third integral of H, the differential of order $k+2$ of X at any point $x_0 \in P$ is obtained by integrating the differential of order $k-1$ of H at x_0 three times. But we observed that $X_{(j)}(x_0) = 0$ for $0 \leqslant j \leqslant k+2$ and almost all $x_0 \in P$. Hence

$$H_{(j)}(x_0) = 0 \quad \text{for} \quad 0 \leqslant j \leqslant k-1 \text{ and almost all } x_0 \in P.$$

This, together with the fact that H satisfies condition Λ_k at each point of P, shows that for almost all points $x \in P$ we have

$$H(x+t) + H(x-t) = O(t^k). \tag{39}$$

(For k even we would get, instead, $H(x+t) - H(x-t) = O(t^k)$.) Hence, by Lemma 11, $H_{(k)}(x)$ exists almost everywhere in P. It follows that $F_{(k)}(x)$ exists almost everywhere in P, and so also almost everywhere in E.

(6) The continuity of F in E is not really indispensable in the argument. The conclusion $\Phi''' = F$ in P can be reached if F is continuous in the mean at each $x \in E$, i. e., if $\int_0^h |F(x+t) - F(x)|\,dt = o(h)$, and this holds, anyway, for almost all x.

Suppose, therefore, that for each $x_0 \epsilon E$ we have (cf. (2))

$$\varphi_{x_0}(t; F) = a_0(x_0) + \frac{a_2(x_0)}{2!} t^2 + \ldots + \frac{a_{k-1}(x_0)}{(k-1)!} t^{k-1} + \frac{\varepsilon_{x_0}(t)}{k!} t^k,$$

with $\varepsilon_{x_0}(t)$ bounded as $t \to 0$ and $\varepsilon_{x_0}^2(t)/t$ integrable near $t = 0$. We have to show that F has a k-th differential almost everywhere in E. The boundedness of $\varepsilon_{x_0}(t)$ as $t \to 0$ implies that $F(x_0+t)+F(x_0-t)-2F(x_0)$ tends to 0 as $t \to 0$, so that F is continuous almost everywhere in E (Lemma 9), and since the problem of the differentiability of F is local, we may assume that F is periodic (of period 2π), bounded, and continuous in E. Let $\tilde{F}(x)$ be the function conjugate to F, $F(\varrho, x)$ the Poisson integral of F, and $\tilde{F}(\varrho, x)$ the conjugate Poisson integral of F.

By Lemma 1, for each $x_0 \epsilon E$ the function

$$\left| \frac{\partial^{k+1}}{\partial x^{k+1}} F(\varrho, x_0+\xi) + \frac{\partial^{k+1}}{\partial x^{k+1}} F(\varrho, x_0-\xi) \right|^2$$

is integrable over any domain $|\xi| \leqslant C(1-\varrho)$, $\frac{1}{2} \leqslant \varrho < 1$, and by Lemma 8 the same can be said, for almost all $x_0 \epsilon E$, about the function

$$\left| \frac{\partial^{k+1}}{\partial x^{k+1}} F(\varrho, x_0+\xi) \right|^2$$

and so also, by Lemma 2, about the function

$$\left| \frac{\partial^{k+1}}{\partial x^{k+1}} \tilde{F}(\varrho, x_0+\xi) \right|^2.$$

This implies, by Lemma 3, that the function

$$\frac{\partial^k}{\partial x^k} \{F(\varrho, x) + i\tilde{F}(\varrho, x)\},$$

which is regular inside the unit circle, has a non-tangential limit at almost all points $x_0 \epsilon E$. In particular, the radial limit

$$\lim_{\varrho \to 1} \frac{\partial^k}{\partial x^k} \tilde{F}(\varrho, x)$$

exists an is finite almost everywhere in E.

Denote by S the Fourier series of F, by $\tilde{S}$ the series conjugate to S, and by $S^{(k)}$ and $\tilde{S}^{(k)}$, respectively, the series S and $\tilde{S}$ differentiated termwise k times. We have just proved that $\tilde{S}^{(k)}$ is Abel summable almost everywhere in E. By Lemmas 4 and 5, $\tilde{S}^{(k)}$ is summable $(C, k+1)$ almost everywhere in E, and, by Lemma 7, the same holds for $S^{(k)}$. The latter series has coefficients $o(n^k) = o(n^{k+1})$.

10. This completes the proof of Theorem 1. Assuming the validity of Theorem 2, we may also consider Theorems 3 and 3′ as established.

We now pass to the proof of Theorems 4 and 4′. The theorems say essentially the same thing and we may confine our attention to Theorem 4 and assume for the sake of definiteness that k is odd; we may also assume that $0 < |E| < \infty$. We begin with the second part of the theorem which asserts that if for each $x_0 \epsilon E$ the function F has a k-th differential in L^2, if this differential is

$$P_{x_0}(t) = \sum_{j=0}^{k} \frac{a_j(x_0)}{j!} t^j$$

and if $\varepsilon_{x_0}(t)$ is defined by the equation

$$F(x_0+t) = P_{x_0}(t) + \varepsilon_{x_0}(t)\frac{t^k}{k!},$$

then $\varepsilon_{x_0}(t) - \varepsilon_{x_0}(-t)$ satisfies condition N at almost all points of E.

By Lemma 16, we can find a closed subset P of E, with $|E-P| < \varepsilon$ and a decomposition $F = G+H$ such that $G \epsilon C^k$, $G = F$ in P, and if $\{(a_n, b_n)\}$ is the sequence of intervals contiguous to P we have

$$\int_{a_n}^{b_n} H^2(t)\,dt \leqslant M(b_n - a_n)^{2k+1}, \tag{40}$$

except possibly for a finite number of these intervals; the constant M is independent of n. In view of Theorem 2, it is enough to prove the required result of $F = H$ and $E = P$. Since $H = 0$ on P, the last inequality implies that for every x_0 which is a point of density of P we have

$$\int_{-h}^{h} H^2(x_0+t)\,dt = o(h^{2k+1})$$

so that the k-th differential $P_{x_0}(t)$ of H in L^2 is identically 0 and $\varepsilon_{x_0}(t) = k!H(x_0+t)t^{-k}$. It is therefore enough to show that for almost all $x_0 \epsilon P$ the function $H^2(x_0+t)|t|^{-(2k+1)}$ is integrable near $t = 0$.

Take any point x_0 of density of P and $\eta > 0$ so small that for all the intervals (a_n, b_n) situated in $I = (x_0 - \eta, x_0 + \eta)$ we have (40). By reducing η still more we may assume that if $d_n = d_n(x_0)$ is the distance of (a_n, b_n) from x_0, then $d_n \geqslant b_n - a_n$ for all (a_n, b_n) in I. We may also assume that $x_0 \pm \eta \epsilon P$. Let $\chi(x)$ be the function equal to 0 in P and to $b_n - a_n$ in the intervals (a_n, b_n).

Then

$$\int_{-\eta}^{\eta} \frac{H^2(x_0+t)}{|t|^{2k+1}}\,dt = \sum \int_{a_n}^{b_n} \frac{H^2(x)}{|x-x_0|^{2k+1}}\,dx \leqslant \sum d_n^{-2k-1} \int_{a_n}^{b_n} H^2(x)\,dx$$

$$\leqslant M\sum d_n^{-2k-1}(b_n-a_n)^{2k+1} \leqslant M\cdot 2^{2k+1}\sum\int_{a_n}^{b_n}\frac{\chi^{2k}(x)}{|x-x_0|^{2k+1}}\,dx$$

$$= M\int_{x_0-\eta}^{x_0+\eta}\frac{\chi^{2k}(x)}{|x-x_0|^{2k+1}}\,dx$$

and, by Lemma 15, the last integral is finite for almost all x_0 in P.

11. It remains to prove the first part of Theorem 4, namely that if at each point of a set E (assuming. e. g., k odd) the function $\varepsilon_{x_0}(t)$ which appears in

$$\varphi_{x_0}(t;F) = a_0(x)+\frac{a_2(x)}{2!}t^2+\ldots+\frac{a_{k-1}(x_0)}{(k-1)!}t^{k-1}+\varepsilon_{x_0}(t)\frac{t^k}{k!}$$

satisfies condition N, then F has a k-th differential in L^2 at almost all points of E.

The proof is to a considerable degree parallel to that of Theorem 1. The integrability of $\varepsilon_{x_0}^2(t)/|t|$ near $t=0$ implies, firstly, that $\int_0^h|\varepsilon_{x_0}(t)|\,dt = o(h)$ and, secondly, that $\{F(x_0+t)+F(x_0-t)\}^2$ is integrable near $t=0$. By Lemma 13, F^2 is integrable near almost all points of E. We may therefore assume from the start that $F(x)$ is periodic and of the class L^2, and arguing as in the case of Theorem 1 we have the decomposition (37) with Ψ and X having the same properties as before (cf. also the footnote in the proof of Theorem 1). By differentiating we again obtain (38) with G in C^k and $H = X'''$ in P. As before, the $(k-1)$-th differential of H is identically 0 at almost all points of P, and, by Theorem 2 applied to G, we have

$$\int_0^{\pi}\frac{[H(x+t)+H(x-t)]^2}{t^{2k+1}}\,dt<\infty \tag{35a}$$

(instead of (39)) for almost all points $x\in P$. Hence, a fortiori,

$$\int_0^h\{H(x+t)+H(x-t)\}^2dt = o(h^{2k+1})$$

almost everywhere in P, and so also, by Lemma 14 ([6a]),

$$\int_{-h}^{h}H^2(x+t)\,dt = o(h^{2k+1}) \tag{35b}$$

([6a]) Added in proof. This is the only place where Lemma 14 is used, but we could do without it and use the key Lemma 8 instead, with $f(x,y) = H(x)y^{-k-1}$. For then (35a) implies the integrability of functions $H^2(x\pm t)t^{-2k-1}$ near $t=0$, and so also (35b), almost everywhere in P.

almost everywhere in P. It follows that H, and so also $F = G+H$, has a k-th differential in L^2 at almost all points of P, and therefore also at almost all points of E. This completes the proof of Theorem 4.

Chapter III

1. Let us return to the definition of the derivative in L^p and suppose that $1 \leqslant p < \infty$. It can be shown (7) that if, say, k is even and for each $x_0 \in E$ the even part $\varphi_{x_0}(t)$ of F has at $t = 0$ a k-th differential in L^p, then F itself has a k-th differential in L^p at almost all points of E; the same conclusion holds if k is odd and we replace $\varphi_{x_0}(t)$ by $\psi_{x_0}(t)$. If we interchange the roles of φ_{x_0} and ψ_{x_0}, without changing the parity of k, we are led to the notions of conditions λ_k and Λ_k in the metric L^p.

Suppose that F belongs to L^p in the neighborhood of x_0 and that the even part $\varphi_{x_0}(t)$ has at $t = 0$ a differential in L^p of odd order k, or that the odd part $\psi_{x_0}(t)$ has a differential of even order k. It is not difficult to see that in either case the k-th derivative must be 0. For if, for example, k is odd (the definitions and arguments which follow are analogous for k even) and $U(t)$ is the k-th differential of $\varphi_{x_0}(t)$ at $t = 0$, the hypothesis

$$\left\{h^{-1} \int_{-h}^{h} |\varphi_{x_0}(t) - U(t)|^p dt\right\}^{1/p} = o(h^k)$$

and the even character of φ_{x_0} imply that $U(t)$ is also even, and so is of degree $k-1$; hence the k-th derivative of $\varphi_{x_0}(t)$ at $t = 0$ is actually 0. We therefore have

$$\varphi_{x_0}(t, F) = U(t) + \varepsilon_{x_0}(t)\frac{t^k}{k!}, \tag{1}$$

where

$$\left\{h^{-1} \int_{0}^{h} |\varepsilon_{x_0}(t)|^p t^{kp} dt\right\}^{1/p} = o(h^k),$$

a condition which is easily seen to be equivalent to

$$\left\{h^{-1} \int_{0}^{h} |\varepsilon_{x_0}(t)|^p dt\right\}^{1/p} = o(1). \tag{2}$$

If (2) holds we shall say that F *satisfies condition* λ_k^p *at* x_0, and by replacing here o by O we define condition Λ_k^p.

If F satisfies condition Λ_k^p at each point of E, it does not necessarily follow that that F has a k-th derivative in L^p almost everywhere in E,

---

(7) See [14].

though the existence of the $(k-1)$-th derivative is assured by the remark made at the beginning of the section. We have, however, the following

THEOREM 5. *If F satisfies condition Λ_k^p, $1 \leqslant p < \infty$, in a set E, then the necessary and sufficient condition for F to have a k-th differential in L^p almost everywhere in E is that the function*

$$\varepsilon_{x_0}^*(t) = \frac{1}{t}\int_0^t \varepsilon_{x_0}(s)\,ds \tag{2a}$$

(see (1)) *satisfies condition* N *almost everywhere in E.*

The definition of condition Λ_k^p presupposes (if k is odd) that $\varepsilon_{x_0}(t)$ is in L^p near $t=0$. By an analogue of Lemma 9, with exponent p, F itself is in L^p near almost all points of E. Hence, without any loss of generality, we may assume that F is in L^p over the whole interval of definition.

Let F_1 be the indefinite integral of F. Integrating (1) with respect to t we get (omitting the subscript x_0 on the right)

$$\psi_{x_0}(t;F_1) = U_1(t)+\eta(t)\frac{t^{k+1}}{(k+1)!}, \tag{3}$$

where

$$U_1(t) = \int_0^t U(s)\,ds, \qquad \eta(t) = \frac{k+1}{t^{k+1}}\int_0^t \varepsilon(s)s^k\,ds.$$

Hence

$$|\eta(t)| \leqslant \frac{k+1}{t^{k+1}}\int_0^t |\varepsilon(s)|s^k\,ds \leqslant (k+1)\frac{1}{t}\int_0^t |\varepsilon(s)|\,ds \tag{3a}$$

$$\leqslant (k+1)\left\{\frac{1}{t}\int_0^t |\varepsilon(s)|^p\,ds\right\}^{1/p},$$

so that if F satisfies condition Λ_k^p at x_0, F_1 satisfies condition Λ_{k+1}.

Next, $\varepsilon^*(t)$ satisfies condition N if and only if $\eta(t)$ does. This follows from the two formulas

$$(k+1)\varepsilon^*(t) = \eta(t)+\frac{k}{t}\int_0^t \eta(s)\,ds,$$

$$\frac{\eta(t)}{k+1} = \varepsilon^*(t)-\frac{k}{t^{k+1}}\int_0^t \varepsilon^*(s)s^k\,ds,$$

which are easily obtainable by integration by parts. Both integrals on the right are absolutely convergent, as may be easily seen from the first inequality (3a) and (2) (with O instead of o).

We shall now prove the sufficiency of the condition in Theorem 5. Suppose that F satisfies condition Λ_k^p in E and that $\varepsilon_{x_0}^*(t)$ satisfies condition N there. Hence the indefinite integral F_1 of F satisfies condition Λ_{k+1} in E and we have (3) with η satisfying condition N. By Theorem 1, F_1 has a Peano $(k+1)$-th derivative almost everywhere in E. By Lemma 6, we have a decomposition $F_1 = G_1 + H_1$, where $G_1 \in C^{k+1}$, $G_1 = F_1$ in a closed set $P \subset E$ with $|E-P|$ arbitrarily small, and the derivatives of order $\leqslant k+1$ of H_1 vanish almost everywhere on P. By differentiating the equation $F_1 = G_1 + H_1$ we obtain $F = G + H$, where $G = G_1' \in C^k$ and $H = H_1'$ vanishes almost everywhere on P.

Consider now any point $x_0 \in P$ where F_1 has a differential of order $k+1$ and H_1 has a differential of order $k+1$ vanishing identically; at x_0 the differentials of order $k+1$ of F_1 and G_1 are equal. At such an x_0 we must have $\eta_{x_0}(t) = o(1)$ since the left-hand side of (3) is an odd function of t, $U_1(t)$ is an odd polynomial of degree $\leqslant k$, and $k+1$ is even. If, for the same x_0, we write $\varphi_{x_0}(t, G_1) = V_1(t) + o(t^{k+1})$, then $U_1(t) = V_1(t)$. It follows that the differential of order k of $\varphi_{x_0}(t, G)$ at $t = 0$ is $V_1'(t) = U_1'(t) = U(t)$. Hence, by (1),

$$\varphi_{x_0}(t, H) = \varphi_{x_0}(t, F) - \varphi_{x_0}(t, G) = \varepsilon_{x_0}(t)\frac{t^k}{k!} + o(t^k),$$

$$\left\{\frac{1}{h}\int_0^h |\varphi_{x_0}(t, H)|^p\,dt\right\}^{1/p} \leqslant h^k \left\{\frac{1}{h}\int_0^h |\varepsilon_{x_0}(t)|^p\,dt\right\}^{1/p} + o(h^k)$$

$$= O(h^k) + o(h^k) = O(h^k).$$

Since $H = 0$ almost everywhere in P, an application of the analogue of Lemma 14 with exponent p instead of 2, shows that at almost all points $x \in P$ we have

$$\left\{\frac{1}{2h}\int_{-h}^{h} |H(x+t)|^p\,dt\right\}^{1/p} = o(h^k).$$

At such a point x the function $F = G + H$ has a k-th differential in L^p. Hence F has such a differential almost everywhere in P, and so also almost everywhere in E. This completes the proof of the sufficiency part of Theorem 5.

The proof of the necessity of the condition in Theorem 5 is simple. Suppose that F has a k-th differential in L^p in E and satisfies condition Λ_k^p there. It is immediate that the indefinite integral F_1 of F has a $(k+1)$-th Peano differential in E. Moreover, we have (3), where $U_1(t) = \int_0^t U(s)\,ds$ is a polynomial of degree k and $\eta(t) = o(1)$. By Theorem 2, $\eta(t)$ satisfies condition N almost everywhere in E and this implies, as we indicated above, that $\varepsilon_{x_0}^*(t)$ satisfies condition N almost everywhere in E.

The following result is a corollary of Theorem 5:

THEOREM 6. *Suppose that $F \in L^p$, $1 \leqslant p < \infty$. The necessary and sufficient condition for F to have a k-th derivative in L^p almost everywhere in a set E is that, almost everywhere in E,*

a) *F satisfies condition Λ_k^p;*

b) *the function $\varepsilon_{x_0}^*(t)$ defined by* (2a) *satisfies condition* N.

That condition Λ_k^p is satisfied at each point where the k-th derivative in L^p exists, is clear, and then the necessity of condition b) follows from Theorem 5. The latter theorem also implies the sufficiency of conditions a) and b).

2. In the case $p \geqslant 2$ Theorem 6 can be stated in a different form

THEOREM 7. *Suppose that $F \in L^p$, $2 \leqslant p < \infty$. The necessary and sufficient condition for F to have a k-th derivative almost everywhere in a set E is that, almost everywhere in E,*

a) *F satisfies condition Λ_k^p;*

b) *the function $\varepsilon_{x_0}(h)$ in* (1) *(for k odd, with a corresponding modification for k even) satisfies condition* N.

The necessity of condition b) (condition a) here is the same as in Theorem 5) follows from Theorem 4 and the fact that, since $p \geqslant 2$, differentiability in L^p implies differentiability in L^2. The sufficiency of the conditions follows from Theorem 6 observing that if $\varepsilon_{x_0}(t)$ satisfies condition N so does $\varepsilon_{x_0}^*(t)$ (a simple consequence of Schwarz's inequality).

3. For our next theorem we need the following lemma:

LEMMA 17. *Suppose that $F \in L^p$, $p \geqslant 2$, and that (for k odd, with the corresponding modification for k even) the function $\varepsilon_{x_0}(t)$ in* (1) *satisfies condition* N *in a set E. Then the following two conditions are equivalent almost everywhere in E:*

a) *F satisfies condition Λ_k^p;*

b) *the integral $\int_0 t^{-1}|\varepsilon_{x_0}(t)|^p dt$ is finite.*

It is clear that condition b) implies a) at each point. It is the converse that requires proof. It will be convenient to assume that F is periodic of period 2π.

Since $\varepsilon_{x_0}(t)$ satisfies condition N in E, F has a k-th derivative in L^2 almost everywhere in E (Theorem 4). Let us apply to F Lemma 16 and keep the notation of that lemma. Without loss of generality we may assume that the function G is not only in C^k but also of period 2π. Then the function $\varepsilon_x(t)$ of (1) but corresponding to G ($\varepsilon_x(t) = \varepsilon_x(t, G)$) satisfies the inequality

$$\left\{\int_{-\pi}^{\pi}\int_{-\pi}^{\pi}\varepsilon_x^2(t,G)\frac{dx\,dt}{t}\right\}^{1/2}\leqslant A\left\{\int_{-\pi}^{\pi}g^2(x)\,dx\right\}^{1/2}\qquad (g=G^{(k)}),$$

with A independent of G. This is proved in [13], and in the case $k=1$ pretty familiar (see [4]). We claim that if $2<p<\infty$, then

$$\left\{\int_{-\pi}^{\pi}\int_{-\pi}^{\pi}|\varepsilon_x(t,G)|^p\frac{dx\,dt}{t}\right\}^{1/p}\leqslant A\left\{\int_{-\pi}^{\pi}|g|^p\,dx\right\}^{1/p}.$$

For $p=2$ this is the preceding inequality, and it also holds for $p=\infty$ if we interpret is as $\operatorname{Ess\,sup}_{x,t}|\varepsilon_x(t,G)|\leqslant A\operatorname{Ess\,sup}_x|g(x)|$. Hence, by M. Riesz' convexity theorem, it is valid for $2<p<\infty$ (for $k=1$, the result is already in [4]). It follows that

$$\int_0^{\pi}|\varepsilon_x(t,G)|^p t^{-1}\,dt<\infty$$

for almost all x.

In particular, G satisfies condition Λ_k^p almost everywhere. Hence, if F satisfies condition Λ_k^p almost everywhere in P, so does the function H in the decomposition $F=G+H$. Hence, by the Remark to Lemma 14 (with $a=kp+1$),

$$\left\{\frac{1}{h}\int_0^h|H(x\pm t)|^p\,dt\right\}^{1/p}=o(h^k)$$

for almost all $x\in P$. This shows (since $H=0$ in P) that the function $|H(x\pm t)|^p t^{-(kp+1)}$, and so also the function $|\varepsilon_x(t,H)|^p t^{-1}$, is integrable near $t=0$ for almost all $x\in P$. Hence the same holds for $\varepsilon_x(t,F)$ and the lemma is established.

THEOREM 8. *Suppose that $F\in L^p$, $p\geqslant 2$, and k is odd (the corresponding result holds for k even), then F has a k-th derivative in L^p almost everywhere in E if and only if the two integrals*

$$\int_0 \{\varepsilon_x(t)\}^2\frac{dt}{t},\qquad \int_0|\varepsilon_x(t)|^p\frac{dt}{t}$$

are finite almost everywhere in E.

If the two integrals are finite almost everywhere in E, then, by Theorem 7, F has a k-th derivative in L^p at almost all points of E. The converse follows from Theorem 7 and Lemma 17.

If $k=1$, Theorem 8 asserts that, for $p\geqslant 2$, the simultaneous existence of the two integrals

$$\int_0\frac{[F(x+t)+F(x-t)-2F(x)]^2}{t^3}\,dt,\qquad \int_0\frac{|F(x+t)+F(x-t)-2F(x)|^p}{t^{p+1}}\,dt$$

almost everywhere in E is both necessary and sufficient for the differentiability of F in L^p almost everywhere in E.

4. The following result is an immediate corollary of Theorem 4' and of known results:

THEOREM 9. *Suppose that $F(x)$ is in $L^2(-\infty, +\infty)$ and has a $(k-1)$-th Peano derivative at each point of a set E. Suppose also that the function $\omega_x(t)$ defined by the equation*

$$F(x+t) = \sum_{j=0}^{k-1} \frac{1}{j!} F_{(j)}(x) t^j + \omega_x(t) \frac{t^k}{k!} \qquad (x \in E)$$

satisfies at each point $x \in E$ the condition

$$\int_0^\infty \frac{[\omega_x(t) - \omega_x(-t)]^2}{t} dt < \infty. \tag{4}$$

Then the integral

$$\int_0^\infty \frac{\omega_x(t) - \omega_x(-t)}{t} dt = \lim_{\varepsilon \to +0} \int_\varepsilon^\infty \tag{5}$$

exists almost everywhere in E.

Proof. The finiteness of the integral (4) implies that F has a k-th differential in L^2 at almost all points of E. At each point at which this occurs, and so almost everywhere in E, the function $\Phi(x) = \int_0^x F(t)\,dt$ has a $(k+1)$-th Peano derivative. By the main result of [10], the integral (5) exists almost everywhere in E.

In the case $k = 1$, Theorem 9 asserts that if $F \in L^2(-\infty, +\infty)$, the existence of the integral of Marcinkiewicz

$$\int_0^\infty \frac{[F(x+t)+F(x-t)-2F(x)]^2}{t^3} dt$$

for $x \in E$, implies the existence of

$$\int_0^\infty \frac{F(x+t)+F(x-t)-2F(x)}{t^2} dt$$

almost everywhere in E. The converse is not true; to see this, it is enough to construct an $F \in L^2(-\infty, +\infty)$, such that $\Phi(x) = \int_0^x F(t)\,dt$ has a second Peano derivative almost everywhere while F is almost everywhere without a first derivative in L^2. The construction is not difficult and is omitted here. A similar argument shows that the converse of Theorem 9 is false for each k.

5. Some of the results proved in this paper have extensions to functions of several variables. We give here one of such extensions which is an analogue of Theorem 3′.

THEOREM 10. *Suppose that* $F(x) = F(x_1, x_2, \ldots, x_n) \in L^2(E^n)$, *and that it satisfies condition* Λ *for each* $x_0 \in E \subset E^n$, *i. e., we have*

$$F(x_0+h)+F(x_0-h)-2F(x_0) = O(|h|), \tag{6}$$

where $h = (h_1, h_2, \ldots, h_n)$, $|h| = (\sum h_j^2)^{1/2}$. *Then the necessary and sufficient condition for* F *to have a first total differential almost everywhere in* E *is that*

$$\int_{E^n} \frac{[F(x_0+h)+F(x_0-h)-2F(x_0)]^2}{|h|^{n+2}}\, dh < \infty \tag{7}$$

almost everywhere in E.

The integral in (7) is the n-dimensional analogue of the integral of Marcinkiewicz. The part of it extended over $|h| \geqslant \varepsilon > 0$ is always finite.

The necessity of the condition is proved exactly as in the one-dimensional case, by decomposing F into a "good" and "bad" part (see [1], p. 189). We have here even a somewhat stronger result: we have (7) almost everywhere in E if we merely assume that F has at each point of E a total differential in the sense of L^2.

The sufficiency of the condition is a relatively simple consequence of the corresponding result for the one-dimensional case. We sketch the proof, and we omit routine arguments involving the measurability of the sets which occur in the proof. Without loss of generality we may replace (6) by

$$|F(x_0+h)+F(x_0-h)-2F(x_0)| \leqslant M|h| \quad (x_0 \in E;\ |h| \leqslant \delta).$$

Suppose we have (7) for a fixed x_0 and let $e_1, e_2, \ldots, e_n$ be a system of n mutually orthogonal unit vectors. Then, by Fubini's theorem, we have

$$\sum_{k=1}^{n} \int_0^{\infty} t^{-3}[F(x_0+e_k t)+F(x_0-e_k t)-2F(x_0)]^2\, dt < \infty \tag{8}$$

for almost all choices of the frame $e_1, e_2, \ldots, e_n$. Using Fubini's theorem again, we obtain the existence of a *fixed* frame $e_1, e_2, \ldots, e_n$ such that (8) holds for almost all $x_0 \in E$. Applying a rotation, we may assume that this frame lies along the $x_1, x_2, \ldots, x_n$ axes.

By our one-dimensional theorem, the partial derivative $(\partial/\partial x_k)F$ $(k = 1, \ldots, n)$ exist almost everywhere in E. This implies that F has an approximate total differential almost everywhere in E, i. e., for almost all $x_0 \in E$ there is a set H_{x_0} having 0 as a point of density and such that

$F(x_0+h) = \sum h_k(\partial/\partial x_k)F(x_0)+o(|h|)$, provided h tends to 0 through the set H_{x_0}. Take a fixed point x_0 at which this occurs. We may assume without loss of generality that $x_0 = 0$ and that $(\partial/\partial x_k)F(x_0) = 0$, $k = 1, 2, \ldots, n$. Write H_0 for H_{x_0}. Since 0 is a point of density for H_0, for any h with $|h|$ sufficiently small, and for any $\varepsilon > 0$, we can find two points h_1 and h_2 in H_0 such that, in the first place, h_1 is the mid-point of the segment (h, h_2) and, second, $|h-h_1| \leqslant \varepsilon|h|$. Since $F(h_1) = o(|h_1|) = o(|h|)$ and, similarly, $F(h_2) = o(|h|)$, and since

$$|F(h)-2F(h_1)+F(h_2)| \leqslant M|h-h_1| \leqslant M\varepsilon|h|,$$

it follows that $|F(h)| \leqslant 2\varepsilon M|h|$ for h arbitrary and sufficiently small. Hence $F(h) = o(|h|)$ and the theorem is established.

References

[1] A. P. Calderón and A. Zygmund, *Local properties of solutions of elliptic partial differential equations*, Studia Math. 20 (1961), p. 171-225.

[2] K. O. Friedrichs, *An inequality for potential functions*, American Journal for Math. 68 (1946), p. 581-592.

[3] A. Khintchine, *Recherches sur la structure des fonctions mesurables*, Fund. Math. 9 (1923), p. 212-279.

[4] J. Marcinkiewicz, *Sur quelques intégrales du type de Dini*, Annales de la Soc. Pol. de Math. 17 (1938), p. 42-50.

[5] — *Sur les séries de Fourier*, Fund. Math. 27 (1936), p. 38-69.

[6] — *Sur quelques intégrales du type de Dini*, Annales de la Soc. Pol. de Math. 17 (1938), p. 42-50.

[7] — and A. Zygmund, *On the differentiability of functions and summability of trigonometric series*, Fund. Math. 26 (1936), p. 1-43.

[8] — and A. Zygmund, *On the behavior of trigonometric and power series*, Trans. Amer. Math. Soc. 50 (1941), p. 407-453.

[9] E. M. Stein and A. Zygmund, *Smoothness and differentiability of functions*, Ann. Univ. Sci. Budapest, Sectio Math., III-IV (1960-1961), p. 295-307.

[10] M. Weiss and A. Zygmund, *On the existence of conjugate functions of higher order*, Fund. Math 48 (1960), p. 175-187.

[11] A. Zygmund, *Trigonometric series*, 2nd ed., Cambridge 1959, 2 vols.

[12] — *Smooth functions*, Duke Math. Journal 12 (1945), p. 47-76.

[13] — *A theorem on generalized derivatives*, Bull. Amer. Math. Soc. 49 (1943), p. 917-923.

[14] M. Weiss, *Symmetric differentiation in L^p* (to appear in the Studia Mathematica).

Reçu par la Rédaction le 13. 5. 1963

STUDIA MATHEMATICA. T. XXIV. (1964)

On higher gradients of harmonic functions

by

A. P. CALDERÓN and A. ZYGMUND (Chicago)*

Chapter I

1. Let $U(x) = U(x_1, x_2, \ldots, x_n)$ be a real-valued harmonic function defined in a domain D of the n-dimensional Euclidean space ($n \geqslant 2$). Consider the norm $W(x)$ of the gradient of $U(x)$,

$$W(x) = |\mathrm{grad}\, U| = \left\{ \sum_{j=1}^{n} \left(\frac{\partial U}{\partial x_j} \right)^2 \right\}^{1/2}.$$

It is a classical fact that $W(x)$ is subharmonic in D, and therefore $\{W(x)\}^p$ is also subharmonic for any $p \geqslant 1$. E. M. Stein and G. Weiss [3] established a remarkable fact that $\{W(x)\}^p$ is subharmonic in D for some values of p less than 1, more precisely, subharmonic for any

(1.1.1) $$p \geqslant \frac{n-2}{n-1}.$$

The example $U(x) = \left(\sum x_j^2\right)^{-(n-2)/2}$ shows that the result is false for p less than $(n-2)/(n-1)$. The case $n = 2$ is, of course, classical if we interpret the result as the subharmonicity of $\log W$.

In this chapter we extend the Stein-Weiss result to higher gradients.

2. Let $a = (a_1, a_2, \ldots, a_n)$ be any multi-index of weight m, that is, $a_1, a_2, \ldots, a_n$ are non-negative integers and $m = |a| = a_1 + a_2 + \ldots + a_n$. We write $a! = a_1! a_2! \ldots a_n!$ and

$$D^a = \left(\frac{\partial}{\partial x_1} \right)^{a_1} \left(\frac{\partial}{\partial x_2} \right)^{a_2} \ldots \left(\frac{\partial}{\partial x_n} \right)^{a_n}.$$

Given any harmonic function $U(x)$ we consider its gradient of order m, that is, the set of all distinct derivatives of order m (arranged in any fixed way)

$$\mathrm{grad}_m U(x) = \{D^a U\}_{|a|=m}$$

* Research resulting in this paper was partly supported by the Air Force contract AF-AFOSR-62-118 and the NSF contract GP-574.

and its norm

$$W(x) = |\mathrm{grad}_m U| = \Big\{ \sum_{|\alpha|=m} (D^\alpha U)^2 (\alpha!)^{-1} \Big\}^{1/2}. \tag{1.2.1}$$

We have then the following result which for $m = 1$ reduces to the one stated above:

THEOREM 1. *If* $U(x) = U(x_1, x_2, \ldots, x_n)$ *is harmonic, the function* $\{W(x)\}^p = |\mathrm{grad}_m U|^p$ *is subharmonic for*

$$p \geqslant \frac{n-2}{m+n-2}. \tag{1.2.2}$$

We note that for fixed n the right-hand side here is arbitrarily small if m is large enough. If $n = 2$ the result should again be interpreted as the subharmonicity of $\log W(x)$ (and also easily follows from classical facts).

3. The rest of the chapter will be devoted to the proof of Theorem 1. In this chapter (but not in the other two) we shall also use another notation for the derivatives of order m. If $\beta = (\beta_1, \beta_2, \ldots, \beta_m)$ is a multi-index *of m components* such that $1 \leqslant \beta_i \leqslant n$ for $i = 1, 2, \ldots, m$, we will write

$$U_\beta = \frac{\partial}{\partial x_{\beta_1}} \frac{\partial}{\partial x_{\beta_2}} \cdots \frac{\partial}{\partial x_{\beta_m}} U$$

and the set $\{U_\beta\}$ of such derivatives designate by a single letter u. The numbers m and n are kept fixed throughout. Thus u is a vector function with m^n components and the norm $|u|$ of u will be

$$\Big\{\sum U_\beta^2\Big\}^{1/2}.$$

In this sum each component $D^\alpha U$ of $\mathrm{grad}_m U$ occurs exactly $m!/\alpha!$ times so that

$$|u| = (m!)^{1/2} |\mathrm{grad}_m U|.$$

By u_i we shall denote the derivatives $\partial u/\partial x_i$ of the vector function u. Let

$$f = \sum U_\beta^2 = |u|^2,$$

where U is harmonic. We are interested in functions $\varphi(t)$, $t \geqslant 0$, increasing and such that $\varphi(f)$ is subharmonic. We may restrict ourselves to functions φ that are concave; hence $\varphi' \geqslant 0$, $\varphi'' \leqslant 0$. We begin by computing the Laplacian of $\varphi(f)$.

We have

$$\Delta\varphi(f) = \varphi''(f)\,|\mathrm{grad} f|^2 + \varphi'(f)\,\Delta f.$$

Clearly,

$$|\mathrm{grad} f|^2 = 4\sum_i (u\cdot u_i)^2,$$

where the dot designates scalar product, and

$$\Delta f = 2\sum_\beta [U_\beta \Delta U_\beta + |\mathrm{grad}\, U_\beta|^2] = 2\sum_{\beta,i}\left(\frac{\partial U_\beta}{\partial x_i}\right)^2 = 2\sum_i |u_i|^2.$$

Let now

$$M = \max\frac{\sum_i (u\cdot u_i)^2}{|u|^2\sum_i |u_i|^2}. \tag{1.3.1}$$

Since $\varphi'' \leqslant 0$, we have

$$\begin{aligned}\Delta\varphi(f) &= \varphi''(f)4\sum_i (u\cdot u_i)^2 + \varphi'(f)2\sum_i |u_i|^2 \\ &\geqslant \varphi''(f)4M|u|^2\sum |u_i|^2 + 2\varphi'(f)\sum |u_i|^2.\end{aligned} \tag{1.3.2}$$

It follows that if

$$2Mt\varphi''(t) + \varphi'(t) \geqslant 0, \tag{1.3.3}$$

the function $\varphi(f)$ is subharmonic. In particular, we have subharmonicity if

$$2Mt\varphi''(t) + \varphi'(t) = 0,$$

which can be written $\varphi'(t) = Ct^{-1/2M}$. Here $C > 0$ since $\varphi' \geqslant 0$. Taking the second constant of integration 0, we find $\varphi(t) = C't^{1-1/2M}$, and the main problem now is finding the value of M. If we show that

$$M = \frac{m+n-2}{2m+n-2}, \tag{1.3.4}$$

the function $\varphi(f) = f^{(n-2)/2(m+n-2)} = |u|^{(n-2)/(m+n-2)}$ will be subharmonic, and Theorem 1 established. If $n = 2$, then $M = \frac{1}{2}$ and the preceding argument leads to $\varphi(t) = \log t$.

4. In the lemma that follows the index β has the meaning explained in Section 3.

LEMMA 1. *Let $U(x) = U(x_1, x_2, \ldots, x_n)$ be a solid harmonic of degree m. Then*

$$|u|^2 = \sum_\beta U_\beta^2 = C \int_{|x|\leqslant 1} U^2(x)\,dx, \tag{1.4.1}$$

where the constant C depends on m and n only. More generally, if U and U' are any two solid harmonics of degree m, then

$$\sum_\beta U_\beta U'_\beta = C \int_{|x|\leqslant 1} U(x)\,U'(x)\,dx. \tag{1.4.2}$$

Proof. It is enough to prove (1.4.1). By homogeneity, we have

$$\int_{|x|\leqslant 1} U^2(x)\,dx = C \int_{|x|=1} U \frac{\partial U}{\partial \nu}\,d\sigma,$$

C denoting a constant depending on m and n only. On the other hand, Green's formula gives

$$\int_{|x|=1} U \frac{\partial U}{\partial \nu}\,d\sigma = \int_{|x|\leqslant 1} \sum_i \left(\frac{\partial U}{\partial x_i}\right)^2 dx,$$

so that

$$\int_{|x|\leqslant 1} U^2(x)\,dx = C \int_{|x|\leqslant 1} \sum_i \left(\frac{\partial U}{\partial x_i}\right)^2 dx,$$

with the same C. Successive application of this gives (1.4.1).

5. We now pass to the calculation of M.

Let $U(x)$ be a harmonic function and x_0 a point of its definition. Expand U in spherical harmonics at x_0. If V and W are the terms of the development of degrees m and $m+1$ respectively, then at the point x_0 the derivatives of order m of U are the same as the derivatives of order m of V, and the derivatives of order $m+1$ of U are the same as those of W. On account of (1.4.2) we have in (1.3.1)

$$u \cdot u_i = C \int_{|x|\leqslant 1} V \frac{\partial W}{\partial x_i}\,dx,$$

$$|u|^2 = C \int_{|x|\leqslant 1} V^2\,dx,$$

$$\sum |u_i|^2 = C \int_{|x|\leqslant 1} \sum \left(\frac{\partial W}{\partial x_i}\right)^2 dx.$$

Hence M is simply the maximum of

$$I = \sum_{i=1}^{n}\left[\int_{|x|\leqslant 1} V\frac{\partial W}{\partial x_i}\,dx\right]^2 = \sum_{i=1}^{n}\left(V\cdot\frac{\partial W}{\partial x_i}\right)^2, \tag{1.5.1}$$

where V and W are all possible solid harmonics of degrees m and $m+1$ respectively, satisfying the conditions

$$\int_{|x|\leqslant 1} V^2 dx = \int_{|x|\leqslant 1}\sum\left(\frac{\partial W}{\partial x_i}\right)^2 dx = 1,$$

the dot product in (1.5.1) being the integral over $|x| \leqslant 1$ of product.

Let us now fix W and maximize with respect to V. Then

$$\tfrac{1}{2}\delta I = \left[\sum_{i=1}^{n}\left(V\cdot\frac{\partial W}{\partial x_i}\right)\frac{\partial W}{\partial x_i}\right]\cdot\delta V = 0,$$

provided

$$\delta(V\cdot V) = 2(V\cdot\delta V) = 0.$$

But this implies that

$$\sum_i\left(V\cdot\frac{\partial W}{\partial x_i}\right)\frac{\partial W}{\partial x_i} = \lambda V.$$

The largest value of λ is the maximum. Write $\xi_i = (V\cdot\partial W/\partial x_i)$ and multiply the last equation by $\partial W/\partial x_j$. We obtain

$$\sum_i \xi_i\left(\frac{\partial W}{\partial x_i}\cdot\frac{\partial W}{\partial x_j}\right) = \lambda\xi_j,$$

and if we now multiply by ξ_j and sum,

$$\sum_{i,j}\xi_i\xi_j\left(\frac{\partial W}{\partial x_i}\cdot\frac{\partial W}{\partial x_j}\right) = \lambda\sum_{i,j}\xi_j^2.$$

Now if we assume that the $\partial W/\partial x_i$ are linearly independent, then the quantities $t\xi_i$ are arbitrary. It follows that λ is simply the maximum of

$$\sum\xi_i\xi_j\left(\frac{\partial W}{\partial x_i}\cdot\frac{\partial W}{\partial x_j}\right) = \left(\sum\xi_i\frac{\partial W}{\partial x_i}\right)^2$$

with the condition that $\Sigma\xi_i^2 = 1$. Let us denote by ξ the unit vector with components ξ_i. Then

$$\lambda = \max\left(\frac{\partial W}{\partial \xi}\right)^2 = \max_{\xi}\left(\frac{\partial W}{\partial \xi}\cdot\frac{\partial W}{\partial \xi}\right),$$

and

$$M = \max_{\xi, W} \left(\frac{\partial W}{\partial \xi}\right)^2, \quad \sum \left(\frac{\partial W}{\partial x_i}\right)^2 = 1.$$

(We use here systematically the notation: $(F)^2 = (F \cdot F) =$ the integral of F^2 over $|x| \leqslant 1$.)

We have excluded sofar the W for which the derivatives $\partial W / \partial x_i$ are linearly dependent. But since any such W can be approximated in norm ($\|W\|^2 = (W)^2$) by a W with linearly independent derivatives, the maximum of I remains the same if one imposes on W the last condition.

Finally, since the space of W is rotation invariant we may replace ξ by any unit vector, so that

$$M = \max \left(\frac{\partial W}{\partial x_1}\right)^2 \quad \text{with} \quad \sum \left(\frac{\partial W}{\partial x_i}\right)^2 = 1.$$

Now

$$\sum \left(\frac{\partial W}{\partial x_i}\right)^2 = \int_{|x|=1} W \frac{\partial W}{\partial \nu} d\sigma = (m+1) \int_{|x|=1} W^2 d\sigma,$$

and

$$(W)^2 = \int_0^1 \varrho^{n-1+2m+2} d\varrho \left[\int_{|x|=1} W^2 d\sigma\right] = \frac{1}{n+2m+2} \int_{|x|=1} W^2 d\sigma.$$

It follows that

$$(W)^2 = (n+2m+2)^{-1}(m+1)^{-1}$$

or

$$M = \max \left(\frac{\partial W}{\partial x_1}\right)^2 \quad \text{with} \quad (W)^2 = (n+2m+2)^{-1}(m+1)^{-1},$$

and finally we arrive at the formula

(1.5.2) $$M = \max_W \frac{(\partial W / \partial x_1)^2}{(m+1)(n+2m+2)(W)^2}.$$

6. Consider now a solid harmonic W of degree $m+1$. We have (see [1], p. 239)

(1.6.1) $$W = \sum_{\mu=0}^{m+1} r^{m+1-\mu} C_{m+1-\mu}^{\mu+(n-2)/2} \left(\frac{x_1}{r}\right) H_\mu(x_2, \ldots, x_n),$$

where H_μ is a solid harmonic of degree μ in $x_2, \ldots, x_n$, C_k^ν is an ultraspherical polynomial defined by the equation

$$(1-2xt+t^2)^{-\nu} = \sum_{k=0}^{\infty} C_k^\nu(x)t^k,$$

and $r^2 = x_1^2+x_2^2+\ldots+x_n^2$.

We want to maximize $(\partial W/\partial x_1)^2/(W)^2$. We observe that the terms of the sum in (1.6.1) are orthogonal over the unit sphere $x_1^2+x_2^2+\ldots+x_n^2 = 1$, for if we fix x_1 and r the functions resulting from the terms are orthogonal over the sphere $x_2^2+\ldots+x_n^2 = r^2-x_1^2$. If we differentiate with respect to x_1 we obtain a similar expression for $\partial W/\partial x_1$ whose terms will likewise be orthogonal. It follows from this that in order to maximize it is enough to assume that the right-hand side of (1.6.1) consists of a single term. For if we denote by W_μ the μ-th term in the sum (1.6.1), then

$$\frac{(\partial W/\partial x_1)^2}{(W)^2} = \frac{\sum_\mu (\partial W_\mu/\partial x_1)^2}{\sum_\mu (W_\mu)^2} \leqslant \max_\mu \frac{(\partial W_\mu/\partial x_1)^2}{(W_\mu)^2},$$

and if the maximum on the right is attained for $\mu = \mu_0$ and a suitable W_{μ_0}, then the left-hand side attains its maximum for W consisting of a single term W_{μ_0}.

We now fix μ and calculate $(\partial W_\mu/\partial x_1)^2$ and $(W_\mu)^2$. We have

$$(W_\mu)^2 = \int\limits_{|x|\leqslant 1} r^{2m+2-2\mu}\left[C_{m+1-\mu}^{\mu+(n-2)/2}\left(\frac{x_1}{r}\right)\right]^2 H_\mu(x_2, \ldots, x_n)^2\, dx.$$

We may assume that H_μ^2 is normalized; say, its integral over the surface of the unit sphere in the space $x_2, \ldots, x_n$ is equal to 1. Then

$$(W_\mu)^2 = \int_{-1}^{+1} dx_1 \int_0^{\sqrt{1-x_1^2}} r^{2m+2-\mu}\left[C_{m+1-\mu}^{\mu+(n-2)/2}\left(\frac{x_1}{r}\right)\right]^2 \varrho^{2\mu+n-2} d\varrho,$$

where $\varrho^2+x_1^2 = r^2$, or setting $x_1 = r\cos\theta$, $\varrho = r\sin\theta$,

$$(W_\mu)^2 = \int_0^\pi d\theta \int_0^1 r^{2m+n+1}[C_{m+1-\mu}^{\mu+(n-2)/2}(\cos\theta)]^2(\sin\theta)^{2\mu+n-2}dr$$

$$= \frac{1}{2m+n+2}\int_0^\pi [C_{m+1-\mu}^{\mu+(n-2)/2}(\cos\theta)]^2(\sin\theta)^{2\mu+n-2}d\theta.$$

Finally, substituting $\cos\theta = x$, we obtain

$$(1.6.2)\qquad (W_\mu)^2 = \frac{1}{2m+n+2}\int_{-1}^{+1}[C_{m+1-\mu}^{\mu+(n-2)/2}(x)]^2(1-x^2)^{\mu+(n-3)/2}dx.$$

7. From the definition of W_μ,

$$\frac{\partial W_\mu}{\partial x_1} = \left\{(m+1-\mu)\frac{x_1}{r}C_{m+1-\mu}^{\mu+(n-2)/2}(x_1/r)+\left[C_{m+1-\mu}^{\mu+(n-2)/2}\left(\frac{x_1}{r}\right)\right]'\left(1-\frac{x_1^2}{r^2}\right)\right\}r^{m-\mu}H_\mu.$$

But (see [1], p. 175, formula (15))

$$(1-x^2)[C_k^\nu(x)]'+kxC_k^\nu(x) = (k+2\nu-1)C_{k-1}^\nu(x),$$

which shows that

$$\frac{\partial W_\mu}{\partial x_1} = (m+\mu+n-2)r^{m-\mu}C_{m-\mu}^{\mu+(n-2)/2}\left(\frac{x_1}{r}\right)H_\mu(x_2, \dots, x_n).$$

Comparing this with (1.6.2) we obtain

$$(1.7.1)\qquad \left(\frac{\partial W_\mu}{\partial x_1}\right)^2 = \frac{(m+\mu+n-2)^2}{2m+n}\int_{-1}^{+1}[C_{m-\mu}^{\mu+(n-2)/2}(x)]^2(1-x^2)^{\mu+(n-3)/2}dx.$$

From the classical formula (see [1], p. 236, formula (26))

$$J_k^\nu = \int_{-1}^{+1}[C_k^\nu(x)]^2(1-x^2)^{\nu-1/2}dx = \frac{2^{1-2\nu}\Gamma(k+2\nu)\pi}{k!(k+\nu)\Gamma(\nu)^2}$$

we deduce that

$$\frac{J_{k-1}^\nu}{J_k^\nu} = \frac{k(k+\nu)}{(k-1+\nu)(k-1+2\nu)}.$$

From this, (1.6.2), (1.7.1), (1.5.2) and also the observation that in computing $(\partial W/\partial x_1)^2/(W)^2$ we may restrict ourselves to the W's of the form W_μ, we obtain after elementary computation that

$$M = \max_{0\leqslant\mu\leqslant m+1}\frac{(m+\mu+n-2)(m+1-\mu)}{(m+1)(2m+n+2)}$$
$$= \max_{0\leqslant\mu\leqslant m+1}\frac{(m+1)^2-\mu^2+(n-3)[(m+1)-\mu]}{(m+1)(2m+n+2)}.$$

If $n\geqslant 3$ the maximum is clearly attained when $\mu = 0$; if $n = 2$ it is attained when $\mu = 0$ or $\mu = 1$. In all cases, therefore,

$$M = \frac{m+n-2}{2m+n-2}.$$

This proves the formula (1.3.4) and so establishes Theorem 1.

8. Theorem 1 asserts that, for any harmonic U and $u = \operatorname{grad}_m U$ the function $\psi_0(|u|)$ is subharmonic, where

$$\psi_0(t) = t^{p_0}, \quad p_0 = \frac{n-2}{m+n-2} \quad (t \geqslant 0).$$

One may ask whether this is a best possible result. Of course, if $\omega(t)$ is convex and increasing, then the subharmonicity of $\psi_0(|u|)$ implies that of $\omega[\psi_0(|u|)]$, and a positive answer to our question is given by the following

THEOREM 2. *Suppose that $\psi(t)$ is continuous for $t \geqslant 0$ and $\psi(|u|)$, $u = \operatorname{grad}_m U$, is subharmonic for any harmonic U. Then $\psi(t) = \omega(t^{p_0})$, where $\omega(t)$ is increasing and convex and $p_0 = (n-2)/(m+n-2)$. If $n = 2$, we replace here t^{p_0} by* $\log t$.

Write $|u|^2 = f$, $\varphi(t) = \psi(t^{1/2})$. We have to show that $\psi(t)$ is a convex function of t^{p_0}, i. e., $\varphi(t)$ is a convex function of $t^{p_0/2}$. Suppose first that ψ is continuously twice differentiable for $t > 0$. Take a fixed point x_0 and any U such that $|u| > 0$, $\sum u_i^2 > 0$ at x_0, and denote by M_U the ratio under the max sign in (1.3.1). Since $\Delta\varphi(f) \geqslant 0$ at x_0, the first equation (1.3.2) shows that $2M_U f\varphi''(f)+\varphi'(f) \geqslant 0$. Taking the upper bound of M_U and observing that by replacing U by λU we may give f any preassigned positive value without changing M_U, we see that we have $2Mt\varphi''(t) +\varphi'(t) \geqslant 0$ for all $t > 0$, a fact which as can be easily verified expresses the convexity of $\varphi(t)$ with respect to $t^{p_0/2}$.

To dispose of the hypothesis that ψ is twice differentiable for $t > 0$ we use the method of regularization. Let $\{\chi_n(t)\}$ be a sequence of functions defined for $t \geqslant 0$, non-negative, in C'', satisfying the condition $\int_0^\infty \chi_n(t)\,dt = 1$ and having support shrinking to the point 1 as $n \to \infty$. Let

$$\psi_n(t) = \int_0^\infty \psi(st)\chi_n(s)\,ds.$$

The functions $\psi_n(t)$ are in C'' for $t > 0$. Moreover, as easily seen, if $\psi(|u|)$ is subharmonic for all U, so is $\psi_n(|u|)$. Hence $\psi_n(t)$ is a convex function of t^{p_0} for $t > 0$. But $\psi_n(t)$ tends to $\psi(t)$ for $t > 0$. It follows that $\psi(t)$ is a convex function of t^{p_0} for $t > 0$, and so also for $t \geqslant 0$ since it is continuous for $t = 0$.

To show that $\psi(t)$ is increasing (i. e., non-decreasing) for $t \geqslant 0$, observe that if it were not so, then we would have $\psi(0) > \psi(t)$ for all t positive and sufficiently small. Take any U such that $|u| = 0$ at x_0. Then $\psi(|u|)$ would have a strict maximum at x_0 which is incompatible with the hypothesis that $\psi(|u|)$ is subharmonic.

Remark. Further extensions of theorem 1 have been obtained by E. M. Stein and G. Weiss in a paper not yet published.

Chapter II

1. It is a familiar fact that a system of n functions of n variables is the gradient of a harmonic function if and only if both the divergence and the curl of the system vanish. In this chapter we investigate the problem when is a given system of functions a gradient of order m of a harmonic function. We shall need the results in Chapter III. Some of the arguments below are borrowed from [2].

We recall the notation. We consider functions of a variable $x = (x_1, x_2, \ldots, x_n)$ and we write $|x| = (\sum x_j^2)^{1/2}$. By α we denote *multi-indices* $(\alpha_1, \alpha_2, \ldots, \alpha_n)$, where the α_j are non-negative integers, and by the *weight* of α we mean the number $|\alpha| = \sum \alpha_j$. We write

$$\alpha! = \alpha_1!\alpha_2!\ldots\alpha_n!, \qquad x^\alpha = x_1^{\alpha_1}x_2^{\alpha_2}\ldots x_n^{\alpha_n}, \qquad \left(\frac{\partial}{\partial x}\right)^\alpha = \left(\frac{\partial}{\partial x_1}\right)^{\alpha_1}\ldots\left(\frac{\partial}{\partial x_n}\right)^{\alpha_n}.$$

If $P(x)$ is a polynomial $\sum a_\alpha x^\alpha$, we mean by $P(\partial/\partial x)$ the operator $\sum a_\alpha (\partial/\partial x)^\alpha$.

By Π_m we shall denote the linear space of all homogeneous polynomials of degree m in x. By h_m we shall mean the subclass of Π_m consisting of all harmonic polynomials of degree m.

If P and Q are in Π_m, we set

$$(P, Q) = P\left(\frac{\partial}{\partial x}\right)Q.$$

It is easy to see that (P, Q) *is an inner product on* Π_m. For suppose that $|\alpha| = |\beta| = m$. Then $(\partial/\partial x)^\alpha x^\beta = 0$ if $\beta \neq \alpha$, and $(\partial/\partial x)^\alpha x^\alpha = \alpha!$. Thus if $P = \sum_{|\alpha|=m} a_\alpha x^\alpha$, $Q = \sum_{|\beta|=m} b_\beta x^\beta$, then

$$(P, Q) = P\left(\frac{\partial}{\partial x}\right)Q = \sum a_\alpha b_\alpha \alpha! = (Q, P). \tag{2.1.1}$$

2. Lemma 1. *Suppose that $Q \in \Pi_m$. Then $(Q, P) = 0$ for all $P \in h_m$ if and only if Q is divisible by $x_1^2 + x_2^2 + \ldots + x_n^2$.*

Let $\Delta = \sum \partial^2/\partial x_j^2$. If Q is divisible by $x_1^2 + \ldots + x_n^2$, then $Q = R\cdot(x_1^2 + \ldots + x_n^2)$ and

$$(Q, P) = R\left(\frac{\partial}{\partial x}\right)\Delta P = 0 \quad \text{for all} \quad P \in h_m.$$

Suppose, conversely, that $(Q, P) = 0$ for all $P \in h_m$. Consider the mapping $\varphi: P \to \Delta P$ of Π_m into Π_{m-2}; we claim that the mapping is "onto". In fact, if $R\in\Pi_{m-2}$ and R is orthogonal to all polynomials of the form ΔP, $P\in\Pi_m$, then $R(\partial/\partial x)\Delta P=0$ for all $P\in\Pi_m$. Setting $P(x)=$ $=(x_1^2+\ldots+x_n^2)R(x)$ we obtain

$$R\left(\frac{\partial}{\partial x}\right)\left[\frac{\partial^2}{\partial x_1}+\ldots+\frac{\partial^2}{\partial x_n}\right]R(x)(x_1^2+\ldots+x_n^2)=0,$$

which, in view of the fact that the operation (P,Q) is an inner product (see (2.1.1)) implies that $R(x)(x_1^2+\ldots+x_n^2)=0$, i. e. $R=0$. Thus the mapping φ is actually "onto".

The kernel of the mapping φ is precisely h_m. Hence $\dim h_m=\dim\Pi_m$ $-\dim\Pi_{m-2}$, and the orthogonal complement $h_m^\perp$ has dimension

$$\dim\Pi_m-\dim h_m=\dim\Pi_{m-2}.$$

Consider now the mapping $\psi\colon\Pi_{m-2}\to\Pi_m$ given by $\psi(Q)=$ $(x_1^2+\ldots+x_n^2)Q(x)$, $Q\in\Pi_{m-2}$. The mapping is one-one and so the image $\psi(\Pi_{m-2})$ of Π_{m-2} has dimension $\dim\Pi_{m-2}$. Furthermore, $\psi(\Pi_{m-2})\in h_m^\perp$, for f $P\in h_m$, then

$$\big(\psi(Q),P\big)=\big((x_1^2+\ldots+x_n^2)Q,P\big)=Q\left(\frac{\partial}{\partial x}\right)\Delta P(x)=0.$$

Consequently, since $\dim\psi(\Pi_{m-2})=\dim h_m^\perp$, we have $h_m^\perp=\psi(\Pi_{m-2})$, that is, every $P\in h_m^\perp$ is of the form $(x_1^2+\ldots+x_n^2)Q$, $Q\in\Pi_{m-2}$, and the lemma is established.

3. THEOREM 1. *Let $\{P_a\}$ be a set of homogeneous polynomials of degree k, where a runs through all multi-indices of weight m. Then $P_a=(\partial/\partial x)^aP$, where $P\in h_{m+k}$ if and only if $\sum Q_a(\partial/\partial x)P=0$ for all sets of polynomials Q_a of degree k such that $\sum x^aQ_a(x)$ is divisible by $x_1^2+\ldots+x_n^2$.*

The necessity of the condition is clear. To prove the sufficiency, consider the set of polynomials $R_a=(\partial/\partial x)^aP$, where $P\in h_{m+k}$. They form a linear subspace of the space of the vectors $\{S_a\}$, $S_a\in\Pi_k$. In the space of vectors $S=\{S_a\}$ we have an inner product $(S_1,S_2)=\sum(S_{1a},S_{2a})$. If $\{Q_a\}$ is a vector orthogonal to all R_a, $R_a=(\partial/\partial x)^aP$, then we have

$$\sum\left(Q_a,\left(\frac{\partial}{\partial x}\right)^aP\right)=0$$

for all $P\in h_{m+k}$. But

$$\sum\left(Q_a,\left(\frac{\partial}{\partial x}\right)^aP\right)=\sum(x^aQ_a,P)=\left(\sum x^aQ_a,P\right)=0$$

for all $P\in h_{m+k}$. According to Lemma 1 this implies that $\sum x^aQ_a$ is divisible by $x_1^2+\ldots+x_n^2$ and thus, by hypothesis, $\sum(Q_a,P_a)=\sum Q_a(\partial/\partial x)P_a(x)=0$.

Consequently, if Q_a is orthogonal to the space of vectors $\{R_a\}$, it is also orthogonal to the vector $\{P_a\}$, i. e., $\{P_a\}$ is among the $\{R_a\}$.

4. THEOREM 2. *Let $\{u_a\}$ be a set of C^∞ functions in the sphere $|x| < R$ where a runs through all multi-indices of weight m. Then $u_a = (\partial/\partial x)^a u$, where u is a harmonic function if and only if $\sum Q_a(\partial/\partial x) u_a = 0$ whenever Q_a are homogeneous polynomials of the same degree such that $\sum x^a Q_a(x)$ is divisible by $x_1^2 + \ldots + x_n^2$.*

Remark. The condition that the u_a are in C^∞ can be dropped, but then $Q_a(\partial/\partial x)$ must be taken in the sense of distributions.

Proof. The necessity of the condition is obvious as before.

In the proof of sufficiency, observe, first of all that the u_a are necessarily harmonic. For set $Q_a(x) = x_1^2 + \ldots + x_n^2$ if $a = \beta$ and $Q_a = 0$ if $a \neq \beta$. Then our hypotheses imply that $\Delta u_\beta = 0$.

Let now $u_a = \sum a_\nu^a P_\nu(x)$ be the expansion of u_a into normalized spherical harmonics. We observe that a series $\sum a_\nu P_\nu(x)$ of normalized spherical harmonics converges for $|x| < R$ if and only if $\sum |a_\nu| \varrho^\nu < \infty$ for all $\varrho < R$. Consequently, we have $\sum |a_\nu^a| \varrho^\nu < \infty$ for $\varrho < R$.

Suppose now that $Q_a(x)$ are homogeneous polynomials of the same degree such that $\sum x^a Q_a(x)$ is divisible by $x_1^2 + x_2^2 + \ldots + x_n^2$. Then

$$\sum_a Q_a\left(\frac{\partial}{\partial x}\right) u_a = \sum_\nu \sum_a a_\nu^a Q_a\left(\frac{\partial}{\partial x}\right) P_\nu^a(x) = 0.$$

Since the inner sum on the right represents a harmonic polynomial of degree $\nu - |a|$, the vanishing of the series implies the vanishing of each of the terms. Thus we have

$$\sum_a Q_a\left(\frac{\partial}{\partial x}\right) [a_\nu^a P_\nu^a(x)] = 0$$

whenever $\sum x^a Q_a(x)$ is divisible by $x_1^2 + \ldots + x_n^2$. By the preceding theorem there exist harmonic polynomials, which we will denote by $b_\nu P_\nu$ such that P_ν is a normalized spherical harmonic and

$$a_\nu^a P_\nu^a(x) = \left(\frac{\partial}{\partial x}\right)^a b_{\nu+m} P_{\nu+m} \qquad (m = |a|).$$

These polynomials $b_\nu P_\nu$ are uniquely determined for $\nu \geqslant m$. For if also $a_\nu^a P_\nu^a = (\partial/\partial x)^a b'_{\nu+m} P'_{\nu+m}$, then

$$\left(\frac{\partial}{\partial x}\right)^a [b'_{\nu+m} P'_{\nu+m} - b_{\nu+m} P_{\nu+m}] = 0$$

for all a, $|a| = m$, which would imply that $b'_{\nu+m} P'_{\nu+m} - b_{\nu+m} P_{\nu+m}$ is a polynomial of degree $\leqslant m-1$. If we show that the series $\sum b_\nu P_\nu(x)$ converges for $|x| < R$, then denoting its sum by $u(x)$ we shall have

$$\left(\frac{\partial}{\partial x}\right)^a u(x) = \sum_\nu \left(\frac{\partial}{\partial x}\right)^a b_\nu P_\nu(x) = \sum_\nu a_\nu^a P_\nu^a(x) = u_a(x).$$

Now, for a normalized spherical harmonic P_ν of degree ν we have

$$|P_\nu(x)| \leqslant C\nu^{n-2}|x|^\nu,$$

where C depends on the dimension n only. If $\partial/\partial\varrho$ denotes differentiation in the direction of the unit vector $(\mu_1, \mu_2, \ldots, \mu_n)$, we have

$$\frac{\partial}{\partial\varrho} = \sum_{j=1}^{m} \mu_j \frac{\partial}{\partial x_j}$$

and

$$\left|\left(\frac{\partial}{\partial\varrho}\right)^m b_{\nu+m} P_{\nu+m}(x)\right| = \left|\left[\sum \mu_j \frac{\partial}{\partial x_j}\right]^m b_{\nu+m} P_{\nu+m}(x)\right| \leqslant C \sum |a_\nu^a|\,|P_\nu^a(x)|,$$

where C is a sufficiently large constant, and this is majorized by $C\sum\limits_a |a_\nu^a|\nu^{n-2}|x|^\nu$.

Integrating along the ray we obtain

$$|b_{\nu+m}P_{\nu+m}(x)| \leqslant C\,|x|^{\nu+m} \sum_a |a_\nu^a|\,\nu^{n-2}.$$

Since $\sum |a_\nu^a|\,\varrho^\nu < \infty$ for $\varrho < R$, it follows that $\sum |b_{\nu+m}P_{\nu+m}(x)| < \infty$ for $|x| < R$ and Theorem 2 is established.

Chapter III

1. Let $f(x) = f(x_1, x_2, \ldots, x_n) \in L^p = L^p(E_n)$. We consider its Poisson integral

$$P_t f = \frac{2}{\omega_{n+1}} \int_{E_n} \frac{t}{[(x-z)^2 + t^2]^{(n+1)/2}} f(z)\,dz, \qquad t > 0.$$

Young's inequality implies that $P_t f \in L^q$, $p \leqslant q \leqslant \infty$, for each $t > 0$. If $t = t_1 + t_2$, then $P_t f = P_{t_1} P_{t_2} f$.

We consider Riesz transforms $R_j f$, $j = 1, 2, \ldots, n$, of f. There are a number of definitions of Riesz transforms. Using Fourier transforms we may define $R_j f$ by the equation

$$(R_j f)^\wedge = -i \frac{x_j}{|x|} \hat{f}.$$

This definition is legitimate for $f \in L^2$, and it then turns out that for any f in C^∞ and having finite support we have $\|R_j f\|_p \leqslant A_p \|f\|_p$, where $1 < p < \infty$, and A_p depends on p only. Thus R_j can be extended by continuity to all of $L_p(E_n)$. This extension defines $R_j f$ only almost everywhere, but there is another definition of $R_j f$, given by means of singular integrals, which shows that $R_j f$ can be defined everywhere, and is pointwise continuous if f is continuously differentiable (and in L^p). In the arguments below, where we consider Riesz transforms of Poisson integrals the transforms are assumed to be continuous.

If $a = (a_1, a_2, \ldots, a_n)$ is a multi-index and the a_j are non-negative integers, we set $R^a = R_1^{a_1} R_2^{a_2} \ldots R_n^{a_n}$. If $f \in L^p$, $1 < p < \infty$, then $R^a f$ is defined and $\|R^a f\|_p \leqslant A_p^{|a|} \|f\|_p$.

Suppose that $f \in L^2$. Then by differentiating under the integral sign one sees that $P_t f$, as a function of x and t, is in C^∞. Furthermore, all derivatives of $P_t f$ are in $L^2(E_n)$. The Fourier transform of the Poisson kernel is $e^{-|x|t}$. Consequently, we have $(P_t f)^\wedge = \hat{f}(x) e^{-|x|t}$.

Let again $f \in L^2$. Taking Fourier transforms we see that $R^a P_t f = P_t R^a f$ and consequently $R^a P_t f$ and all its derivatives are in C^∞ and $L^2(E_n)$. If D is a monomial differential operator in x and t, then, since $DP_t f$ is in L^2 and $R^a P_t f$ is in C^∞, both $R^a D P_t f$ and $D R^a P_t f$ are well defined and by taking their Fourier transforms we see that

$$R^a D P_t f = D R^a P_t f. \tag{3.1.1}$$

Finally, by again taking Fourier transforms, we see that

$$\frac{\partial}{\partial x_j} R^a D P_t f = R_j \frac{\partial}{\partial t} R^a D P_t f, \tag{3.1.2}$$

i. e., the operators $\partial/\partial x_j$ and $R_j \partial/\partial t$ coincide on all functions $R^a D P_t f$, $f \in L^2$.

2. THEOREM 1. *Let*

$$\beta = (a_1, a_2, \ldots, a_n, k) = (a, k)$$

be the multi-indices of weight m,

$$m = |\beta| = |a| + k,$$

and $f_\beta(x, t)$ *a system of functions of* x *and* t *given by*

$$f_\beta(x, t) = R^a P_t f, \qquad \beta = (a, k), \tag{3.2.1}$$

where f *is real-valued and in* $L^p(E_n)$, $1 \leqslant p < \infty$. *Then the* $f_\beta(x, t)$ *are harmonic functions and*

$$\Big\{ \sum_\beta f_\beta^2(x, t) (\beta!)^{-1} \Big\}^{1/2} = \Big\{ \sum_{k=0}^{m} \frac{1}{k!} \sum_{|a|=m-k} (R^a P_t f)^2 (a!)^{-1} \Big\}^{1/2} \tag{3.2.2}$$

is subharmonic for

(3.2.3) $$l \geqslant \frac{n-1}{m+n-1}.$$

Proof. We assume first that f is bounded and has bounded support; hence $f \in L^2$. The function $R^a P_t f$ is in C^∞, and in view of (3.1.1) and the fact that $P_t f$ is harmonic, the functions $f_\beta(x, t)$ are harmonic.

To prove the subharmonicity of $\left(\Sigma f_\beta^2(\beta!)^{-1}\right)^{l/2}$ we apply Theorem 1 of Chapter I and Theorem 2 of Chapter II. It is enough to show that if $Q_\beta(x, t)$ are homogeneous polynomials of the same degree N, such that

$$\sum x^a t^k Q_\beta(x, t), \quad \beta = (a, k), \quad |\beta| = m,$$

is divisible by $x_1^2+\ldots+x_n^2+t^2$, then $\sum Q_\beta(\partial/\partial x, \partial/\partial t) f = 0$. Now, since $f_\beta = R^a P_t f$ and the operators $\partial/\partial x_j$ and $R_j(\partial/\partial t)$ coincide for all functions of the form $R^a D P_t f$, $f \in L^2$ (see (3.1.2)), we have

$$\sum_\beta Q_\beta\left(\frac{\partial}{\partial x}, \frac{\partial}{\partial t}\right) f_\beta = \sum_\beta Q_\beta\left(R\frac{\partial}{\partial t}, \frac{\partial}{\partial t}\right) R^a P_t f$$

$$= \sum_\beta R^a Q_\beta(R, 1)\left(\frac{\partial}{\partial t}\right)^N P_t f.$$

Since

$$\sum_\beta x^a t^k Q_\beta(x, t) = (x_1^2+\ldots+x_n^2+t^2)\, L(x, t),$$

we have

$$\sum_\beta R^a Q_\beta(R, 1) = L(R, 1)(R_1^2+R_2^2+\ldots+R_n^2+I) = 0,$$

in view of the identity $\sum R_j^2 = -I$, which is an immediate consequence of the definition of the R_j. Thus $\sum Q_\beta(\partial/\partial x, \partial/\partial t) f_\beta = 0$, as we wished to show.

Suppose now that $1 < p < \infty$, $f \in L^p$, and let f_n be bounded, of finite support and tend to f in L^p. Then $P_t f_n$ converges to $P_t f$ in L^p for each $t > 0$, and thus $f_\beta^{(n)} = R^a P_t f_n$ converges to $R^a P_t f$ in $L^p(E_n)$. On the other hand,

$$f_\beta^{(n)} = R^a P_t f_n = P_t R^a f_n,$$

and since $R^a f_n$ converges to $R^a f$ in $L^p(E_n)$, it follows that $f_\beta^{(n)} = P_t R^a f_n$ converges uniformly for $t \geqslant \varepsilon > 0$ to $P_t R^a f = R^a P_t f = f_\beta$. Thus the $f_\beta(x, t)$ are harmonic and $\{\sum f_\beta^2(x, t)(\beta!)^{-1}\}^{l/2}$ subharmonic for $t > 0$ and satisfying (3.2.2).

It remains to consider the case $p = 1$ of Theorem 1. Observe that if $t > \varepsilon$, then $P_t f = P_{t-\varepsilon}(P_\varepsilon f)$, and that $P_\varepsilon f \in L_p$ for all $p \geqslant 1$. This reduces the case to the previous cases.

3. If p is strictly greater than 1, then the functions (3.2.1) are the Poisson integrals of the functions $R^\alpha f$. This is in general not true if $p = 1$ even though $R^\alpha f$ can be defined in that case. It is however not integrable, even locally, so that $P_t R^\alpha f$ has no meaning.

4. The significance of the theorem of this chapter is as follows. If f is in L^p, $p \geqslant 1$, then $|P_t f|^r$ is subharmonic for $r \geqslant 1$. The Stein-Weiss result quoted in Chapter I asserts that if we adjoin to $P_t f$ its Riesz transforms, we obtain a harmonic vector $P_t f, R_1 P_t f, \ldots, R_n P_t f$ whose norm is subharmonic when raised to the power $(n-1)/n$. By the theorem of this chapter, if we keep adding to the last system higher and higher Riesz transforms we obtain harmonic vectors whose norms remain subharmonic when raised to smaller and smaller powers.

Perhaps a change in notation will make this a little clearer. In defining the norm of $\operatorname{grad}_m U$ we considered only distinct derivatives of order m. If, however, we define $u = \operatorname{grad}_m U$ successively as the first gradient of the $(m-1)$-st (which is in a way more natural, as the argument of Chapter I shows) and set

$$R_\gamma f = R_{\gamma_1}, R_{\gamma_2}, \ldots, R_{\gamma_m} f$$

for any multi-index $\gamma = (\gamma_1, \gamma_2, \ldots, \gamma_n)$ *of m components*, where now $1 \leqslant \gamma_j \leqslant n$ for all j, then the factorials in (3.2.2) can be dropped and our theorem asserts that the function

$$\Big\{\sum_{k=0}^{m} \sum_{\gamma_1,\gamma_2,\ldots,\gamma_k=1}^{n} (R_{\gamma_1} R_{\gamma_2} \ldots R_{\gamma_k} P_t f)^2\Big\}^{l/2}$$

is subharmonic for $l \geqslant (n-1)/(m+n-1)$.

References

[1] H. Bateman, *Higher transcendental functions*, vol. 2, 1953.

[2] A. P. Calderón, *Integrales singulares*, Buenos Aires 1961.

[3] E. M. Stein and G. Weiss, *On the theory of harmonic functions of several variables I*, Acta Math. 103 (1960), p. 25-62.

Reçu par la Rédaction le 9. 9. 1963

ON THE FRACTIONAL DIFFERENTIABILITY OF FUNCTIONS

By E. M. STEIN *and* A. ZYGMUND

[Received 24 November 1964]

To J. E. LITTLEWOOD *on his 80th birthday*

1. In this paper we give certain necessary and sufficient conditions for a function to have a fractional derivative almost everywhere in a given set.

We begin by recalling some familiar definitions. Let α be positive and non-integral, say $k < \alpha < k+1$ (k integral). Then the fractional derivative of order α of f, which we shall denote by $f^{(\alpha)}$, is defined by the formula

(1) $$f^{(\alpha)}(x) = \frac{d^{k+1}}{dx^{k+1}} f_\beta(x) \quad (\beta = k+1-\alpha),$$

where f_β designates the integral of f of (fractional) order β. This definition requires further explanation since the fractional integration of order β, as well as the derivative d^{k+1}/dx^{k+1}, may be given several meanings.

There are a number of definitions of fractional integration. Here we adopt M. Riesz's definition,

(2) $$f_\beta(x) = \int_{-\infty}^{\infty} f(t)\,|x-t|^{\beta-1}\,dt = (f * K_{1-\beta})(x) \quad (0 < \beta < 1),$$

where $K_\gamma(x) = |x|^{-\gamma}$.† If $f \in L(-\infty, \infty)$, then f_β exists almost everywhere and is locally integrable.

The results we are going to prove remain valid, the proofs essentially unchanged, if we define f_β as $f_\beta * K_{1-\beta}{}^*$, where $K_\gamma{}^*(x) = \operatorname{sign} x\,|x|^{-\gamma}$, or if we define f_β by

$$\int_{-\infty}^{x} f(t)\,|x-t|^{\beta-1}\,dt = \{\tfrac{1}{2} f * (K_{1-\beta} + K_{1-\beta}{}^*)\}(x);$$

however, we shall systematically adhere to the definition (2).

In the context of this paper a function f defined in a neighbourhood of the point x_0 will be said to have a kth derivative if it has it in the sense

† This agrees with the usual definition except for the multiplicative factor $2\cos(\pi\beta/2)/\Gamma(\beta)$. We omit this factor because it is of no significance for our purposes.

Proc. London Math. Soc. (3) 14A (1965) 249–64

Reprinted from *Proc. London Math. Soc.* (3) 14A, 249–264 (1965).

of Peano, that is if

$$f(x_0+t) = \sum_{j=0}^{k} \frac{\alpha_j t^j}{j!} + R(t), \tag{3}$$

where $R(t) = o(|t|^k)$ as $t \to 0$.

It will also be necessary to consider a generalization of this notion. We shall say that f has kth derivative in the L^2 sense at x_0 if we have (3) with

$$\left\{\frac{1}{t}\int_0^t |R(u)|^2 du\right\}^{1/2} = o(|t|^k) \quad (t \to 0).\dagger$$

Keeping these definitions in mind we say that $f^{(\alpha)}$ exists at x_0 if $f_\beta = f * K_\beta$ has a $(k+1)$th derivative in the sense defined above, with $\beta = k+1-\alpha$. Similarly, we say that $f^{(\alpha)}$ exists in the L^2 sense at x_0 if f_β has a $(k+1)$th derivative in the L^2 sense at x_0.

If α is fractional, $k < \alpha < k+1$, we shall say that f satisfies a Lipschitz condition of order α at x_0, that is f belongs to Λ_α at x_0, if

$$R(t) = O(|t|^\alpha) \quad \text{as} \quad t \to 0.$$

Similarly, we shall say that f belongs to $\Lambda_\alpha{}^2$ at x_0 if

$$\left\{\frac{1}{t}\int_0^t [R(u)]^2 du\right\}^{1/2} = O(|t|^\alpha) \quad \text{as} \quad t \to 0.$$

Now it is known that if f belongs to Λ_α for each point x_0 of a set E, then $f^{(\gamma)}$ exists a.e. if $0 < \gamma < \alpha$; see (**5**). However, $f^{(\alpha)}$ need not exist at any point. The characterizing condition for fractional differentiability turns out to be a condition involving quadratic integrability, and not an 'O' condition. Thus we shall say that f *satisfies condition* N_α *at* x_0 if

$$\int_{-\delta}^{\delta} \frac{[R(t)]^2}{|t|^{1+2\alpha}} dt < \infty \quad \text{for some} \quad \delta > 0,$$

where $R(t)$ is the remainder in the development (3). Our results are then as follows.

THEOREM 1. *Suppose that f satisfies the condition Λ_α at every point of a set E of positive measure.‡ Then $f^{(\alpha)}(x)$ exists almost everywhere in E if and only if f satisfies condition N_α almost everywhere in E.*

There is also a variant of this theorem, where the condition Λ_α is not assumed at the outset, which characterizes the functions satisfying condition N_α.

† See (**1**), where analogous notions for L^p, $1 \leqslant p \leqslant \infty$, are treated systematically, and in the context of several variables.

‡ We always make the assumption that f is integrable on $(-\infty, \infty)$.

THEOREM 2. *The necessary and sufficient condition that f satisfies the condition N_α almost everywhere in a set E is that f satisfies the condition Λ_α^2, and $f^{(\alpha)}$ exists in the L^2 sense, almost everywhere in this set.*

These theorems are consequences, although not immediate ones, of corresponding theorems involving differentiation of integral order which we considered previously in (**4**). To state these results consider

$$\tfrac{1}{2}\{f(x_0+t)+(-1)^{k+1}f(x_0-t)\} = P_{x_0}(t)+\varepsilon_{x_0}(t)\,|t|^k.$$

Here $P(t)$ is the even or odd part of the polynomial $\sum_{j=0}^{k} \alpha_j t^j/(j!)$ appearing in the development (3), depending on whether k is odd or even. Then f is said to satisfy the condition $\Lambda_k{}^*$ at x_0 if $\varepsilon(t) = O(1)$ as $t \to 0$. Moreover, ε is said to satisfy the condition N if $\varepsilon^2(t)/t$ is integrable in a neighbourhood of $t = 0$.

The results analogous to Theorems 1 and 2 above are

THEOREM A. *Suppose that f satisfies condition $\Lambda_k{}^*$ at every point of a set E. Then f has a kth derivative almost everywhere in the set if and only if ε_{x_0} satisfies condition N for almost every x_0 in E.*

THEOREM B. *The necessary and sufficient condition that ε_{x_0} satisfies condition N for almost every point of E is that f has a kth derivative in the L^2 sense at almost every point of E.*

Before we begin the proofs we shall make a clarifying comment. The definition of the fractional derivative $f^{(\alpha)}$ is 'semi-local' in the sense that if f is modified by setting it equal to zero outside an interval containing x_0, the existence of $f^{(\alpha)}(x_0)$ is unchanged. The value of $f^{(\alpha)}(x_0)$ is, however, changed, but this is of no concern to us since we are only interested in the problem of the existence of $f^{(\alpha)}$.

2. We begin by stating a series of lemmas needed below.

LEMMA 1. *Suppose that $\Phi(x) = \int_0^\infty K(x,y)\varphi(y)\,dy$, where K is homogeneous of degree -1, and $\int_0^\infty |K(1,y)|\,y^{-\frac{1}{2}}\,dy < \infty$. Then if φ is in $L^2(0,\infty)$, so is Φ.*

The familiar argument may be found in ((**2**) Chapter 9).

LEMMA 2. *Suppose $0 < \alpha$, α is non-integral, $k < \alpha < k+1$, and $\beta = k+1-\alpha$. Let $g = G_1 * K_\beta{}^*$, where G_1 belongs to $C^{(k)}$ and has finite support. Then g satisfies the condition Λ_α uniformly, and $g \in N_\alpha$ for almost every x_0, $-\infty < x_0 < \infty$.*

Proof. $g^{(k)} = G_1{}^{(k)} * K_\beta{}^*$, and hence

$$|g^{(k)}(x_0+t) - g^{(k)}(x_0)| \leqslant \|G_1{}^{(k)}\|_\infty \int_{-\infty}^{\infty} |K_\beta{}^*(x+t) - K_\beta{}^*(x)|\, dx.$$

The integral is easily seen to be $O(|t|^\beta)$, and thus

$$|g^{(k)}(x_0+t) - g^{(k)}(x_0)| \leqslant A(|t|^\beta).$$

Therefore, by Taylor's theorem, g satisfies the condition Λ_α uniformly.

Next, $g^{(k)}$ is a convolution of an L^2 function $(G_1{}^{(k)})$ with $K_\beta{}^*$, and since the Fourier transform of $K_\beta{}^*$ is $K_{1-\beta}{}^*$ (up to a constant factor), we see that

$$g^{(k)}(x) = \int_{-\infty}^{\infty} \gamma(u)\, |u|^{\beta-1} e^{ixu}\, du, \quad \text{with} \quad \gamma(u) \in L^2(-\infty, \infty).$$

By Plancherel's formula,

$$\frac{1}{2\pi}\int_{-\infty}^{\infty} |g^{(k)}(x+t) - g^{(k)}(x)|^2\, dx = \int_{-\infty}^{\infty} |\gamma(u)|^2 |u|^{2\beta-2} \left[2^2 \sin\left(\frac{tu}{2}\right)^2\right] du.$$

Let us define $\omega_x(t)$ by

$$g^{(k)}(x+t) - g^{(k)}(x) = \omega_x(t)\, |t|^{1-\beta}.$$

Inserting this in the above, and integrating with respect to t, we get

$$\frac{1}{2\pi}\int_{-\infty}^{\infty}\left\{\int_{-\infty}^{\infty} \frac{[\omega_x(t)]^2}{|t|}\, dt\right\} dx = \int_{-\infty}^{\infty} |\gamma(u)|^2 \left\{\int_{-\infty}^{\infty} \frac{2^2 \sin^2 \dfrac{tu}{2}}{|t|^{3-2\beta}}\, |u|^{2\beta-2}\, dt\right\} du.$$

However, the inner integral is a finite constant, $\displaystyle\int_{-\infty}^{\infty} \frac{2^2 \sin^2 \frac{t}{2}}{|t|^{3-2\beta}}\, dt$, and $\gamma(u)$ is square-integrable; thus

$$\int_{-\infty}^{\infty} \frac{[\omega_{x_0}(t)]^2}{|t|}\, dt < \infty \quad \text{for almost every } x_0.$$

Now write $g(x+t) = P_x(t) + R_x(t)$, with $P_x(t)$ the polynomial of degree k in t corresponding to the Taylor development of g about x, and $R_x(t)$ the remainder. From the fact that $g^{(k)}(x+t) - g^{(k)}(x) = \omega_x(t)\, |t|^{1-\beta}$, and the remainder formula, it follows that

$$R_{x_0}(t) = \frac{1}{k!}\int_0^t (t-u)^{k-1} \omega_{x_0}(u)\, |u|^{1-\beta}\, du,$$

and hence

$$\int_{-\infty}^{\infty} \frac{|R_{x_0}(t)|^2}{|t|^{1+2\alpha}}\, dt < \infty,$$

by an easy application of Lemma 1. For these x_0, g satisfies the condition N_α.

We shall also need the following inversion formula for fractional integrals. (See also (**3**).)

LEMMA 3. *Suppose that* $f \in L^2(-\infty,\infty)$ *and has finite support, and that* $F = f * K_{1-\beta}$, $0 < \beta < 1$. *Then*

$$\frac{1}{A_\beta}\int_{|t|\geq\varepsilon} \frac{F(x+t)-F(x)}{|t|^{1+\beta}}\,dt$$

converges in L^2 *norm to* $f(x)$, *as* $\varepsilon \to 0$. *(Here* A_β *is a suitable constant.)*

Proof. The function $K_{1-\beta}(x+t)-K_{1-\beta}(x)$ belongs, for each t, to $L^1(-\infty,\infty)$, and its Fourier transform is $B_\beta(e^{ixt}-1)|x|^{-\beta}$, since the Fourier transform of $K_{1-\beta}$ is K_β (up to a constant multiple). Thus the Fourier transform of $F(x+t)-F(x)$ is $B_\beta \hat{f}(x)(e^{ixt}-1)|x|^{-\beta}$, and therefore the Fourier transform of $\int_{|t|\geq\varepsilon} \frac{F(x+t)-F(x)}{|t|^{1+\beta}}\,dt$ is $\hat{f}(x)M_\varepsilon(x)$ with

$$M_\varepsilon(x) = B_\beta|x|^{-\beta}\int_{|t|\geq\varepsilon} \frac{e^{ixt}-1}{|t|^{1+\beta}}\,dt.$$

Now it is apparent that $M_\varepsilon(x)$ is uniformly bounded in ε and x, and that

$$\lim_{\varepsilon\to 0} M_\varepsilon(x) = B_\beta\int_{-\infty}^{\infty} \frac{e^{it}-1}{|t|^{1+\beta}}\,dt = \text{constant}$$

if $x \neq 0$. It is also apparent from this expression that the constant is different from zero. If we take the reciprocal of this constant to be $1/A_\beta$, and use Plancherel's theorem, we obtain the lemma. (In fact, $A_\beta = -(2\pi\cot(\pi\beta/2))/\beta$.)

The following lemma and its variants are basic. We assume $\lambda > 0$.

LEMMA 4. (See ((**6**) 130).) *Let* P *be a closed set, and* $\Delta(x) =$ *distance of* x *from* P, *with* $\Delta(x)$ *put equal to zero outside some interval. Then*

$$\int_{-\infty}^{\infty} \frac{\Delta^\lambda(x_0+t)}{|t|^{\lambda+1}}\,dt$$

is finite for almost every x_0 *in* P.

An alternative form is as follows (see e.g. ((**1**) 189)).

LEMMA 4′. *Suppose that* $\frac{1}{h}\int_0^h |H(x_0+t)|\,dt \leqslant Ah^\lambda$, $0 < |h| < \infty$, *for every* x_0 *in* P. *Then*

$$\int_{-\infty}^{\infty} \frac{|H(x_0+t)|}{|t|^{\lambda+1}}\,dt < \infty$$

for almost every x_0 *in* P.

We shall also use still another form, a discrete variant, which arises from Lemma 4 by considering points of density of P.

LEMMA 4″. *Suppose that $\{\delta_n\} = \{(a_n, b_n)\}$ are those open intervals of the complement of P which are situated in a fixed finite interval. Let δ_n also denote the length of the interval (a_n, b_n). Then*

$$\sum_n \frac{\delta_n^{\lambda+1}}{|x_0 - a_n|^{\lambda+1}} < \infty$$

for almost every x_0 in P.

LEMMA 5. *Suppose that $\alpha > 0$, with $k < \alpha < k+1$, and that h has the development*

$$h(x_0+t) = \sum_{j=0}^{k} h_j(x_0)t^j + \rho_{x_0}(t)$$

for each x_0 in P, with $\frac{1}{t}\int_0^t |\rho_{x_0}(u)|^2\,du \leqslant A|t|^{2\alpha}$, $0 < |t| \leqslant \delta$. Then with x_0 in P, x_0+t in P,

$$h_l(x_0+t) = \sum_{j=0}^{k-l} \frac{t^j}{j!} h_{j+l}(x_0) + O(|t|^{\alpha+l}) \quad (l = 0, 1, \dots, k),$$

and O is uniform for x_0, x_0+t in P, and $|t| \leqslant \delta$.

See ((**1**) Theorem 8 and Definition 3).

3. In this section we prove that the conditions stated in Theorems 1 and 2 are sufficient to ensure the required differentiability.

Suppose, for example, that $k = [\alpha]$ is even, and, as we may, that f is in L^2 and has bounded support. Let $\beta = k+1-\alpha$. It is enough to show that at each point x_0 of E we have

$$\tfrac{1}{2}\{f_\beta(x_0+t) + f_\beta(x_0-t)\} = \alpha_0' + \alpha_2' \frac{t^2}{2!} + \dots + \alpha_k' \frac{t^k}{k!} + \eta(t)t^{k+1}, \tag{4}$$

where $\alpha_j' = \alpha_j'(x_0)$, and $\eta(t)$ satisfies condition N if f satisfies condition N_α at x_0; also that $\eta(t)$ is bounded near $t = 0$ if f satisfies condition Λ_α at x_0. It will then suffice to apply Theorems A and B. A parallel argument works for k odd.

We may assume that $x_0 = 0$. Suppose that the support of f is contained in an interval $(-a, a)$. Let $\lambda(t)$ be a function infinitely differentiable, equal to 1 for $|t| \leqslant a$ and to 0 for $|t| \geqslant b > a$. Write $f = \lambda f + (1-\lambda)f$. Correspondingly, the integral (2) splits into two, generated respectively by λf and $(1-\lambda)f$, both integrals extended over the interval $|t| \leqslant b$. Since $(1-\lambda)f = 0$ in $(-a, a)$, the second integral represents in the interior of $(-a, a)$ an infinitely differentiable function. Let $P(t)$ be the polynomial on the right of (3) (with $x_0 = 0$), and $R(t)$ the remainder, so that $f = P + R$.

The integral

$$\int_{-b}^{b} \lambda(t)P(t)\,|x-t|^{\beta-1}\,dt = \int_{-\infty}^{\infty} |t|^{\beta-1}\lambda(x-t)P(x-t)\,dt$$

represents an infinitely differentiable function, so that the whole problem reduces to showing that the function $\int_{-b}^{b} \lambda(t)R(t)\,|x-t|^{\beta-1}\,dt$ satisfies at x_0 the required conditions. To sum up, we may assume that $x_0 = 0$, that all the α_j in (3) vanish, and that the support of $f(t) = R(t)$ is in an interval $(-b, b)$.

Hence, for $0 < h < \frac{1}{2}b$,

$$\text{(5)}\qquad \tfrac{1}{2}\{f_\beta(h)+f_\beta(-h)\} = \int_{-b}^{b} R(t)\tfrac{1}{2}\{|h-t|^{\beta-1}+|h+t|^{\beta-1}\}\,dt$$

$$= \int_{0}^{b} \rho(t)\tfrac{1}{2}\{|h-t|^{\beta-1}+|h+t|^{\beta-1}\}\,dt,$$

where

$$\rho(t) = R(t)+R(-t) = \{\varepsilon(t)+\varepsilon(-t)\}\,|t|^{\alpha} = \omega(t)\,|t|^{\alpha},$$

say. We shall first verify that f_β satisfies condition $\Lambda_{k-1}{}^*$ at x_0, that is that the last integral is the sum of a polynomial of degree k in h and a remainder $O(h^{k+1})$.

We split the integral into two parts, $\int_0^{2h}$ and $\int_{2h}^{b}$. The first part is

$$\text{(6)}\qquad \frac{1}{2}\int_0^{2h} \rho(t)(|h-t|^{\beta-1}+|h+t|^{\beta-1})\,dt = O(h^{\alpha})\int_0^{2h} |h-t|^{\beta-1}\,dt = O(h^{\alpha+\beta})$$

$$= O(h^{k+1}),$$

and the second part is

$$\int_{2h}^{b} \rho(t)t^{\beta-1}\frac{1}{2}\left\{\left(1-\frac{h}{t}\right)^{\beta-1}+\left(1+\frac{h}{t}\right)^{\beta-1}\right\}dt$$

$$= \int_{2h}^{b} \rho(t)t^{\beta-1}\left\{1+A_2\left(\frac{h}{t}\right)^2+A_4\left(\frac{h}{t}\right)^4+\ldots+A_k\left(\frac{h}{t}\right)^k+O\left(\frac{h^{k+2}}{t^{k+2}}\right)\right\}dt.$$

Since $\rho(t) = O(t^{\alpha})$, it is immediate that

$$\text{(7)}\qquad \int_{2h}^{b} \rho(t)t^{\beta-1}O\left(\frac{h^{k+2}}{t^{k+2}}\right)dt = O(h^{k+2})\int_{2h}^{b} t^{-2}\,dt = O(h^{k+1}).$$

On the other hand, for $j = 0, 1, 2, \ldots, \frac{1}{2}k$ we have

$$\text{(8)}\qquad h^{2j}\int_{2h}^{b} \rho(t)t^{\beta-1}t^{-2j}\,dt = h^{2j}\int_{0}^{b} \rho(t)t^{\beta-1}t^{-2j}\,dt + h^{2j}\int_0^{2h} O(t^{\alpha+\beta-1-2j})\,dt$$

$$= h^{2j}\int_0^{b} \rho(t)t^{\beta-1}t^{-2j}\,dt + O(h^{k+1}).$$

Collecting the results, we see that the right-hand side of (5) is actually a polynomial of degree k plus a term $O(h^{k+1})$.

It remains to show that the function $\eta(t)$ in (**4**) satisfies condition N. Taking into account the origin of η, and considering the expressions (6), (7), (8), we see that it is enough to show that each of the following functions of h satisfies condition N at $h = 0$:

$$\text{(a)} \qquad h^{-(k+1)}\int_0^{2h} |\rho(t)|\,|h-t|^{\beta-1}\,dt, \qquad \text{(b)} \quad h\int_{2h}^{b} |\rho(t)|\,t^{\beta-k-3}\,dt,$$

$$\text{(c)} \qquad h^{2j-k-1}\int_0^{2h} |\rho(t)|\,t^{\beta-2j-1}\,dt \quad (j = 0, 1, 2, \ldots, \tfrac{1}{2}k).$$

(In (a) we omitted the term $|h+t|^{\beta-1}$ since it is majorized by $|h-t|^{\beta-1}$.)

We write $|\omega(t)|\,|t|^{\alpha} = \rho(t)$, and $\varphi(t) = |\omega(t)|/|t|^{\frac{1}{2}}$. Recall that by our discussion $\varphi(t)$ is even, and satisfies the condition

$$\int_0^{\infty} |\varphi(t)|^2\,dt = \int_0^{\infty} \frac{\omega^2(t)}{t} < \infty,$$

since this integrability condition is satisfied near $t = 0$, and ω vanishes outside a finite interval. We apply Lemma 1 in the cases (a), (b), and (c) above, calling $\Phi(h)h^{\frac{1}{2}}$ in turn each of these expressions. Thus in case (a) we have

$$\Phi(h) = h^{-k/2-3/2}\int_0^{2h} t^{\alpha+\frac{1}{2}}\,|h-t|^{\beta-1}\varphi(t)\,dt,$$

and here

$$K(x,y) = \begin{cases} x^{-k/2-3/2}y^{\alpha+\frac{1}{2}}\,|x-y|^{\beta-1} & \text{if} \quad 0 < y \leqslant 2x, \\ 0 \text{ otherwise.} \end{cases}$$

The condition for the lemma is satisfied since $\int_0^2 y^{\alpha}\,|1-y|^{\beta-1}\,dy < \infty$.

For the case (b) we take $\Phi(h) = h^{\frac{1}{2}}\int_{2h}^{\infty} \varphi(t)t^{-3/2}\,dt$, and for case (c),

$$\Phi(h) = h^{2j-k-3/2}\int_0^{2h} \varphi(t)t^{k-2j+\frac{1}{2}}\,dt \quad \left(j = 0,\ldots,\frac{k}{2}\right).$$

The lemma applies equally well in the latter two cases because

$$\int_2^{\infty} t^{-2}\,dt < \infty \quad \text{and} \quad \int_0^2 t^{k-2j}\,dt < \infty, \quad j = 0,\ldots,\frac{k}{2}.$$

Thus we have shown that if f satisfies condition Λ_α at x_0, then f_β satisfies condition $\Lambda_{k+1}{}^*$ at x_0; also if f satisfies condition N_α at x_0, then $\eta(t)$ satisfies condition N.

4. We shall now begin to prove that if f satisfies the condition Λ_α^2 at each point of a set E, and $f^{(\alpha)}$ exists in the L^2 sense in E, then f satisfies condition N_α at almost every point of E; that is, the function $R(t)$ in the development (3) satisfies the condition that $R^2(t)/|t|^{1+2\alpha}$ is integrable in a two-sided neighbourhood of $t = 0$, for almost every x_0 in E.

The assumption that

$$\frac{1}{t}\int_0^t |f(x_0+u)-P_{x_0}(u)|^2\,du = O(|t|^{2\alpha+1})$$

clearly implies that f is in L^2 in a neighbourhood of each point of E. Thus, modifying f outside a finite interval, we may assume that f has finite support and is in L^2. In addition, by restricting consideration to an appropriate closed subset P of E, with $|E-P| < \varepsilon$, we may assume that $F = f_\beta$ has a $(k+1)$th derivative in the L^2 sense uniformly on P, and that f satisfies condition Λ_α^2 uniformly on P.

Now by the basic decomposition lemma (see ((**1**) 189), where an n-dimensional form may be found), we can write

$$F(x) = G(x)+H(x), \tag{9}$$

where $G \in C^{(k+1)}$ and $G^{(j)}(x_0) = F_{(j)}(x_0) = j$th derivative of F in the L^2 sense, for each x_0 in P, and $0 \leqslant j \leqslant k+1$. In particular, $H(x_0) = 0$ for x_0 in P. We may also assume that G has finite support. In view of the uniform differentiability of F on P we get

$$\frac{1}{t}\int_0^t |H(x_0+u)|^2\,du \leqslant A\,|t|^{2k+2}, \quad 0 < |t| < \infty, \quad x_0 \in P. \tag{10}$$

This inequality holds for t near 0, say $|t| \leqslant \delta$, by virtue of the differentiability of F. For large t it is an immediate consequence of the boundedness of $H = F-G$ outside a finite interval; since G is everywhere bounded, and F is the fractional integral of f, which is in L^2 and has, as assumed, finite support.

Let us now apply the inversion formula, Lemma 3, to F: we obtain

$$f(x) = \lim_{\varepsilon\to 0}\frac{1}{A_\beta}\int_{|t|\geqslant\varepsilon}\frac{F(x+t)-F(x)}{|t|^{1+\beta}}\,dt,$$

the limit existing in the L^2 norm. Consider also $g(x)$ given by

$$g(x) = \lim_{\varepsilon\to 0}\frac{1}{A_\beta}\int_{|t|\geqslant\varepsilon}\frac{G(x+t)-G(x)}{|t|^{1+\beta}}\,dt.$$

Since $G \in C^{(k+1)}$ and has finite support, it is clear that this limit exists also in the L^2 norm, and additionally uniformly in x. Let us calculate $g(x)$. Since

$$G(x+t)-G(x) = \int_0^t G'(x+u)\,du,$$

we have

$$g(x) = \lim_{\varepsilon\to 0} \frac{1}{A_\beta} \int_{|t|\geqslant\varepsilon} |t|^{-1-\beta} \int_0^t G'(x+u)\,du\,dt.$$

Interchanging the orders of integration and passing to the limit, which we may since G' has finite support, we get

$$(11) \qquad g(x) = \frac{1}{\beta A_\beta} \int_{-\infty}^{\infty} \operatorname{sign} t \frac{G'(x+t)}{|t|^\beta}\,dt = -\frac{1}{\beta A_\beta}(G' * K_\beta^*)(x).$$

If we now set $h(x) = f(x) - g(x)$ we have

$$(12) \qquad f(x) = g(x) + h(x),$$

where $h(x) = \lim_{\varepsilon\to 0} \frac{1}{A_\beta} \int_{|t|\geqslant\varepsilon} \frac{H(x+t)-H(x)}{|t|^{1+\beta}}\,dt$, in the L^2 norm. However, $H(x) = 0$ if $x \in P$, and the integral $\int_{-\infty}^{\infty} \frac{|H(x+t)|}{|t|^{1+\beta}}\,dt$ converges for almost every x in P, by Lemma 4′ and (10) (recall that H is bounded outside a finite interval). Thus

$$(13) \qquad h(x_0) = \frac{1}{A_\beta} \int_{-\infty}^{\infty} \frac{H(x_0+t)}{|t|^{1+\beta}}\,dt, \quad \text{for almost every } x_0 \text{ in } P.$$

Looking back at (11), we see that we may apply Lemma 2 to g with $G_1 = -G'/\beta A_\beta$. Hence g satisfies the condition Λ_α uniformly, and since f satisfies the condition Λ_α^2 uniformly in P, then so does h satisfy the condition Λ_α^2 uniformly in P. Moreover, g satisfies the condition N_α for almost every x, and therefore the problem of proving that f satisfies the condition N_α for almost every x in P reduces to the same problem for h.

We shall now show that, in fact, $h(x)$ satisfies the condition N_α at every point of P which is simultaneously a point of density of P, for which the conclusions of Lemma 4 (Lemmas 4′ and 4″) hold, and for which the representation (13) holds. Let us assume that x_0 is such a point, and for simplicity of notation set $x_0 = 0$. Then we have

$$(14) \qquad h(t) = \sum_{j=0}^{k} \frac{\alpha_j t^j}{j!} + \xi(t)|t|^\alpha = P(t) + \xi(t)|t|^\alpha,$$

and we must show that

$$(15) \qquad \int_{|t|\leqslant\delta} \frac{\xi^2(t)}{|t|}\,dt < \infty.$$

We break up the range of integration in (15) into P and its complement Q. For simplicity we drop the factor $1/A_\beta$ in (13), and assume that $0 \leqslant x$. The case $x \leqslant 0$ is treated similarly. Now if $x \in P$,

$$h(x) = \int \frac{|H(t)|}{|x-t|^{1+\beta}}\,dt = \int_{|t|\geqslant 2x} + \int_{|t|\leqslant 2x} = S + T.$$

If $|t| \geqslant 2x$, then $|x-t|^{-1-\beta} = |t|^{-1-\beta}\{1+A_1(x/t)+\ldots+A_k(x/t)^k+R\}$, with $|R| \leqslant A|x/t|^{k+1}$ Thus

$$S = \sum_{j=0}^{k} A_j x^j \int_{-\infty}^{\infty} \frac{H(t)}{t^j|t|^{1+\beta}}\,dt + R_1, \tag{16}$$

with

$$R_1 = \sum_{j=0}^{k} A_j x^j \int_{|t|\leqslant 2x} \frac{H(t)}{t^j|t|^{1+\beta}}\,dt + O\left(x^{k+1}\int_{|t|\geqslant 2x} \frac{|H(t)|}{|t|^{k+2+\beta}}\,dt\right).$$

Clearly

$$|T| \leqslant \int_{|t|\leqslant 2x} \frac{|H(t)|\,dt}{|t-x|^{1+\beta}} \leqslant \int_{x/2}^{2x} \frac{|H(t)|\,dt}{|t-x|^{1+\beta}} + Ax^{-\beta-1}\int_{-2x}^{2x} |H(t)|\,dt.$$

Hence, *if we restrict ourselves to x in P*, it is enough to show that each of the three expressions

$$(17)\quad \begin{cases} I_1(x) = x^{\beta}\displaystyle\int_{|t|\geqslant 2x} \frac{|H(t)|\,dt}{|t|^{k+\beta+2}}, \quad I_2(x) = x^{\beta-1}\int_{|t|\leqslant 2x} \frac{|H(t)|}{|t|^{k+\beta+1}}\,dt, \\ I_3(x) = x^{-k-1+\beta}\displaystyle\int_{x/2}^{2x} \frac{|H(t)|}{|t-x|^{\beta+1}}\,dt \\ \text{(we set } I_j(x) = 0 \text{ if } x \leqslant 0) \end{cases}$$

satisfies condition N, that is $\displaystyle\int_P \frac{I_j^2(x)\,dx}{x} < \infty$, $j = 1, 2, 3$. Now

$$\int_P \frac{I_1^2(x)\,dx}{x} \leqslant \int_0^{\infty} \frac{I_1^2(x)\,dx}{x} \leqslant A\int_{-\infty}^{\infty} \frac{|H(t)|^2}{|t|^{2k+3}}\,dt,$$

since $\dfrac{I_1(x)}{x^{\frac{1}{2}}} = \displaystyle\int_{-\infty}^{\infty} K(x,y)\frac{|H(y)|}{|y|^{k+3/2}}\,dy$, with $K(x,y)$ and $K(x,-y)$ satisfying the conditions of Lemma 1.

However, by (10) and Lemma 4′ we have $\displaystyle\int_{-\infty}^{\infty} \frac{|H(t)|^2}{|t|^{2k+3}}\,dt < \infty$. Therefore $\displaystyle\int \frac{I_1^2(x)}{x}\,dx < \infty$. The parallel argument applies to I_2.

Let us now consider I_3. Let $\{(a_n, b_n)\}$ be the open intervals contiguous to P. Then if $x \in P$,

$$\int_{a_n}^{b_n} \frac{|H(t)|\,dt}{|x-t|^{\beta+1}} = \int_{a_n}^{c_n} \frac{|H(t)|\,dt}{|x-t|^{\beta+1}} + \int_{c_n}^{b_n} \frac{|H(t)|\,dt}{|x-t|^{\beta+1}},$$

with c_n the mid-point of (a_n, b_n). An integration by parts shows that

$$\int_{a_n}^{c_n} \frac{|H(t)|\,dt}{|x-t|^{\beta+1}} = (\beta+1)\int_{a_n}^{c_n} \frac{\operatorname{sign}(x-t)}{|x-t|^{\beta+2}}\left\{\int_{a_n}^{t} |H(u)|\,du\right\}dt + \int_{a_n}^{c_n} \frac{|H(t)|\,dt}{|x-c_n|^{\beta+1}}.$$

Thus by (10), since $\Delta(x) \leqslant |x-t|$,

$$\int_{a_n}^{c_n} \frac{|H(t)|\,dt}{|x-t|^{\beta+1}} \leqslant A\int_{a_n}^{c_n} \frac{\Delta^{k+1}(t)}{|x-t|^{\beta+1}}\,dt,$$

where $\Delta(t)$ is the distance of t from P. A similar estimate holds for the integral over (c_n, b_n). This gives

$$\int_{a_n}^{b_n} \frac{|H(t)|\,dt}{|x-t|^{\beta+1}} \leqslant A \int_{a_n}^{b_n} \frac{\Delta^{k+1}(t)}{|x-t|^{\beta+1}}\,dt. \tag{18}$$

Next, since 0 is a point of density of P, there exists a positive δ so small that for each x in $(0, \delta)$ there is a pair x_1 and x_2 with the property that $x/3 < x_1 \leqslant x/2$, $2x < x_2 < 3x$, and x_1 and x_2 are both in P. Therefore, by (18),

$$\begin{aligned}\int_{x/2}^{2x} \frac{|H(t)|\,dt}{|x-t|^{\beta+1}} \leqslant \int_{x_1}^{x_2} \frac{|H(t)|\,dt}{|x-t|^{\beta+1}} &= \sum_{(a_n,b_n)\subseteq(x_1,x_2)} \int_{a_n}^{b_n} \frac{|H(t)|\,dt}{|x-t|^{\beta+1}} \\ &\leqslant A \sum_{(a_n,b_n)\subseteq(x_1,x_2)} \int_{a_n}^{b_n} \frac{\Delta^{k+1}(t)}{|x-t|^{\beta+1}}\,dt \\ &\leqslant A \int_{x/3}^{3x} \frac{\Delta^{k+1}(t)}{|x-t|^{\beta+1}}\,dt.\end{aligned}$$

Hence we get

$$\begin{aligned} I_3(x) &\leqslant A x^{-k-1+\beta} \int_{x/3\leqslant t\leqslant 3x} \frac{\Delta^{k+1}(t)}{|x-t|^{\beta+1}}\,dt \\ &\leqslant A x^{-k-1+\beta} \int_{x/3\leqslant t\leqslant 3x} \frac{\Delta^{k+\frac{1}{2}-\beta/2}(t)}{|x-t|^{(\beta+1)/2}}\,dt, \end{aligned} \tag{19}$$

since $\Delta(x) \leqslant |x-t|$. Let us now set $\Delta(t) = 0$ if $t \geqslant 3\delta$ (since we restrict x to $(0, \delta)$). Another application of Lemma 1 then gives

$$\int_0^{\delta} \frac{I_3^2(x)}{x}\,dx \leqslant A \int_0^{3\delta} \frac{\Delta^{2k+1-\beta}(t)}{t^{2k+2-\beta}}\,dt,$$

since $\int_{1/3}^{3} |1-t|^{-\beta+\frac{1}{2}} t^{k+\frac{1}{2}-\beta/2}\,dt < \infty$. The last integral is finite by Lemma 4, and so we have proved that

$$\int_{P\cap(0,\delta)} \frac{\xi^2(x)}{x}\,dx < \infty.$$

5. We must now show that

$$\int_{Q\cap(0,\delta)} \frac{\xi^2(x)}{x}\,dx = \int_{Q\cap(0,\delta)} \frac{\rho^2(x)\,dx}{x^{2\alpha+1}} < \infty, \tag{20}$$

with Q the complement of P, and $\rho(x) = |x|^{\alpha}\xi(x)$.

Let us denote by $\delta_n = (a_n, b_n)$ $(n = 1, 2, \ldots)$ the disjoint open intervals of Q situated in $(0, \delta)$; we may assume that the point δ itself is in P. The length of δ_n will also be denoted by δ_n. The finiteness of (20) is a consequence of the finiteness of the two sums

$$\sum_n \int_{a_n}^{b_n} \frac{[\rho(x)-\rho(a_n)]^2}{x^{2\alpha+1}}\,dx, \tag{21}$$

$$\sum_n \int_{a_n}^{b_n} \frac{[\rho(a_n)]^2}{x^{2\alpha+1}}\,dx. \tag{22}$$

We begin with (22). Since 0 is a point of density of P, we have $b_n \leqslant 2a_n$, for all n, provided that δ was taken sufficiently small, and then (22) is majorized by a constant multiple of

$$\int_{a_n}^{b_n} \frac{[\xi(a_n)]^2\,dx}{x}.$$

Also, since $a_n \in P$, $\xi(a_n)$ is majorized by a fixed multiple of the sums of the expressions (17), taken with $x = a_n$. Consider first

$$a_n{}^{\beta} \int_{|t|\geqslant 2a_n} \frac{|H(t)|\,dt}{|t|^{k+\beta+2}}.$$

For $a_n \leqslant x \leqslant b_n$ this integral is increased if we replace it by

$$x^{\beta} \int_{|t|\geqslant x} \frac{|H(t)|}{|t|^{k+\beta+2}}\,dt = I_1{}^*(x).$$

Similarly

$$a_n{}^{\beta-1} \int_{|t|\leqslant 2a_n} \frac{|H(t)|}{|t|^{k+\beta+1}}\,dt$$

is majorized by $2^{\beta} I_2(x)$. Thus the part of (22) arising from these two contributions is majorized by a constant multiple of

$$\int_Q \frac{[I_1{}^*(x)]^2\,dx}{x} + \int_Q \frac{I_2{}^2(x)}{x}\,dx \leqslant \int_0^{\infty} \frac{[I_1{}^*(x)]^2 + I_2{}^2(x)}{x}\,dx < \infty.$$

The last inequality is proved in the same way as the analogous inequalities proved earlier for $I_1(x)$ and $I_2(x)$.

Therefore in order to prove the finiteness of the sum (22) it remains, in view of (19), to prove the finiteness of

$$\sum_n \int_{a_n}^{b_n} \frac{dx}{x} \left\{ a_n^{-k-1+\beta} \int_{a_n/3}^{3a_n} \frac{\Delta^{k+1}(t)}{|t-a_n|^{1+\beta}}\,dt \right\}^2.$$

However by the inequalities $\Delta(t) \leqslant |t-a_n|$ and $b_n \leqslant 2a_n$, this is majorized by a constant multiple of

$$\sum \int_{a_n}^{b_n} \frac{dx}{x^{\alpha+1}} \int_{x/6}^{3x} \frac{\Delta^{k+1}(t)}{|t-a_n|^{1+\beta}}\,dt \quad (k+1-\beta = \alpha). \tag{23}$$

Let now δ_n' be the interval concentric with δ_n but twice as large. Since

$$\int_{\delta_n'} \frac{\Delta^{k+1}(t)}{|t-a_n|^{\beta+1}}\,dt \leqslant C\delta_n{}^{k-\beta+1} = C\delta_n{}^{\alpha},$$

the contribution on δ_n' of the sum (23) is majorized by a multiple of $\sum (\delta_n/a_n)^{1+\alpha} < \infty$, by Lemma 4". Hence in estimating (23) it is enough to take the inner integrals over that part of $(x/6, 3x)$ which is outside δ_n'. For such t, $|t-a_n| \geqslant \frac{1}{2}|t-x|$ $(x \in (a_n, b_n))$. Denote by $\psi_Q(x)$ the characteristic function of Q; and for x in Q let $\lambda_x(t)$ denote the function of t which equals 0 when t is in the component of Q which contains x, and is 1 elsewhere. We then see that the part of (23) over $(x/6, 3x) - \delta_n'$ is majorized by a multiple of

$$(24)\quad \int_Q x^{-\alpha-1}\left\{\int_{x/6}^{3x} \frac{\Delta^{k+1}(t)\lambda_x(t)}{|t-x|^{1+\beta}}\,dt\right\}dx = \int_0^\infty \Delta^{k+1}(t)\left\{\int_{t/3}^{6t} x^{-\alpha-1}\frac{\lambda_x(t)\psi_Q(x)}{|x-t|^{1+\beta}}\,dx\right\}dt$$
$$\leqslant C\int_0^\infty \frac{\Delta^{k+1}(t)}{t^{1+\alpha}}\left\{\int_0^\infty \frac{\lambda_x(t)\psi_Q(x)\,dx}{|x-t|^{1+\beta}}\right\}dt.$$

Observe now that for each fixed t in Q we need integrate internally only over the components of Q which do not contain t. Hence the inner integral is majorized by

$$\int_{|t-x|\geqslant\Delta(t)} \frac{dx}{|x-t|^{\beta+1}} \leqslant C\{\Delta(t)\}^{-\beta} = C\{\Delta(t)\}^{-k+\alpha-1}.$$

Inserting in the above shows that the double integrals of (24) are majorized by a multiple of $\int_0^\infty \frac{(\Delta(t))^\alpha}{t^{1+\alpha}}\,dt$, which is finite, by Lemma 4. Therefore the finiteness of (23), and hence of (22), has been established.

6. To finish the proof of (20), we must show the finiteness of (21).

Let us recall that h satisfies the condition $\Lambda_\alpha{}^2$ uniformly in the set P. We have in fact, for any x_0 in P,

$$h(x_0+t) = P_{x_0}(t) + \rho_{x_0}(t),$$

where $P_{x_0}(t)$ is the polynomial of degree k in t given by

$$P_{x_0}(t) = \sum_{j=0}^{k} h_j(x_0)t^j,$$

and

$$(25)\qquad \frac{1}{t}\int_0^t |\rho_{x_0}(u)|^2\,du = O(|t|^{2\alpha}) \quad (t\to 0),\ \text{uniformly for } x_0 \text{ in } P.$$

In (14), $P(t) = P_0(t)$, $\xi(t)|t|^\alpha = \rho_0(t)$.

Now we can apply Lemma 5 (§ 2) to h. The meaning of the lemma relevant to us is that, in particular when restricting to P, the polynomial part of the development of h_j is the jth derivative of the polynomial part of the development of h.

Since $a_n \in P$,

$$h(a_n+t) = \sum_{j=0}^{k} \frac{h_j(a_n)t^j}{j!} + \rho_{a_n}(t).$$

Hence $h(x)-h(a_n) = \sum_{j=1}^{k} \frac{h_j(a_n)t^j}{j!} + \rho_{a_n}(x-a_n)$. However,

$$P(x)-P(a_n) = \sum_{j=1}^{k} \frac{P^{(j)}(a_n)}{j!}(x-a_n)^j.$$

Moreover, by the lemma and (14), $P^{(j)}(x)$ is the principal part of $h_j(x)$, for x in P, in the development at the origin. Thus

$$h_j(a_n)-P^{(j)}(a_n) = O(a_n^{\alpha-j}).$$

Finally,

$$\rho(x)-\rho(a_n) = h(x)-h(a_n)-[P(x)-P(a_n)].$$

Therefore, for x in (a_n, b_n),

$$|\rho(x)-\rho(a_n)| \leqslant |\rho_{a_n}(x-a_n)| + A\sum_{j=1}^{k} a_n^{\alpha-j}\delta_n^{\,j}. \tag{26}$$

If we insert (26) in (21), and use the estimate (25) for $\rho_{a_n}(x-a_n)$, we get

$$\sum_n \int_{a_n}^{b_n} \frac{[\rho(x)-\rho(a_n)]^2}{x^{2\alpha+1}}\,dx \leqslant A\sum_n \left(\frac{\delta_n^{\,3}}{a_n^{\,3}} + \frac{\delta_n^{\,1+2\alpha}}{a_n^{\,1+2\alpha}}\right).$$

(Recall that since $b_n \leqslant 2a_n$, $\delta_n = b_n - a_n \leqslant a_n$.) Thus the finiteness of (21) is proved by appealing to Lemma 4″.

7. The arguments of §§ 4–6 show that if, on E, f satisfies the condition Λ_α^2, and $f^{(\alpha)}$ exists in the L^2 sense there, then f satisfies condition N_α almost everywhere in E. However, we saw in § 3 that if f satisfies condition N_α in E, then $f^{(\alpha)}$ exists almost everywhere in E in the L^2 sense, while if f satisfies in addition the condition Λ_α in E, $f^{(\alpha)}$ exists (in the ordinary sense) almost everywhere in E. This proves Theorem 1. To conclude the proof of Theorem 2 let us notice that if f satisfies the condition N_α at a point x_0, it *a fortiori* satisfies the condition Λ_α^2 there. In fact suppose that

$$\int_{|t|<\delta} \frac{|f(x_0+t)-P_{x_0}(t)|^2}{|t|^{1+2\alpha}}\,dt < \infty;$$

then clearly

$$\frac{1}{t}\int_0^t |f(x_0+u)-P_{x_0}(u)|^2\,du = o(|t|^{2\alpha}), \quad \text{as} \quad t \to 0,$$

which is as asserted.

REFERENCES

1. A. P. CALDERÓN and A. ZYGMUND, 'Local properties of solutions of elliptic differential equations', *Studia Math.* 20 (1961) 171–225.
2. G. H. HARDY, J. E. LITTLEWOOD, and G. POLYA, *Inequalities* (Cambridge, 1952).
3. E. M. STEIN, 'The characterization of functions arising as potentials', *Bull. American Math. Soc.* 67 (1961) 102–4.
4. E. M. STEIN and A. ZYGMUND, 'On the differentiability of functions', *Studia Math.* 23 (1964) 247–83.
5. A. ZYGMUND, 'A theorem on fractional derivatives', *Duke Math. J.* 12 (1945) 455–64.
6. —— *Trigonometric series*, 2nd edn, Vol. 1 (Cambridge, 1959).

Princeton University *University of Chicago*

Boundedness of translation invariant operators on Hölder spaces and L^p-spaces

By E. M. Stein and A. Zygmund

Many linear operators T occurring in analysis enjoy one or both of the following properties:

(a) T maps the space of Hölder (Lipschitz) continuous functions with exponent α, Λ_α, to the space Λ_β, for appropriate α and β.

(b) T maps the L^p space to L^q, for appropriate p and q.

It is therefore of interest to know what is the relation between property (a) and (b). This problem may have further significance in view of the results concerning functions of bounded mean oscillation of John and Nirenberg [3], and their variants, the spaces $\mathfrak{L}^{p,\lambda}$ considered by several authors.[1] While we shall not describe the spaces $\mathfrak{L}^{p,\lambda}$ here, we shall be content with recalling that, for one range of λ, these reduce to the Λ_α spaces, and for another value of λ, these reduce to the L^p spaces. Finally for an intermediate value of λ, these reduce to the space of functions of bounded mean oscillation. In this context it is natural to ask whether the spaces $\mathfrak{L}^{p,\lambda}$ are stable with respect to some method of interpolation.

In answering these questions we shall deal with translation invariant operators T on euclidean n-space. The following results will be proved.

(1) If T maps the space Λ_α to the space Λ_β, with $\beta > \alpha$, then T maps L^p to L^q, where $1/q = 1/p - (\beta - \alpha)/n$, and $1 < p, q < \infty$.

This can be viewed as the Hardy-Littlewood Soboleff theorem on fractional integration in its general setting.

(2) However, there is an example of an operator T, mapping Λ_α to Λ_α, but so that T is not bounded on any L^p space, $p \neq 2$.[2]

The example (2) shows that for this T, constructed in the setting of one-dimensional Fourier series, there is a bounded function whose image under T does not belong to any L^q, $q > 2$. Thus this T does not take functions of bounded mean oscillation into functions of bounded mean oscillation. Hence

(3) The spaces $\mathfrak{L}^{p,\lambda}$ are not stable under interpolation.

[1] See e.g. [4], where references to the literature may be found.

[2] T must automatically be bounded on L^2.

The origin of those results is a little known and less understood paper of Hardy and Littlewood [1]. In that paper, they give sufficient conditions for a translation invariant operator on one-dimensional Fourier series to map L^p to L^q. However the real force of their conditions appears only in a rather disguised form (in fact, the space of Hölder continuous functions does not arise). Thus the proof of Theorem 1 requires a re-interpretation of their conditions, together with a different proof which avoids complex methods, and works in n-dimension. The example (2) is also based on a construction given in the Hardy and Littlewood paper, but the details given here are simpler, and the conclusion is more decisive.

This paper is organized as follows. Section 1 contains the proof of the general result stated in (1) above. In § 2 we digress to show how the space of functions of bounded mean oscillation occur as limiting cases for the operators T of the type above. Section 3 deals with the counter-example which proves (2) and (3). An appendix states some related results.

1. The general theorem

We begin by defining the spaces Λ_α. (For details see [7, Ch. II, III], and [5a]). Suppose first $0 < \alpha < 1$, then Λ_α is the space of continuous and bounded functions in E_n which are Hölder continuous and with norm

$$\sup_x |f(x)| + \sup_{x\neq y} \frac{|f(x) - f(y)|}{|x-y|^\alpha} .$$

When $\alpha = 1$, the appropriate class, which we here call Λ_1, is defined by a now well understood modification. Thus we define Λ_1 as the class of continuous and bounded functions for which the norm

$$\sup_x |f(x)| + \sup_{x\neq y} \frac{\left|f(x) + f(y) - 2f\left(\frac{x+y}{2}\right)\right|}{|x-y|}$$

is finite. (In [7], this class is refered to as Λ_*). Next if $1 < \alpha < 2$, we can give two equivalent definitions: $f \in \Lambda_\alpha$ if f is bounded and continuously differentiable and each $\partial f/\partial x_k \in \Lambda_{\alpha-1}$; or equivalently, if $f(x)$ is continuous and the norm

$$\sup_x |f(x)| + \sup_{x\neq y} \frac{\left|f(x) + f(y) - 2f\left(\frac{x+y}{2}\right)\right|}{|x-y|^\alpha}$$

is finite. Finally for any positive α, $f \in \Lambda_\alpha$ if for the largest integer k which is less than α, the partial derivatives of total order k belong to $\Lambda_{\alpha-k}$, and f is bounded. (The classes Λ_α just defined are called $\Lambda(\alpha, \infty, \infty)$ in [5]).

A basic feature of the classes Λ_α, $0 < \alpha < \infty$, is the potential operator J_γ defined for any real γ. J_γ is the mapping given by convolution with the distribution whose Fourier transform is $(1 + |x|^2)^{-\gamma/2}$. The fact is that J_γ is a bicontinuous isomorphism of Λ_α to $\Lambda_{\alpha+\gamma}$ (as long as $\alpha + \gamma > 0$), (See [5a], and [6]).

Now suppose that T is a linear transformation mapping Λ_α to Λ_β, for some $\alpha, \beta > 0$. Then clearly T is defined on smooth functions of bounded support, and we may ask if there is for such functions an inequality of the type

$$\| T(f) \|_q \leqq A \| f \|_p .$$

If so we say that T is of type (p, q).

THEOREM 1.[3] *Let T be a linear transformation mapping Λ_α to Λ_β boundedly for some α and β with $\beta > \alpha$. Suppose that T commutes with translations. Then T is of type (p, q) where*

$$1/q = 1/p - \frac{(\beta - \alpha)}{n}, \qquad \text{and } 1 < p, q < \infty .$$

PROOF. In view of the isomorphism given by J_γ and the fact that T must commute with J_γ, i.e., $J_\gamma T J_{-\gamma} = T$, it follows that T is bounded from $\Lambda_{\alpha+\gamma}$ to $\Lambda_{\beta+\gamma}$. Thus only the difference $\delta = \beta - \alpha$, which is the degree of smoothing, is significant. Our assumption is that $\beta - \alpha > 0$, and we shall assume temporarily that $\delta = \beta - \alpha < 1$. The translation invariant operators with this smoothing property have been characterized (see [6] and [5]). The operator is given by

$$T(f) = \int k(x - y) f(y) dy ,$$

where the kernel k satisfies the following two conditions

(1) $$\int_{E_n} | k(x) | dx < \infty$$

(2) $$\int_{E_n} | k(x - y) - k(x) | dx \leqq A | y |^\delta .$$

The proof of the theorem is based on the following simple lemma.

LEMMA 1. *Suppose k satisfies conditions (1) and (2) above. Then, given any $\alpha > 0$, there exists a splitting $k(x) = k_1(x) + k_\infty(x)$ so that*

(a) $| k_\infty(x) | \leqq \alpha$ *all* x

(b) $\int_{E_n} | k_1(x) | dx \leqq A\alpha^{\delta/(\delta - n)}$.

This lemma can be proved by using suitable approximations to $k(x)$. In

[3] See also the further remarks in the appendix.

this case, it is advantageous to use harmonic functions in the half-space $t > 0$; i.e., we consider the Poisson integral $k(x, t) = P_t(x) * k(x)$, and

$$P_t(x) = \frac{ct}{(|x|^2 + t^2)^{\frac{n+1}{2}}} .$$

Then conditions (1) and (2) imply easily (see [5a])

$$\text{(1')} \qquad \int_{E_n} |k(x, t)| dx \leqq A$$

$$\text{(2')} \qquad \int_{E_n} \left| \frac{\partial k}{\partial t}(x, t) \right| dx \leqq At^{-1+\delta} .$$

However since $k(x, t)$ is the Poisson integral of an L^1 function $\lim_{t\to 0} k(x, t) = k(x)$ for almost every x, and thus

$$k(x, t) - k(x) = \int_0^t \frac{\partial k(x, s)}{\partial s} ds ,$$

and hence

$$\int_{E_n} | k(x, t) - k(x) | dx \leqq A \int_0^t s^{-1+\delta} ds \leqq At^\delta .$$

Next

$$\frac{\partial k}{\partial t}(x, t) = P_{t/2} * \frac{\partial k}{\partial t}(x, t/2) ,$$

and therefore

$$\sup_x \left| \frac{\partial k}{\partial t}(x, t) \right| \leqq \sup_x | P_{t/2}(x) | \int_{E_n} \left| \frac{\partial k}{\partial t}(x, t/2) \right| dx$$

and so

$$\sup \left| \frac{\partial k}{\partial t}(x, t) \right| \leqq At^{-n} \cdot t^{-1+\delta} .$$

Also $k(x, t) = P_{t/2} * k(x, t/2)$, and so $\sup_x | k(x, t) | \leqq At^{-n} \to 0$, as $t \to \infty$. Therefore $k(x, t) = -\int_t^\infty \frac{\partial k}{\partial s}(x, s) ds$, and so

$$\sup_x | k(x, t) | \leqq At^{-n+\delta} .$$

We now write $k(x) = \{k(x) - k(x, t)\} + k(x, t) = k_1(x) + k_\infty(x)$, and choose t so that $\alpha = At^{-n+\delta}$. Then $| k_\infty(x) | \leqq \alpha$, and $\int | k_1(x) | dx \leqq A\alpha^{\delta/\delta - n}$, which proves the lemma.

We shall now show that the operator T is of weak type (p_0, q_0), where $p_0 = 1$, and $1/q_0 = 1 - \delta/n$, i.e., $q_0 = n/n - \delta$. In order to do this, assume $\int | f(x) | dx \leqq 1$ and let us estimate the set $E_{2\alpha} = \{| Tf(x) | > 2\alpha\}$. Since $Tf =$

$k*f$ and $k = k_1 + k_\infty$, this set is clearly in the union of the sets where $|k_\infty * f| > \alpha$ and where $|k_1 * f| > \alpha$. However by our construction $|k_\infty * f| \leqq \sup_x |k_\infty(x)| \, \|f\|_1 \leqq \alpha$, and so the first set is empty. The measure of the second set is not more than

$$\frac{1}{\alpha}\int |k_1 * f|dx \leqq \frac{1}{\alpha}\|k_1\|_1 \|f\|_1 \leqq A\alpha^{-1}\alpha^{-\delta/\delta-n} = A\alpha^{-n/n-\delta} = A\alpha^{-q_0} .$$

This shows that $f \rightarrow T(f)$ is of weak type $(1, q_0)$. In fact a very similar argument, which we leave to the reader, shows that the operator T is actually of weak type (p, q) where $1 \leqq p, q < \infty$, $1/q = 1/p - \delta/n$. The reader should note that the splittings of k into L^1 and L^∞ parts as given in the lemma imply easily corresponding splittings into L^1 and $L^{p'}$ parts.

Now an application of the Marcinkiewicz interpolation theorem proves Theorem 1, still under the assumption $\delta < 1$. For general $\delta > 0$, we proceed as follows. Let r be a non-negative integer chosen so that $0 < \delta - r/2 \leqq 1/2$. Write $T = J_{1/2} \cdot J_{1/2} \cdots J_{1/2} \cdot (J_{-r/2}T)$. In view of what has been said above about the action of the potential operators T_γ on the Λ_α spaces, we get that T is a product of operators of the same kind but for which the order of smoothing is either 1/2 or $\gamma - r/2$, $(\leqq 1/2)$. Successive applications of the case discussed above then prove the general theorem.

2. Remarks on functions of bounded mean oscillation

We wish now to deal with the limiting cases of these results, i.e., when p is near $n/\beta - \alpha$, and thus q is near infinity; also with the case when p is near infinity. The results, which are intimately connected with functions of bounded mean oscillation may be stated as follows:

THEOREM 2. *Suppose that T satisfies the conditions of Theorem* 1.

(a) *Suppose that f is of weak-type $p = n/(\beta - \alpha)$*, i.e., *the measure of the set $E_\lambda = \{|f(x)| > \lambda\}$ is $\leqq A\lambda^{-p}$. Then $T(f)$ is of bounded mean oscillation.*

(b) *Conversely, suppose that f is the sum of an L^1 and an L^∞ function, and that f is of bounded mean oscillation on the whole space E_n. Then $T(f) \in \Lambda_{\beta-\alpha}$.*

Before we come to the proof, we need to make the following remarks. If we assumed that f were actually in L^p, then the fact that $T(f)$ is of bounded mean oscillation actually follows Theorem 1 and an interpolation theorem of Stampacchia [4]. Next for the proof of both (a) and (b), it suffices to consider the special case when T is actually $J_{\beta-\alpha}$. In fact, we write as before $T = J_\gamma(J_{-\gamma}T) = (TJ_{-\gamma})J_\gamma$. For part (a), we use the fact $J_{-\gamma}T$ maps L^p to L^q, where

$1/q = 1/p - (\beta - \alpha + \gamma)/n$, with $1 < p, q < \infty$. Thus it also maps weak L^p to weak L^q for the same range of p's and q's. (This can be proved by an elementary argument, or one can appeal to the general form of the Marcinkiewicz interpolation theorem, as in [2]). For part (b), we merely use the fact that $TJ_{-\gamma}$ maps $\Lambda_{\alpha+\gamma}$ to Λ_β.

PROOF OF (a). It suffices to show that there exists a constant M and an exponent q_0, $1 \leqq q_0$, so that if f satisfies the condition $f^*(x) \leqq x^{-1/p}$ (here $f^*(x)$ denotes the equi-measurable, non-increasing rearrangement of f on the positive half-line) then for every cube Q, there exists a constant a_Q so that

$$\int_Q | T(f) - a_Q |^{q_0} dx \leqq Mm(Q) . \tag{2.1}$$

Since we are dealing with translation invariant operators let us fix the cube Q to be centered at the origin, and write $f(x) = f_1(x) + f_2(x)$, where $f_1(x) = f(x)$ in the cube Q^* twice the size of Q and also centered at the origin, and $f_1(x) = 0$ for x outside Q^*. Then $T(f) = F_1(x) + F_2(x)$, where $F_i(x) = T(f_i)$. Pick $1 < p_0 < p$, and determine q_0 by $1/q_0 = 1/p_0 - (\beta - \alpha)/n$. Then we know that

$$\int_Q | F_1(x) |^{q_0} dx \leqq \int | F_1(x) |^{q_0} dx \leqq A\Big(\int | f_1(x) |^{p_0} dx\Big)^{q_0/p_0}$$

$$\leqq A\Big(\int_0^{2^n m(Q)} x^{-p_0/p} dx\Big)^{q_0/p_0} = A'\big(m(Q)\big)^{(-p_0/p+1)(q_0/p_0)} = A'm(Q) ,$$

since $1/p = (\beta - \alpha)/n$, $1/q_0 = 1/p_0 - (\beta - \alpha)/n$. Therefore we have

$$\int_Q | F_1(x) |^{q_0} dx \leqq A'm(Q) .$$

Next

$$F_2(x) = \int G_{\beta-\alpha}(x - y) f_2(y)\, dy ,$$

since $T = J_{\beta-\alpha}$. Set

$$a_Q = \int G_{\beta-\alpha}(y) f_2(y)\, dy .$$

Then

$$| F_2(x) - a_Q | \leqq \int_{cQ^*} | G_{\beta-\alpha}(x - y) - G_{\beta-\alpha}(y) | \, | f_2(y) | dy ,$$

and for $x \in Q$, this gives

$$| F_2(x) - a_Q | \leqq A\big(m(Q)\big)^{1/n} \int_{cQ^*} | y |^{-n+\beta-\alpha-1} | f(y) | dy ,$$

since $| \nabla G_{\beta-\alpha}(x) | \leqq A' | x |^{-n+\beta-\alpha-1}$.[4]

Therefore if $x \in Q$,

[4] For the required properties of the Bessel kernel $G_{\beta-\alpha}$, see N. Aronszajn and K. T. Smith, *Theory of Bessel potentials*: I, Ann. Inst. Fourier 11 (1961), p. 417.

$$|F_2(x) - a_Q| \leqq Al \int_{|y|>l/2} |y|^{-n+\beta-\alpha-1} |f(y)|dy ,$$

where l is the length of the edges of Q. Now let us split f again; $f(y) = f'(y) + f''(y)$, where $f'(y) = f(y)$ when $|f(y)| \geqq k$, and $f'(y)$ is zero otherwise. (The constant k will be fixed in terms of l momentarily). This gives us

$$|F_2(x) - a_Q| \leqq A'[l^{-n+\beta-\alpha}\|f'\|_1 + l^{\beta-\alpha}\|f''\|_\infty] , \qquad \text{if } x \in Q .$$

However, $\|f''(y)\|_\infty \leqq k$, and $\|f'(y)\|_1 = \int_{x^{-1/p}\geqq k} x^{-1/p}dx = (p/(p-1))k^{1-p}$. Now choose k so that $l^{\beta-\alpha}\cdot k = 1$. Then $k = l^{\alpha-\beta}$ and

$$l^{-n+\beta-\alpha}\|f'\|_1 \leqq (p/(p-1))l^{-n+\beta-\alpha}l^{(\alpha-\beta)(1-p)} = p/(p-1) ,$$

since $p = n/(\beta - \alpha)$.

Therefore $|F_2(x) - a_Q| \leqq A''$ if $x \in Q$, and (2.1) is proved.

PROOF OF (b). We shall need the following observation about functions f of bounded mean oscillation. Suppose E_1 and E_2 are two sets of comparable measure and both are contained in a cube of comparable measure; the difference between the mean values of f over E_1 and E_2 is bounded. We shall be more explicit in the following special case we need. Suppose Q_1 and Q_2 are two cubes, $m(Q_1) = m(Q_2)$, and $Q_1 \cup Q_2$ is contained in a cube Q so that $m(Q_i) \geqq cm(Q)$. Then

$$\left| \frac{1}{m(Q_1)} \int_{Q_1} f(x)dx - \frac{1}{m(Q_2)} \int_{Q_2} f(x)dx \right| \leqq A ,$$

where A depends on c and the bound arising in the definition of functions of bounded mean oscillation. In fact this difference equals

$$1/m(Q_1) \int_Q \varphi(x) f(x)dx, \qquad \text{where } \varphi(x) = \chi_{Q_1}(x) - \chi_{Q_2}(x) ,$$

and χ_{Q_i} are the characteristic functions of the Q_i. But this difference also equals $1/m(Q_1) \int_Q \varphi(x)[f(x) - a_Q]dx$, so is bounded in absolute value by $2/c\cdot \int_Q |f(x) - a_Q|dx$.

Let us remark that in one dimension, if we take Q_1 and Q_2 to be two adjacent intervals of equal length, we get $|F(x+h) + F(x-h) - 2F(x)| \leqq A|h|$, where $F(x) = \int^x f(t)dt$, i.e., $F \in \Lambda_1$.

Let now $u(x, t)$ denote the Poisson integral of f. From the assumption $f \in L^1 + L^\infty$, we get

(2.2) $$\sup_x |u(x, t)| \leqq A , \qquad \text{if } t \geqq 1 .$$

The main burden of the proof will be to show

(2.3) $\sup_x |u_{x_i}(x, t)| \leq A/t\,,$ if $t > 0,\ i = 1, \cdots n\,.$

By translation invariance, it suffices to prove (2.3) at the origin. Now $u_{x_i}(0, t) = -\int(\partial P_t/\partial x_i)(x)f(x)dx$, and we make the following easily verified assertions about the differentiated Poisson kernel $(\partial P_t/\partial x_i)(x)$.

(a) $\dfrac{\partial P_t}{\partial x_i}(x)$ is odd (in x)

(b) $\left|\dfrac{\partial P_t}{\partial x_i}(x)\right| \leq At^{-n-1}$

(c) $\left|\dfrac{\partial P_t}{\partial x_i}(x)\right| \leq At|x|^{-n-2}\,.$

For each fixed t, $0 < t < \infty$, we decompose the euclidean space into a union of cube whose interiors are disjoint, as follows. We first consider the increasing family of cubes $Q^0 \subset Q^1 \subset Q^2 \cdots \subset Q^m \subset \cdots$, where Q^m is the cube centered at the origin and whose sides have length $3^m t$. Next when $m \geqq 1$ we write $Q^m - Q^{m-1} = \bigcup_\alpha Q^m_\alpha$, where the decomposition is in terms of the obvious $3^n - 1$ parallel cubes, each congruent to Q^{m-1}, and the length of the sides of each Q^m_α is $3^{m-1}t$ (The index α ranges from 1 to $3^n - 1$. Thus

$$E_n = Q^0 \cup \bigcup_{m=1}^\infty \bigcup_\alpha Q^m_\alpha\,.$$

Let us notice that Q^0 is symmetric with respect to the origin. Therefore since $(\partial P_t/\partial x_i)(x)$ is odd, its integral over Q^0 vanishes. Similarly for each cube Q^m_α, let $Q^m_{\alpha'}$ denote the cube symmetric to Q^m_α with respect to the origin. (That there is such a cube in our decomposition is clear). Then the integral of $(\partial P_t/\partial x_i)(x)$ over $Q^m_\alpha \cup Q^m_{\alpha'}$ is again zero. Hence for any constants a^0, $\{a^m_\alpha\}$, we have

$$\begin{aligned} u_{x_i}(0, t) &= \int_{Q^0}\frac{\partial}{\partial x_i}P_t(x)\,[f(x) - a^0]dx \\ &\quad + \frac{1}{2}\sum_{m\geq 1}^{\alpha}\int_{Q^m_\alpha \cup Q^m_{\alpha'}}\frac{\partial}{\partial x_i}P_t(x)\,[f(x) - a^m_\alpha]dx\,. \end{aligned} \tag{2.4}$$

Let us denote for a^0 the mean value of f on Q^0, and for a^m_α the average of the mean values of f on Q^m_α and $Q^m_{\alpha'}$. Thus by what was said above

$$\int_{Q^0}|f(x) - a^0|dx \leq Am(Q^0)\,, \quad \text{and} \quad \int_{Q^m_\alpha \cup Q^m_{\alpha'}}|f(x) - a^m_\alpha|dx \leq Am(Q^m_\alpha)\,. \tag{2.5}$$

Next on the cube Q^0, we use the estimate (b) for $(\partial P_t/\partial x_i)(x)$, and for the cubes Q^m_α, $m \geqq 1$, we use estimate (c), and the fact that $|x| \geqq C3^m t$, when $x \in Q^m_\alpha$. Thus $|(\partial P_t/\partial x_i)(x)| \leq A'(3^m)^{-n-2}t^{n-1}$ for $x \in Q^m_\alpha$. Inserting this with (2.5) in (2.4) gives

$$|u_{x_i}(0, t)| \leq At^{-n-1}\cdot t^n + A\sum_{m=1}(3^m)^{-n-2}t^{n-1}(3^m t)^n \leq A't^{-1},$$

and (2.3) is proved.

From (2.2) and (2.3), it follows that f is in the class called $\Lambda(0, \infty, \infty)$ in Taibleson [5a], and thus its potential of order $\beta - \alpha$ is in the class $\Lambda(\beta - \alpha, \infty, \infty)$ which is the class $\Lambda_{\beta-\alpha}$. This proves part (b) of the theorem.

3. Counter-examples

We now come to the main counter-example which is constructed in the context of one-dimensional Fourier series. Consider a multiplier transformation $f \to T(f)$ given by

$$(3.1) \qquad Tf(x) \sim \sum a_k \mu_k e^{ikx}, \qquad \text{where } f(x) \sim \sum a_k e^{ikx}.$$

This transformation which commutes with translations is, of course, convolution with the distribution whose trigonometric expansion is $\sum \mu_k e^{ikx}$. The example is as follows:

THEOREM 3. *There exists a transformation T of the type above with the following properties*

(a) *T maps Λ_α to Λ_α, and L^2 to L^2.*

(b) *There exists a bounded f, so that $T(f)$ is not in any L^q, for $q > 2$.*

We need the following simple lemma.

LEMMA 2. *Suppose $\| \sum_{2^r \leq |k| < 2^{r+1}} \mu_k e^{ikx} \|_1 \leq A$, $r = 0, 1, 2, \cdots$. Then the transformation $f \to T(f)$ maps Λ_α to Λ_α, $0 < \alpha < 1$.*

PROOF OF THE LEMMA. Let $K(x) = \sum'(\mu_k/ik)e^{ikx}$. By [6], it suffices to show that $\int |K(x+t) + K(x-t) - 2K(x)|dx = O(t)$, as $t \to 0$. For this purpose write

$$\sum \mu_k e^{ikx} = \mu_0 + \sum_{r=0}^{\infty} \Delta_r(x),$$

and

$$K(x) = \sum_{r=0}^{\infty} \tilde{\Delta}_r(x).$$

Here $\Delta_r(x) = \sum_{2^r \leq |k| < 2^{r+1}} \mu_k e^{ikx}$, $\tilde{\Delta}_r(x) = \sum_{2^r \leq |k| < 2^{r+1}} (\mu_k/ik)e^{ikx}$. Since $\|\Delta_r(x)\|_1 \leq A$, it follows by a well known argument that $\|\tilde{\Delta}_r(x)\|_1 \leq A'2^{-r}$, and by a variant of Bernstein's inequality that $\|\Delta_r'(x)\|_1 \leq A2^r$. Now

$$K(x+t) + K(x-t) - 2K(x) = \sum_{r=0}^{\infty}\{\tilde{\Delta}_r(x+t) + \tilde{\Delta}_r(x-t) - 2\tilde{\Delta}_r(x)\}.$$

In view of the above, we can make the following two estimates

$$\|\tilde{\Delta}_r(x+t) + \tilde{\Delta}_r(x-t) - 2\tilde{\Delta}_r(x)\|_1 \leq A2^r t^2, \quad \text{or} \quad \leq 3A'2^{-r}.$$

Therefore $\|K(x+t) + K(x-t) - 2K(x)\|_1 \leq A\sum_1^N 2^r t^2 + 3A'\sum_{N+1}^{\infty} 2^{-r}$. If we set N to be the largest integer so that $2^N \leq 1/t$, we get the desired result, and the lemma is proved. We now describe the transformation more explicity. We write

$$\sum \mu_k e^{ikx} = \sum_{r=2} \frac{D_r(x+\alpha_r)}{\log r}\, e^{i2^r(x+\alpha_r)} \tag{3.2}$$

where

$$D_r(x) = \sum_{k=0}^{r} e^{ikx} = \frac{1 - e^{i(r+1)x}}{1 - e^{ix}}\,,$$

and $\alpha_1, \alpha_2 \cdots \alpha_r, \cdots$ are translations which for the moment are arbitrary but fixed.

When written out in full

$$\sum \mu_k e^{ikx} = \sum \Delta_r(x)\,, \qquad \text{where } \Delta_r(x) = \frac{D_r(x+\alpha_r)e^{i2^r(x+\alpha_r)}}{\log r}\,.$$

Since it is clear that $\int |D_r(x)|dx \leqq A \log r$, it follows by Lemma 2 that the transformation T given by (3.2) maps Λ_α to Λ_α (and of course L^2 to L^2). Next we construct f and $T(f)$.

We write

$$f(x) \sim \sum_{r=2}^{\infty} \frac{1}{r(\log r)^3}\, D_r(x-\alpha_r)e^{i2^r(x-\alpha_r)}$$

(3.3) and

$$T(f)(x) \sim \sum_{r=2}^{\infty} \frac{1}{r(\log r)^4}\, D_r(x)e^{i2^r x}\,.$$

In order to obtain a bounded f we shall show that, for a particular choice of $\alpha_1, \alpha_2, \cdots, \alpha_r \cdots$,

$$\sum \frac{1}{r(\log r)^3}\, |D_r(x-\alpha_r)| \leqq A < \infty\,,$$

with A independent of x. In fact, if r is between 2^k and 2^{k+1}, i.e., $2^k \leqq r < 2^{k+1}$, define $\alpha_r = (r - 2^k)/2^k \cdot 2\pi$. Thus the α_r are 2^k equi-distant points on $[-\pi, \pi]$. Now $|D_r(x)| \leqq r + 1$, and $|D_r(x)| \leqq A/|x|$. Therefore

$$\sum_{2^k \leqq r < 2^{k+1}} |D_r(x-\alpha_r)| \leqq 2^{k+1} + A2^k \sum_{r=1}^{2^k} 1/r \leqq A' 2^k \cdot k\,.$$

Hence

$$\sum_{r=2}^{\infty} 1/r(\log r)^3\, |D_r(x-\alpha_r)| \leqq A' \sum 2^{-k} k^{-3} \cdot 2^k \cdot k = A' \sum 1/k^2 < \infty\,,$$

which shows that f is bounded.

To consider $T(f)$, we use the Littlewood-Paley theory (see [7, Ch. 15]). Thus $T(f) \in L^q$ if and only if

(3.4) $$\left(\sum \frac{|D_r(x)|^2}{r^2(\log r)^8}\right)^{1/2} \in L^q .$$

However, clearly $|D_r(x)| \geq Ar$, if $|x| \leq 1/r$, and so the expression (3.4) is larger than

$$A\left(\sum_{r<1/|x|} \frac{1}{(\log r)^8}\right)^{1/2} \geq A' \frac{|x|^{-1/2}}{(\log 1/x)^4}, \qquad \text{as } |x| \to 0 .$$

This last function does not belong to L^q, if $q > 2$. Therefore we see that the operator T is not bounded on any L^q space $q > 2$, and hence by duality, it is not bounded on any L^p space $p < 2$.

Finally we wish to remark that if T is of the form (3.1) and takes L^p in L^q, it need not take Λ_α to Λ_β. More specifically let $\mu_k = 1$ if $k = 2^r$, $\mu_k = 0$ otherwise. Then it follows from the theory of lacunary series that, if $f \in L^p$ $1 < p$, then $Tf \in L^q$ for every $q < \infty$. (See [7, Ch. 5]). However the function $\sum 2^{-k\alpha} e^{i2^k x}$ belongs to exactly Λ_α and not to Λ_β, if $\beta > \alpha$. So T does not map Λ_α to Λ_β if $\beta > \alpha$.

A last example is as follows. Let $\mu_k = 1$ if $2^r \leq k < 2^{r+1}$ and r is even, but let $\mu_k = 0$ if $2^r \leq k < 2^{r+1}$ if r is odd. Then it follows by the Marcinkiewicz multiplier theorem (see [7, Ch. 15]) that T maps L^p to L^p $1 < p < \infty$. However T does not map Λ_α to Λ_α for any $\alpha > 0$. Otherwise it would follow by [6], that $K(x) \sim \sum' (\mu_k/ik)e^{ikx}$ should satisfy

$$\int |K(x+t) + K(x-t) - 2K(x)|dx = O(t) ,$$

and this would easily imply that $\|\sum_{2^r \leq k < 2^{r+1}} \mu_k e^{ikx}\|_1 \leq A$. However, this contradicts the fact that $\|\sum_{2^r \leq k < 2^{r+1}} e^{ikx}\|_1 \geq Ar$.

Appendix

We wish to describe certain variants of the results obtained above. Theorem 1 is in part an extension of [1, Th. 13] (for the case $\sigma = 1$). An extension corresponding to the case $1 \leq \sigma < s$ can also be proved by the methods of this paper, since for this class of kernels an analogue of the decomposition of Lemma 1 holds. In fact the argument of Lemma 1 shows that if $k \in \Lambda(\alpha_0, p_0, \infty)$,[5] where $\alpha_0 > 0$ and $1/q_0 = 1/p_0 - \alpha/n$, then k is of weak type L^{q_0}; i.e., $m\{|k(x)| > \lambda\} \leq A\lambda^{-q_0}$. In the limiting case ($\sigma = s$, in [1, Th. 14]) we claim the following[6]

(1) *Suppose that* $k \in \Lambda(0, q_0, \infty)$.* *Then* $T(f) = k * f$ *takes* L^p *to* L^q,

[5] In the notation of Taibleson [5].

[6] We are informed that the result (1) of this appendix is already contained in a paper of Taibleson to appear in J. Math. Mech. (1966).

* See footnote 5.

where $1/q = 1/p + 1/q_0 - 1$, *provided that* $1 < p \leqq 2 \leqq q < \infty$.

Next, in the case of one dimensional power series we consider

$$
\begin{aligned}
k(e^{i\theta}) &= \sum_{n\geqq 0} \mu_n e^{in\theta} \\
f(e^{i\theta}) &= \sum_{n\geqq 0}^{\infty} a_n e^{in\theta} \qquad \text{(A.1)} \\
T(f) = k*f &= \sum_{n\geqq 0} a_n \mu_n e^{in\theta} .
\end{aligned}
$$

The analogue of the condition $k \in \Lambda(0, q_0, \infty)$ is then

$$
\left(\int_0^{2\pi} |k'(\rho e^{i\theta})|^{q_0} d\theta\right)^{1/q_0} \leqq A/1 - \rho . \qquad \text{(A.2)}
$$

(2) *The condition* (A.2) *is a necessary condition that the transformation maps* H^1 *to* H^{q_0} *boundedly,* $1 \leqq q_0 \leqq \infty$.

If we combine this with [1, Th. 14], we get

(3) *The necessary and sufficient condition that the transformation* (A.1) *maps* H^1 *to* H^{q_0}, *with* $2 \leqq q_0 < \infty$, *is that condition* (A.2) *be satisfied.*

As far as (1) is concerned, one proves that, if $k \in \Lambda(0, q_0, \infty)$, then $f \to k*f$ maps $\Lambda(0, p, 2)$ to $\Lambda(0, q, 2)$ where $1/p = 1/p + 1/q_0 - 1$. Next, one uses the inclusion relations

$$
L^p \subset \Lambda(0, p, 2) \qquad \text{if } 1 < p \leqq 2, \text{ and } \Lambda(0, q, 2) \subset L^q
$$

if $2 \leqq q < \infty$. See [5a], [5b]. As far as (2) is concerned, we use the function $f_\rho(z) = \sum_{n=1}^{\infty} n\rho^n z^n = \rho z/(1 - \rho z)^2$. Then $\int_0^{2\pi} |f_\rho(e^{i\theta})| d\theta \leqq A/1 - \rho$ and $k*f_\rho = k'(\rho e^{i\theta})$.

If one uses Parseval's relation, then the condition (A.2) becomes in the case $q_0 = 2$, $\sum n^2 |\mu_n|^2 \rho^{2n} = O((1 - \rho)^{-2})$, $0 < \rho < 1$, which is equivalent to

$$
\sum_N^{2N} |\mu_n|^2 = O(1), \ N \to \infty . \qquad \text{(A.3)}
$$

Hence,

(4) *The necessary and sufficient condition that the transformation* (A.1) *maps* H^1 *to* H^2 *is that the condition* (A.3) *is satisfied.*

We now consider the special case when the transformation (A.1) is a projection, i.e., when all μ_n are either 0 or 1. Then we can write $T(f) = \sum a_{n_k} e^{in_k\theta}$, for an appropriate subsequence $\{n_k\}$. It follows easily from (4) that

(5) *A necessary and sufficient condition that a projection of the above kind maps* H^1 *to* H^2 *is that the sequence* $\{n_k\}$ *is a union of finitely many lacunary sequences.*

That this fact, which contains the converse of a theorem of Paley, follows, was pointed out to us by W. Rudin.

PRINCETON UNIVERSITY
UNIVERSITY OF CHICAGO

References

1. Hardy, G. H. and Littlewood, J. E., *Theorems concerning the mean values of analytic or harmonic functions*, Quart. J. Math., 12 (1941), 221-256.
2. Hunt, R. A., *An extension of the Marcinkiewicz interpolation theorem to Lorentz spaces*, Bull. Amer. Math. Soc., 70 (1964), 803-807.
3. John, F. and Nirenberg, L., *On functions of bounded mean oscillation*, Comm. Pure Appl. Math., 14 (1961), 415-426.
4. Stampacchia, G., *The spaces* $\mathfrak{L}^{(p,\lambda)}$ *and* $N^{(p,\lambda)}$ *and interpolation*, Annali di Pisa 29 (1965), 443-462.

5a. Taibleson, M., *On the theory of Lipschitz spaces and distributions on euclidean n-space*: I, J. Math. Mech., 13 (1964), 407-479.

5b. ———: II, 14 (1965), 821-829.

6. Zygmund, A., *On the preservation of classes of functions*, J. Math. Mech., 8 (1959), 889-896.
7. ———, Trigonometric Series, 2nd ed., Cambridge, 1959, 2 vols.

(Received July 18, 1966)

STUDIA MATHEMATICA, T. XXX. (1968)

On multipliers preserving convergence of trigonometric series almost everywhere

by

MARY WEISS and ANTONI ZYGMUND (Chicago)

1. Consider a trigonometric series $\sum_{-\infty}^{+\infty} c_n e^{in\theta}$, which in the case $c_{-n} = \bar{c}_n$ can also be written in the form

$$\frac{1}{2}a_0 + \sum_{n=1}^{\infty}(a_n \cos n\theta + b_n \sin n\theta) = \sum_{0}^{\infty} A_n(\theta),$$

say. Its conjugate is

$$\sum_{-\infty}^{+\infty}(-i \operatorname{sign} n) c_n e^{in\theta} = \sum_{1}^{\infty}(a_n \sin n\theta - b_n \cos n\theta) = \sum_{0}^{\infty} B_n(\theta),$$

say (with $B_0 = 0$).

One of the topics of the theory of trigonometric series that enjoyed popularity a few decades ago was the problem of the behavior (convergence or summability, at individual points or almost everywhere) of the series $\sum A_n(\theta) n^a$, $\sum B_n(\theta) n^a$, where a is a constant. The problem has obvious connections with differentiability or integrability (in general, of fractional order) of functions, and a was almost exclusively real. In this note we consider complex values of a, $a = \beta + i\gamma$, but in view of the fact that the case of real a has been exhaustively dealt with we limit ourselves to a purely imaginary, $a = i\gamma$, which shows some novel features. The problem we are discussing here arose out of some concrete applications but the latter are not considered here.

The main result of the paper is the following

THEOREM. *If the series $\sum_{n=0}^{\infty} A_n(\theta)$ is summable (C, k), $k > -1$, at each point of a set E of positive measure, then the series $\sum_{0}^{\infty} A_n(\theta) n^{i\gamma}$ is summable (C, k) almost everywhere in E. In particular, the convergence of $\sum A_n(\theta)$ in E implies the convergence of $\sum_{0}^{\infty} A_n(\theta) n^{i\gamma}$ almost everywhere in E.*

Here γ is a real number distinct from 0. Without loss of generality we may assume, wherever needed, that $a_0 = 0$, in which case successive termwise integrations of the series $\sum A_n(\theta)$ lead again to trigonometric series. We also assume that the a's and b's are real numbers.

2. The proof which follows systematically uses the notion of differentiability in the metric L^p, $1 \leqslant p < \infty$, and we recall the definition (for more details see [1]). We say that the function $f(\theta)$, defined almost everywhere in the neighborhood of the point θ_0, has an *m-th differential at* θ_0, *in* L^p, if there is a polynomial

$$P(t) = \sum_{j=0}^{m} a_j \frac{t^j}{j!}$$

of degree $\leqslant m$ such that

$$\left(h^{-m} \int_{-h}^{h} |f(\theta_0+t)-P(t)|^p dt\right)^{1/p} = o(h^m) \quad (h \to 0). \tag{2.1}$$

The polynomial $P(t)$ is called the *m-th differential* (*in* L^p) *of* f *at* θ_0, and the number a_m is the *m-th derivative* (*in* L^p) *of* f *at* θ_0. The differentials in L^p have a number of properties missing in the classical case $p = \infty$ (when $f(\theta_0+t) = P(t)+o(t^m)$) and for this reason are both interesting and useful to consider.

The proof our theorem is based on a few lemmas which we now state.

LEMMA 1. *If a trigonometric series* $\sum A_n(\theta)$ *is summable* (C, k), $k = 0, 1, \ldots$, *to sum* $s(\theta)$ *in a set* E *of positive measure* (*thus, in particular*, $|a_n|+|b_n| = o(n^k)$), *then the function* $G(\theta)$ *obtained by integrating* $\sum A_n(\theta)$ *termwise* $k+1$ *times has almost everywhere in* E *a* $(k+1)$*-st derivative in the metric* $L^p, p < \infty$, *equal to* $s(\theta)$.

This lemma is known (see [4]) and we take it for granted here.

LEMMA 2. *Suppose that a trigonometric series* $\sum A_n(\theta)$ *is the Fourier series of a function* $F(\theta) \in L^p$, $1 \leqslant p < \infty$, *and that* F *has an m-th derivative in* L^p *at the point* θ_0 *equal to* s. *Then the series obtained by differentiating* $\sum A_n(\theta)$ *termwise m times is summable* $(C, m+2)$ *at* θ_0 *to sum* s.

This lemma holds even with $(C, m+2)$ replaced by $(C, m+\varepsilon)$, $\varepsilon > 0$, but the index of summability is of no importance and, in the form stated, the lemma is a simple corollary of known results.

For, in the first place, since differentiability in L^p clearly implies differentiability in L^{p_1} if $p_1 < p$, we may assume that $p = 1$. Thus we have (2.1) with f replaced by F and $p = 1$. Omitting the sign of absolute value we see that the indefinite integral G of F has an $(m+1)$-st derivative at θ_0 equal to s, in the classical sense. But then, by a very well known result (see [6_{II}], p. 60) the series obtained by differentiating the

Fourier series of G termwise $m+1$ times — or what is the same thing differentiating the Fourier series of F termwise m times — is summable $(C, m+1+1) = (C, m+2)$ to s, and this is our lemma 2.

LEMMA 3. *If $F(\theta) \sim \sum A_n(\theta)$, $F \in L^p$, $1 < p < \infty$, then $\sum A_n(\theta) n^{i\gamma}$ is the Fourier series of a function $\tilde{F}(\theta)$ which is also in L^p. Moreover, if F has an m-th derivative in L^p at each point of a set E of positive measure, then $\tilde{F}$ has an m-th derivative in L^p almost everywhere in E.*

The first part of this lemma is well known (see $[6_{\mathrm{II}}]$, p. 232, Example; another proof is contained implicitly in [2]). To the proof of the second part we return in the next section. Here we only observe that we require p to be strictly greater than 1.

LEMMA 4. *If a (numerical) series $\sum_0^\infty u_n$ is summable (C, k), $k > -1$, and the series $\sum_0^\infty u_n n^{i\gamma}$ (γ real) is Abel summable, then it is also summable (C, k).*

The proof of lemma 3 is given in §§ 3, 4 below. That of lemma 4 is briefly discussed in § 5. We shall now deduce our theorem from the lemmas above.

Suppose that $\sum A_n(x)$, with $a_0 = 0$, is summable (C, k), $k > -1$, at each point of a set E, $|E| > 0$; in particular, it is summable (C, k'), where k' is the least integer $\geqslant k$. By lemma 1, the sum $G(\theta)$ of the series obtained by integrating $\sum A_n(\theta)$ termwise $k'+1$ times has almost everywhere in E a $(k'+1)$-st derivative, in the metric L^p, $p < \infty$. Suppose, e.g. that $k'+1$ is even, so that

$$G(\theta) = (-1)^{\frac{1}{2}(k'+1)} \sum A_n(\theta) n^{k'+1}.$$

By lemma 3, the function

$$\tilde{G}(\theta) = (-1)^{\frac{1}{2}(k'+1)} \sum A_n(\theta) n^{k'+1} n^{i\gamma}$$

has a $(k'+1)$-st derivative in L^p, $1 < p < \infty$, at almost all points of E. By lemma 2, the last series differentiated termwise $k'+1$ times, that is the series $\sum A_n(\theta) n^{i\gamma}$, is summable $(C, k'+3)$ almost everywhere in E; in particular, it is Abel summable almost everywhere in E. Finally, by lemma 4, at each point where $\sum A_n(\theta) n^{i\gamma}$ is Abel summable, and so almost everywhere in E, it is summable (C, k). This completes the proof of the theorem provided we supply the proofs of lemmas 3 and 4.

3. In this and next sections we prove lemma 3, which is of independent interest, for general $1 < p < \infty$. It should however be observed that for the proof of theorem 1 we need only some fixed p, e.g. $p = 2$, in which case the fact that $\sum A_n(\theta) n^{i\gamma}$ is, like $\sum A_n(\theta)$, in L^2 is obvious.

The proof of lemma 3 resembles that of the fact that if a function has an m-th derivative in L^p, $1 < p < \infty$, at each point of a set E, then its Hilbert transform (conjugate function) has the same property almost everywhere in E (see [1], theorems 6, 7). The main idea is in both cases the same but unfortunately there is enough formal difference, of a not completely trivial nature, not to leave it to the reader to take care of the required changes. Thus we must go through certain details of computation.

Let C_n^a be the Cesàro numbers defined by the generating function

$$\sum_{n=0}^{\infty} C_n^a z^n = (1-z)^{-a-1}.$$

We have $C_n^a \simeq n^a/\Gamma(a+1)$ $(a \neq -1, -2, \ldots)$, $n \to \infty$, and in particular

$$C_n^{i\gamma} \simeq n^{i\gamma}/\Gamma(i\gamma+1).$$

We first prove lemma 3 with factors $n^{i\gamma}$ replaced by $C_n^{i\gamma}$ and we shall see later that this easily leads to lemma 3 as stated.

Suppose therefore that $F \sim \sum A_n(\theta) \in L^p$, $1 < p < \infty$, that F has an m-th derivative in L^p at each point of a set E, $|E| > 0$. It is known (see [1], theorems 9, 10) that given any $\varepsilon > 0$ we can find a closed set $P \subset E$, $|E-P| < \varepsilon$, and a decomposition

$$F = G+H,$$

where G and H are again periodic, G is in $C^{(m)}$ and $G^{(j)}(\theta)$, $j = 0, 1, \ldots, m$, coincides with the j-th derivative of F (in L^p); thus the derivatives of order $j = 0, 1, \ldots, m$ of H are all 0 on P. Moreover,

$$\left(h^{-1} \int_{-h}^{h} |H(\theta+t)|^p dt\right)^{1/p} \leqslant Mh^m \qquad (\theta \in P)$$

with M independent of θ and

$$\text{(3.1)} \qquad \left(h^{-1} \int_{-h}^{h} |H(\theta+t)|^p dt\right)^{1/p} = o(h^m) \qquad (h \to 0)$$

uniformly in $\theta \in P$. In particular, also

$$h^{-1} \int_{-h}^{h} |H(\theta+t)| dt = o(h^m) \qquad (h \to 0,\ \theta \in P).$$

Basic for the proof of theorem 1 is also the fact (see [1], theorem 10) that

$$\text{(3.2)} \qquad \int_{-\pi}^{\pi} \frac{H(t)}{|t-\theta|^{m+1}}\, dt < \infty$$

for almost all $\theta \in P$.

If we write

$$\tilde{F}(\theta) \sim \sum_{n=0}^{\infty} A_n(\theta) C_n^{i\gamma},$$

then $\tilde{F} = \tilde{G} + \tilde{H}$. Since the multipliers $C_n^{i\gamma}$ preserve the class L^p, $1 < p < \infty$ (see [6_{II}], p. 232, theorem 4.14), it is clear that $\tilde{G}(\theta)$ has almost everywhere an m-th derivative in L^p and it is enough to prove that $\tilde{H}(\theta)$ has an m-th derivative in L^p almost everywhere in P.

Let

$$K(r, t) = \frac{1}{2} + \sum_{n=1}^{\infty} C_n^{i\gamma} r^n \cos nt = \frac{1}{2}\left\{\frac{1}{(1-re^{it})^{1+i\gamma}} + \frac{1}{(1-re^{-it})^{1+i\gamma}} - 1\right\}.$$

Clearly, for almost all θ, $\tilde{H}(\theta)$ is the limit of the Abel means of its Fourier series, i.e.,

(3.3) $$\tilde{H}(\theta) = \lim_{r\to 1} \frac{1}{\pi} \int_{-\pi}^{\pi} H(\theta+t) K(r, t)\, dt$$

$$= \lim_{r\to 1} \frac{1}{\pi} \int_{-\pi}^{\pi} [H(\theta+t) - H(\theta)] K(r, t)\, dt + H(\theta).$$

On the other hand, it is well known (see [2]) that if $f(x) \in L^p(-\infty, +\infty)$, $1 \leqslant p < \infty$, then the expression

(3.4) $$g(x) = \int_0^1 \frac{f(x+t) - f(x)}{t^{1+i\gamma}}\, dt + \int_1^{\infty} \frac{f(x+t)}{t^{1+i\gamma}}\, dt = \lim_{\varepsilon\to 0} \int_{\varepsilon}^{1} + \int_1^{\infty}$$

exists almost everywhere (clearly, it is only the existence of $\int_0^1$ that re-requires proof; the integral $\int_1^{\infty}$ converges absolutely and uniformly); moreover, if $1 < p < \infty$, we have

(3.5) $$\|g\|_p \leqslant A_p \|f\|_p,$$

where A_p depends on p only. All these facts remain essentially unchanged if in the definition of $g(x)$ instead of the decomposition $\int_0^1 + \int_1^{\infty}$ we use $\int_0^{\omega} + \int_{\omega}^{\infty}$, $0 < \omega < \infty$, the constant A_p in (3.5) remaining the same (this follows by a change of variable). Also, each integral $\int_0^{\omega}$ and $\int_{\omega}^{\infty}$ satisfies an inequality analogous to (3.5).

Using these facts it is easy to deduce from (3.3) that for periodic functions $H(\theta)$ that are merely integrable we have

$$\tilde{H}(\theta) = \frac{1}{\pi}\int_{-\pi}^{\pi}[H(\theta+t)-H(\theta)]K(1,t)\,dt+H(\theta) \tag{3.6}$$

$$= \frac{1}{\pi}\int_{-\pi}^{\pi}[H(t)-H(\theta)]K(1,t-\theta)\,dt+H(\theta),$$

where

$$K(1,t) = \lim_{r\to 1} K(r,t) = \frac{1}{2}\left\{\frac{1}{(1-e^{it})^{1+i\gamma}}+\frac{1}{(1-e^{-it})^{1+i\gamma}}-1\right\},$$

and that if $H \in L^p$, $1 < p < \infty$, then

$$\|\tilde{H}\|_p \leqslant A_p\|H\|_p,$$

where the norms are over the interval $(0, 2\pi)$.

For example, in order to deduce (3.6) from (3.3) it is enough to verify that $|K(r,t)| \leqslant A/\delta$, where $\delta = 1-r$ and that $|K(r,t)-K(1,t)| \leqslant A\delta/t^2$ for $\delta \leqslant |t| \leqslant \pi$, so that the expression equal to $K(r,t)$ for $|t| \leqslant \delta$ and equal to $K(r,t)-K(1,t)$ for $\delta \leqslant |t| \leqslant \pi$ is majorized by a fixed multiple of the Poisson kernel.

We also add that since $K(t)$ is infinitely differentiable for $t \neq 0$, the second integral (3.6) shows that we do not affect the differentiability properties of H at the point θ if we modify H away from θ.

4. If $\varphi(z)$ denotes the function $\{(1-e^{iz})/iz\}^{-1-i\gamma}$, regular for $|z| < 2\pi$, we have

$$K(1,t) = \frac{1}{2}\left\{\frac{\varphi(t)}{(it)^{1+i\gamma}}+\frac{\varphi(-t)}{(-it)^{1+i\gamma}}\right\},$$

and taking a sufficiently large number of terms of the Taylor series of φ we easily see that the problem reduces to showing that the functions

$$\int_{\theta}^{\pi}[H(t)-H(\theta)]\frac{(t-\theta)^k}{|t-\theta|^{1+i\gamma}}\,dt, \qquad \int_{-\pi}^{\theta}[H(t)-H(\theta)]\frac{(t-\theta)^k}{|t-\theta|^{1+i\gamma}}\,dt \tag{4.1}$$

have m-th derivatives in L^p at almost all points of P; the exponent k here takes a finite number of values 0 ,1, 2, ... We may restrict ourselves to the first integral (4.1). We shall discuss only the case $k = 0$ which is, in some sense, the most subtle (it corresponds to the constant term of the Taylor development of φ); for other values of k the proof is parallel.

Thus we will prove that the function

$$S(\theta) = \int_{\theta}^{\pi} [H(t)-H(\theta)] \frac{dt}{(t-\theta)^{1+i\gamma}} \tag{4.2}$$

has an m-th derivative in L^p at almost all points of P.

Let us consider any point $\theta_0 \epsilon P$ at which the integral (3.2) is finite. We shall prove that S has an m-th derivative in L^p at $\theta = \theta_0$. Assume, as we may, that $\theta_0 = 0$. We fix a small $h > 0$ and we consider the function $S(\theta)$ in the interval $(-h, h)$. We write

$$S(\theta) = \int_{\theta}^{\theta+3h} \frac{H(t)-H(\theta)}{(t-\theta)^{1+i\gamma}} dt + \int_{\theta+3h}^{\pi} \frac{H(t)}{(t-\theta)^{1+i\gamma}} dt - H(\theta) \int_{\theta+3h}^{\pi} \frac{dt}{(t-\theta)^{1+i\gamma}} dt$$

$$= S_1+S_2+S_3,$$

say. Since, for $|\theta| \leqslant h$,

$$S_1(\theta) = \int_0^{3h} \{H(t+\theta)-H(\theta)\} t^{-1-i\gamma} dt = \int_0^{3h} [H_1(t+\theta)-H_1(\theta)] t^{-1-i\gamma} dt,$$

where $H_1(t) = H(t)$ for $|t| \leqslant 4h \leqslant 1$ and $H_1(t) = 0$ elsewhere, an application of (3.5) (and the remarks that follow it) shows that

$$\left(h^{-1} \int_{-h}^{h} |S_1|^p d\theta\right)^{1/p} \leqslant \left(h^{-1} \int_{-\infty}^{+\infty} |S_1|^p dt\right)^{1/p} \leqslant A_p \left(h^{-1} \int_{-\infty}^{+\infty} |H_1|^p d\theta\right)^{1/p} \tag{4.3}$$

$$\leqslant A_p \left(h^{-1} \int_{-4h}^{4h} |H|^p dt\right)^{1/p} = o(h^m),$$

by (3.1). Also, since $|S_3(\theta)| \leqslant A_\gamma |H(\theta)|$,

$$\left(h^{-1} \int_{-h}^{h} |S_3|^p d\theta\right)^{1/p} = o(h^m). \tag{4.4}$$

Now observe that $2h \leqslant \theta+3h \leqslant 4h$, so that

$$S_2 = \int_{2h}^{\pi} \frac{H(t)}{(t-\theta)^{1+i\gamma}} dt - \int_{2h}^{\theta+3h} \frac{H(t)\,dt}{(t-\theta)^{1+i\gamma}} = S_{2,1}+S_{2,2}.$$

Clearly,

$$|S_{2,2}| \leqslant h^{-1} \int_0^{4h} |H(t)|\,dt = o(h^m). \tag{4.5}$$

On the other hand,

$$S_{2,1}(\theta) = \int_{2h}^{\pi} \frac{H(t)}{t^{1+i\gamma}(1-\theta/t)^{1+i\gamma}}\,dt \tag{4.6}$$

$$= \int_{2h}^{\pi} H(t)\frac{1}{t^{1+i\gamma}}\left\{\sum_{j=0}^{m}\lambda_j\left(\frac{\theta}{t}\right)^j + O\left(\frac{\theta}{t}\right)^{m+1}\right\}dt$$

$$= \sum_{j=0}^{m}\lambda_j\theta^j\int_{2h}^{\pi}\frac{H(t)}{t^{j+1+i\gamma}}\,dt + O\left(|\theta|^{m+1}\int_{2h}^{\pi}\frac{|H(t)|}{t^{m+2}}\,dt\right).$$

Observe now that, by hypothesis, the integral $\int_0^{\pi}\frac{|H(t)|}{t^{m+1}}\,dt$ is finite, so that

$$\int_{2h}^{\pi}|H(t)|\,t^{-m-2}\,dt = o(h^{-1})$$

and the last term on the right of $o(h^m)$, also, if $j = 0, 1, \ldots, m$,

$$\int_{2h}^{\pi}\frac{H(t)}{t^{j+1+i\gamma}}\,dt = \int_0^{\pi} - \int_0^{2h} = \int_0^{\pi} + O(h^{m-j})$$

as easily seen by integration by parts and applying (3.1). Thus $S_{2,1}(\theta)$ is a fixed polynomial of degree $\leqslant m$ in θ plus an error term $o(h^m)$. Collecting results we obtain that $S(\theta)$ does indeed have an m-th derivative in L^p at $\theta = 0$.

We have thus proved lemma 3 with factors $n^{i\gamma}$ replaced by $C_n^{i\gamma}$. But the result holds also for the factors $C_n^{i\gamma-s}$, where s is any positive integer. For the new generating function is

$$\sum_{n=0}^{\infty} C_n^{i\gamma-s} z^n = \frac{(1-z)^s}{(1-z)^{i\gamma+1}}$$

and it is clear that the argument above holds for kernels $z^j/(1-z)^{i\gamma+1}$, $j = 0, 1, 2, \ldots$ Since the asymptotic formula

$$C_n^{i\gamma} = n^{i\gamma}\{\lambda_0 + \lambda_1 n^{-1} + \ldots + \lambda_s n^{-s} + O(n^{-s-1})\}$$

leads to

$$n^{i\gamma} = \mu_0 C_n^{i\gamma} + \mu_1 C_n^{i\gamma-1} + \ldots + \mu_s C_n^{i\gamma-s} + O(n^{-s-1});$$

taking s sufficiently large we are easily led to lemma 3 as stated.

5. It remains to consider lemma 4. We do not give a proof here since it is implicitly contained in paper [5] (Satz 2, p. 317). It is shown there that if the Cesàro means σ_n^k of a series $\sum_0^\infty u_n$ satisfy the condition $\sigma_n^k = o(\mu_n)$, where μ_n is a sequence monotonically tending to $+\infty$ and such that

$$\Delta^j \frac{1}{\mu_n} = O\left(\frac{1}{n^j \mu_n}\right) \quad (j = 0, 1, \ldots, k'),$$

k' being the least integer $\geqslant k$, then $\sum u_n/\mu_n$ is either summable (C, k) or else is not Abel summable. Actually the proof without any change gives the following result: if $\{1/\mu_n\}$ is a bounded sequence satisfying $\Delta^j(1/\mu_n) = O(n^{-j})$ for $j = 0, 1, \ldots, k'$ and $\sum u_n$ is summable (C, k), then $\sum u_n/\mu_n$ is also summable (C, k), provided it is Abel summable.

The result clearly applies in the case when $\mu_n = n^{-i\gamma}$ $(n > 0)$.

Since the case $k = 0$ of theorem 1 is of special interest, it may be worth pointing out that in this case lemma 4 is immediate. For it is very well known that if a series $\sum_0^\infty v_\nu$ is Abel summable and

$$\sum_{\nu=0}^{n} \nu v_\nu = o(n), \tag{5.1}$$

then Σv_ν converges. But (5.1) is immediate, by summation by parts, if $v_\nu = u_\nu \nu^{i\gamma}$ and $\sum u_\nu$ converges.

6. The theorem of this paper has connection with some recent results of E. M. Stein. As he pointed out to us, the methods of his paper [3] give the following theorem:

If a function

$$F(z) = \sum_0^\infty c_n z^n$$

analytic in $|z| < 1$ has a non-tangential limit at each point of a set E situated on $|z| = 1$, then the function

$$G(z) = \sum_0^\infty c_n n^{i\gamma} z^n$$

has a non-tangential limit almost everywhere in E.

References

[1] A. P. Calderón and A. Zygmund, *Local properties of solutions of elliptic partial differential equations*, Studia Math. 20 (1961), p. 171-225.

[2] B. Muckenhoupt, *On certain singular integrals*, Pacific J. Math. 10 (1961), p. 239-261.

[3] E. M. Stein, *Classes H^p, multiplicateurs et fonctions de Littlewood-Paley*, C. R. Acad. Sci. Paris 263 (1966), p. 716-719.

[4] Mary Weiss, *On symmetric derivatives in L^p*, Studia Math. 24 (1964), p. 89-100.

[5] A. Zygmund, *Über einige Sätze aus der Theorie der divergenten Reihen*, Bull. Intern. Acad. Pol. Sci. et Lettres, 1927, p. 309-331.

[6] — *Trigonometric series*, 2nd ed., Cambridge 1959, vol. I and II.

UNIVERSITY OF ILLINOIS AT CHICAGO CIRCLE
UNIVERSITY OF CHICAGO

Reçu par la Rédaction le 22. 8. 1967

JOURNAL OF APPROXIMATION THEORY 2, 249–257 (1969)

On Certain Lemmas of Marcinkiewicz and Carleson*

A. ZYGMUND

University of Chicago, Chicago, Illinois 60637

1. In this lecture I am going to discuss things that are known, but probably not *too* well known. Correspondingly, the character of the presentation be more expository than exploratory, though there may be some elements of novelty here. I shall discuss certain metric properties of sets and functions, properties that are important for applications of the Lebesgue integral to classical analysis.

Two such properties are fundamental, and I shall begin with the one which is familiar to all: it is the fact that the derivative of the indefinite integral of an integrable function exists almost everywhere, and almost everywhere is equal to the integrand, More specifically, the result is as follows: if $f(x) = f(x_1, x_2, \ldots, x_n)$ is an integrable function defined over the n-dimensional Euclidean space E_n, and if

$$F(E) = \int_E f$$

is the indefinite integral of f, then at almost all points x we have

$$\frac{F(Q)}{|Q|} \to f(x), \tag{1.1}$$

where Q designates a cube containing the point x and shrinking to x, and $|Q|$ is the measure of Q. The result holds if Q is an n-dimensional interval (parallelepiped) containing x, provided the ratio of the largest and the smallest edge of Q remains bounded (we consider only intervals with edges parallel to the co-ordinate axes).

Also, and this easily follows from the preceding, we have at almost all points x the somewhat stronger result, namely

$$\frac{\int_Q |f(y) - f(x)|\, dy}{|Q|} \to 0. \tag{1.2}$$

The points at which (1.2)—or a suitable generalization of it—holds, are usually called *Lebesgue points* of the function f.

* A lecture delivered at the Second Symposium on Inequalities at USAF Academy, Colorado, August 1967.

Reprinted from *J. Approx. Theory* 2, 249–257 (1969).

One need not stress the importance of the fact; we all know that without it the present-day analysis would be impossible. What is, however, less known is that, in a number of problems, the result does not suffice and must be supplemented by another result whose general importance seems to have been recognized for the first time by Marcinkiewicz.

Some of us who worked in Fourier series in the period between the two world wars remember a number of problems which remained unsolved for a long time and which were rather tantalizing; tantalizing, because, without appearing to be out of reach (unlike, e.g., the problem of convergence almost everywhere of Fourier series of continuous functions), they were still quite elusive. I shall mention three examples as illustrations.

(a) In the early nineteen twenties Carleman proved that if $f(x)$ is a periodic function of the class L^2 and $s_k(x)$ is the kth partial sum of the Fourier series of f, then for almost all values of x we have the relation

$$\frac{1}{n+1}\sum_{k=0}^{n}|s_k(x)-f(x)|^2 \to 0 \tag{1.3}$$

("strong summability" of Fourier series). The question naturally arose whether this result, or a suitable modification of it, holds for functions that are merely integrable. During a number of years much effort was being spent on it and a number of generalizations were obtained. For example, it was shown that the result holds for f in any class L^p, provided p is strictly greater than 1, and that in this case we can even replace the exponent 2 in (1.3) by any positive q, arbitrarily large. (See, e.g., [*10*, II, p. 180] and references there.) The relation (1.3) and its generalizations were usually shown to hold at the Lebesgue points of f, so that when Hardy and Littlewood [*3*] showed that, for f merely integrable, (1.3) need not hold at such points the possibility arose (cf. [*3*]) that perhaps, after all, strong summability need not hold almost everywhere for functions that are merely integrable. The question remained unsolved until 1939 when Marcinkiewicz showed that (1.3) is indeed true almost everywhere for f integrable (see [*4*] or [*10*, II, p. 184]).

(b) A very well-known result asserts that the Fourier series of any integrable function $f(x)$ is summable (C,δ), $\delta > 0$, almost everywhere; more precisely, at each Lebesgue point of f. It is also very well known that the result fails for $\delta = 0$: there are integrable functions whose Fourier series diverge at each point. Thus the result of Hardy and Littlewood that the termwise differentiated Fourier series of a function f is summable $(C, 1+\delta)$, $\delta > 0$, at each point where f' exists and is finite, appeared final, the more so as they showed by examples that the conclusion fails for $\delta = 0$. But Marcinkiewicz showed that though the conclusion may fail at individual points it is nevertheless valid almost everywhere; more precisely, if $f'(x)$ exists at each point of a set E, then the termwise differentiated Fourier series of f is summable $(C,1)$ almost everywhere in E (see [*5*] or [*10*, II, p. 81]).

(c) One of the classical results of the theory of Fourier series asserts that if a periodic and continuous function $f(x)$ satisfies the condition

$$f(x+h)-f(x)=o\left\{\frac{1}{\log 1/|h|}\right\} \qquad (h\to 0) \tag{1.4}$$

uniformly in x, then the Fourier series of f converges uniformly (the Dini–Lipschitz test). It is easy to show by examples that a continuous function f may satisfy the condition (1.4) at *some* point x without its Fourier series converging at that point. The question remained: if a periodic and merely integrable f satisfies (1.4) at each point x of a set E, does the Fourier series of f necessarily converge almost everywhere in E? It was again Marcinkiewicz (see [*6*] or [*10*, II, p. 170]) who showed that it is actually so. He even proved a stronger result: the conclusion holds if at each point $x\in E$ we have instead of (1.4) the obviously weaker relation:

$$\frac{1}{h}\int_0^h |f(x+t)-f(x)|\,dt=O\left\{\frac{1}{\log 1/|h|}\right\} \qquad (h\to 0). \tag{1.5}$$

(Observe that we have "O" here). Incidentally, he also showed that the result is best possible: the conclusion fails if the expression $1/(\log 1/h)$ on the right of (1.5) is replaced by any function of h tending to 0 more slowly [*5*].

In all three cases the solution was made possible by an application of the same theorem which expresses a certain metric property of sets and functions and which succeeds where the theorem about the differentiability of integrals seems to be insufficient. And it is a curious fact that this property, in a somewhat modified form, plays an important role in Carleson's proof of his fundamental theorem on the convergence almost everywhere of Fourier series of functions of the class L^2 (see [*2*], Lemma 5). I shall now describe that property.

2. Given any closed set P situated in the Euclidean space E_n we shall call the distance of any point x from P the *distance function*; it will be denoted by $\delta(x;P)$, or simply by $\delta(x)$. Thus $\delta(x)=0$ if and only if x is in P. If $n=1$ and (a,b) is any interval contiguous to P and situated between the terminal points of P, then the graph of $\delta(x)$ over (a,b) is an isosceles triangle with base (a,b) and altitude $\frac{1}{2}(b-a)$; outside the terminal points of P the graph of $\delta(x)$ is a linear function. If $n > 1$, the graph of $\delta(x)$ is in general much less simple, but since if we move from a point x to another point y the distance from P does not increase by more than $|x-y|$, it is clear that

$$|\delta(x)-\delta(y)| \leqslant |x-y|,$$

that is $\delta(x)$ satisfies a Lipschitz condition of order 1.

Marcinkiewicz's lemma (or theorem) may be stated as follows (see [*10*, I, pp. 129–131 and p. 377].)

(A) *Let P be a closed subset of E_n and $\delta(x)=\delta(x;P)$ the corresponding distance function. Let λ be a positive number and $f(x)$ a non-negative function integrable over the complement Q of P. Then for almost all points $x \in P$ the integral*

$$J_\lambda(x) = J_\lambda(x;f,P) = \int_{E_n} \frac{\delta^\lambda(y)\, f(y)}{|x-y|^{n+\lambda}}\, dy \tag{2.1}$$

is finite.

In particular, if P is bounded and K is any finite sphere containing P, the integral

$$\int_K \frac{\delta^\lambda(y)}{|x-y|^{n+\lambda}}\, dy \tag{2.2}$$

is finite almost everywhere in P.

The usual proof of (A) actually gives a little more (see [*10*, I], pp. 129–131), namely *the function $J_\lambda(x)$ is integrable over P*. A few years ago Professor R. O'Neil pointed out to the author that if $f \in L^p(E_n - P)$, $1 \leqslant p < \infty$, then $J_\lambda \in L^p(P)$ and we have the obvious inequalities for the norms. His proof was based on Hardy–Littlewood maximal theorems. In what follows we give a slightly different proof of (A) and its generalization by using a modification of the integral J_λ.

Whatever the behavior of the integral $J_\lambda(x)$ in P, it generally diverges outside P; this is certainly true of the integral (2.2) which is infinite at the points x interior to $K-P$. Let us however consider the following modification of J_λ:

$$H_\lambda(x) = \int_{E_n} \frac{\delta^\lambda(y) f(y)}{|x-y|^{n+\lambda} + \delta^{n+\lambda}(x)}\, dy. \tag{2.3}$$

It has two obvious properties: (a) it coincides with $J_\lambda(x)$ for $x \in P$; (b) it is finite at each point x not in P, provided $f \in L^p(E_n - P)$, $1 \leqslant p \leqslant \infty$. To prove the latter we consider separately the y's close to x, in which case the denominator stays away from 0, and the more distant y's to which we can apply Hölder's inequality.

We shall also consider another modification of J_λ, namely

$$H_\lambda'(x) = \int_{E_n} \frac{\delta^\lambda(y) f(y)}{|x-y|^{n+\lambda} + \delta^{n+\lambda}(y)}\, dy, \tag{2.4}$$

which, like H_λ, is finite at each point not in P. In view of the inequality $\delta(y) \leqslant |x-y| + \delta(x)$ we have, by Jensen's inequality,

$$\delta^{n+\lambda}(y) \leqslant 2^{n+\lambda-1}\{|x-y|^{n+\lambda} + \delta^{n+\lambda}(x)\}$$

and a similar inequality with x and y interchanged. We immediately deduce from this that

$$A^{-1} H_\lambda'(x) \leqslant H_\lambda(x) \leqslant A H_\lambda'(x)\,(A = 2^{n+\lambda-1} + 1), \tag{2.5}$$

so that inequalities for H_λ' immediately lead to inequalities for H_λ, but H_λ' is sometimes easier to deal with than H_λ.

Also, since the values of H_λ and H_λ' are independent of the values of f on P, we shall assume for the sake of simplicity of enunciation that f is defined over the whole of E_n and, say, is 0 in P.

(B) *If $f \in L^p(E_n)$, $1 \leqslant p < \infty$, then $H_\lambda \in L^p(E_n)$ and*

$$\Big\{\int_{E_n} H^p(x)\, dx\Big\}^{1/p} \leqslant Ap\Big(\int_{E_n} f^p\, dx\Big)^{1/p} \qquad (A = A_{n,\lambda}). \tag{2.6}$$

If f is bounded, say $0 \leqslant f \leqslant 1$, and has support in a sphere $K \supset P$, then

$$\int_K \exp\{\gamma H_\lambda(x)\}\, dx \leqslant A|K| \qquad (A = A_{n,\lambda}) \tag{2.7}$$

provided γ is small enough, $\gamma \leqslant A'_{n,\lambda}$.

3. We first prove (2.6), with H_λ' instead of H_λ.

Let $g(x)$ be any non-negative locally integrable function and let $\bar{g}(x)$ denote the corresponding Hardy–Littlewood maximal function

$$\bar{g}(x) = \operatorname*{Sup}_\rho \Big\{\rho^{-n} \int_{|z| \leqslant \rho} g(x+z)\, dz\Big\}.$$

It is a familiar fact that if $g \in L^r(E_n)$, $1 < r < \infty$, then $\bar{g}$ is likewise in $L^r(E_n)$, and

$$\|\bar{g}\|_r \leqslant A_n \frac{r}{r-1} \|g\|_r.^{1} \tag{3.1}$$

[1] The fact that $g \in L^r(E_n)$ implies $\bar{g} \in L^r(E_n)$, $1 < r < \infty$, and the inequality $\|\bar{g}\|_r \leqslant \{2r/(r-1)\}^n \|g\|_r$ is, using repeated integration, a simple corollary of the Hardy–Littlewood classical result for $n = 1$. The estimate (3.1), where we have $r/(r-1)$ in the first power, is slightly deeper and is due to Wiener [9]. See also [1], where it is deduced from the case $n = 1$ by the "method of rotation."

Also the following observation is very well known (and immediate): if λ and δ are positive numbers, $g(x)$ is non-negative and in $L^r(E_n)$, $1 \leqslant r \leqslant \infty$, then

$$\int_{E_n} \frac{\delta^\lambda g(x+z)}{|z|^{n+\lambda} + \delta^{n+\lambda}} dz \leqslant A\bar{g}(x) \qquad (A = A_{n,\lambda}). \tag{3.2}$$

For, decomposing the integral into two, extended, respectively, over $|z| \leqslant \delta$ and $|z| \geqslant \delta$, we see that the first is majorized by

$$\delta^{-n} \int_{|z| \leqslant \delta} g(x+z)\, dz \leqslant \bar{g}(x),$$

and the second by

$$\int_{|z| \geqslant \delta} \frac{g(x+z)}{|z|^{n+\lambda}} dz \leqslant A_{n,\lambda} \bar{g}(x),$$

as a simple integration by parts shows. This proves (3.2).

Let now $g(x)$ be any non-negative function such that $\|g\|_{p'} = 1$, where $p' = p/(p-1)$. Then

$$\begin{aligned}
\int_{E_n} H_\lambda'(x) g(x)\, dx &= \int_{E_n} f(y) \delta^\lambda(y)\, dy \left\{ \int_{E_n} \frac{g(x)\, dx}{|x-y|^{n+\lambda} + \delta^{n+\lambda}_{(y)}} \right\} \\
&\leqslant A_{n,\lambda} \int_{E_n} f(y) \bar{g}(y)\, dy \\
&\leqslant A_{n,\lambda} \|f\|_p \|\bar{g}\|_{p'} \leqslant A_{n,\lambda} \|f\|_p \cdot A_n \frac{p'}{p'-1} \|g\|_{p'} \\
&= p \,.\, A_{n,\lambda} \|f\|_p,
\end{aligned}$$

and since the least upper bound of the left-hand side here for all such g is the left-hand side of (2.6) with H_λ' for H_λ, this proves the first part of (B).

Passing to (2.7) we observe that the left-hand side there is

$$\begin{aligned}
|K| + \sum_{p=1}^{\infty} \frac{\gamma^p}{p!} \int_K H_\lambda{}^p\, dx &\leqslant |K| + \sum_{p=1}^{\infty} \frac{\gamma^p}{p!} A^p p^p \int_K f^p\, dx \\
&\leqslant |K| \left(1 + \sum_{p=1}^{\infty} \frac{(\gamma A p)^p}{p!} \right).
\end{aligned}$$

Since the last series converges for $\gamma Ae < 1$, (2.7) follows.

4. Let us consider (2.7) in the special case $f \equiv 1$ in K, and let $\omega(\eta) = \omega(\eta; K)$ be the distribution function of H_λ in K, that is, the measure of the set of the points $x \in K$ such that $H_\lambda(x) > \eta > 0$. An immediate corollary of (2.7) is

(C) *If γ is sufficiently small, $0 < \gamma \leqslant A'_{n,\lambda}$, then the distribution function of $H_\lambda(x;1,P)$ in K satisfies an inequality*

$$\omega(\eta) < A_{n,\lambda}|K|\, e^{-\gamma\eta}. \tag{4.1}$$

It is clear that, conversely, (4.1) gives (2.7) with any smaller value of γ.

5. Let K be any finite closed sphere in E_n, and let K_1, K_2, ... be a sequence, finite or not, of non-overlapping spheres contained in K. The center of K_j we denote by ξ_j, the radius by r_j. Let K_j^* be the sphere concentric with K_j of radius $\frac{1}{2}r_j$. Let K_j^0 be the interior of K_j and $P = K - \bigcup K_j^0$. Let $\delta(x)$ be the distance of x from P. If $x \in K_j^*$, then $\frac{1}{2}r_j \leqslant \delta(x) \leqslant r_j$. Consider the function $H_\lambda'(x)$ for f equal to the characteristic function of the set $\bigcup K_j^*$. Thus

$$H_\lambda'(x) = \sum_j \int_{K_j^*} \frac{\delta^\lambda(y)}{|x-y|^{n+\lambda} + \delta^{n+\lambda}(y)}\, dy$$

and an elementary argument (we consider separately the cases when x is or is not in K_j^*) shows that $H_\lambda'(x)$ is contained between two positive multiples, depending on n and λ only, of the sum

$$S_\lambda(x) = \sum_j \frac{r_j^{n+\lambda}}{|x-\xi_j|^{n+\lambda} + r_j^{n+\lambda}}. \tag{5.1}$$

Hence, using (C), or rather its analog for H'_λ, we obtain the following result:

(D) *With the notation just introduced, the distribution function on K of the sum $S_\lambda(x)$ satisfies the inequality* (4.1)

For $n = \lambda = 1$ this is Lemma 5 of Carleson's paper [*2*]. In his proof, which is very short, he uses properties of harmonic functions. We now see that his lemma has close connection with the results of Marcinkiewicz, and indicating this was one of the purposes of this lecture. Obviously the result holds if the spheres K, K_j are replaced by cubes.

6. In all the foregoing the parameter λ was a strictly positive number. If $\lambda = 0$ the arguments break down, and one can also show by examples that the theorems are false. However, already Marcinkiewicz considered in this case the substitute function

$$J_0(x;f,P) = \int_K \frac{\log\{1/\delta(y)\}^{-1} f(y)}{|x-y|^n}\, dy \tag{6.1}$$

which has a number of properties in common with J_λ, $\lambda > 0$. Since the function $|x|^{-n}$ is not integrable at infinity it is convenient to integrate in (6.1) over a finite sphere K, which is supposed to contain our closed set P. Morevoer, it will be convenient to assume that the diameter of the sphere is $\leqslant \frac{1}{2}$, so that the integrand in (6.1) is non-negative. Correspondingly, we shall also consider the function

$$H_0(x;f,P) = \int_K \frac{f(y)\log\{1/\delta(y)\}^{-1}}{|x-y|^n + \delta^n(x)}\,dy \tag{6.2}$$

which for $x \in P$ reduces to J_0, and the function $H_0'(x)$ which is obtained from $H_0(x)$ by replacing the term $\delta^n(x)$ in the denominator by $\delta^n(y)$. The inequality (2.5) holds also in this case. We shall only consider the behaviour of H_0 and H_0' on K and we have then the following theorem, in which the diameter of K is $\leqslant \frac{1}{2}$.

(E) *If $f \in L^p(K)$, $1 \leqslant p < \infty$, then $H_0 \in L^p(K)$ and*

$$\left\{\int_K H_0^p(x)\,dx\right\}^{1/p} \leqslant Ap\left\{\int_K f^p\,dx\right\}^{1/p} \qquad (A = A_n) \tag{6.3}$$

and if $f \equiv 1$ in K, then

$$\int_K \exp \gamma H_0(x)\,dx \leqslant A|K| \qquad (A = A_n, \gamma \leqslant A_n') \tag{6.4}$$

The proof is parallel to that of B. In the case of (6.3) it is enough to observe that if $g(x)$ is non-negative in K and the integral of $g^{p'}$ over K is 1, then the integral of $H_0'(x)g(x)$ over K can be written

$$\int_K f(y)\left\{(\log 1/\delta(y))^{-1}\int_K \frac{g(x)\,dx}{|x-y|^n + \delta^n(y)}\right\}dy,$$

and that the expression in curly brackets is majorized by the sum

$$\{\log 1/\delta(y)\}^{-1}\int_{|z|\leqslant\delta(y)} \frac{g(y+z)}{\delta^n(y)}\,dz + \{\log 1/\delta(y)\}^{-1}\int_{\delta(y)\leqslant|z|\leqslant 1} \frac{g(y+z)}{|z|^n}\,dz,$$

of which the first term does not exceed $\bar{g}(y)(\log 2)^{-1}$, and the second does not exceed $A_n\bar{g}(y)$. The proof of (6.4) is identical with that of (2.7). We also have analogous results for the function

$$S_0(x) = \sum_j \frac{r_j^n\{\log(1/r_j)\}^{-1}}{|x-\xi_j|^n + r_j^n}$$

[cf. (5.1)].

We add that, in the case $p = 1$, the integrability of H_0 over K implies the finiteness of J_0 almost everywhere in P. The latter result is, essentially, one of the original results of Marcinkiewicz.[2]

[2] Marcinkiewicz [7] himself considered only the case $n = 1$ and instead of $\delta(x)$ the function $\delta^*(x)$ equal to 0 in P and to $b_i - a_i$ in each interval (a_i, b_i) contiguous to P. But the function δ seems to be more natural than δ^* and extensions to higher dimensions more routine.

7. We conclude with an incomplete result. Following ideas of Ostrow and Stein [*8*], we may consider instead of H_λ the somewhat more general integral

$$T_\lambda(x) = \int_{E_n} \frac{f(y)\,\delta^\lambda(y)\,\phi(x-y)}{|x-y|^{n+\lambda} + \delta^{n+\lambda}(x)}\,dy,$$

where ϕ is a non-negative locally integrable function satisfying the condition

$$\int_{|x|\leqslant\rho} \phi(x)\,dx \leqslant M\rho^n \qquad (0<\rho<\infty). \tag{7.1}$$

It is very easy to show that if $f \in L$ then T_λ is likewise in L and $\|T_\lambda\|_1 \leqslant MA_{n,\lambda}\|f\|_1$; in particular the integral

$$\int_{E_n} \frac{f(y)\,\delta^\lambda(y)\,\phi(x-y)}{|x-y|^{n+\lambda}}\,dy$$

generalizing J_λ, is finite almost everywhere in P. Whether, however, $f \in L^p$, $1<p<\infty$, implies $T_\lambda \in L^p$ seems to be an open problem.

References

1. A. P. Calderón and A. Zygmund, On singular integrals. *Am. J. Math.* **78** (1956), 289–309.

2. L. Carleson, On convergence and growth of partial sums of Fourier series. *Acta Math.* **116** (1966), 135–157.

3. G. H. Hardy and J. E. Littlewood, On the strong summability of Fourier series. *Fund. Math.* **25** (1935), 162–189.

4. J. Marcinkiewicz, Sur la sommabilité forte des séries de Fourier. *J. London Math. Soc.* **14** (1939), 162–168.

5. J. Marcinkiewicz, Sur les séries de Fourier. *Fund. Math.* **27** (1936), 38–69.

6. J. Marcinkiewicz, On the convergence of Fourier series. *J. London Math. Soc.* **10** (1935), 264–268.

7. J. Marcinkiewicz, Sur quelques intégrales du type de Dini. *Ann. Soc. Polon. Math.* **17** (1938) 42–50.

8. E. M. Ostrow and E. M. Stein, A generalization of lemmas of Marcinkiewicz and Fine with applications to singular integrals. *Ann. Scuola Normale Sup. Pisa* **11** (1957), 117–135.

9. N. Wiener, The ergodic theorem. *Duke Math. J.* **5** (1935), 1–18.

10. A. Zygmund, "Trigonometric Series," Vols. I and II. Cambridge University Press, 1959.

STUDIA MATHEMATICA, T. XLIII. (1972)

A Cantor–Lebesgue theorem for double trigonometric series

by

A. ZYGMUND (Chicago)

Abstract. Let $\xi = (x, y)$ be points of the plane, $\nu = (m, n)$ – lattice points, and $\langle\nu\cdot\xi\rangle = mx+ny$. It is shown that given any set E of positive measure situated in the square $0 < x < 1$, $0 < y < 1$, there is a constant $A = A_E$ such that for any trigonometric polynomial $T(\xi)$ of the form $\sum_{|\nu|=R} c_\nu e^{2\pi i(\nu\cdot\xi)}$ we have

$$\sum |c_\nu|^2 < A \int_E |T(\xi)|^2 d\xi.$$

In particular, if an infinite series $\sum c_\nu e^{2\pi i(\nu,\xi)}$ converges circularly in a set of positive measure, then $\sum_{|\nu|=R} |c_\nu|^2 \to 0$ as $R \to \infty$.

1. Let $\xi = (x, y) \in R^2$ and let $\nu = (m, n)$ denote lattice points in R^2. Consider a double trigonometric series

$$\sum_\nu c_\nu e^{2\pi i\langle\nu\cdot\xi\rangle}, \tag{T}$$

where $\langle\nu\cdot\xi\rangle = mx+ny$, and its circular partial sums

$$T_R(\xi) = \sum_{|\nu|\leqslant R} c_\nu e^{2\pi i\langle\nu\cdot\xi\rangle}.$$

We shall also write

$$A_R(\xi) = \sum_{|\nu|=R} c_\nu e^{2\pi i\langle\nu\cdot\xi\rangle}.$$

Recently, R. L. Cooke proved the following result (see [1]).

THEOREM 1. *If $A_R(\xi) \to 0$ almost everywhere as $R \to \infty$ (and, in particular, if T converges almost everywhere), then $c_\nu \to 0$ as $|\nu| \to \infty$. More generally, we then have*

$$\sum_{|\nu|=R} |c_\nu|^2 \to 0 \quad (R \to \infty). \tag{1.1}$$

In this note we prove a somewhat more general result.

THEOREM 2. *If $A_R(\xi) \to 0$ at each point ξ of a set of positive measure, we have* (1.1).

2.

LEMMA 1. *Let* $E \subset R^2$, $0 < |E| < \infty$, *and let*

$$\chi^{\wedge}(\nu) = \int_E e^{-2\pi i \langle \nu \cdot \xi \rangle} d\xi .$$

Then there is a strictly positive ε *such that*

$$|\chi^{\wedge}(\nu)| < |E| - \varepsilon \quad \textit{for all } \nu \neq 0 .$$

Proof. Since $\chi^{\wedge}(\nu) \to 0$ as $|\nu| \to \infty$, it is enough to show that each $|\chi^{\wedge}(\nu)|$ is strictly less than $|E|$ if $\nu \neq 0$.

This is a consequence of the elementary fact that if, for any complex-valued f, we have $|\int_E f| = \int_E |f|$, and $f \neq 0$ on E, then $\arg f$ is constant, $\bmod 2\pi$, almost everywhere on E. Hence, in our case, if we had $|\chi^{\wedge}(\nu)| = |E|$ for some $\nu \neq 0$, $\langle \nu \cdot \xi \rangle$ would have to be constant, mod1, almost everywhere on E, almost all points of E would be situated on a finite or denumerable family of straight lines, and we would have $|E| = 0$, contrary to hypothesis.

LEMMA 2. *For any three distinct lattice points* λ, μ, ν *situated on a circumference of radius* R *we have*

$$|\lambda - \mu|\,|\mu - \nu|\,|\nu - \lambda| \geqslant 2R. \tag{2.1}$$

This is a corollary of the classical theorem of Elementary Geometry which asserts that if a triangle with sides a, b, c has area S, and if R is the radius of the circle circumscribed on the triangle, then

$$R = abc/4S. \tag{2.2}$$

For in our case a, b, c are the factors on the left in (2.1), and if we write $\lambda = l_1 + il_2$, $\mu = m_1 + im_2$, $\nu = n_1 + in_2$, then S is the absolute value of

$$\frac{1}{2}\begin{vmatrix} 1 & l_1 & l_2 \\ 1 & m_1 & m_2 \\ 1 & n_1 & n_2 \end{vmatrix},$$

and so $S \geqslant 1/2$. This leads to (2.1). Lemma 2, which is essential for the proof of Theorem 2, was first proved (in reply to a question put by the author) by A. Schinzel, in an elementary and purely analytical way. That it is a corollary of (2.2) was later pointed out by A. Pełczyński.

3. Passing to the proof of the theorem we may assume (by Egoroff's theorem) that T_R converges uniformly in E, and that E is contained in the square $0 \leqslant x < 1$, $0 \leqslant y < 1$. Then $A_R(\xi) \to 0$ uniformly in E.

We have

$$(3.1)\quad \int_E |A_R(\xi)|^2 d\xi = |E| \sum_{|\mu|=R} |c_\mu|^2 + \sum_{\substack{|\mu|=\nu=R\\ \mu\neq\nu}} c_\mu \bar{c}_\nu \chi\hat{\ }(\nu-\mu) = P+Q,$$

say.

Let $\Delta = \Delta_R$ be the set of all the differences $\nu-\mu \neq 0$, for $|\nu| = |\mu| = R$. Let R_0 be so large that

$$(3.2)\quad \Big(2 \sum_{|\lambda|>R_0} |\chi\hat{\ }(\lambda)|^2\Big)^{1/2} < \tfrac{1}{2}\varepsilon,$$

where ε is that of Lemma 1 corresponding to our set E. Write $\Delta = \Delta' \cup \Delta''$, where Δ' consists of the elements of Δ of modulus $\leqslant R_0$ and Δ'' is the remainder of Δ.

Correspondingly,

$$(3.3)\quad Q = Q'+Q''.$$

Clearly,

$$(3.4)\quad |Q''| \leqslant \Big(\sum |c_\mu \bar{c}_\nu|^2\Big)^{1/2} \Big(\sum |\chi\hat{\ }(\nu-\mu)|^2\Big)^{1/2} (|\mu| = |\nu| = R,\ |\mu-\nu| > R_0)$$
$$\leqslant \sum_{|\lambda|=R} |c_\lambda|^2 \cdot \Big(\sum_{|\lambda|>R_0} 2|\chi\hat{\ }(\lambda)|^2\Big)^{1/2} \leqslant \tfrac{1}{2}\varepsilon \sum_{|\lambda|=R} |c_\lambda|^2,$$

by (3.2). Here we used the fact that a circle can have at most two chords of prescribed length and direction.

Let $C(0, R)$ be the circumference of center 0 and radius R. The meaning of Lemma 2 is that if two lattice points on $C(0, R)$ are 'close' to each other, then any other lattice point on $C(0, R)$, should it exist, is necessarily 'distant' from those two.

Having fixed R_0 we take R so large that any pair (μ, ν) on $C(0, R)$ with $|\mu-\nu| \leqslant R_0$ is distant by more than R_0 from any other lattice point on $C(0, R)$. Hence the lattice points on $C(0, R)$ can be split into 'distant' pairs (μ, ν) with $|\mu-\nu| \leqslant R_0$. For each such pair (μ, ν), writing $\nu-\mu = \lambda$ we have, by Lemma 1,

$$|c_\mu \bar{c}_\nu \chi\hat{\ }(\lambda) + c_\nu \bar{c}_\mu \chi\hat{\ }(-\lambda)|$$
$$\leqslant \tfrac{1}{2}(|c_\mu|^2 + |c_\nu|^2)(|E|-\varepsilon)\cdot 2 = (|c_\mu|^2 + |c_\nu|^2)(|E|-\varepsilon).$$

It follows that

$$(3.5)\quad |Q'| \leqslant \sum_{|\lambda|=R} |c_\lambda|^2 (|E|-\varepsilon).$$

Collecting the results (see (3.1), (3.3), (3.4), (3.5)) we obtain

$$\int_E |A_R(\xi)|^2 d\xi = P+Q'+Q''$$

$$\geqslant \sum_{|\lambda|=R} |c_\lambda|^2 \{|E| - \tfrac{1}{2}\varepsilon - (|E|-\varepsilon)\} = \tfrac{1}{2}\varepsilon \sum_{|\lambda|=R} |c_\lambda|^2.$$

Thus

$$\sum_{|\lambda|=R} |c_\lambda|^2 \leqslant 2\varepsilon^{-1} \int_E |A_R(\xi)|^2 d\xi, \tag{3.6}$$

and if $A_R(\xi)$ tends uniformly to 0 on E, $\sum_{|\lambda|=R} |c_\lambda|^2 \to 0$. This completes the proof of Theorem 2.

4. We conclude with a few observations.

a) The proof of Theorem 2 is essentially two-dimensional. Whether an analogue of Theorem 2 (or even only of Theorem 1) holds in higher dimensions remains an open problem.

b) Strictly speaking, the proof of Theorem 2 has little to do with the relation $A_R(\xi) \to 0$ $(\xi \in E, |E| > 0)$. Analyzing the proof (see, in particular, (3.6)) we see that it gives the following result.

THEOREM 3. *Given any set E of positive measure situated in the unit square $0 \leqslant x < 1$, $0 \leqslant y < 1$, we can find a positive number A_E such that*

$$\sum_{|\lambda|=R} |c_\lambda|^2 \leqslant A_E \int_E \Big| \sum_{|\lambda|=R} c_\lambda e^{2\pi i(\lambda\cdot\xi)}\Big|^2 d\xi. \tag{4.1}$$

That (4.1) holds for R sufficiently large, $R \geqslant R_E$, is implicit in the proof of Theorem 2, and for $R < R_E$ follows from the equivalence of norms in spaces of the same finite dimension.

c) One may ask for an estimate analogous to (4.1) for sums $\sum c_\nu e^{2\pi i\langle\nu\cdot\xi\rangle}$ extended over lattice points situated on some plane curve Γ.

The argument of Section 3, where $\Gamma = C(0, R)$, utilizes two properties of the circle: α) it has at most two chords of prescribed length and direction; β) given R_0, for any lattice point $\lambda \in C(0, R)$ there is at most one neighbor $\mu \in C(0, R)$ with $|\lambda - \mu| \leqslant R_0$, provided R is large enough (there may actually be such neighbors; take e.g. the points $(n, n+1)$ and $(n+1, n)$ on $C(0, R)$ with $R^2 = 2n^2+2n+1$).

Property α) was needed to estimate the term Q'' in (3.3) (see (3.4)), property β) — for Q'. As to α), it is certainly satisfied for any strictly convex curve Γ, but it is easily seen to be unnecessarily restrictive: if Γ has at most k chords of prescribed length and direction, the factor 2 in the next to last term of (3.4) can be replaced by k and the argument still works.

Property β) is a little more subtle and we do not propose to study it. The special case of an ellipse is both of independent interest and sufficient simplicity to be considered here and we limit ourselves to it. (Even simpler, though less interesting, is the case when Γ is a convex polygone whose sides make with the x-axis angles incommensurable with π; it is obvious that each side of Γ can contain at most one lattice point.)

Let E_{ab} denote an ellipse with semi-axes $a, b, a > b$, and center 0 (it is obvious that the latter condition is no restriction of generality); the direction of the axes is not specified. Let $C(0, a)$ be the circle circumscribed on E_{ab}. Let λ, μ, ν be three distinct lattice points on E_{ab}, and λ', μ', ν' their projections parallel to the minor axis out to the circumscribed circle (λ', μ', ν' need not be lattice points).

Let S and S' denote respectively the areas of the triangles $\lambda\mu\nu$ and $\lambda'\mu'\nu'$; thus $S \geqslant \frac{1}{2}$. In view of (2.2) we have the relation

$$|\mu'-\lambda'|\,|\mu'-\nu'|\,|\lambda'-\nu'| = 4S'a = 4\cdot\frac{a}{b}S\cdot a \geqslant 2\frac{a^2}{b}. \tag{4.2}$$

Since the passage from S to S' increases the sides by factors $< a/b$, (4.2) leads to the following analogue of (2.1):

$$|\mu-\lambda|\,|\mu-\lambda|\,|\lambda-\nu| \geqslant 2\frac{b^2}{a}. \tag{4.3}$$

If the distance of μ from both λ and ν does not exceed R_0, we deduce from (4.3) that

$$R_0\,.\,R_0\,.\,2R_0 \geqslant 2\frac{b^2}{a},$$

that is,

$$b \leqslant R_0^{3/2}a^{1/2}.$$

It follows that if $b > R_0^{3/2}a^{1/2}$, then for any lattice point $\mu \in E_{ab}$ there can exist at most one lattice point $\lambda \in E_{ab}$ with $0 < |\lambda-\mu| \leqslant R_0$. Hence, as in the case of a circle, the lattice points on E_{ab} of distance $\leqslant R_0$ can be split into 'distant' pairs, the estimate for Q' holds and we arrive at the following generalization of Theorem 3.

THEOREM 4. *Let E_{ab} be an ellipse with center 0, semi-axes $a, b, a > b$, their direction arbitrary. Then for any set E of positive measure situated in the square $0 \leqslant x < 1, 0 \leqslant y < 1$ we can find constants $A = A_E, K = K_E$ such that if*

$$b > Ka^{1/2}, \tag{4.4}$$

then

$$\sum_{\nu\in E_{ab}} |c_\nu|^2 \leqslant A\int_E \Big|\sum_{\nu\in E_{ab}} c_\nu e^{2\pi i\langle\nu\cdot\xi\rangle}\Big|^2 d\xi. \tag{4.5}$$

Observe that initially we obtain (4.5) under the additional condition that a is large, $a \geqslant L_E$, which later, as in the case of the circle, may be dropped.

Of course, under (4.4) the eccentricity of E_{ab} may tend to 1 as $a \to \infty$.

I am indebted to Dr. M. Jodeit for some clarifying observations.

References

[1] R. L. Cooke, *A Cantor–Lebesgue theorem for two dimensions*, Preliminary Report, Notices of the Amer. Math. Soc. 17 (1970), p. 933.

Received September 21, 1971 (375)

STUDIA MATHEMATICA, T. LXV. (1979)

A note on singular integrals

by

A. P. CALDERÓN and A. ZYGMUND* (Chicago, Ill.)

Abstract. The purpose of the paper is to further investigate relationships between various conditions on singular kernels K which imply continuity of the corresponding operator.

1. In the study of the existence and properties of singular integrals

$$\int_{R^n} f(y)K(x-y)\,dy$$

various hypotheses about the kernel K can be made, in addition to the basic properties that $K(x)$ is homogeneous of degree $-n$ (n the dimension of the space) and that the mean value of K over the surface

$$(\Sigma) \qquad |x| = 1$$

of the unit sphere is 0.

One of the earlierst assumptions used was (see e.g., [2]) that the kernel K satisfies the Dini condition on Σ, that is to say that the modulus of continuity $\omega(t)$ of K on Σ be such that

$$(1.1) \qquad \int_0^1 \frac{\omega(t)}{t}\,dt < \infty.$$

This implicitly presupposes the continuity of K on Σ. If this holds then the transformation

$$\tilde{f}(x) = Tf(x) = \lim_{\varepsilon\to 0} \int_{|x-y|\geqslant\varepsilon} f(y)K(x-y)\,dy = \text{P.V.} \int_{R^n} f(y)K(x-y)\,dy$$

is of type (p, p) for $1 < p < \infty$, and of weak type $(1, 1)$ (see [2]).

It may also be noted that condition (1.1) was merely used to show that

$$(1.2) \qquad \int_{|x|\geqslant 2|y|} |K(x-y)-K(x)|\,dx \leqslant C \quad (y \neq 0)$$

from which the properties of T just stated were derived (see also [1], [5]).

* This research was supported by NSF Grant MCS 75-05567.

In view of the importance of singular integrals any weakening or significant modification of assumptions about the kernel K may be of interest. For example, the theorem just stated about the operation $\tilde{f} = Tf$ holds if the modulus of continuity $\omega(t)$ is replaced by the integral modulus of continuity $\omega_1(t)$ (see below), that is, if

$$\int_0^1 \frac{\omega_1(t)}{t}\, dt < \infty \tag{1.3}$$

because, as was shown in [4], (1.3) implies (1.2). In that paper, it was also shown that (1.3) implies

$$\int_{\Sigma} |K(x)| \log^+ |K(x)|\, d\sigma_x, \tag{1.4}$$

which had been previously known to guarantee that T is of type (p, p), $1 < p < \infty$ (see [3]).

2. In this paper we want to establish some additional relations between the conditions (1.2), (1.3) and (1.4). As we said, (1.3) implies both (1.2) and (1.4). Here we shall show that, conversely, (1.2) implies (1.3) and (1.4). We recall the definition of $\omega_1(t)$ (see [4]). Let ϱ be a proper rotation of $\boldsymbol{R}^n$ about the origin and let

$$|\varrho| = \sup_{|x|=1} |x - \varrho x|.$$

Then

$$\omega_1(t) = \sup_{|\varrho| \leqslant t} \int_{\Sigma} |K(\varrho x) - K(x)|\, d\sigma_x,$$

where $d\sigma_x$ denotes the surface area element of $\Sigma = \{|x| = 1\}$. We shall also consider two more moduli of continuity of the kernel $K(x)$, namely

$$\omega_2(t) = \omega_2(t, a, b, \bar{y}) = \int_{a \leqslant |x| \leqslant b} |K(x - t\bar{y}) - K(x)|\, dx, \qquad |\bar{y}| = 1, \tag{2.1}$$

$$\omega_3(t) = \omega_3(t, a, b) = \sup_{|y| \leqslant t} \int_{a \leqslant |x| \leqslant b} |K(x - y) - K(x)|\, dx, \tag{2.2}$$

where $0 < a < b$ and $|\bar{y}| = 1$.

Let $a > 1$. Setting

$$I_a(y) = \int_{|x| \geqslant a|y|} |K(x - y) - K(x)|\, dx, \qquad J_a(K) = \sup_y I_a(y) \tag{2.3}$$

(notice that $I_a(y)$ is a homogeneous function of degree zero of y), our main result can be formulated as follows:

THEOREM. *Let $K(x)$ be positively homogeneous of degree $-n$ and locally integrable in $|x| \neq 0$. Then, if $1 < \alpha < \beta$,*

(i) $I_\beta(y) \leqslant I_\alpha(y) \leqslant 2\dfrac{\beta-1}{\alpha-1} I_\beta(y)$;

(ii) $\dfrac{a}{b-a}\int_0^{a/\alpha} \dfrac{\omega_2(t)}{t}\,dt \leqslant I_\alpha(y) \leqslant \dfrac{b}{b-a}\int_0^{b/\alpha}\dfrac{\omega_2(t)}{t}\,dt$, *where a and b are as in (2.1) and $\bar{y} = y/|y|$;*

(iii) $\dfrac{1}{c}\int_0^{\delta(a/\alpha)} \dfrac{\omega_3(t)}{t}\,dt \leqslant J_\alpha(K) \leqslant c\int_0^{b/\alpha} \dfrac{\omega_3(t)}{t}\,dt$, $\delta > 0$, *where δ depends only on the dimension n, and c depends on n, a and b;*

(iv) $\omega_1(t) \leqslant c\omega_3(t)$, $0 < t \leqslant 2$, *where c depends on a, b and n; and finally, if*

$$\lambda = J_\alpha(K) + \int_{a\leqslant|x|\leqslant b} |K(x)|\,dx < \infty,$$

then

(v) $\displaystyle\int_{a\leqslant|x|\leqslant b} \frac{|K(x)|}{\lambda}\log\left(1+\frac{|K(x)|}{\lambda}\right)dx \leqslant c$ *where c depends on a, b, α and n.*

3. We begin proving (i). Let $\beta-1 \leqslant 2(\alpha-1)$. Then

$$\int_{|x|\geqslant\alpha|y|} |K(x-y)-K(x)|\,dx \tag{3.1}$$
$$\leqslant \int_{|x|\geqslant\alpha|y|} |K(x-y)-K(x-y/2)|\,dx + \int_{|x|>\alpha|y|} |K(x-y/2)-K(x)|\,dx.$$

Setting $\bar{x} = x - y/2$ in the first integral on the right above and observing that $|\bar{x}+y/2| \geqslant \alpha|y|$ implies $|\bar{x}| \geqslant (2\alpha-1)\left|\dfrac{y}{2}\right| \geqslant \beta\left|\dfrac{y}{2}\right|$, we see that this integral is majorized by

$$\int_{|\bar{x}|\geqslant\beta\left|\frac{y}{2}\right|} |K(\bar{x}-y/2)-K(\bar{x})|\,d\bar{x}.$$

Now, because $\beta \leqslant 2\alpha$, this also majorizes the second integral on the right of (3.1). Consequently,

$$\int_{|x|\geqslant\alpha|y|} |K(x-y)-K(x)|\,dx \leqslant 2\int_{|x|\geqslant\beta\frac{|y|}{2}} |K(x-y/2)-K(x)|\,dx$$

and

$$I_\alpha(y) \leqslant 2I_\beta(y/2) = 2I_\beta(y)$$

which implies (i) for $\beta-1 \leqslant 2(a-1)$. In particular, we have

$$I_a(y) \leqslant 2I_\beta(y), \quad \beta-1 = 2(a-1)$$

and from this we obtain

$$I_a(y) \leqslant 2^k I_\beta(y), \quad 2^{k-1}(a-1) \leqslant \beta-1 \leqslant 2^k(a-1), \quad k = 1, 2, \dots$$

whence (i) follows in the general case.

To prove (ii) we set $x = t\bar{x}, y = s\bar{y}, |\bar{x}| = |\bar{y}| = 1$ in the integral defining $I_a(y)$ in (2.3) and obtain

$$I_a(y) = \int\limits_{|x|\geqslant a|y|} |K(x-y)-K(x)|\,dx = \int\limits_{\Sigma} d\sigma_{\bar{x}} \int\limits_{as}^{\infty} |K(t\bar{x}-s\bar{y})-K(t\bar{x})|\,t^n \frac{dt}{t},$$

where Σ denotes the unit sphere $|x| = 1$ and $d\sigma_{\bar{x}}$ the surface area element. We replace now t by the variable $\bar{t} = \tau\dfrac{s}{t}$, where τ is a constant for the moment, and, using the homogeneity of $K(x)$, we find that

$$I_a(y) = \int\limits_0^{\tau/a} \frac{d\bar{t}}{\bar{t}}\,\tau^n \int\limits_{\Sigma} |K(\tau\bar{x}-\bar{t}\bar{y})-K(\tau\bar{x})|\,d\sigma_{\bar{x}}.$$

If we integrate this equation with respect to τ over the interval (a, b), $0 < a < b$, and write t for $\bar{t}$, we obtain

$$\text{(3.2)} \qquad I_a(y) \geqslant \frac{1}{b-a}\int\limits_0^{a/a} \frac{dt}{t}\int\limits_a^b \tau^{n-1}a\,d\tau \int\limits_{\Sigma} |K(\tau\bar{x}-t\bar{y})-K(\tau\bar{x})|\,d\sigma_{\bar{x}}$$

$$= \frac{a}{b-a}\int\limits_0^{a/a}\frac{dt}{t}\int\limits_{a\leqslant|x|\leqslant b} |K(x-t\bar{y})-K(x)|\,dx.$$

According to (2.1), this is the first inequality in (ii). Clearly, the second inequality in (ii) can be obtained by a similar argument which we leave to the reader.

The proof of (iii) is more elaborate and depends on the following lemma.

LEMMA 1. *Let A_λ be the annulus $\{x|\ \lambda \leqslant |x| \leqslant 2\lambda\}$ and E a subset of A_λ such that $|E| \geqslant |A_\lambda|c$, where $|E|$ and $|A_\lambda|$ denote the measures of E and A_λ respectively and $c > \frac{1}{2}$. Then the set*

$$E+E = \{x|\ x = x_1+x_2, x_1 \in E, x_2 \in E\}$$

contains a sphere $|x| \leqslant \delta\lambda$, δ being a positive number which depends only on c.

It evidently suffices to prove the lemma in the case when $\lambda = 1$. Let $|y| \leqslant \delta$ and consider the sets E and $y-E = \{y-x|\ x \in E\}$. They are both contained in the annulus $\bar{A} = \{x|\ 1-\delta \leqslant |x| \leqslant 2+\delta\}$. But $|\bar{A}| = \frac{(2+\delta)^n-(1-\delta)^n}{2^n-1}|A_1|$, so that, if δ is chosen so small that $|\bar{A}| < 2c|A_1|$, we will have

$$|\bar{A}| < 2c|A_1| \leqslant 2|E| = |E|+|y-E|,$$

that is, $|\bar{A}| < |E|+|y-E|$. Because the sets E and $y-E$ are contained in $\bar{A}$, this implies that their intersection is non-empty or, equivalently, that $y \in E+E$. Consequently, every point of $\{|y| \leqslant \delta\}$ is contained in $E+E$, as we wished to show.

Returning to (iii), integrating (3.2) with respect to $\bar{y}$ over the unit sphere $\Sigma = \{|x| = 1\}$ we obtain

$$(3.3)\qquad |\Sigma| J_a(K) \geqslant \frac{a}{b-a}\int_0^{a/a}\frac{dt}{t}\int_{a\leqslant|x|\leqslant b}|K(x-t\bar{y})-K(x)|\,dx\,d\sigma_{\bar{y}}$$

$$= \frac{a}{b-a}\int_{|y|\leqslant a/a}|y|^{-n}\int_{a\leqslant|x|\leqslant b}|K(x-y)-K(x)|\,dx\,dy,$$

where $|\Sigma|$ denotes the surface area of Σ.

Let now $0 < a_1 < a < b < b_1$, $a_1 = a/2$, $b_1 = b+a/2$, and set

$$\theta(y) = \int_{a\leqslant|x|\leqslant b}|K(x-y)-K(x)|\,dx,$$

$$\theta_1(y) = \int_{a_1\leqslant|x|\leqslant b_1}|K(x-y)-K(x)|\,dx.$$

Then, if $|y_1| \leqslant a/2$, $|y_2| \leqslant a/2$, we have

$$(3.4)\qquad \theta(y_1+y_2) \leqslant \int_{a\leqslant|x|\leqslant b}|K(x-y_1-y_2)-K(x-y_1)|\,dx +$$

$$+ \int_{a\leqslant|x|\leqslant b}|K(x-y_1)-K(x)|\,dx \leqslant \theta_1(y_1)+\theta_1(y_2).$$

If A_λ denotes the annulus $\lambda \leqslant |x| \leqslant 2\lambda$, and

$$a_\lambda = \frac{1}{|A_\lambda|}\int_{A_\lambda}\theta_1(y)\,dy,\qquad E_\lambda = \{y|\ y \in A_\lambda,\ \theta_1(y) \leqslant 4a_\lambda\},$$

then $\theta_1(y) > 4a_\lambda$ on $A_\lambda - E_\lambda$, which clearly implies that $|A_\lambda - E_\lambda| < \frac{1}{4}|A_\lambda|$ and, consequently, $|E_\lambda| > \frac{3}{4}|A_\lambda|$. Now, if $y = y_1+y_2$, $y_1, y_2 \in E_\lambda$, then according to (3.4) we have

$$\theta(y) \leqslant \theta_1(y_1)+\theta_1(y_2) < 8a_\lambda.$$

But Lemma 1 asserts that $E_\lambda + E_\lambda$ contains the sphere $|y| \leqslant \delta\lambda$ and we find that the preceding inequality holds for $y \leqslant \delta\lambda$. Recalling the definition of $\omega_3(t)$, this shows that $\omega_3(t) < 8a_\lambda$ for $t \leqslant \delta\lambda$. Thus we have

$$\int_{\delta\frac{\lambda}{2}}^{\delta\lambda} \omega_3(t)\frac{dt}{t} < 8a_\lambda \log 2 = 8\log 2 \frac{1}{|A_\lambda|}\int_{A_\lambda}\theta_1(y)\,dy \leqslant c\int_{A_\lambda}\theta_1(y)\,|y|^{-n}dy.$$

Setting $\lambda = 2^{-h}a_1/a = 2^{-h}a/2a$, $h = 1, 2, \ldots$, and adding the corresponding inequalities we obtain

$$\int_0^{\delta\frac{a}{4a}} \omega_3(t)\frac{dt}{t} \leqslant c \int_{|y|\leqslant a_1/a} \theta_1(y)\,|y|^{-n}\,dy,$$

which combined with (3.3) yields

$$\int_0^{\delta\frac{a}{4a}} \omega_3(t)\frac{dt}{t} \leqslant cJ_a(K),$$

where c depends on a and b. Thus the first half of (iii) is established. The second half follows from the second inequality in (ii) by observing that $\omega_2(t) \leqslant \omega_3(t)$.

We pass now to the proof of (iv). Our argument depends on Lemma 6 in [4], which is also valid in the following slightly different situation.

Lemma 2. *There exist positive constants c, η depending only on the dimension n such that if*

$$A = \{x|\ a \leqslant |x| \leqslant b\}, \qquad a > 2\delta_0,$$
$$A' = \{x|\ a-\delta_0 \leqslant |x| \leqslant b+\delta_0\},$$
$$A'' = \{x|\ a-2\delta_0 \leqslant |x| \leqslant b+2\delta_0\},$$

and $h(x) = \big(h_1(x), \ldots, h_n(x)\big)$ is a C^1 vector-valued function satisfying

$$\text{(a)}\quad |h(x)| \leqslant \delta \leqslant \delta_0, \quad \left|\frac{\partial h_i}{\partial x_j}(x)\right| \leqslant \eta \text{ for all } x \text{ in } A',$$

then

$$\int_A \big|f\big(x+h(x)\big)-f(x)\big|\,dx \leqslant c \sup_{|y|\leqslant\delta}\int_{A'}|f(x+y)-f(x)|\,dx \tag{3.5}$$

for every function f integrable in A''.

To prove the lemma we argue as follows. Choosing η sufficiently small, the matrices with entries $\delta_{ij}+\partial h_i/\partial x_j$, which is the functional matrix of the change of variables $z = x+h(x)$, will have a determinant of absolute value larger than $1/2$. Consequently, for any function $g(x)$

we shall have

$$\int_{A'} |g(x+h(x))|\,dx \leqslant 2 \int_{A''} |g(z)|\,dz, \tag{3.6}$$

as is readily seen by changing variables in the integral on the left. Let now $\varphi \geqslant 0$ be a function in C_0^∞ with support in $|x| \leqslant 1$ and such that $\int \varphi\,dx = 1$, and let

$$\varphi_\delta(x) = \delta^{-n}\varphi\left(\frac{x}{\delta}\right), \quad f_\delta(x) = f * \varphi_\delta.$$

Let us also denote the supremum on the right-hand side of (3.5) by $\omega(\delta)$. Then, since $\varphi_\delta(y)$ has support in $|y| \leqslant \delta$ and $\int \varphi_\delta\,dx = 1$, we have

$$\int_{A'} |f(x)-f_\delta(x)|\,dx = \int_{A'} \left| \int [f(x)-f(x-y)]\varphi_\delta(y)\,dy \right| dx$$
$$\leqslant \int \varphi_\delta(y) \int_{A'} |f(x)-f(x-y)|\,dx\,dy \leqslant \omega(\delta),$$

that is,

$$\int_{A'} |f(x)-f_\delta(x)|\,dx \leqslant \omega(\delta), \tag{3.7}$$

which combined with (3.6) gives

$$\int_{A} |f(x+h(x))-f_\delta(x+h(x))|\,dx \leqslant 2\omega(\delta). \tag{3.8}$$

On the other hand, because

$$\int \frac{\partial}{\partial y_j}\varphi_\delta(y)\,dy = 0 \quad \text{and} \quad \int \left|\frac{\partial}{\partial y_j}\varphi_\delta(y)\right| dy \leqslant c\delta^{-1},$$

we have

$$\int_{A'} \left|\frac{\partial f_\delta}{\partial x_j}(x)\right| dx = \int_{A'} \left| \int f(x-y)\frac{\partial}{\partial y_j}\varphi_\delta(y)\,dy \right| dx$$
$$= \int_{A'} \left| \int [f(x-y)-f(x)]\frac{\partial}{\partial y_j}\varphi_\delta(y)\,dy \right| dx \leqslant c\delta^{-1}\omega(\delta).$$

From this and (3.6) which is also valid with $th(x)$, $0 \leqslant t \leqslant 1$, replacing $h(x)$, we obtain

$$\int_{A} |f_\delta(x+h(x))-f_\delta(x)|\,dx = \int_{A} \left| \int_0^1 \sum_{j=1}^{n} \frac{\partial f_\delta}{\partial x_j}(x+th(x))h_j(x)\,dt \right| dx$$
$$\leqslant \delta \sum_j \sup_t \int_A \left|\frac{\partial f_\delta}{\partial x_j}(x+th(x))\right| dx$$
$$\leqslant 2\delta \sum_j \int_{A'} \left|\frac{\partial f_\delta}{\partial x_j}(x)\right| dx \leqslant 2nc\omega(\delta),$$

that is,

$$\int_A \left| f_\delta(x+h(x)) - f_\delta(x) \right| dx \leqslant 2nc\,\omega(\delta),$$

and this combined with (3.7) and (3.8) gives the desired result.

Returning to the proof of (iv), let A be the annulus $\{x|\ a+\delta_0 \leqslant |x| \leqslant b-\delta_0\}$, where $\delta_0 < a$, $2\delta_0 < b-a$. Then, if ϱ denotes a rotation of $\boldsymbol{R}^n$ about the origin, we have

$$\int_\Sigma |K(\varrho x) - K(x)|\, d\sigma_x = \frac{1}{\log \dfrac{b-\delta_0}{a+\delta_0}} \int_A |K(\varrho x) - K(x)|\, dx.$$

This is readily seen if one takes into account the fact that $K(x)$ is homogeneous of degree $-n$. Setting $h(x) = \varrho(x) - x$, or $x + h(x) = \varrho(x)$, and using the preceding lemma we find that

$$\text{(3.9)}\quad \sup_{|\varrho|\leqslant\delta} \int_\Sigma |K(\varrho x) - K(x)|\, d\sigma_x \leqslant \frac{c}{\log \dfrac{b-\delta_0}{a+\delta_0}} \sup_{|y|\leqslant\delta} \int_{a\leqslant|x|\leqslant b} |K(x+y) - K(x)|\, dx,$$

provided that δ is sufficiently small, say $\delta \leqslant \varepsilon$, where ε depends only on the dimension n. But according to the definitions of ω_1 and ω_3, this inequality is the same as

$$\omega_1(t) \leqslant c\,\omega_3(t), \qquad t \leqslant \varepsilon.$$

In order to extend this inequality to the interval $\varepsilon \leqslant t \leqslant 2$, we observe that the group of proper rotations is compact and connected and, consequently, there exists a finite collection of rotations $\varrho_1, \varrho_2, \ldots, \varrho_k$ such that for every ϱ there exists an element ϱ_{j_1} of this collection with the property that

$$|\varrho x - \varrho_{j_1} x| \leqslant \varepsilon,$$

for all x with $|x| = 1$. Furthermore, there exist $\varrho_{j_1}, \varrho_{j_2}, \ldots, \varrho_{j_l} = I$, where I is the identity rotation, such that

$$|\varrho_{j_i} x - \varrho_{j_{i+1}} x| \leqslant \varepsilon, \qquad |x| = 1, \qquad i = 1, 2, \ldots, l-1.$$

In other words, we have

$$|\varrho_{j_1}^{-1} \varrho| \leqslant \varepsilon, \qquad |\varrho_{j_{i+1}}^{-1} \varrho_{j_i}| \leqslant \varepsilon, \qquad |\varrho_l| < \varepsilon.$$

Now

$$\int_{\Sigma} |K(\varrho x)-K(x)|\,d\sigma_x \leqslant \int_{\Sigma} |K(\varrho x)-K(\varrho_{j_1} x)|\,d\sigma_x +$$

$$+\sum_{i=1}^{l-1} \int_{\Sigma} |K(\varrho_{j_i} x)-K(\varrho_{j_{i+1}} x)|\,d\sigma_x + \int_{\Sigma} |K(\varrho_l x)-K(x)|\,d\sigma_x$$

$$= \int_{\Sigma} |K(\varrho_{j_1}^{-1}\varrho x)-K(x)|\,d\sigma_x + \sum_{i=1}^{l-1} \int_{\Sigma} |K(\varrho_{j_{i+1}}^{-1}\varrho_{j_i} x)-K(x)|\,d\sigma_x +$$

$$+\int_{\Sigma} |K(\varrho_l x)-K(x)|\,d\sigma_x,$$

and since all rotations in this last expression have modulus less than or equal to ε, we find that

$$\int_{\Sigma} |K(\varrho x)-K(x)|\,d\sigma_x \leqslant (l+1)\,\omega_1(\varepsilon) \leqslant (k+1)\,\omega_1(\varepsilon)\,d\sigma_x$$

which implies that for $t \geqslant \varepsilon$

$$\omega_1(t) \leqslant (k+1)\,\omega_1(\varepsilon) \leqslant c(k+1)\,\omega_3(t).$$

Now k evidently depends only on the dimension n. Thus (iv) is valid for all t.

Now there only remains to prove (v). Clearly it will suffice to prove this inequality for the positive part K^+ of K (without loss of generality we may assume that K is real). Evidently we have

$$|K^+(x-y)-K^+(x)| \leqslant |K(x-y)-K(x)|,$$

and, on account of (3.3),

$$\text{(3.10)}\qquad \frac{a}{b-a}\int_{|y|\leqslant a/a} |y|^{-n} \int_{a\leqslant |x|\leqslant b} |K^+(x-y)-K^+(x)|\,dx \leqslant |\Sigma|\,J_a(K).$$

Consider now the maximal function of K^+:

$$\overline{K}(x) = \sup_{t\leqslant a/a} \frac{1}{t^n}\int_{|y|<t} K^+(x-y)\,dy.$$

Then

$$\overline{K}(x) \leqslant \sup_{t\leqslant a/a} \frac{1}{t^n}\int_{|y|<t} |K^+(x-y)-K^+(x)|\,dy + \Omega K^+(x),$$

where Ω is the measure of $\{|x|\leqslant 1\}$, and

$$\overline{K}(x) \leqslant \int_{|y|\leqslant a/a} \frac{|K^+(x-y)-K^+(x)|}{|y|^n}\,dy + \Omega K^+(x).$$

Integrating with respect to x and using (3.10) we find that

$$\int_{a\leqslant |x|\leqslant b} \overline{K}(x)\,dx \leqslant |\Sigma|\frac{b-a}{a}J_a + \Omega \int_{a\leqslant |x|\leqslant b} K^+(x)\,dx.$$

But a theorem of E. M. Stein (see [6]) asserts that if the maximal function $\bar{K}$ of K^+ is integrable, then so is $K^+\log(1+K^+)$, and since the same argument applies to the negative K^- of K, we conclude that

$$\int_{a\leqslant|x|\leqslant b} |K(x)|\log[1+|K(x)|]\,dx < \infty$$

for every function $K(x)$ which is homogeneous of degree $-n$, is locally integrable in $|x|>0$ and for which $J_a(K)<\infty$ for some a, $a>1$. But this implies the inequality in (v). To prove this implication consider the convex function $\Phi(t) = t\log(1+t)$, $t\geqslant 0$, and the space L_Φ of functions F in $a\leqslant|x|\leqslant b$ with the property that $\Phi(|F|)$ is integrable, and define a norm in L_Φ by (see [7], Chapter IV, Section 10)

$$\int_{a\leqslant|x|\leqslant b} \Phi\left(\frac{|F|}{\|F\|_\Phi}\right) dx = 1.$$

On the other hand, consider also the space B of functions $K(x)$ which are homogeneous of degree $-n$ and for which $J_a(K)<\infty$, $a>1$, with the norm

$$\|K\|_B = J_a(K) + \int_{a\leqslant|x|\leqslant b} |K(x)|\,dx.$$

As is readily verified, B is a Banach space and its embedding in $L^1\{a\leqslant|x|\leqslant b\}$ is continuous.

Now, what we have shown above is that $B\subset L_\Phi$. Consequently, we have

$$B\subset L_\Phi\subset L^1\{a\leqslant|x|\leqslant b\}.$$

But the embedding of B in $L^1\{a\leqslant|x|\leqslant b\}$ is continuous and, as is readily seen, so is that of L_Φ. Thus, according to the closed graph theorem, the embedding of B in L_Φ is also continuous, that is, there exists a constant c such that

$$\|K\|_\Phi \leqslant c\left(J_a(K) + \int_{a\leqslant|x|\leqslant b} |K|\,dx\right) \tag{3.11}$$

Clearly, we may assume that $c\geqslant 1$. Now, as is readily verified, the function $\Phi(t)/t^2$ is a decreasing function of t and therefore, since $c\geqslant 1$, we have

$$\frac{\Phi(t/c)}{(t/c)^2} \geqslant \frac{\Phi(t)}{t^2},$$

that is,

$$\Phi(t/c) \geqslant \frac{1}{c^2}\Phi(t), \qquad c\geqslant 1.$$

Thus, setting

$$\lambda = J_a(K) + \int_{a \leqslant |x| \leqslant b} |K(x)|\, dx,$$

(3.11) becomes $\|K\|_\Phi \leqslant c\lambda$, and we obtain

$$1 = \int_{a \leqslant |x| \leqslant b} \Phi\left(\frac{|K|}{\|K\|_\Phi}\right) dx \geqslant \int_{a \leqslant |x| \leqslant b} \Phi\left(\frac{|K|}{c\lambda}\right) dx \geqslant \frac{1}{c^2} \int_{a \leqslant |x| \leqslant b} \Phi\left(\frac{K}{\lambda}\right) dx,$$

that is

$$\int_{a \leqslant |x| \leqslant b} \Phi\left(\frac{K}{\lambda}\right) dx \leqslant c^2,$$

which is the inequality (v). This concludes the proof of our theorem.

References

[1] A. Benedek, A. P. Calderón and R. Panzone, *Convolution operators on Banach space valued functions*, Proc. Nat. Acad. Sci. U. S. A. 48 (1962), pp. 356–365.

[2] A. P. Calderón and A. Zygmund, *On singular integrals*, Acta Math. 88 (1952), pp. 85–139.

[3] — *On singular integrals*, Amer. J. Math. 78 (1956), pp. 289–309.

[4] A. P. Calderón, M. Weiss and A. Zygmund, *On the existence of singular integrals*, Proc. Symposia Pure Math., Amer. Math. Soc., vol. 10, pp. 56–73.

[5] L. Hörmander, *Estimates for translation invariant operators in L^p spaces*, Acta Math. 104 (1960), pp. 93–139.

[6] E. M. Stein, *Interpolation of linear operators*, Trans. Amer. Math. Soc. 83 (1956), pp. 482–492.

[7] A. Zygmund, *Trigonometric series*, 2nd. ed., Vol. I and II, Cambridge Univ. Press, New York 1959.

Received April 5, 1977 (1287)